Springer Monographs in Mathematics

More information about this series at http://www.springer.com/series/3733

R. Lowen

Index Analysis

Approach Theory at Work

Springer

R. Lowen
Department of Mathematics and
Computer Science
University of Antwerp
Antwerp
Belgium

ISSN 1439-7382 ISSN 2196-9922 (electronic)
ISBN 978-1-4471-7266-6 ISBN 978-1-4471-6485-2 (eBook)
DOI 10.1007/978-1-4471-6485-2

Mathematics Subject Classification: 06B35, 06D22, 18B30, 18B99, 45A03, 46A19, 46B04, 46M15, 54A05, 54B20, 54B30, 54D30, 54D35, 54E35, 54E99, 60B10, 60F05, 68N30, 68Q99

Springer London Heidelberg New York Dordrecht

Printed on acid-free paper

Springer is part of Springer Science+Business Media (www.springer.com)

I dedicate this book to my dear friend and colleague Horst Herrlich for all his support and encouragement and for the inspiration I have always found in his fundamental and beautiful mathematical work

Preface

The essence of mathematics lies in its freedom.
(Georg Cantor)

The moving power of mathematical invention is not reasoning but imagination.
(Augustus De Morgan)

Approach theory was introduced in a series of papers which appeared between 1988 and 1995, and I refer to the bibliography for details. In 1997 a first book, Approach Theory: the Missing Link in the Topology-Uniformity-Metric Triad, appeared with Oxford University Press (Lowen 1997). With the maturing of the theory and the many further developments in applications since then, the time was ripe to write a more definitive account.

The work presented in this book could not have been completed without the enthusiastic collaboration of many colleagues and students.

A very special thought goes to my wife, Eva Colebunders for the exciting times we had when developing so many fundamental aspects together in our many joint papers on approach theory, metrically generated theories, and lax-algebraic theories.

Further, I have collaborated on approach theory, either in its own right or related to metrically generated theories, lax-algebraic theories, or approach frames, with many colleagues: Maria-Manuel Clementino, Guiseppe Di Maio, Eraldo Giuli, Horst Herrlich, Dirk Hofmann, Som Naimpally, Sevda Sagiroglu, Gavin Seal, Walter Tholen, Jan Van Casteren, David Vaughan and especially Bernhard Banaschewski and Piet Wuyts.

Many of my Ph.D. students, over the years, have helped develop parts of the theory: Rony Baekeland, Ben Berckmoes, Marc Nauwelaerts, Kristin Robeys, Wannes Rosiers, Mark Sioen, Anneleen Van Geenhoven, Christophe Van Olmen, Francis Verbeeck, Stijn Verwulgen, and Bart Windels. Several Ph.D. students of Eva Colebunders too contributed to the development of parts of the theory: Veerle Claes, Sarah De Wachter, An Gerlo, Gert Sonck, and Eva Vandersmissen. Many of their results are present in the text and references to their work in Ph.D. theses and in joint publications can be found in the bibliography.

Most of these students were supported by Research Foundation Flanders (FWO) doctoral and/or post-doctoral grants, and the FWO also funded the research project on Metrically Generated Theories in cooperation with Eva Colebunders which has had a fundamental impact on approach theory. Other students were supported by doctoral grants from the University of Antwerp. Both the FWO and the University of Antwerp also supported the series of conferences "Aspects of Contemporary Topology" where approach theory always played an important role. Thus the FWO and the University of Antwerp have indirectly also contributed to the coming into existence of the present work and I wish to express my thanks for those many years of considerable financial support.

Besides the colleagues mentioned above, I have been fortunate also to have had interesting exchanges on approach theory with, and enjoy the support and encouragement of many other colleagues, including the following, several of whom have also worked on approach theory individually: Gerald Beer, Lamar Bentley, Guillaume Brümmer, Peter Collins, Ákos Császár, Dikran Dikranjan, Szymon Dolecki, Paul Embrechts, Marcel Erné, David Holgate, Mirek Hušek, George Janelidze, Max Kelly (†), Darrell Kent, Hans-Peter Künzi, Bill Lawvere, Sandro Levi, Geert Molenberghs, Frédéric Mynard, Louis Nel, Gerhard Preuss (†), Ales Pultr, Dieter Pumplün, Gary Richardson, and Jerry Vaughan.

To all the aforementioned colleagues and students I would like to express my sincere appreciation for the many years of cooperation and support.

I thank Ben Berckmoes, Nieves Blasco, and Mark Sioen for proofreading parts of the manuscript and especially Piet Wuyts who proofread a complete first draft. Of course any mistakes which remain are entirely my own responsibility.

Finally I would like to thank Springer, in particular Lynn Brandon and Catherine Waite for the professional and pleasant cooperation during the writing and the final production of this book.

Antwerp, 2014 R. Lowen

Reference

Lowen, R.: Approach spaces: The Missing Link in the Topology-Uniformity-Metric Triad Oxford Mathematical Monographs. Oxford University Press, Oxford (1997)

Introduction

> *Beauty is the first test: there is no permanent place in the world for ugly mathematics.*
> (Godfrey Harold Hardy)

1. The basics of approach theory The genesis of approach spaces and more comprehensively, the whole of approach theory, finds itself in a very simple ascertainment. We can produce a canonical metric for finite products of metrizable (topological or uniform) spaces, only ad hoc metrics for countable products, and no metric at all for uncountable products. These simple facts lie at the basis of a vast history of development of important parts of mathematics. It was one of the main reasons for the apparition initially of topological spaces and later of uniform spaces (Weil 1937; Dieudonné 1939), the former in order to be able to deal with the known local properties of metrizable spaces in a more general context, which as history shows was entirely unavoidable, and the latter in order to be able to deal with uniform aspects in an equally unavoidable more general context. Top (respectively CReg), the category of topological spaces (respectively completely regular spaces) and continuous maps, and qUnif (respectively Unif) the category of quasi-uniform (respectively uniform) spaces and uniformly continuous maps both allow for all usual constructions such as subspaces, products, quotients, and coproducts, and constructions in one category concord well with those in the other category. Those same constructions in Met (the category of metric spaces and non-expansive maps) however, in general, do not concord with, e.g., either topological or uniform initial structures and hence are virtually useless. In view of the importance of initial structures, this explains why in many areas of mathematics, out of necessity, one often has to abandon an original metric setup and migrate to the settings of topological and/or uniform spaces.

In the wake of this development there appeared many other mathematical theories in their own right which could not have existed without topology and/or uniformity, such as, e.g., topological vector spaces, locally convex spaces, topological groups, and a host of specific (mostly non-metrizable) topologies in various fields, such as the weak topology on probability measures or the topology of convergence in measure of random variables (see e.g. Billingsley 1968; Parthasarathy 1967) the Wijsman and Vietoris topology or any of several proximal topologies on hyperspaces (see e.g. Beer 1993; Lechicki and Levi 1987), various topologies of function spaces such as pointwise convergence or uniform convergence on compacta (see e.g. Bourbaki 1961), many auxiliary non-metrizable topologies in the theory of Banach spaces such as the weak and weak* topologies (see e.g. Brezis 2011). In many of these examples one starts with a metric setting and then auxiliary structures (topologies and/or uniformities) are introduced which are no longer metrizable. The reason for this can always be traced back to the ascertainment mentioned higher up: uncountable products or more generally uncountable initial structures of metrizable spaces, be it topological or uniformizable, are no longer metrizable. Hence one drops from a numerical setup to a non-numerical topological, respectively uniform, setup.

Approach theory completely solves this by introducing precisely those two new types of numerically structured spaces which are required: approach spaces on the local level and uniform gauge spaces on the uniform level. Approach spaces formalize exactly the numerical information which is preserved when making arbitrary products of metrizable topological spaces and likewise uniform gauge spaces formalize exactly the numerical information, which is preserved when making arbitrary products of metrizable uniform spaces. In both cases this is achieved by a type of structure which generalizes metrics respectively in a topological way and in a uniform way.

The basic concepts of the local theory of approach spaces are largely explained in the first two chapters. Because of the many different structures which characterize approach spaces and of which we have to show the equivalence and for which we have to prove transition formulas, this takes quite some work. In Chap. 5, we elaborate on the uniform counterpart, so-called (quasi-)uniform gauge spaces. Here we basically only give two different structural characterizations.

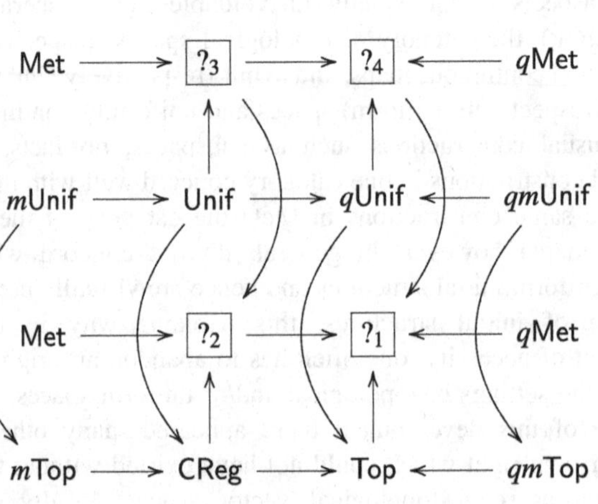

Referring to the diagram above, these local and uniform theories relate to each other as topological spaces relate to quasi-uniform spaces and as completely regular topological spaces relate to uniform spaces. In Chap. 2, we see that the category App of approach spaces (first question mark) contains both Top and qMet as full subcategories and the category UAp (second question mark) of uniform approach spaces contains both CReg and Met as full subcategories.

Hence it was indeed to be expected that similarly, well-behaved topological categories should exist on the uniform level and which extend the categorical relation that exists between Top and qUnif on the one hand and between CReg and Unif on the other hand. In Chap. 5, we see that qUG (fourth question mark) contains both qUnif and qMet as full subcategories and that UG (third question mark) contains both Unif and Met as full subcategories. We note that in the diagram, mUnif, qmUnif, mTop and qmTop stand respectively for the categories of metrizable uniform spaces, quasi-metrizable uniform spaces, metrizable topological spaces and quasi-metrizable topological spaces, all of course with their obvious morphisms. All horizontal arrows as well as all straight vertical arrows are embeddings and all curved arrows are forgetful functors. The question marks are numbered in the order in which the categories are introduced in the text.

In order to be self-contained we include the foundational parts from Lowen (1997) in the present work in the first two chapters. Concerning the basic structures of approach spaces, some minor changes have been made to the definitions and several new characterizations have been added. In particular, the notion of a basis for a gauge has been improved. This also required a rewrite of most proofs involving this concept. Further, three new structures are introduced bringing the total number of fundamentally and conceptually different structures characterizing approach spaces to ten. The new structures are upper regular function frames, upper hull operators, and functional ideal convergence. The former two are the logical counterparts to lower regular function frames and lower hull operators (previously simply called regular function frames and hull operators) and the latter is an entirely new and alternative way to describe the notion of convergence in approach spaces. Of course all required and/or interesting new transition formulas are also contained and proved in the present work.

Since App has both a topological and a metric side to it, both the notions of completeness and of compactness are simultaneously meaningful for general approach spaces. Hence, we also have both a construction of completion and a construction of compactification in a suitable subcategory of App. Moreover, we obviously also have notions of completeness and completion on the uniform level in UG. Thus in Chap. 6, we will study three types of extensions of spaces and morphisms. This chapter recaptures part from Lowen (1997) while adding completion of uniform gauge spaces together with the uniform aspects of the Čech-Stone compactification.

2. The basics of index analysis Topology and analysis, like all mathematical theories usually deal with "good" objects such as for instance contractions, continuous functions, compact sets, convergent sequences, etc, ... Barring contraposition formulations, there are hardly any theorems of type: if X does not fulfill good property A and $f : X \to Y$ does not fulfill good property B then .... However, how many objects in mathematical theories fulfill good properties compared to the number of objects available. Let us look at some simple examples. Of course it is easy to create situations where only finite sets are compact or where only constant functions are continuous, but that usually requires pathological or uninteresting setups. Therefore, let us focus on natural non-pathological situations, specifically let us consider the real line $\mathbb{R}$ with its usual topology and metric. There are $\mathfrak{c}$ continuous functions from $\mathbb{R}$ to $\mathbb{R}$ and so there are $2^{\mathfrak{c}}$ non-continuous functions. The same situation arises for non-expansive maps. There are $\mathfrak{c}$ non-expansive maps and hence $2^{\mathfrak{c}}$ maps which are not non-expansive. There are $\mathfrak{c}$ compact subsets of $\mathbb{R}$ and $2^{\mathfrak{c}}$ non-compact ones. There are only $\mathfrak{c}$ compact metric spaces up to homeomorphism. Hence there is a class of metric spaces which are not compact. A connected set in $\mathbb{R}$ is an interval, there are $\mathfrak{c}$ such sets but only four nonempty ones up to homeomorphism and there are $2^{\mathfrak{c}}$ non-connected sets. Do all these not deserve some consideration?

Let us compare some aspects of topological versus metric spaces. In a topological space there is no notion of "approximate convergence," whereas in a metric space we have notions like asymptotic center and radius (see e.g. Edelstein 1972; Lim 1980). In a topological space we have no notion of "approximate compactness," whereas in a metric space we have the notion of measure of noncompactness (see e.g. Banaś 1997; Banaś and Goebel 1980). In topological spaces we do not have a notion of "approximate homeomorphism," whereas in the theory of Banach spaces we do have the notion of near-isometry (see e.g. Hyers and Ulam 1945, 1947; Bourgin 1946). So what we see is that the presence of metric information allows for a more powerful and discerning analysis of various otherwise topological phenomena. Each of the above concepts however was introduced and developed in a fairly ad hoc way independent of the other concepts. Whereas convergence, compactness, and homeomorphisms are topological concepts fully and canonically embedded in the theory of topology, the approximate versions were not embedded in any well-founded and unifying theory. This however is exactly what approach theory provides.

The theory of approach spaces and index analysis set forth in this book is a first step in the direction of a comprehensive mathematical framework whereby, as much as possible and as far as meaningful, the extent to which properties are fulfilled is measured by means of indices, and theorems that involve indexed concepts contain as few conditions as possible and consist mainly of inequalities involving indices.

Of course not all concepts lend themselves to being indexed. Therefore, in Chap. 3, we first study some such non-indexed properties of approach spaces. These are mainly classical concepts, of either a topological nature, a metric nature or a pure approach nature, which are invariant over equivalence classes of isomorphic objects. In particular we study (1) uniformity, which is the counterpart to the topological notion of complete regularity and the metric notion of symmetry, (2) weak adjointness, which actually is a pure approach property (although it also once appeared in the theory of quasi-metric spaces), (3) some lower separation properties, which are the counterparts for the topological notions of T_0, T_1, T_2 and regularity, (4) countability properties which come in three flavors, two of which are the counterparts to the topological notions of first and second countability and a third one which again is a pure approach property, and (5) completeness, which of course is the counterpart to metric completeness.

In Chap. 4, we then develop the basics of what we have called index analysis. We deal with, in the first place, spaces and functions, such as for instance an index of compactness of spaces and an index of contractivity of functions, but also, as the need arises, with indices of other mathematical objects, certain properties of which can naturally be measured. A common property of indices (which are $[0, \infty]$-valued functions) is that they measure a "distance" from satisfying an ideal property, and consequently the interpretation is that the smaller the value of an index is, the better the object satisfies the ideal property which is being measured. As mentioned, these indices are not restricted to spaces and functions, many other mathematical objects can be indexed. Thus, for instance, the basic defining structures of approach spaces, namely distance and limit operator themselves can be considered as being indices, the first as an index of closure and the second as an index of convergence. In this chapter we will define and study basic properties of (1) indices of contractivity, closed expansiveness, open expansiveness and properness for functions, (2) indices of compactness, relative compactness, sequential compactness, relative sequential compactness, countable compactness and the Lindelöf index, (3) index of local compactness, and (4) index of connectedness. Making use of the above indices, we will prove many basic "indexed theorems" which mainly consist of inequalities involving indices. If χ_{P_1} and χ_{P_2} are indices of properties P_1 and P_2, then the interpretation of a simple basic inequality $\chi_{P_1}(O_1) \leq \chi_{P_2}(O_2)$ is that the better O_2 satisfies property P_2 the better O_1 will satisfy property P_1. Moreover this is not a vague heuristic claim, it is an exact statement with numerical indices, the canonicity and appropriateness of which will always be evident from the body of results.

3. Traces in mathematical theories In Chaps. 7–11 on applications in topology, functional analysis, probability theory, hyperspaces, and domain theory, we exhibit the abundance of situations where approach theory actually is hidden within those theories and how bringing it to the forefront and using it systematically enriches the theory and meets the goals we set forth. The applications we deal with in Chaps. 7–11 are of three different kinds.

(1) A first case is where the basic setup is actually built with numerical information given typically by a metric or norm, and in the course of the development of the theory it is required to construct auxiliary spaces and structures. Either the auxiliary structure can be obtained employing countable operations, such as, e.g., countable products of metric spaces, and then one can again obtain a metric, which however is necessarily ad hoc, or, which is more usual, no such construction is possible and the auxiliary space can only be endowed with a uniformity or a topology. This means that an original setup with rich numerical information, which makes use of the additive semigroup structure of the positive reals, is abandoned and replaced by a considerably weaker setup at a purely qualitative level.

Remarkably, in many cases the necessary numerical information is actually available, but it was never recognized as such and only a fraction of it was used to define a uniformity or a topology in the same way that the quantitative information of a metric may be discarded and used merely to define the underlying topology. So we go from an isometric level [e.g., a metric space (E, d)], the level of structures where numerical information is preserved and used, to the isomorphic level [e.g., a derived topological and/or uniform space $(D(E), \mathcal{T}, \mathcal{U})$] where only derived qualitative information on a topological (or uniform) level is available.

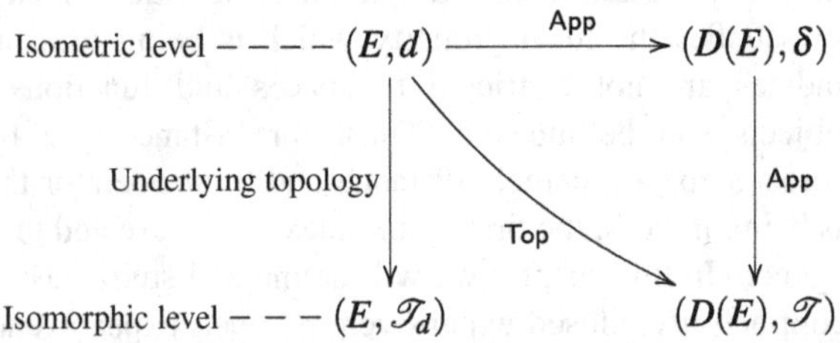

This is the case of the applications in functional analysis where we start with a normed space (Chap. 8), probability theory where we start with a separable Polish space (Chap. 9) and hyperspace theory where we start with a metric space (Chapter 10; Sects. 10.1 and 10.2). In all these cases, instead of going straight from (E, d) to $(D(E), \mathcal{T})$ we construct in a canonical way an approach space $(D(E), \delta)$ the underlying topological space of which is $(D(E), \mathcal{T})$. Analysis is now performed in $(D(E), \delta)$ on the same isometric level as (E, d) giving indexed results of which all classical results are simple consequences.

(2) A second case is where the basic setup (a topology or uniformity) contains no numerical information (the isomorphic level) as such but is actually endowed with a canonical or natural metric or normed structure (the isometric level), such as, e.g., the real line where the Euclidean topology invariable comes linked to its usual norm. Here, we take the stance that we extend what happens on the topological or uniform level to the approach and uniform gauge level and then apply this to the canonical metric or normed situation.

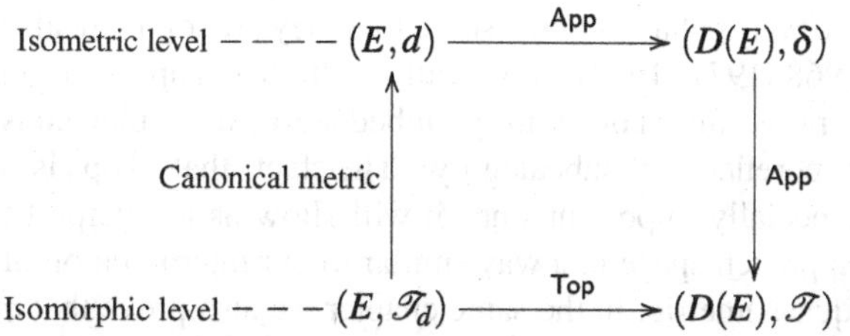

This is the situation in applications to topology (Chap. 7). Function spaces often involve the real line with its natural normed structure, or they involve another given metric or normed space, which we can take as a starting point. Similarly, when we study the Čech-Stone compactification of $\mathbb{N}$ we again use the canonical metric structure which $\mathbb{N}$ inherits from $\mathbb{R}$. So instead of staying with the topological situation $(E, \mathscr{T}_d)$ we go to the canonical metric space (E, d) and derive $(D(E), \delta)$ again in a canonical way and such that the underlying topological space is $(D(E), \mathscr{T})$. Again analysis is performed in $(D(E), \delta)$ and classical results for $(D(E), \mathscr{T})$ are simple consequences.

(3) A third case is where classically one actually also stays in the isometric realm. Again one starts with a metric (or in our particular case quasi-metric) setup and out of necessity has to restrict the development of the theory to countably constructed auxiliary spaces since one explicitly wants to obtain an, again ad hoc, quasi-metric also in the auxiliary situation.

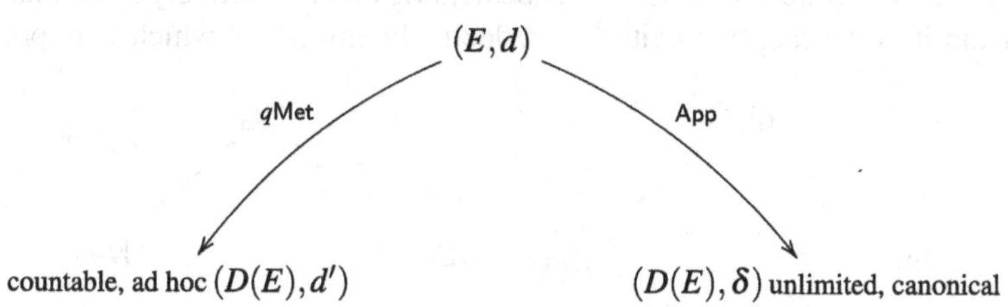

This for instance is the case in the application to DCPO's and domains (Chap. 11), where going the approach way not only does away with the countability limitation but also produces a canonical approach structure giving the required numerical information in the auxiliary spaces. In this case, if available, i.e., in the countable case, the canonically constructed approach space $(D(E), \delta)$ will have $(D(E), d')$ as underlying quasi-metric space.

It is of course not our intention, and would also not be feasible in this work, to develop each and every application to a very large extent. What we will do in each case however is demonstrate the canonicity of the structures, unveil their links with existing structures and give a body of evidence consisting of basic results which apply the new structures and make use of index analysis.

4. The categorical connection The categories of approach spaces and of uniform gauge spaces are well-behaved in the sense that they are topological categories (see e.g. Herrlich 1968, 1971, 1983) as we will see in the Chap. 1 and in the Chap. 5. In Chap. 2 we prove that Top is fully embedded as a simultaneously concretely reflective and coreflective subcategory. The fact that Top is coreflectively embedded is especially important since it will allow us to interpret the Top-coreflection of an approach space in a way similar to our interpretation of the topology underlying a (quasi-)metric. In the same chapter we also prove that qMet and Met are fully embedded as concretely coreflective subcategories of App. Here it is the fact that neither of these two categories is reflectively embedded, which is important as it will imply that, e.g., arbitrary products of metric or quasi-metric spaces in the category of approach spaces are hardly ever again metric or quasi-metric spaces, they are, in general, genuine non-metric and non-topological approach spaces.

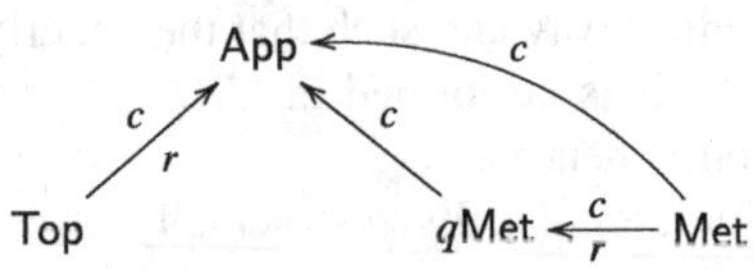

Similar results are shown in Chap. 5 for (quasi-)uniform gauge spaces: qUnif and Unif are fully embedded as simultaneously concretely reflective and coreflective subcategories of respectively qUG and UG. Also, qMet and Met are fully embedded as concretely coreflective subcategories of respectively qUG and UG. Here again it is the fact that neither is reflectively embedded which is important.

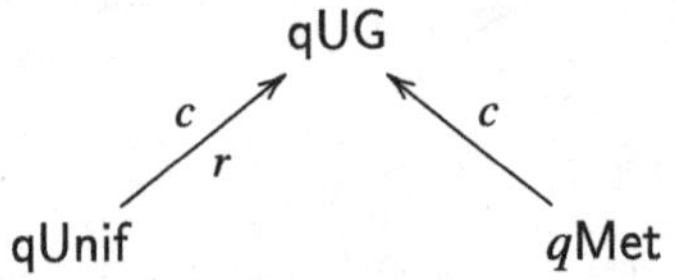

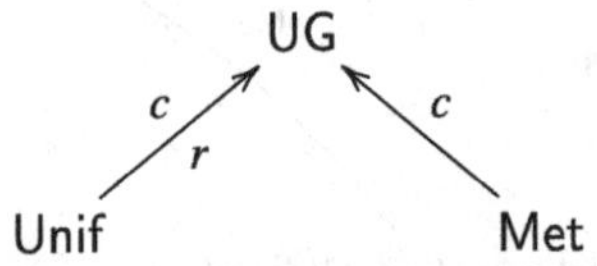

The situation for App is studied in more detail in the last chapter.

In the first place, it has the remarkable property of having very many simultaneously concretely reflectively and coreflectively embedded subcategories, a situation it shares with qMet and Met, and in Chap. 12 we characterize all of these subcategories.

In the second place, App is neither extensional (see e.g. Herrlich 1988a, b) nor cartesian closed (see e.g. Herrlich 1974), and hence also not a quasi-topos. In Chap. 12, we construct and completely describe the extensional (PrAp), cartesian closed (EpiAp) and quasi-topos (PsAp) hulls of App and see how they relate to similar hulls of Top. The situation is depicted in the diagram below.

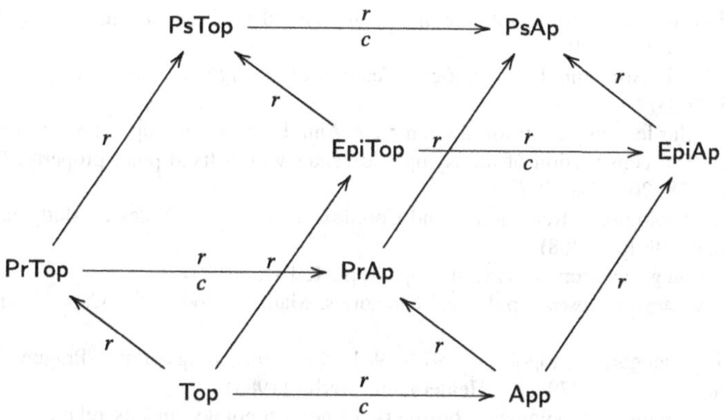

The above is an overview of the main topics treated in the last chapter showing in which way category theory has influenced the development of the theory of approach spaces and uniform gauge spaces.

However, there also is another connection between approach theory and category theory, going in the opposite direction. In Hofmann et al. (2014) it is abundantly shown that approach spaces are the penultimate example of "lax-algebraic categories," and many developments in that book are inspired by the situation in App. In a private communication to Tholen in 2000, Lawvere suggested that, in the same way as topological spaces generalize ordered sets, approach spaces should be describable as generalized metric spaces using multicategories instead of just categories. Simultaneously, following a suggestion by Janelidze, in 2003, Clementino and Hofmann gave a lax-algebraic description of approach spaces using a numerical extension of the ultrafilter monad (Clementino and Hofmann 2003). In the final section of the last chapter we revisit this and give a more direct proof of that description. These developments also led the way to the comprehensive research which is presented in the book "Monoidal Topology" (Hofmann et al. 2014) and which includes several other lax-algebraic characterizations of App.

References

Banás, J. Goebel, K.: Measures of noncompactness in banach spaces. Lecture Notes in Pure and Applied Mathematical. vol. 60. Marcel Dekker, New York (1980)

Banás, J.: Applications of measures of weak noncompactness and some classes of operators in the theory of functional equations in the Lebesgue space. Nonlinear Anal. **30**, 3283–3293 (1997)

Beer, G.: Topologies on Closed and Closed Convex Sets. Kluwer Academic Publishers, Dordrecht-Boston-London (1993)

Billingsley, P.: Convergence of Probability Measures. Wiley, New York (1968)

Bourbaki, N.: Topologie Générale, chapitre 10: Espaces fonctionnels. Hermann, Paris (1961)

Bourgin, D.G.: Approximate isometries Bull. Am. Math. Soc. **52**, 704–714 (1946)
Brezis, H.: Functional Analysis, Sobolev Spaces and Partial Differential Equations. Universitext Springer, New York (2011)
Clementino, M.M., Hofmann, D.: Topological features of lax algebras. Appl. Categor. Struct. **11,** 267–286 (2003)
Dieudonné, J.: Sur les espaces uniformes complets. Ann. Ecol. Norm. Supér. **56**, 276–291 (1939)
Edelstein M.: The construction of an asymptotic center with a fixed-point property. Bull. Am. Math. Soc. **78**, 206–208 (1972)
Herrlich, H.: Topologische Reflexionen und Coreflexionen. Lecture Notes in Mathematics, vol. 78, Springer, Berlin (1968)
Herrlich, H.: Categorical topology. Gen. Top. Appl. **1**, 1–15 (1971)
Herrlich, H.: Cartesian closed topological categories. Math. Colloq. Univ. Cape Town **9**, 1–16 (1974)
Herrlich, H.: Categorical topology 1971–1981. In: Proceedings Fifth Prague Topology Symposium 1981, pp. 279–383. Heldermann, Berlin (1983)
Herrlich, H.: Hereditary topological constructs. General topology and its relations to modern analysis and algebra VI. In: Proceedings of Sixth Prague Topological Symposium 1986, pp. 249–262. Heldermann (1988a).
Herrlich, H.: On the representability of partial morphisms in Top and in related constructs categorical algebra and its applications, Proceedings of a Conference, Held in Louvain-la-Neuve 1987, Lecture Notes in Mathematics, vol. 1348, Springer, Berlin 143–153 (1988b)
Hofmann, D., Seal, G.J., Tholen, W. (eds.): Monoidal Topology: A Categorical Approach to Order, Metric and Topology. Cambridge University Press, Cambridge (2014)
Hyers, D.H., Ulam, S.M.: On approximate isometries. Bull. Am. Math. Soc. **51**, 288–292 (1945)
Hyers, D.H., Ulam, S.M.: On approximate isometries of the space of continuous functions. Ann. Math. **48**, 285–289 (1947)
Lechicki, A., Levi, S.: Wijsman convergence in the hyperspace of a metric space. Boll. Un. Mat. Ital. **5-B**, 435–452 (1987)
Lim, T.-C.: On asymptotic centers and fixed points of nonexpansive mappings. Can. J. Math. **32**, 421–430 (1980)
Lowen, R.: Approach spaces: The Missing Link in the Topology-Uniformity-Metric Triad Oxford Mathematical Monographs, Oxford University Press, Oxford (1997)
Parthasarathy, K.R.: Probability Measures on Metric Spaces. Academic Press, New York (1967)
Weil, A.: Sur les espaces à structure uniforme et sur la topologie générale. Actual. Sci. Ind. **551** (1937)

Contents

Chapter 1
Approach Spaces

> *All the other vehicles of mathematical rigor are secondary to definitions, even that of rigorous proof.*
>
> (Yuri Manin)
>
> *Structures are the weapons of the mathematician.*
>
> (Bourbaki)

In this chapter we define the basic structures which determine what is called an approach space. One of the powerful features of approach spaces is that they can be determined by conceptually totally different but nevertheless equivalent structures. This is not unlike the situation in topology, where the structure of a topological space can be determined by a number of equivalent concepts, such as, for example, open sets, closed sets, closure operator, neighbourhood system, and convergence structure. In the case of approach spaces there are even more different basic structures, namely ten in total. These structures can have a topological side and/or a metric side to them, and the reason for this is made abundantly clear in the second chapter.

We will not only introduce these various structures but obviously we will also prove that they are indeed equivalent. This will provide us with transition formulas as to how one structure unambiguously determines another. Moreover we will of course also define the morphisms and other types of functions that are most naturally linked to approach spaces and we will characterize them in terms of the various structures.

We will also show that with the right morphisms (contractions) approach spaces constitute a topological category. The main aspect of this being that both initial and final structures exist. We then go on to describe these making use of several of the defining structures.

1.1 The Structures

Concerning the concept of a metric, we should warn the reader right from the start that in this work we will adopt a terminology which differs from the one from our former work in this field and from the usual conventions. Given a set X a map

R. Lowen, *Index Analysis*, Springer Monographs in Mathematics,
DOI 10.1007/978-1-4471-6485-2_1

$d : X \times X \longrightarrow [0, \infty]$ which vanishes on the diagonal and satisfies the triangle inequality will be called a *quasi-metric*. If the map is moreover symmetric then it is called a *metric*, if the map is finite the metric will simply be called a *finite metric* and if the underlying topology is Hausdorff the metric will be called *separated*. So in this work a quasi-metric (or metric) need neither be finite nor separated, and if it satisfies any of these conditions then it will explicitly be mentioned.

Throughout this work, given a set X, we denote the set of all subsets of X by 2^X and for conciseness in notation the set of all finite subsets of X will sometimes be denoted by $2^{(X)}$. Further we put $\mathbb{R}^+ := [0, \infty[$ and $\mathbb{R}_0^+ :=]0, \infty[$. The unbounded closed interval $[0, \infty]$ will play a prominent role and therefore we also reserve a special notation for it, $\mathbb{P} := [0, \infty]$. We consider $\mathbb{P}$ with its natural quantale structure basically consisting of the usual order and complete lattice structure and the additive semigroup. Hence we will freely use the symbols $+$ and $-$ also for the natural "extensions" of these operations to $\mathbb{P}$. More precisely, $+$ and $-$ stand for the usual addition and subtraction in the case of real numbers and further, for any $x \in [0, \infty[$ we have $x + \infty = \infty + x = \infty + \infty = \infty$, $\infty - x = \infty$, and $\infty - \infty = 0$. In order to have a "subtraction" interior to $\mathbb{P}$, when required, we will use *truncated subtraction* which is defined and denoted as $a \ominus b := (a - b) \vee 0$ for all $a, b \in \mathbb{P}$. Given a function $\mu \in \mathbb{P}^X$, for simplicity in notation we will often denote $\inf_{x \in X} \mu(x)$ simply by $\inf \mu$ and likewise $\sup_{x \in X} \mu(x)$ simply by $\sup \mu$.

When proving an inequality $a \leq b$, since the values 0 and ∞ may appear, we, often silently, assume that $a \neq 0$ and that $b \neq \infty$ since otherwise there is of course nothing to show.

Distances

Intuitively, probably the most appealing structure which we will consider is that of a distance between points and sets. Whereas in a metric space (X, d) a distance between pairs of points is given, and a distance between points and sets can then be derived from this according to the usual formula which says that for all $x \in X$ and $A \subseteq X$

$$\delta_d(x, A) := \inf_{a \in A} d(x, a)$$

here we consider the latter as a primitive structure, i.e. we start from a concept of distance between points and sets. Before giving the precise definition we need to introduce the following notation.

If X is a set, and we have a function $\delta : X \times 2^X \longrightarrow \mathbb{P}$, then for any subset $A \subseteq X$ and any $\varepsilon \in \mathbb{P}$, we define

$$A^{(\varepsilon)} := \{x \in X \mid \delta(x, A) \leq \varepsilon\}.$$

1.1.1 Definition (Distance) A function

$$\delta : X \times 2^X \longrightarrow \mathbb{P}$$

is called a *distance* if it satisfies the following properties.

(D1) $\forall x \in X, \forall A \subseteq X : x \in A \Rightarrow \delta(x, A) = 0.$
(D2) $\forall x \in X : \delta(x, \emptyset) = \infty.$
(D3) $\forall x \in X, \forall A, B \subseteq X : \delta(x, A \cup B) = \min(\delta(x, A), \delta(x, B)).$
(D4) $\forall x \in X, \forall A \subseteq X, \forall \varepsilon \in \mathbb{P} : \delta(x, A) \leq \delta(x, A^{(\varepsilon)}) + \varepsilon.$

In the same way as in a metric space the value $\delta(x, A)$ is interpreted as the distance from the point x to the set A. We wish to emphasize, however, that this distance cannot necessarily be derived from the distances $\delta(x, \{a\})$ for $a \in A$. We will see numerous examples of this fact throughout this work.

The following proposition contains some simple but fundamental properties which we will use implicitly in the sequel.

1.1.2 Proposition *If $\delta : X \times 2^X \longrightarrow \mathbb{P}$ is a distance then the following properties hold.*

1. $\forall x \in X, \forall A, B \subseteq X : A \subseteq B \Rightarrow \delta(x, B) \leq \delta(x, A).$
2. $\forall x \in X, \forall \mathscr{A} \subseteq 2^X, \mathscr{A}$ *finite* $: \delta(x, \bigcup \mathscr{A}) = \min_{A \in \mathscr{A}} \delta(x, A).$
3. $\forall x \in X, \forall A, B \subseteq X : \delta(x, A) \leq \delta(x, B) + \sup_{b \in B} \delta(b, A).$

Proof The first and second properties follow immediately from (D3). To prove the third property let $x \in X$ and $A, B \subseteq X$, then with

$$\varepsilon := \inf \left\{ \theta \in \mathbb{P} \mid B \subseteq A^{(\theta)} \right\}$$

the result follows from (D4). □

The third property in the foregoing result, modulo the difference in domains for κ-metrics and for distances, is the same as the so-called regularity condition for κ-metrics in Shchepin (1980).

Although a distance is defined as a function of two variables, points and sets, it will sometimes be useful also to consider the following associated functions on points. For a given subset $A \subseteq X$ we define

$$\delta_A : X \longrightarrow \mathbb{P} : x \mapsto \delta(x, A).$$

As is usually done in the context of hyperspaces of metric spaces (see e.g. Lechicki and Levi 1987) we will also call such functions *distance functionals*.

Limit Operators

The basic set-theoretical tools needed to define convergence in a topological space are sequences, nets and filters. Sequences suffice in metrizable spaces, but not in general topological spaces and as it will soon turn out, neither do they suffice in our case. Although in particular cases we will sometimes give results for sequences, for the general theory we have to resort to either filters or nets. We have chosen filters, as introduced by Choquet (1947), because of their conceptual beauty. All results can, however, be properly rephrased in terms of nets.

Whenever convenient we will use the following notations. $\mathsf{F}(X)$ will stand for the set of all filters on X, and $\mathsf{U}(X)$ will stand for the set of all ultrafilters on X. Sometimes it will be useful to generalize this notation in the following way. If $\mathscr{F}$ is a given filter on X, then we will denote by $\mathsf{F}(\mathscr{F})$ the collection of all filters on X which are finer than $\mathscr{F}$, and by $\mathsf{U}(\mathscr{F})$ the collection of all ultrafilters on X which are finer than $\mathscr{F}$. If $\mathscr{F}$ is the trivial filter on X, i.e. $\mathscr{F} = \{X\}$, then $\mathsf{F}(\mathscr{F})$ reduces to $\mathsf{F}(X)$ and $\mathsf{U}(\mathscr{F})$ reduces to $\mathsf{U}(X)$.

If $\mathscr{A}$ is a collection of subsets of X, then the *stack of* $\mathscr{A}$ is defined as

$$\text{stack}\mathscr{A} := \{B \subseteq X \mid \exists A \in \mathscr{A} : A \subseteq B\}.$$

If $\mathscr{F}$ is a filter on X then the *sec of* $\mathscr{F}$ is defined as

$$\text{sec}\mathscr{F} := \bigcup_{\mathscr{U} \in \mathsf{U}(\mathscr{F})} \mathscr{U} = \{A \subseteq X \mid \forall F \in \mathscr{F} : A \cap F \neq \emptyset\}.$$

If $\mathscr{A}$ is a filter basis then stack$\mathscr{A}$ is the filter generated by $\mathscr{A}$. Not to overload the notations, and if the meaning is self-evident, we often omit writing stack before a filterbasis. In case $\mathscr{A}$ reduces to a single set A we write stack$A$, or even shorter $\dot{A}$, instead of stack $\{A\}$, and in case the single set A furthermore reduces to a single point a we write stack$a$, or $\dot{a}$, instead of stack $\{a\}$. If $A \subseteq X$, $A \neq \emptyset$, then, for notational simplicity, especially in formulas, we will often write $\mathsf{F}(A)$ instead of $\mathsf{F}(\text{stack}A)$, and $\mathsf{U}(A)$ instead of $\mathsf{U}(\text{stack}A)$, i.e., whenever possible and no confusion can occur, we use the same notation for all filters on a (non-empty) set $A \subseteq X$ as for all filters on X containing A.

We refer to Kent (1964), Kowalsky (1954) and Lowen-Colebunders (1989) for more information on the so-called *Kowalsky diagonal operation* which we require in one of the axioms.

Since we will also be using this concept in the sections on regularity (3.3), on functional ideal convergence (1.1) and on lax algebraic descriptions of approach spaces (12.7) we give it in the required generality. Given sets J and X, a filter $\mathscr{F} \in \mathsf{F}(J)$ and a map $\sigma : J \longrightarrow \mathsf{F}(X)$ then the *diagonal filter* of σ *with respect to* $\mathscr{F}$ is defined as

$$\Sigma\sigma(\mathscr{F}) := \bigvee_{A \in \sigma(\mathscr{F})} \bigcap_{\mathscr{G} \in A} \mathscr{G} = \bigvee_{F \in \mathscr{F}} \bigcap_{j \in F} \sigma(j).$$

The symbol $\bigvee$ in this notation stands for the supremum of a collection of filters. Note that in the particular case of a diagonal filter this supremum always exists. We refer to $(\sigma(j))_{j\in J}$ as being a *selection of filters on* X.

We will need the following results concerning this diagonal operation.

1.1.3 Proposition *Let* J, L *and* X *be sets, let* $\sigma : J \longrightarrow \mathsf{F}(X)$ *and* $\gamma : L \longrightarrow \mathsf{F}(J)$ *and let* $\mathscr{F} \in \mathsf{F}(J)$. *Then the following properties hold.*

1. $\Sigma\sigma(\mathscr{F}) = \bigcup_{F\in\mathscr{F}} \bigcap_{j\in F} \sigma(j)$.
2. $\Sigma\sigma(\bigcap_{l\in L} \gamma(l)) = \bigcap_{l\in L} \Sigma\sigma(\gamma(l))$.
3. $\Sigma\sigma(\mathscr{F}) = \bigcap_{\rho\in\prod_{j\in J} \mathsf{U}(\sigma(j))} \Sigma\rho(\mathscr{F})$.
4. $\Sigma\sigma(\mathscr{F}) = \bigcap_{\rho\in\prod_{j\in J} \mathsf{U}(\sigma(j))} \bigcap_{\mathscr{U}\in\mathsf{U}(\mathscr{F})} \Sigma\rho(\mathscr{U})$.
5. *If all filters involved are ultrafilters, then so is* $\Sigma\sigma(\mathscr{F})$.

Proof We only prove the third property, leaving the remaining ones to the reader. One inclusion is clear; to show the other one, suppose that $A \notin \Sigma\sigma(\mathscr{F})$. For any $j \in J$, we now choose an ultrafilter in the following way: if $A \notin \sigma(j)$ then choose $\rho(j) \in \mathsf{U}(\sigma(j))$ such that $A \notin \rho(j)$ and if $A \in \sigma(j)$ then choose $\rho(j) \in \mathsf{U}(\sigma(j))$ arbitrarily. Then it follows from the first property that, for any $F \in \mathscr{F}$, there exists $j \in F$ such that $A \notin \sigma(j)$ and hence, $A \notin \rho(j)$. Consequently, $A \notin \Sigma\rho(\mathscr{F})$. □

Before moving on we give some preliminary properties related to filters. The following is an extremely useful, purely filter-theoretic result which we will require numerous times throughout the book.

1.1.4 Lemma *If* $\mathscr{F}$ *is a filter, and for each ultrafilter* $\mathscr{U} \in \mathsf{U}(\mathscr{F})$ *we have selected a set* $S(\mathscr{U}) \in \mathscr{U}$, *then there exists a finite set* $\mathsf{U}_S \subseteq \mathsf{U}(\mathscr{F})$ *such that*

$$\bigcup_{\mathscr{U}\in\mathsf{U}_S} S(\mathscr{U}) \in \mathscr{F}.$$

Proof Suppose the conclusion does not hold. Then this implies that the family

$$\mathscr{F} \cup \{X \setminus S(\mathscr{U}) \mid \mathscr{U} \in \mathsf{U}(\mathscr{F})\}$$

has the finite intersection property, and thus is contained in some ultrafilter $\mathscr{U} \in \mathsf{U}(\mathscr{F})$. This, however, implies that both $S(\mathscr{U}) \in \mathscr{U}$ and $X \setminus S(\mathscr{U}) \in \mathscr{U}$ which is a contradiction. □

The following result is a kind of *minimax* formula which will allow us to interchange liminf and limsup in several instances.

1.1.5 Lemma *If $\mathcal{U}$ is an ultrafilter on X and $f : X \longrightarrow \mathbb{P}$ is an arbitrary function, then*

$$\sup_{U\in\mathcal{U}} \inf_{y\in U} f(y) = \inf_{U\in\mathcal{U}} \sup_{y\in U} f(y).$$

Proof It is immediately clear that, for any $U, V \in \mathcal{U}$, we have $\inf_{y\in U} f(y) \leq \sup_{y\in V} f(y)$. Consequently

$$\sup_{U\in\mathcal{U}} \inf_{y\in U} f(y) \leq \inf_{U\in\mathcal{U}} \sup_{y\in U} f(y).$$

To prove the other inequality, suppose that $\inf_{U\in\mathcal{U}} \sup_{y\in U} f(y) > r$. Then, for all $U \in \mathcal{U}$, we have $U \cap \{f > r\} \neq \emptyset$ and consequently $\{f > r\} \in \mathcal{U}$. This then also implies that $r \leq \inf_{y\in\{f>r\}} f(y)$, from which the converse inequality follows. □

1.1.6 Definition (Limit operator) A function

$$\lambda : \mathsf{F}(X) \longrightarrow \mathbb{P}^X$$

is called a *limit operator* if it satisfies the following properties.

(L1) $\forall x \in X : \lambda\dot{x}(x) = 0$.
(L2) For any (non-empty) family $(\mathcal{F}_j)_{j\in J}$ of filters on X

$$\lambda(\bigcap_{j\in J} \mathcal{F}_j) = \sup_{j\in J} \lambda\mathcal{F}_j.$$

(L3) For any $\mathcal{F} \in \mathsf{F}(X)$ and any selection of filters $(\sigma(x))_{x\in X}$ on X

$$\lambda\Sigma\sigma(\mathcal{F}) \leq \lambda\mathcal{F} + \sup_{x\in X} \lambda\sigma(x)(x).$$

The value $\lambda\mathcal{F}(x)$ is interpreted as the distance that the point x is away from being a limit point of the filter $\mathcal{F}$. The word distance here is meant generically and not in the sense of 1.1.1. However, apart from the general formulas which exist involving distances and limits and which we will prove in this chapter, in the third and fourth chapters some more specific relationships will be shown to hold between these two notions.

The smaller the value of $\lambda\mathcal{F}(x)$, the closer x comes to being a limit point of $\mathcal{F}$. Notice also that it immediately follows from (L2) that

$$\forall \mathcal{F}, \mathcal{G} \in \mathsf{F}(X) : \mathcal{G} \subseteq \mathcal{F} \Rightarrow \lambda\mathcal{F} \leq \lambda\mathcal{G}.$$

Condition (L3) has been generalized in various interesting ways by Brock and Kent (1997a, b) in order to investigate a notion of regularity for approach spaces and we will come back to this in the third chapter.

One of the consequences of their investigations is the following alternative combination of (L2) and (L3) which instead of a selection of filters requires an extra variable, namely a function $\psi : J \longrightarrow X$ for (L3) and a weaker form of (L2).

1.1.7 Theorem *Given a function* $\lambda : \mathsf{F}(X) \longrightarrow \mathbb{P}^X$ *satisfying* (L1), λ *is a limit operator if and only if it satisfies the following properties.*

(L2w) *For any* $\mathscr{G} \subseteq \mathscr{F}$: $\lambda\mathscr{F} \leq \lambda\mathscr{G}$.

(L) *For any set* J, *for any* $\psi : J \longrightarrow X$, *for any* $\sigma : J \longrightarrow \mathsf{F}(X)$ *and for any* $\mathscr{F} \in \mathsf{F}(J)$

$$\lambda\Sigma\sigma(\mathscr{F}) \leq \lambda\psi(\mathscr{F}) + \sup_{j\in J} \lambda\sigma(j)(\psi(j)).$$

Proof We first show the if-part. Letting $J := X$ and $\psi := 1_X$ we see that (L) implies (L3). To show (L2), let $(\mathscr{F}_j)_{j\in J}$ be a family of filters and let $x \in X$. Define σ, ψ and $\mathscr{F}$ as follows

$$\begin{aligned}
&\sigma : J \longrightarrow \mathsf{F}(X) : j \mapsto \mathscr{F}_j,\\
&\psi : J \longrightarrow X : j \mapsto x,\\
&\mathscr{F} := \{J\}.
\end{aligned}$$

Then it follows that $\Sigma\sigma(\mathscr{F}) = \bigcap_{j\in J}\mathscr{F}_j$, $\lambda\psi(\mathscr{F})(x) = 0$ and $\sup_{j\in J}\lambda\sigma(j)(\psi(j)) = \sup_{j\in J}\lambda\mathscr{F}_j(x)$, which shows from (L) that $\lambda(\bigcap_{j\in J}\mathscr{F}_j) \leq \sup_{j\in J}\lambda\mathscr{F}_j$. The other inequality follows from (L2w).

Conversely, if λ is a limit operator, then let J be any set, let $\psi : J \longrightarrow X$, $\sigma : J \longrightarrow \mathsf{F}(X)$ and $\mathscr{F} \in \mathsf{F}(J)$. Put

$$\rho : X \longrightarrow \mathsf{F}(X) : x \mapsto \begin{cases} \dot{x} & x \notin \psi(J),\\ \bigcap_{j\in\psi^{-1}(x)} \sigma(j) & x \in \psi(J), \end{cases}$$

and consider $\psi(\mathscr{F})$ on X. If $A \in \Sigma\rho(\psi(\mathscr{F}))$ then there exists $F \in \mathscr{F}$ such that $A \in \bigcap_{x\in\psi(F)}\rho(x)$. For $j \in F$ we then have $A \in \rho(\psi(j)) = \bigcap_{\psi(k)=\psi(j)}\sigma(k)$, in particular $A \in \sigma(j)$. Hence $\Sigma\rho(\psi(\mathscr{F})) \subseteq \Sigma\sigma(\mathscr{F})$ and thus

$$\begin{aligned}
\lambda\Sigma\sigma(\mathscr{F}) &\leq \lambda\Sigma\rho(\psi(\mathscr{F}))\\
&\leq \lambda\psi(\mathscr{F}) + \sup_{x\in X}\lambda\rho(x)(x)\\
&= \lambda\psi(\mathscr{F}) + \sup_{x\in\psi(J)}\lambda\Big(\bigcap_{k\in\psi^{-1}(x)}\sigma(k)\Big)(x)\\
&= \lambda\psi(\mathscr{F}) + \sup_{j\in J}\lambda\sigma(j)(\psi(j)).
\end{aligned}$$

□

In spite of the fact that (L) (together with (L1) and (L2w)) is all that is required in a complete axiom system for a limit operator actually an even stronger version holds.

1.1.8 Theorem *Given a function* $\lambda : \mathsf{F}(X) \longrightarrow \mathbb{P}^X$ *satisfying* (L1) *and* (L2w), λ *is a limit operator if and only if it satisfies the following properties.*

(L*) *For any set* J, *for any* $\psi : J \longrightarrow X$, *for any* $\sigma : J \longrightarrow \mathsf{F}(X)$ *and for any* $\mathscr{F} \in \mathsf{F}(J)$

$$\lambda \Sigma\sigma(\mathscr{F}) \leq \lambda\psi(\mathscr{F}) + \inf_{F\in\mathscr{F}} \sup_{j\in F} \lambda\sigma(j)(\psi(j)).$$

Proof Obviously (L*) implies (L). Conversely, suppose that J, $\psi : J \longrightarrow X$, $\sigma : J \longrightarrow \mathsf{F}(X)$ and $\mathscr{F} \in \mathsf{F}(J)$ are given and suppose that

$$\inf_{F\in\mathscr{F}} \sup_{j\in F} \lambda\sigma(j)(\psi(j)) < \varepsilon.$$

Choose F_0 such that $\sup_{j\in F_0} \lambda\sigma(j)(\psi(j)) < \varepsilon$ and define

$$\sigma' : J \longrightarrow \mathsf{F}(X) : j \mapsto \sigma'(j) := \begin{cases} \text{stack}\psi(j) & j \notin F_0, \\ \sigma(j) & j \in F_0. \end{cases}$$

From the fact that $\{F \in \mathscr{F} \mid F \subseteq F_0\}$ is a basis for $\mathscr{F}$ it follows that $\sigma'\mathscr{F} = \sigma\mathscr{F}$ and making use of (L1) it further follows from the definition of σ' that

$$\sup_{j\in J} \lambda\sigma'(j)(\psi(j)) = \sup_{j\in F_0} \lambda\sigma(j)(\psi(j)) < \varepsilon.$$

Hence

$$\lambda\Sigma\sigma(\mathscr{F}) \leq \lambda\psi\mathscr{F} + \varepsilon$$

and we are finished. □

In a perfectly analogous way one can prove the following result involving the original axioms where (L3) gets replaced by a similarly stronger version.

1.1.9 Theorem *Given a function* $\lambda : \mathsf{F}(X) \longrightarrow \mathbb{P}^X$ *satisfying* (L1) *and* (L2), λ *is a limit operator if and only if it satisfies the following property.*

(L3*) *For any* $\mathscr{F} \in \mathsf{F}(X)$ *and any selection of filters* $\sigma : X \longrightarrow \mathsf{F}(X)$

$$\lambda\Sigma\sigma(\mathscr{F}) \leq \lambda\mathscr{F} + \inf_{F\in\mathscr{F}} \sup_{x\in F} \lambda\sigma(x)(x).$$

Proof This is analogous to 1.1.8 and we leave this to the reader. □

The foregoing characterizations were formulated in terms of filters. It is however also possible to do this in terms of ultrafilters. Of course (L1) is already a property of ultrafilters and neither (L2) nor (L2w) makes sense when using ultrafilters, so we are basically only looking at conditions (L) and (L*). Further, when we say that a characterization with ultrafilters suffices, then this is of course in the understanding that the limit operator working on arbitrary filters is deduced from the limit operator working on ultrafilters by the formula $\lambda\mathcal{F} := \sup_{\mathcal{U}\in\mathsf{U}(\mathcal{F})} \lambda\mathcal{U}$.

We require the following lemma.

1.1.10 Lemma *If* $\sigma : J \longrightarrow \mathsf{U}(X)$ *is a selection of ultrafilters on* X *then*

$$\mathsf{U}(\Sigma\sigma(\{J\})) = \{\Sigma\sigma(\mathcal{W}) \mid \mathcal{W} \in \mathsf{U}(J)\}.$$

Proof One inclusion follows from 1.1.3. To prove the other inclusion let $\mathcal{V}$ be an ultrafilter finer than $\Sigma\sigma(\{J\})$ and suppose that for every ultrafilter $\mathcal{W}$ on J we have $\Sigma\sigma(\mathcal{W}) \nsubseteq \mathcal{V}$, i.e.

$$\exists\, A_{\mathcal{W}} \in \bigcup_{W\in\mathcal{W}} \bigcap_{j\in W} \sigma(j) \text{ such that } A_{\mathcal{W}} \notin \mathcal{V}.$$

This implies that there exists $W \in \mathcal{W}$ such that $A_{\mathcal{W}} \in \bigcap_{j\in W} \sigma(j)$ but $A_{\mathcal{W}} \notin \mathcal{V}$. From 1.1.4 it then follows that there exist ultrafilters $\mathcal{W}_1, \ldots, \mathcal{W}_n$ on J and sets $W_k \in \mathcal{W}_k$, $k = 1, \ldots, n$ such that $\bigcup_{i=1}^n W_i = J$ and $\bigcup_{i=1}^n A_{\mathcal{W}_i} \notin \mathcal{V}$. Hence, since $\Sigma\sigma(\{J\}) = \bigcap_{j\in J} \sigma(j)$ we also have

$$\bigcup_{i=1}^n A_{\mathcal{W}_i} \notin \bigcap_{j\in J} \sigma(j).$$

This then implies that there exists $j \in J$ such that $\bigcup_{i=1}^n A_{\mathcal{W}_i} \notin \sigma(j)$. Now there further exists $i \in \{1, \ldots, n\}$ such that $j \in W_i$ and from the supposition we have that $A_{\mathcal{W}_i} \in \sigma(j)$ which is a contradiction.

This proves that for every ultrafilter $\mathcal{V}$ finer than $\Sigma\sigma(\{J\})$ there exists an ultrafilter $\mathcal{W} \in \mathsf{U}(J)$ such that $\Sigma\sigma(\mathcal{W}) \subseteq \mathcal{V}$ and hence, by 1.1.3, that $\Sigma\sigma(\mathcal{W}) = \mathcal{V}$. This proves the result. □

We now give the announced result for the ultrafilter version of (L*), the result and proof for (L) are perfectly similar.

1.1.11 Theorem *Given a function* $\lambda : \mathsf{U}(X) \longrightarrow \mathbb{P}^X$ *satisfying* (L1), *the extension to* $\mathsf{F}(X)$ *defined by*

$$\overline{\lambda} : \mathsf{F}(X) \longrightarrow \mathbb{P}^X : \mathcal{F} \mapsto \sup_{\mathcal{U}\in\mathsf{U}(\mathcal{F})} \lambda\mathcal{U}$$

is a limit operator if and only if it satisfies the following property.

(LU*) *For any set J, for any $\psi : J \longrightarrow X$, for any $\sigma : J \longrightarrow \mathsf{U}(X)$ and for any $\mathscr{F} \in \mathsf{U}(J)$*

$$\lambda \Sigma\sigma(\mathscr{F}) \leq \lambda\psi(\mathscr{F}) + \inf_{F\in\mathscr{F}} \sup_{j\in F} \lambda\sigma(j)(\psi(j)).$$

Proof Of course we only need to prove the if-part and hereto we use 1.1.8. (L1) is clear and (L2w) follows from the definition of $\overline{\lambda}$. Because we require this in the sequel of the proof we first prove one inequality of (L2) for the case that all filters involved are ultrafilters. So let $\sigma : J \longrightarrow \mathsf{U}(X)$, fix $x \in X$ and let $\psi : J \longrightarrow X : j \mapsto x$. Then it follows from the foregoing lemma and from (LU*) that

$$\begin{aligned}
\overline{\lambda}(\bigcap_{j\in J} \sigma(j))(x) &= \overline{\lambda}(\Sigma\sigma(\{J\}))(x) \\
&= \sup_{\mathscr{V}\in\mathsf{U}(\Sigma\sigma(\{J\}))} \lambda\mathscr{V}(x) \\
&= \sup_{\mathscr{W}\in\mathsf{U}(J)} \lambda\Sigma\sigma(\mathscr{W})(x) \\
&\leq \sup_{\mathscr{W}\in\mathsf{U}(J)} (\lambda\psi(\mathscr{W})(x) + \inf_{W\in\mathscr{W}} \sup_{j\in W} \lambda\sigma(j)(\psi(j)) \\
&\leq \sup_{\mathscr{W}\in\mathsf{U}(J)} (\lambda\dot{x}(x) + \sup_{j\in J} \lambda\sigma(j)(x)) \\
&= \sup_{j\in J} \lambda\sigma(j)(x).
\end{aligned}$$

Now suppose given $\psi : J \longrightarrow X$, $\sigma : J \longrightarrow \mathsf{F}(X)$ and $\mathscr{F} \in \mathsf{F}(J)$. Then for any $\rho \in \prod_{j\in J} \mathsf{U}(\sigma(j))$ and $\mathscr{U} \in \mathsf{U}(\mathscr{F})$, by supposition, we have

$$\lambda\Sigma\rho(\mathscr{U}) \leq \lambda\psi(\mathscr{U}) + \sup_{U\in\mathscr{U}} \inf_{j\in U} \lambda\rho(j)(\psi(j)),$$

and hence, from the first part, 1.1.3 and 1.1.5, we obtain

$$\begin{aligned}
\overline{\lambda}\Sigma\sigma(\mathscr{F}) &= \overline{\lambda}(\bigcap_{\rho\in\prod_{j\in J}\mathsf{U}(\sigma(j))} \bigcap_{\mathscr{U}\in\mathsf{U}(\mathscr{F})} \Sigma\rho(\mathscr{U})) \\
&\leq \sup_{\rho\in\prod_{j\in J}\mathsf{U}(\sigma(j))} \sup_{\mathscr{U}\in\mathsf{U}(\mathscr{F})} (\lambda\psi(\mathscr{U}) + \sup_{U\in\mathscr{U}} \inf_{j\in U} \lambda\rho(j)(\psi(j))) \\
&\leq \overline{\lambda}\psi(\mathscr{F}) + \sup_{\rho\in\prod_{j\in J}\mathsf{U}(\sigma(j))} \sup_{\mathscr{U}\in\mathsf{U}(\mathscr{F})} \sup_{U\in\mathscr{U}} \inf_{j\in U} \lambda\rho(j)(\psi(j)) \\
&= \overline{\lambda}\psi(\mathscr{F}) + \sup_{\mathscr{U}\in\mathsf{U}(\mathscr{F})} \sup_{U\in\mathscr{U}} \sup_{\rho\in\prod_{j\in J}\mathsf{U}(\sigma(j))} \inf_{j\in U} \lambda\rho(j)(\psi(j)) \\
&= \overline{\lambda}\psi(\mathscr{F}) + \sup_{\mathscr{U}\in\mathsf{U}(\mathscr{F})} \sup_{U\in\mathscr{U}} \inf_{j\in U} \sup_{\mathscr{W}\in\mathsf{U}(\sigma(j))} \lambda\mathscr{W}(\psi(j))
\end{aligned}$$

$$= \overline{\lambda}\psi(\mathscr{F}) + \sup_{\mathscr{U}\in \mathsf{U}(\mathscr{F})} \sup_{U\in\mathscr{U}} \inf_{j\in U} \overline{\lambda}\sigma(j)(\psi(j))$$
$$= \overline{\lambda}\psi(\mathscr{F}) + \sup_{\mathscr{U}\in \mathsf{U}(\mathscr{F})} \inf_{U\in\mathscr{U}} \sup_{j\in U} \overline{\lambda}\sigma(j)(\psi(j)).$$

Now, since obviously for any $\mathscr{U} \in \mathsf{U}(\mathscr{F})$

$$\inf_{U\in\mathscr{U}} \sup_{j\in U} \overline{\lambda}\sigma(j)(\psi(j)) \leq \inf_{F\in\mathscr{F}} \sup_{j\in F} \overline{\lambda}\sigma(j)(\psi(j))$$

this proves that also (L*) is fulfilled. □

Note that in the foregoing theorem, by 1.1.5, the second term on the right-hand side can also be written as $\sup_{F\in\mathscr{F}} \inf_{j\in F} \lambda\sigma(j)(\psi(j))$.

1.1.12 Theorem *Given a function* $\lambda : \mathsf{U}(X) \longrightarrow \mathbb{P}^X$ *satisfying* (L1), *the extension to* $\mathsf{F}(X)$ *defined by*

$$\overline{\lambda} : \mathsf{F}(X) \longrightarrow \mathbb{P}^X : \mathscr{F} \mapsto \sup_{\mathscr{U}\in\mathsf{U}(\mathscr{F})} \lambda\mathscr{U}$$

is a limit operator if and only if it satisfies the following property.

(LU) *For any set* J, *for any* $\psi : J \longrightarrow X$, *for any* $\sigma : J \longrightarrow \mathsf{U}(X)$ *and for any* $\mathscr{F} \in \mathsf{U}(J)$

$$\lambda\Sigma\sigma(\mathscr{F}) \leq \lambda\psi(\mathscr{F}) + \sup_{j\in J} \lambda\sigma(j)(\psi(j)).$$

Proof This is perfectly analogous to the previous theorem and we leave this to the reader. □

Approach Systems

Approach systems can be thought of as a localization of the notion of metric. In each point of the space X we give a collection of $\mathbb{P}$-valued functions, called *local distances*, each of which measures a distance from the given point to any other point of the space. Here too, in order to give the precise definitions, we require some preliminary concepts.

A nonempty subset $\mathscr{A}$ of $\mathbb{P}$-valued functions on a given set X is called an *ideal* in $\mathbb{P}^X$ if it is closed under the operation of taking finite suprema and under the operation of taking smaller functions. In other words, if it is an ideal (dual filter) in the lattice $\mathbb{P}^X$ in the order-theoretic sense, when $\mathbb{P}^X$ is equipped with the pointwise order (see e.g. Birkhoff 1967). Note that a priori we are allowing for the improper ideal $\mathbb{P}^X$, however because of condition (A1) below this will not occur in the present context.

Given a collection of functions $\mathcal{A} \subseteq \mathbb{P}^X$ and a function $\varphi \in \mathbb{P}^X$, we will say that φ is *dominated by* $\mathcal{A}$, or that $\mathcal{A}$ *dominates* φ, if

$$\forall \varepsilon > 0, \forall \omega < \infty : \exists \varphi_\varepsilon^\omega \in \mathcal{A} \text{ such that } \varphi \wedge \omega \leq \varphi_\varepsilon^\omega + \varepsilon.$$

We will then also say that the family $(\varphi_\varepsilon^\omega)_{\varepsilon>0,\omega<\infty}$ dominates φ.

Further we will say that a collection of functions $\mathcal{A} \subseteq \mathbb{P}^X$ is *saturated*, if any function which is dominated by $\mathcal{A}$ already belongs to $\mathcal{A}$.

1.1.13 Definition (Approach system) A collection of ideals $(\mathcal{A}(x))_{x\in X}$ in $\mathbb{P}^X$, indexed by the points of X, is called an *approach system* if for all $x \in X$ the following properties hold.

(A1) $\forall \varphi \in \mathcal{A}(x) : \varphi(x) = 0$.
(A2) $\mathcal{A}(x)$ is saturated.
(A3) $\forall \varphi \in \mathcal{A}(x), \forall \varepsilon > 0, \forall \omega < \infty, \exists (\varphi_z)_{z\in X} \in \prod_{z\in X} \mathcal{A}(z)$ such that

$$\forall z, y \in X : \varphi(y) \wedge \omega \leq \varphi_x(z) + \varphi_z(y) + \varepsilon.$$

For any $x \in X$, a function in $\mathcal{A}(x)$ is called a *local distance* (in x). Note that (A2) implies that $\mathcal{A}(x)$ is closed under the operation of taking smaller functions and that (A1) then implies that the constant function 0 is in $\mathcal{A}(x)$. So the only condition of ideal which, on top of (A1)–(A3), we need to impose explicitly is the one which stipulates that $\mathcal{A}(x)$ must be closed under the formation of finite suprema. (A3) will sometimes be referred to as the *mixed triangular inequality*. The value $\varphi(y)$ of a local distance $\varphi \in \mathcal{A}(x)$ at a point $y \in X$ is interpreted as "the distance from x to y according to φ". The set of local distances in a point can be compared to the set of neighbourhoods of a point in a topological space. Each neighbourhood determines its own set of points which are considered close by (in the neighbourhood of) the given point. In the same way each local distance makes its own measurement of the distance other points in the space lie from the given point.

Often one can determine collections $\mathcal{B}(x), x \in X$, which would be natural candidates to form an approach system, but not all required properties are fulfilled. In particular property (A2) is not often automatically fulfilled. To handle this we introduce a type of basis for approach systems. We recall that a subset $\mathcal{B}$ of $\mathbb{P}^X$ is called an *ideal basis* in $\mathbb{P}^X$ if, for any $\alpha, \beta \in \mathcal{B}$, there exists $\gamma \in \mathcal{B}$ such that $\alpha \vee \beta \leq \gamma$. In other words, if $\mathcal{B}$ is an ideal basis (dual filter basis) in the order-theoretic sense (always considering $\mathbb{P}^X$ to be equipped with the pointwise order).

1.1.14 Definition A collection of ideal bases $(\mathcal{B}(x))_{x\in X}$ in $\mathbb{P}^X$ is called an *approach basis* if, for all $x \in X$, the following properties hold.

(B1) $\forall \varphi \in \mathcal{B}(x) : \varphi(x) = 0$.
(B2) $\forall \varphi \in \mathcal{B}(x), \forall \varepsilon > 0, \forall \omega < \infty, \exists (\varphi_z)_{z\in X} \in \prod_{z\in X} \mathcal{B}(z)$ such that

$$\forall z, y \in X : \varphi(y) \wedge \omega \leq \varphi_x(z) + \varphi_z(y) + \varepsilon.$$

Notice that (B1) is actually (A1) and that (B2) is actually (A3). Hence an approach system is also an approach basis, and any result for approach bases will also hold for approach systems.

In order to derive the set of all local distances from an approach basis we will also require the following *saturation operation*. Given a subset $\mathcal{B} \subseteq \mathbb{P}^X$ we define

$$\widehat{\mathcal{B}} := \left\{ \varphi \in \mathbb{P}^X \mid \mathcal{B} \text{ dominates } \varphi \right\}.$$

We call $\widehat{\mathcal{B}}$ the *saturation* of $\mathcal{B}$.

1.1.15 Definition A collection of ideal bases $(\mathcal{B}(x))_{x \in X}$ is called a *basis for an approach system* $(\mathcal{A}(x))_{x \in X}$, if for all $x \in X$, $\mathcal{A}(x)$ equals the saturation of $\mathcal{B}(x)$, i.e. $\mathcal{A}(x) = \widehat{\mathcal{B}(x)}$. In this case we also say that $(\mathcal{B}(x))_{x \in X}$ *generates* $(\mathcal{A}(x))_{x \in X}$ or that $(\mathcal{A}(x))_{x \in X}$ is *generated by* $(\mathcal{B}(x))_{x \in X}$.

1.1.16 Proposition *If $(\mathcal{B}(x))_{x \in X}$ is an approach basis, then $(\widehat{\mathcal{B}(x)})_{x \in X}$ is an approach system with $(\mathcal{B}(x))_{x \in X}$ as basis and if $(\mathcal{B}(x))_{x \in X}$ is a basis for an approach system $(\mathcal{A}(x))_{x \in X}$, then it is an approach basis.*

Proof To prove the first claim, note that (A1) is trivial. To prove (A2) let $x \in X$ and $\varphi \in \mathbb{P}^X$ be such that, for all $\omega < \infty$ and $\varepsilon > 0$, there exists $\psi \in \widehat{\mathcal{B}(x)}$ such that $\varphi \wedge \omega \leq \psi + \frac{\varepsilon}{2}$. If we choose $\tau \in \mathcal{B}(x)$ such that $\psi \wedge \omega \leq \tau + \frac{\varepsilon}{2}$, then it follows that $\varphi \wedge \omega \leq \tau + \varepsilon$. To prove (A3) let $\varphi \in \widehat{\mathcal{B}(x)}$ and let $\omega < \infty$ and $\varepsilon > 0$. Choose $\psi \in \mathcal{B}(x)$ such that $\varphi \wedge \omega \leq \psi + \frac{\varepsilon}{2}$ and then choose $(\psi_z)_z \in \prod_{z \in X} \mathcal{B}(z)$ such that, for all $z, y \in X$,

$$\psi(y) \wedge \omega \leq \psi_x(z) + \psi_z(y) + \frac{\varepsilon}{2}.$$

Then it follows that, for all $z, y \in X$,

$$\begin{aligned} \varphi(y) \wedge \omega &\leq \psi(y) \wedge \omega + \frac{\varepsilon}{2} \\ &\leq \psi_x(z) + \psi_z(y) + \varepsilon. \end{aligned}$$

To prove the second claim, clearly, for any $x \in X$, $\mathcal{B}(x)$ is a basis for an ideal and (B1) is fulfilled. To prove (B2) let $\psi \in \mathcal{B}(x)$ and let $\omega < \infty$ and $\varepsilon > 0$. Choose $(\varphi_z)_z \in \prod_{z \in X} \mathcal{A}(z)$ such that, for all $z, y \in X$,

$$\psi(y) \wedge \omega \leq \varphi_x(z) + \varphi_z(y) + \frac{\varepsilon}{2}.$$

For each $\varphi_z \in \mathscr{A}(z)$, choose $\psi_z \in \mathscr{B}(z)$ such that $\varphi_z \wedge \omega \leq \psi_z + \frac{\varepsilon}{4}$. Then it follows that, for all $z, y \in X$,

$$\begin{aligned}\psi(y) \wedge \omega &\leq (\varphi_x(z) + \varphi_z(y) + \frac{\varepsilon}{2}) \wedge (\omega + \frac{\varepsilon}{2}) \\ &\leq \varphi_x(z) \wedge \omega + \varphi_z(y) \wedge \omega + \frac{\varepsilon}{2} \\ &\leq \psi_x(z) + \psi_z(y) + \varepsilon.\end{aligned}$$

□

1.1.17 Definition It follows from the saturation condition that the set $\mathscr{A}_b(x)$ of all bounded functions in $\mathscr{A}(x)$ is a particularly interesting basis. It satisfies the saturation condition in a simpler form, which says that for all $\mu \in \mathbb{P}^X$ bounded:

$$\forall \varepsilon > 0, \exists \varphi \in \mathscr{A}_b(x) : \mu \leq \varphi + \varepsilon \Rightarrow \mu \in \mathscr{A}_b(x).$$

We refer to this collection as the *bounded approach basis* or *bounded approach system*. We could have chosen to restrict ourselves to bounded functions in the definition of approach systems, however there are good reasons not do so, as we will see e.g. in the definition of a gauge and in the second chapter (see however the definition of functional ideal convergence).

Gauges

Although we will not be requiring any categorical considerations just yet, for notational convenience we already make the following conventions. The category of all quasi-metric spaces (respectively metric spaces) equipped with non-expansive maps as morphisms is denoted $q\mathsf{Met}$ (respectively Met).

The type of structure which we introduce in this section should be compared with the concept of a uniform space via a family of metrics, called a *uniform gauge*. Note however that in the chapter on uniform gauge spaces we will use this term for a different concept. Whereas a uniform gauge, in the above meaning, satisfies a uniform or global saturation condition, here we require the gauge to fulfil only a local condition with a saturation specific for the theory. We will return to uniform aspects in the chapter on uniform gauge spaces. Given a collection $\mathscr{D} \subseteq q\mathsf{Met}(X)$ and a quasi-metric $d \in q\mathsf{Met}(X)$, we will say that d is *locally dominated by* $\mathscr{D}$, or that $\mathscr{D}$ *locally dominates* d, if for all $x \in X$, $\varepsilon > 0$ and $\omega < \infty$ there exists a $d_x^{\varepsilon,\omega} \in \mathscr{D}$ such that

$$d(x, \cdot) \wedge \omega \leq d_x^{\varepsilon,\omega}(x, \cdot) + \varepsilon.$$

We will then also say that the family $(d_x^{\varepsilon,\omega})_{x \in X, \varepsilon > 0, \omega < \infty}$ locally dominates d.

Further we will say that a collection of quasi-metrics $\mathscr{D}$ is *locally saturated*, if any quasi-metric d which is locally dominated by $\mathscr{D}$ already belongs to $\mathscr{D}$.

1.1.18 Definition (Gauge) A subset $\mathcal{G}$ of $q\mathsf{Met}(X)$ is called a *gauge* if it is an ideal in $q\mathsf{Met}(X)$ which fulfils the following property.

(G1) $\mathcal{G}$ is locally saturated.

As was the case for approach systems, here too, it regularly happens that one has a collection of quasi-metrics which would be a natural candidate to form a gauge but not all conditions are fulfilled. The following type of collection will often be encountered.

1.1.19 Definition A subset $\mathcal{H}$ of $q\mathsf{Met}(X)$ is called *locally directed* if for any $\mathcal{H}_0 \subseteq \mathcal{H}$ finite we have that $\sup_{d \in \mathcal{H}_0} d$ is locally dominated by $\mathcal{H}$.

By definition, a gauge, being an ideal, is locally directed, and similarly to the situation for approach systems and approach bases, here too, any result shown to hold for locally directed sets will also hold for gauges.

In order to derive the gauge from a locally directed set we will also require a *local saturation operation* which is perfectly similar to the one for approach systems. Given a subset $\mathcal{D} \subseteq q\mathsf{Met}(X)$ we define

$$\widehat{\mathcal{D}} := \{d \in q\mathsf{Met}(X) \mid \mathcal{D} \text{ locally dominates } d\}.$$

We call $\widehat{\mathcal{D}}$ the *local saturation* of $\mathcal{D}$.

1.1.20 Definition A set $\mathcal{H}$ in $q\mathsf{Met}(X)$ is called a *basis for a gauge* $\mathcal{G}$ if $\widehat{\mathcal{H}} = \mathcal{G}$. In this case we also say that $\mathcal{H}$ *generates* $\mathcal{G}$ or that $\mathcal{G}$ is *generated by* $\mathcal{H}$.

Note that this definition of basis for a gauge differs from the one we used in Lowen (1997).

1.1.21 Proposition *If $\mathcal{H}$ is locally directed, then $\widehat{\mathcal{H}}$ is a gauge with $\mathcal{H}$ as basis and if $\mathcal{H}$ is a basis for a gauge $\mathcal{G}$, then it is locally directed.*

Proof This goes along the same lines as 1.1.16. □

1.1.22 Definition Here too it is worthwhile to mention that a particularly interesting basis for a gauge $\mathcal{G}$ is given by the set $\mathcal{G}_b$ consisting of all bounded quasi-metrics in $\mathcal{G}$. This set too satisfies the saturation condition in a simpler form, namely for any bounded quasi-metric d

$$\forall x \in X,\ \forall \varepsilon > 0,\ \exists d_x^\varepsilon \in \mathcal{G}_b : d(x, \cdot) \le d_x^\varepsilon(x, \cdot) + \varepsilon \Rightarrow d \in \mathcal{G}_b.$$

We refer to this collection as the *bounded gauge basis*. Note that many quasi-metrics are unbounded and hence restricting the definition of a gauge to bounded functions would not have been appropriate.

Towers

A tower is an ordered family of pre-topologies on X, indexed by the real numbers in $\mathbb{R}^+$, and fulfilling certain coherence conditions. The axioms are presented here in terms of closures, neighbourhood systems and convergence. In the book by Dikranjan and Tholen (1995) it is shown that this also presents some interesting features as a closure operator.

1.1.23 Definition **(Tower)** A family of functions

$$\mathrm{t}_\varepsilon : 2^X \longrightarrow 2^X \quad \varepsilon \in \mathbb{R}^+,$$

is called a *(closure-)tower* if it satisfies the following properties.

(T1) $\forall A \in 2^X, \forall \varepsilon \in \mathbb{R}^+ : A \subseteq \mathrm{t}_\varepsilon(A)$.
(T2) $\forall \varepsilon \in \mathbb{R}^+ : \mathrm{t}_\varepsilon(\emptyset) = \emptyset$.
(T3) $\forall A, B \in 2^X, \forall \varepsilon \in \mathbb{R}^+ : \mathrm{t}_\varepsilon(A \cup B) = \mathrm{t}_\varepsilon(A) \cup \mathrm{t}_\varepsilon(B)$.
(T4) $\forall A \in 2^X, \forall \varepsilon, \gamma \in \mathbb{R}^+ : \mathrm{t}_\varepsilon(\mathrm{t}_\gamma(A)) \subseteq \mathrm{t}_{\varepsilon+\gamma}(A)$.
(T5) $\forall A \in 2^X, \forall \varepsilon \in \mathbb{R}^+ : \mathrm{t}_\varepsilon(A) = \bigcap_{\varepsilon<\gamma} \mathrm{t}_\gamma(A)$.

Note that by (T3) and (T5) we have

$$\forall A \subseteq B \subseteq X, \forall \alpha, \beta \in \mathbb{R}^+ : \alpha \leq \beta \Rightarrow \mathrm{t}_\alpha(A) \subseteq \mathrm{t}_\beta(B).$$

We recall that a *pre-topology* on a set X is determined by an operator

$$\mathrm{cl} : 2^X \longrightarrow 2^X$$

which fulfils the properties. (1) $A \subseteq \mathrm{cl}(A)$, (2) $\mathrm{cl}(\emptyset) = \emptyset$, and (3) $\mathrm{cl}(A \cup B) = \mathrm{cl}(A) \cup \mathrm{cl}(B)$, for all $A, B \in 2^X$. This operator is then called a *pre-topological closure operator*. A set X equipped with a pre-topology is called a *pre-topological space*. Pre-topologies and pre-topological spaces and continuous maps between them were introduced by Choquet (1947). They form the objects and morphisms of a topological category, PrTop. It is of course immediately clear that a tower simply means that for all $\varepsilon \in \mathbb{R}^+$, t_ε is a pre-topological closure operator, such that (T4) and (T5) are fulfilled (whereby we note that t_0 is a topological closure operator).

Topologies and pre-topologies can of course both also be determined by their neighbourhood systems and their convergence structures, and this provides another way to view towers.

1.1.24 Definition A double-indexed family of filters $(\mathcal{V}_\varepsilon(x))_{x\in X, \varepsilon\in\mathbb{R}^+}$ is called a *(neighbourhood-)tower* if it satisfies the following properties.

(T1n) $\forall x \in X, \forall \varepsilon, \gamma \in \mathbb{R}^+, \forall V \in \mathcal{V}_{\varepsilon+\gamma}(x), \exists W \in \mathcal{V}_\varepsilon(x)$ such that $\forall z \in W : V \in \mathcal{V}_\gamma(z)$.
(T2n) $\forall \varepsilon \in \mathbb{R}^+ : \mathcal{V}_\varepsilon(x) = \cup_{\varepsilon<\gamma} \mathcal{V}_\gamma(x)$.

1.1.25 Definition A family of pretopological convergence structures $(\stackrel{\varepsilon}{\longrightarrow})_{\varepsilon\in\mathbb{R}^+}$ is called a *(limit-)tower* if it satisfies the following properties.

(T1c) $\forall \mathcal{F} \in \mathsf{F}(X), \forall x \in X, \forall \varepsilon, \gamma \in \mathbb{R}^+ : \varepsilon \leq \gamma \text{ and } \mathcal{F} \stackrel{\varepsilon}{\longrightarrow} x \Rightarrow \mathcal{F} \stackrel{\gamma}{\longrightarrow} x.$
(T2c) $\forall \mathcal{F} \in \mathsf{F}(X), \forall x \in X, \forall \varepsilon \in \mathbb{R}^+ : \mathcal{F} \stackrel{\varepsilon}{\longrightarrow} x \Leftrightarrow \forall \gamma \in]\varepsilon, \infty[: \mathcal{F} \stackrel{\gamma}{\longrightarrow} x.$
(T3c) $\forall \mathcal{F} \in \mathsf{F}(X), \forall x \in X, \forall$ selection of filters $(\sigma(x))_{x\in X}, \forall \varepsilon, \gamma \in \mathbb{R}^+$:

$$\mathcal{F} \stackrel{\varepsilon}{\longrightarrow} x, \forall y \in X : \sigma(y) \stackrel{\gamma}{\longrightarrow} y \Rightarrow \Sigma\sigma(\mathcal{F}) \stackrel{\varepsilon+\gamma}{\longrightarrow} x.$$

We leave it as an exercise to show that closure-towers, neighbourhood-towers and limit-towers are equivalent concepts. Various ways to go from one to the other are

$$V \in \mathcal{V}_\varepsilon(x) \Leftrightarrow x \notin \mathfrak{t}_\varepsilon(X \setminus V),$$
$$x \in \mathfrak{t}_\varepsilon(A) \Leftrightarrow \forall V \in \mathcal{V}_\varepsilon(x) : V \cap A \neq \emptyset,$$
$$\mathcal{F} \stackrel{\varepsilon}{\longrightarrow} x \Leftrightarrow \mathcal{V}_\varepsilon(x) \subseteq \mathcal{F},$$
$$x \in \mathfrak{t}_\varepsilon(A) \Leftrightarrow \exists \mathcal{F} \in \mathsf{F}(A) : \mathcal{F} \stackrel{\varepsilon}{\longrightarrow} x.$$

A thorough study of limit-towers can be found in Brock and Kent (1997a, 1998). One of their results shows that (T2c) and (T3c) can be captured in one single axiom which has remarkable links to a notion of regularity (see also Brock and Kent 1997b). The description of towers with neighbourhood systems was recently explicitly verified by Jaeger (2012).

In what follows we will simply speak of a tower for any of these descriptions and use whichever one suits us best.

We will encounter towers mainly in the chapter on categorical considerations.

Lower and Upper Hull Operators

The structures which we consider in this section are notions of hull operators for real-valued functions, and they are to be compared with well-known operations such as lower or upper semicontinuous regularization, convex regularization, or nonexpansive regularization of functions, as e.g. in Bourbaki (1960) and in Singer (1986).

1.1.26 Definition (Lower hull operator) A function

$$\mathfrak{l} : \mathbb{P}^X \longrightarrow \mathbb{P}^X$$

is called a *lower hull operator* if it satisfies the following properties.

(LH1) $\forall \mu \in \mathbb{P}^X : \mathfrak{l}(\mu) \leq \mu.$
(LH2) $\forall \mu, \nu \in \mathbb{P}^X : \mathfrak{l}(\mu \wedge \nu) = \mathfrak{l}(\mu) \wedge \mathfrak{l}(\nu).$
(LH3) $\forall \mu \in \mathbb{P}^X : \mathfrak{l}(\mathfrak{l}(\mu)) = \mathfrak{l}(\mu).$
(LH4) $\forall \mu \in \mathbb{P}^X, \forall \alpha \in \mathbb{P} : \mathfrak{l}(\mu + \alpha) = \mathfrak{l}(\mu) + \alpha.$

1.1.27 Proposition *If* $\mathfrak{l} : \mathbb{P}^X \longrightarrow \mathbb{P}^X$ *is a lower hull operator, then the following properties hold.*

1. $\forall \alpha$ *constant:* $\mathfrak{l}(\alpha) = \alpha$.
2. $\forall \mu, \nu \in \mathbb{P}^X : \mu \leq \nu \Rightarrow \mathfrak{l}(\mu) \leq \mathfrak{l}(\nu)$.
3. $\forall \mu \in \mathbb{P}^X, \forall \alpha \in [0, \inf \mu] : \mathfrak{l}(\mu - \alpha) = \mathfrak{l}(\mu) - \alpha$.
4. $\forall \mu \in \mathbb{P}^X, \forall \alpha \in \mathbb{P} : \mathfrak{l}(\mu \ominus \alpha) \geq \mathfrak{l}(\mu) \ominus \alpha$.

Proof The first property follows from (LH4) and (LH1) letting $\mu = 0$. The second property is immediate from (LH2). The third property follows from (LH4) since $\mathfrak{l}(\mu) = \mathfrak{l}(\mu - \alpha + \alpha) = \mathfrak{l}(\mu - \alpha) + \alpha$. The fourth property follows from the third one and the fact that $\mu \ominus \alpha = \mu \vee \alpha - \alpha$. □

In what follows, especially in our investigations concerning hull operators, we will sometimes need to restrict our attention to the set of all bounded functions in $\mathbb{P}^X$ which we will denote by $\mathbb{P}^X_b$.

1.1.28 Proposition *A lower hull operator* $\mathfrak{l}$ *on* X *is completely determined by its restriction to* $\mathbb{P}^X_b$. *In particular, for any* $\mu \in \mathbb{P}^X$ *and any set* $K \subseteq [0, \infty[$ *such that* $\sup K = \infty$, *we have*

$$\mathfrak{l}(\mu) = \sup_{\alpha \in K} \mathfrak{l}(\mu \wedge \alpha).$$

Proof This follows from (LH2) and 1.1.27. □

1.1.29 Proposition *Let* $\mathfrak{l} : \mathbb{P}^X_b \longrightarrow \mathbb{P}^X_b$ *fulfil the following properties.*

(LH1b) $\forall \mu \in \mathbb{P}^X_b : \mathfrak{l}(\mu) \leq \mu$.
(LH2b) $\forall \mu, \nu \in \mathbb{P}^X_b : \mathfrak{l}(\mu \wedge \nu) = \mathfrak{l}(\mu) \wedge \mathfrak{l}(\nu)$.
(LH3b) $\forall \mu \in \mathbb{P}^X_b : \mathfrak{l}(\mathfrak{l}(\mu)) = \mathfrak{l}(\mu)$.
(LH4b) $\forall \mu \in \mathbb{P}^X_b, \forall \alpha < \infty : \mathfrak{l}(\mu + \alpha) = \mathfrak{l}(\mu) + \alpha$.

Then

$$\mathfrak{l}^* : \mathbb{P}^X \longrightarrow \mathbb{P}^X : \mu \mapsto \sup_{\alpha < \infty} \mathfrak{l}(\mu \wedge \alpha)$$

is the unique lower hull operator on X, whose restriction to $\mathbb{P}^X_b$ *coincides with* $\mathfrak{l}$.

Proof (LH1) and (LH2) are immediate. To prove (LH3) let $\mu \in \mathbb{P}^X$. Then

$$\begin{aligned}
\mathfrak{l}^*(\mathfrak{l}^*(\mu)) &= \sup_{\alpha < \infty} \mathfrak{l}((\sup_{\beta < \infty} \mathfrak{l}(\mu \wedge \beta)) \wedge \alpha) \\
&= \sup_{\alpha < \infty} \mathfrak{l}(\sup_{\beta < \infty} (\mathfrak{l}(\mu \wedge \alpha) \wedge \beta)) \\
&= \sup_{\alpha < \infty} \mathfrak{l}(\mu \wedge \alpha) = \mathfrak{l}^*(\mu).
\end{aligned}$$

(LH4) follows from the observation that, for any $\mu \in \mathbb{P}^X$ and any $\alpha < \infty$, we have

$$\begin{aligned} \mathfrak{l}^*(\mu + \alpha) &= \sup_{\alpha \leq \beta < \infty} \mathfrak{l}(\mu \wedge (\beta - \alpha) + \alpha) \\ &= \sup_{\alpha \leq \beta < \infty} \mathfrak{l}(\mu \wedge (\beta - \alpha)) + \alpha \\ &= \mathfrak{l}^*(\mu) + \alpha, \end{aligned}$$

and that $\mathfrak{l}^*(\infty) = \infty$. Uniqueness finally follows from 1.1.28 □

In the sequel we will also require the following type of maps. For any $A \subseteq X$, we define

$$\theta_A : X \longrightarrow \mathbb{P} : x \mapsto \begin{cases} 0 & x \in A, \\ \infty & x \notin A. \end{cases}$$

We will call this map the *indicator of* A. The set of all such indicators on X will be denoted by $\mathsf{Ind}(X)$. If moreover $\omega < \infty$ then we put $\theta_A^\omega := \theta_A \wedge \omega$.

Further we will require to approximate functions in $\mathbb{P}_b^X$ uniformly from below by functions taking only a finite number of values. We will formalize the procedure in the following way.

Let us denote the set of all functions attaining only a finite number of values in $\mathbb{R}^+$ by $\mathsf{Fin}(X)$. Clearly, any function in $\mathsf{Fin}(X)$ can be written as $\inf_{i=1}^{n}(a_i + \theta_{A_i})$, where $(A_i)_{i=1}^n$ is a partitioning of X.

Given $\mu \in \mathbb{P}_b^X$, we will say that the family $(\mu_\varepsilon)_{\varepsilon>0}$ in $\mathsf{Fin}(X)$ is a *development of* μ if for all $\varepsilon > 0$

$$\mu_\varepsilon \leq \mu \leq \mu_\varepsilon + \varepsilon.$$

1.1.30 Proposition *A lower hull operator* $\mathfrak{l}$ *on* X *is completely determined by its restriction to* $\mathsf{Ind}(X)$. *In particular, for any* $\mu \in \mathbb{P}_b^X$ *and any development* $(\mu_\varepsilon := \inf_{i=1}^{n(\varepsilon)}(m_i^\varepsilon + \theta_{M_i^\varepsilon}))_{\varepsilon>0}$ *of* μ, *we have*

$$\mathfrak{l}(\mu) = \sup_{\varepsilon>0}(\inf_{i=1}^{n(\varepsilon)}(m_i^\varepsilon + \mathfrak{l}(\theta_{M_i^\varepsilon}))).$$

Proof Applying (LH2) and (LH4) it follows that, for any $\varepsilon > 0$, we have

$$\mathfrak{l}(\mu_\varepsilon) \leq \mathfrak{l}(\mu) \leq \mathfrak{l}(\mu_\varepsilon) + \varepsilon,$$

and hence, once again applying (LH2) and (LH4), it follows that

$$\mathfrak{l}(\mu) = \sup_{\varepsilon>0} \mathfrak{l}(\mu_\varepsilon) = \sup_{\varepsilon>0}(\inf_{i=1}^{n(\varepsilon)}(m_i^\varepsilon + \mathfrak{l}(\theta_{M_i^\varepsilon}))).$$ □

1.1.31 Corollary *If $\mathfrak{l}$ is a lower hull operator on X, $\mu \in \mathbb{P}^X$, and, for each $\omega < \infty$, $(\inf_{i=1}^{n(\varepsilon,\omega)} (m_i^{\varepsilon,\omega} + \theta_{M_i^{\varepsilon,\omega}}))_{\varepsilon>0}$ is a development of $\mu \wedge \omega$, then*

$$\mathfrak{l}(\mu) = \sup_{\omega<\infty} \sup_{\varepsilon>0}(\inf_{i=1}^{n(\varepsilon,\omega)} (m_i^{\varepsilon,\omega} + \mathfrak{l}(\theta_{M_i^{\varepsilon,\omega}}))).$$

We now define the counterpart of a lower hull operator, namely an upper hull operator. For technical reasons it is best to restrict the definition of this operator to bounded functions. The main reason why we do not equally restrict ourselves to bounded functions for the lower hull operator, in spite of 1.1.28, is because of the relation with distance functionals, which are generally unbounded. These relations will be proved in the sequel, in particular in 1.2.18 and 1.2.19.

1.1.32 Definition (Upper hull operator) A function

$$\mathfrak{u} : \mathbb{P}_b^X \longrightarrow \mathbb{P}_b^X$$

is called an *upper hull operator* if it satisfies the following properties.

(UH0) $\mathfrak{u}(0) = 0$.
(UH1) $\forall \mu \in \mathbb{P}_b^X : \mu \leq \mathfrak{u}(\mu)$.
(UH2) $\forall \mu, \nu \in \mathbb{P}_b^X : \mathfrak{u}(\mu \vee \nu) = \mathfrak{u}(\mu) \vee \mathfrak{u}(\nu)$.
(UH3) $\forall \mu \in \mathbb{P}_b^X : \mathfrak{u}(\mathfrak{u}(\mu)) = \mathfrak{u}(\mu)$.
(UH4) $\forall \mu \in \mathbb{P}_b^X, \forall \alpha < \infty : \mathfrak{u}(\mu + \alpha) = \mathfrak{u}(\mu) + \alpha$.

1.1.33 Proposition *If $\mathfrak{u} : \mathbb{P}_b^X \longrightarrow \mathbb{P}_b^X$ is an upper hull operator, then the following properties hold.*

1. *$\forall \alpha$ constant, $\alpha < \infty$: $\mathfrak{u}(\alpha) = \alpha$.*
2. *$\forall \mu, \nu \in \mathbb{P}_b^X : \mu \leq \nu \Rightarrow \mathfrak{u}(\mu) \leq \mathfrak{u}(\nu)$.*
3. *$\forall \mu \in \mathbb{P}_b^X, \forall \alpha \in [0, \inf \mu] : \mathfrak{u}(\mu - \alpha) = \mathfrak{u}(\mu) - \alpha$.*
4. *$\forall \mu \in \mathbb{P}_b^X, \forall \alpha < \infty : \mathfrak{u}(\mu \ominus \alpha) = \mathfrak{u}(\mu) \ominus \alpha$.*

Proof The first property follows from (UH4) and (UH0) letting $\mu = 0$. The second, third and fourth properties follow as in 1.1.27. □

It is possible to obtain results similar to those of 1.1.30 and 1.1.31 for an upper hull operator. However, since we will not require this in the sequel we leave this to an interested reader.

Lower and Upper Regular Function Frames

The following structures which we consider are those of collections of functions, fulfilling certain stability properties. These structures then are to be compared with collections of lower or upper semicontinuous functions or collections of nonexpansive functions.

1.1.34 Definition (Lower regular function frame) A collection of functions $\mathfrak{L} \subseteq \mathbb{P}^X$ is called a *lower regular function frame*, if it satisfies the following properties.

(LR1) $\forall \mathfrak{R} \subseteq \mathfrak{L} : \sup \mathfrak{R} \in \mathfrak{L}$.
(LR2) $\forall \mathfrak{R} \subseteq \mathfrak{L}$ finite : $\inf \mathfrak{R} \in \mathfrak{L}$.
(LR3) $\forall \mu \in \mathfrak{L}, \forall \alpha \in \mathbb{P} : \mu + \alpha \in \mathfrak{L}$.
(LR4) $\forall \mu \in \mathfrak{L}, \forall \alpha \in [0, \inf \mu] : \mu - \alpha \in \mathfrak{L}$.

The members of $\mathfrak{L}$ are called *lower regular functions*.

1.1.35 Proposition *If $\mathfrak{L} \subseteq \mathbb{P}^X$ is a lower regular function frame, then the following properties hold.*

1. *$\mathfrak{L}$ contains all constant functions.*
2. *$\forall \mu \in \mathfrak{L}, \forall \alpha \in \mathbb{P} : \mu \ominus \alpha \in \mathfrak{L}$.*

Proof Letting $\mathfrak{R} = \emptyset$ in (LR1) it follows that $0 \in \mathfrak{L}$ and then the first property follows from (LR3) letting $\mu = 0$. The second property follows from the first one, (LR1), and (LR4). □

1.1.36 Definition A *basis* for a lower regular function frame $\mathfrak{L}$ is a subset $\mathfrak{B} \subseteq \mathfrak{L}$ which is such that any function in $\mathfrak{L}$ can be obtained as a supremum of functions in $\mathfrak{B}$.

With this definition we follow the spirit of a basis for a topology. It is then, unlike the situation in topology, usually required to add an extra step in the construction of a lower regular function frame starting from an arbitrary collection.

1.1.37 Proposition *Given any subset $\mathfrak{G} \subseteq \mathbb{P}^X$ there exists a smallest lower regular function frame $\mathfrak{L}$ containing $\mathfrak{G}$.*

Proof One first adds all finite translations, then all finite infima and finally all suprema. □

1.1.38 Definition Again a particularly interesting basis for a given lower regular function frame $\mathfrak{L}$ is given by the set $\mathfrak{L}_b$ of bounded functions in $\mathfrak{L}$. We refer to this collection as the *bounded basis for a lower regular function frame* or the *bounded lower regular function frame*. The only difference with the complete lower regular function frame is that the bounded basis satisfies (LR1) only for those suprema which are bounded. The entire lower regular function frame is then obtained by adding all suprema.

1.1.39 Definition (Upper regular function frame) A collection of functions $\mathfrak{U} \subseteq \mathbb{P}_b^X$ is called an *upper regular function frame*, if it satisfies the following properties.

(UR1) $\forall \mathfrak{R} \subseteq \mathfrak{U} : \inf \mathfrak{R} \in \mathfrak{U}$.
(UR2) $\forall \mathfrak{R} \subseteq \mathfrak{U}$ finite : $\sup \mathfrak{R} \in \mathfrak{U}$.

(UR3) $\forall \mu \in \mathfrak{U}, \forall \alpha < \infty : \mu + \alpha \in \mathfrak{U}$.
(UR4) $\forall \mu \in \mathfrak{U}, \forall \alpha \in [0, \inf \mu] : \mu - \alpha \in \mathfrak{U}$.

The members of $\mathfrak{U}$ are called *upper regular functions*.

1.1.40 Proposition *If $\mathfrak{U} \subseteq \mathbb{P}_b^X$ is an upper regular function frame, then the following properties hold.*

1. *$\mathfrak{U}$ contains all finite constant functions.*
2. *$\forall \mu \in \mathfrak{U}, \forall \alpha \in \mathbb{P} : \mu \ominus \alpha \in \mathfrak{U}$.*

Proof Letting $\mathfrak{R} = \emptyset$ in (UR2) it follows that $0 \in \mathfrak{U}$ and then the first property follows from (UR3) letting $\mu = 0$. The second property follows from the first one, (UR2), and (UR4). □

1.1.41 Definition A *basis* for an upper regular function frame $\mathfrak{U}$ is a subset $\mathfrak{B} \subseteq \mathfrak{U}$ which is such that any function in $\mathfrak{U}$ can be obtained as an infimum of functions in $\mathfrak{B}$.

Here too the same remark holds as for lower regular function frames. Any set of bounded functions generates a smallest upper regular function frame containing it.

1.1.42 Proposition *Given any subset $\mathfrak{G} \subseteq \mathbb{P}_b^X$ there exists a smallest upper regular function frame $\mathfrak{U}$ containing $\mathfrak{G}$.*

Proof One first adds all finite translations, then all finite suprema and finally all infima. □

Functional Ideal Convergence

We now define our final structure which we refer to as *functional ideal convergence*. The idea behind this is to embed the numerical information of the theory into ideals of functions in $\mathbb{P}^X$ and to use these rather than filters to describe convergence as we did with limit operators.

1.1.43 Definition An (order theoretic) ideal $\mathfrak{I}$ in $\mathbb{P}^X$ is called a *functional ideal (on X)* if it fulfils the following properties.

(I1) Each function $\varphi \in \mathfrak{I}$ is bounded.
(I2) $\mathfrak{I}$ is *saturated* in the sense that for all $\mu \in \mathbb{P}^X$:

$$\forall \varepsilon > 0, \exists \varphi \in \mathfrak{I} : \mu \leq \varphi + \varepsilon \Rightarrow \mu \in \mathfrak{I}.$$

Note that condition (I1) implies that we are actually considering ideals in $\mathbb{P}_b^X$. Given a functional ideal $\mathfrak{I}$ we define its *characteristic value* as

$$c(\mathfrak{I}) := \sup_{\mu \in \mathfrak{I}} \inf_{x \in X} \mu(x) = \sup\{\alpha \mid \alpha \text{ constant}, \alpha \in \mathfrak{I}\}.$$

It follows immediately from the definition that there is only one functional ideal which has an infinite characteristic value and this is the functional ideal consisting of all bounded functions i.e. $\mathbb{P}_b^X$. We denote this functional ideal $\mathfrak{Z}_X$. For functional ideals it plays more or less the same role which the improper filter consisting of all subsets plays for filters. If necessary to differentiate, a functional ideal with a finite characteristic value will be called a *proper functional ideal* and $\mathfrak{Z}_X$ will be called the *improper functional ideal.*

Given a bounded function μ, the set of all functions which are smaller than or equal to μ is a functional ideal with characteristic value $\inf \mu$ and we say that this functional ideal is *generated by* μ. As with principal filters, we denote this by $\dot{\mu}$.

We have already encountered functional ideals. Given a set X with approach system $(\mathscr{A}(x))_{x \in X}$, then for each $x \in X$, the bounded approach system $\mathscr{A}_b(x)$ is a functional ideal with characteristic value equal to zero. We have also already remarked upon the fact that $\mathscr{A}(x)$ satisfies a stronger saturation property which says that for any function $\mu \in \mathbb{P}^X$

$$\forall \varepsilon > 0,\ \forall \omega < \infty,\ \exists \varphi \in \mathscr{A}(x) : \ \mu \wedge \omega \leq \varphi + \varepsilon \Rightarrow \mu \in \mathscr{A}(x),$$

but that obviously for a bounded function μ this is equivalent to the saturation condition in the definition of a functional ideal, namely condition (I2).

If $\mathfrak{B} \subseteq \mathbb{P}_b^X$ is an ideal then we can *saturate* it in the usual way:

$$\widehat{\mathfrak{B}} := \{\mu \in \mathbb{P}_b^X \mid \forall \varepsilon > 0,\ \exists \varphi \in \mathfrak{B} : \ \mu \leq \varphi + \varepsilon\}.$$

This then is a functional ideal and we say that $\mathfrak{B}$ is a *basis* for $\widehat{\mathfrak{B}}$.

1.1.44 Definition Given a proper functional ideal $\mathfrak{J}$ such that $c(\mathfrak{J}) \leq \alpha < \infty$ we define

$$\mathfrak{f}_\alpha(\mathfrak{J}) := \{\{\mu < \beta\} \mid \mu \in \mathfrak{J},\ \alpha < \beta\}.$$

It is easily verified that this is a filter on X. We will denote $\mathfrak{f}_{c(\mathfrak{J})}(\mathfrak{J})$ simply by $\mathfrak{f}(\mathfrak{J})$. The levels $\alpha \in [c(\mathfrak{J}), \infty[$ will be called $\mathfrak{J}$*-admissible*. If $\mathscr{F}$ is a filter on X then we define

$$\mathfrak{i}(\mathscr{F}) := \{\mu \in \mathbb{P}_b^X \mid \forall \alpha \in]0, \infty[: \{\mu < \alpha\} \in \mathscr{F}\}.$$

This is a proper functional ideal with characteristic value equal to zero and it is generated by $\{\theta_F^\omega \mid F \in \mathscr{F}, \omega < \infty\}$.

If $\mathfrak{J}$ is a functional ideal and $\alpha \in \mathbb{P}$ then we define

$$\mathfrak{J} \oplus \alpha := \begin{cases} \{\nu \mid \exists \mu \in \mathfrak{J} : \nu \leq \mu + \alpha\} & \alpha \text{ finite}, \\ \mathfrak{Z}_X & \alpha = \infty. \end{cases}$$

Obviously $c(\mathfrak{J} \oplus \alpha) = c(\mathfrak{J}) + \alpha$.

The collection of functional ideals is a "conditional" lattice in the following sense. Arbitrary infima always exist and are proper as long as at least one of the functional ideals involved is proper. If $(\mathfrak{I}_j)_{j\in J}$ is a family of functional ideals then the infimum is given by

$$\inf_{j\in J} \mathfrak{I}_j = \bigcap_{j\in J} \mathfrak{I}_j = \{\inf_{j\in J} \mu_j \mid \forall j \in J : \mu_j \in \mathfrak{I}_j\}.$$

In general the union of an arbitrary family of proper functional ideals is no longer a functional ideal. The supremum however always exists. If $(\mathfrak{I}_j)_{j\in J}$ is a family of proper functional ideals the supremum of the family is given by

$$\sup_{j\in J} \mathfrak{I}_j = \widehat{\{\sup_{k\in K} \mu_k \mid K \subseteq J \text{ finite}, \forall k \in K : \mu_k \in \mathfrak{I}_k\}}.$$

This supremum need not be proper as is easily seen from the following example.

1.1.45 Example For any $x \in \mathbb{R}$, let $\mathfrak{I}_x$ be the functional ideal on $\mathbb{R}$ generated by $\mu_x : \mathbb{R} \longrightarrow \mathbb{P} : y \mapsto |x - y|$, i.e. $\mathfrak{I}_x := \dot{\mu}_x$. Then, in spite of the fact that all these functional ideals have characteristic value equal to zero, for any $x, y \in \mathbb{R}$: $\inf_{z\in\mathbb{R}} \mu_x(z) \vee \mu_y(z) = \frac{1}{2}|x - y|$ and hence $\sup_{x\in\mathbb{R}} \mathfrak{I}_x = \mathfrak{Z}_{\mathbb{R}}$.

The improper functional ideal $\mathfrak{Z}_X$ obviously is the unique maximal element in the lattice of functional ideals. If we remove this from the lattice and only consider proper functional ideals then there are no maximal elements. Given any proper functional ideal $\mathfrak{I}$ and any $\alpha < \infty$ obviously $\mathfrak{I} \oplus \alpha$ is a strictly finer proper functional ideal.

The following elementary concepts and facts will be used freely in the sequel.

Let $f : X \longrightarrow X'$ be a function. If $\mu \in \mathbb{P}^X$, then we define and denote the function

$$f(\mu) : X' \longrightarrow \mathbb{P} : x' \mapsto \inf_{x\in f^{-1}(x')} \mu(x).$$

Further if $\mathfrak{I}$ is a functional ideal on X then we define and denote its image by f as

$$f(\mathfrak{I}) := \{\mu \mid \mu \circ f \in \mathfrak{I}\}.$$

It is immediately verified that this is indeed a functional ideal on X'.

1.1.46 Proposition *The following properties hold.*

1. $\forall \mu \in \mathbb{P}^X : f(\mu) \circ f \leq \mu$.
2. $\forall \nu \in \mathbb{P}^{X'} : f(\nu \circ f) \geq \nu$.
3. $\forall \mathfrak{I} \in \mathfrak{F}(X) : c(f(\mathfrak{I})) = c(\mathfrak{I})$.
4. $\forall \mathfrak{I} \in \mathfrak{F}(X), \forall \alpha \in [c(\mathfrak{I}), \infty[: \mathfrak{f}_\alpha(f(\mathfrak{I})) = f(\mathfrak{f}_\alpha(\mathfrak{I}))$.
5. $\forall \mathscr{F} \in \mathsf{F}(X) : f(\mathrm{i}(\mathscr{F})) = \mathrm{i}(f(\mathscr{F}))$.

Proof We leave this to the reader. □

1.1.47 Definition A functional ideal $\mathfrak{I}$ is called *prime* if for all bounded functions μ and ν

$$\mu \wedge \nu \in \mathfrak{I} \Rightarrow \mu \in \mathfrak{I} \text{ or } \nu \in \mathfrak{I}.$$

1.1.48 Proposition *For a proper functional ideal $\mathfrak{I}$, the following properties hold.*

1. $c(\mathfrak{I}) \in \mathfrak{I}$.
2. *If $\mathfrak{I}$ is prime and $A \in \mathfrak{f}(\mathfrak{I})$ then for any $\omega < \infty$: $(c(\mathfrak{I}) + \theta_A) \wedge \omega \in \mathfrak{I}$.*
3. *If $\mathfrak{I}$ is prime and for all $\alpha \in]c(\mathfrak{I}), \infty[$: $\{\mu < \alpha\} \in \mathfrak{f}(\mathfrak{I})$ then $\mu \in \mathfrak{I}$.*
4. *If $\mathfrak{I}$ is prime and $\{\mu \leq c(\mathfrak{I})\} \in \mathfrak{f}(\mathfrak{I})$ then $\mu \in \mathfrak{I}$.*

Proof The first property follows at once from the definition and (I2).

For the second property, if $\omega \leq c(\mathfrak{I})$ there is nothing to prove, hence let $c(\mathfrak{I}) < \omega$. If $(c(\mathfrak{I})+\theta_A)\wedge\omega \notin \mathfrak{I}$ then it follows from the first property that $(c(\mathfrak{I})+\theta_{X\setminus A})\wedge\omega \in \mathfrak{I}$ which implies that $X \setminus A \in \mathfrak{f}(\mathfrak{I})$, a contradiction.

For the third property, for any $\varepsilon > 0$ put

$$\psi_\varepsilon := (c(\mathfrak{I}) + \theta_{\{\mu<c(\mathfrak{I})+\varepsilon\}}) \wedge \sup \mu \text{ and } \phi_\varepsilon := (c(\mathfrak{I}) + \theta_{\{\mu \geq c(\mathfrak{I})+\varepsilon\}}) \wedge \sup \mu.$$

Then $\phi_\varepsilon \wedge \psi_\varepsilon = c(\mathfrak{I}) \in \mathfrak{I}$, and since $\phi_\varepsilon \vee \mu \notin \mathfrak{I}$ it follows that $\psi_\varepsilon \in \mathfrak{I}$. Finally then, since $\mu \leq \psi_\varepsilon + \varepsilon$ for any $\varepsilon > 0$, it follows again from (I2) that $\mu \in \mathfrak{I}$.

The fourth property follows from the third one. □

1.1.49 Theorem *For a proper functional ideal $\mathfrak{I}$ the following properties are equivalent.*

1. *$\mathfrak{I}$ is prime.*
2. *For any $A, B \subseteq X$ and any $\omega < \infty$, if $(c(\mathfrak{I}) + \theta_{A\cup B}) \wedge \omega \in \mathfrak{I}$ then either*

$$(c(\mathfrak{I}) + \theta_A) \wedge \omega \in \mathfrak{I} \text{ or } (c(\mathfrak{I}) + \theta_B) \wedge \omega \in \mathfrak{I}.$$

3. *For any $A \subseteq X$ and any $\omega < \infty$ either*

$$(c(\mathfrak{I}) + \theta_A) \wedge \omega \in \mathfrak{I} \text{ or } (c(\mathfrak{I}) + \theta_{X\setminus A}) \wedge \omega \in \mathfrak{I}.$$

4. *For all $\alpha \in [c(\mathfrak{I}), \infty[$: $\mathfrak{f}_\alpha(\mathfrak{I})$ is ultra.*

Proof 1 ⇒ 2. If $\omega \leq c(\mathfrak{I})$ there is nothing to prove, hence let $c(\mathfrak{I}) < \omega$. The result now follows from the fact that

$$((c(\mathfrak{I}) + \theta_A) \wedge \omega) \wedge ((c(\mathfrak{I}) + \theta_B) \wedge \omega) = (c(\mathfrak{I}) + \theta_{A\cup B}) \wedge \omega \in \mathfrak{I}.$$

2 ⇒ 3. This is trivial.

3 ⇒ 4. Suppose that $X \setminus A \notin \mathfrak{f}_\alpha(\mathfrak{I})$ then it follows that with $\alpha < \omega$

$$(c(\mathfrak{I}) + \theta_{X\setminus A}) \wedge \omega \notin \mathfrak{I}$$

and hence

$$A = \{(c(\mathfrak{I}) + \theta_A) \wedge \omega < \omega\} \in \mathfrak{f}_\alpha(\mathfrak{I}).$$

$4 \Rightarrow 1$. Suppose $\mu \wedge \nu \in \mathfrak{I}$. Put $M := \{\nu \leq \mu\}$ and $N := \{\mu \leq \nu\}$. Now choose $\alpha \geq \sup(\mu \vee \nu)$ and let for instance $M \in \mathfrak{f}_\alpha(\mathfrak{I})$. Then there exists $\beta > \alpha$ and $\xi \in \mathfrak{I}$ such that $\{\xi < \beta\} \subseteq M$. It then follows that $\beta \wedge \theta_M \leq \xi$ and hence $\beta \wedge \theta_M \in \mathfrak{I}$. Consequently also $(\mu \wedge \nu) \vee (\beta \wedge \theta_M) \in \mathfrak{I}$ and since $\nu \leq (\mu \wedge \nu) \vee (\beta \wedge \theta_M)$ also $\nu \in \mathfrak{I}$. □

1.1.50 Theorem *If $\mathfrak{I}$ is a proper prime functional ideal then $\mathfrak{f}(\mathfrak{I})$ is an ultrafilter and moreover, $\mathfrak{f}_\alpha(\mathfrak{I}) = \mathfrak{f}(\mathfrak{I})$ for all $\mathfrak{I}$-admissible α. Conversely, if $\mathcal{U}$ is an ultrafilter then $\mathfrak{i}(\mathcal{U}) \oplus \alpha$ is a proper prime functional ideal with characteristic value equal to α for any $\alpha < \infty$.*

Proof The first claim follows from 1.1.49. For the second claim, if $\mathcal{U}$ is an ultrafilter, $\alpha < \infty$ and $\mu \wedge \nu \in \mathfrak{i}(\mathcal{U}) \oplus \alpha$ then it follows that for any $\beta > \alpha$ we have $\{\mu < \beta\} \in \mathcal{U}$ or $\{\nu < \beta\} \in \mathcal{U}$. Since these collections of sets are decreasing with decreasing values of β it follows that, for instance, $\{\mu < \beta\} \in \mathcal{U}$ for all $\beta > \alpha$ and this, by saturation, shows that $\mu \in \mathfrak{i}(\mathcal{U}) \oplus \alpha$. □

1.1.51 Theorem *A proper functional ideal $\mathfrak{I}$ is prime if and only if there exists an ultrafilter $\mathcal{U}$ and $\beta < \infty$ such that $\mathfrak{I} = \mathfrak{i}(\mathcal{U}) \oplus \beta$.*

Proof That $\mathfrak{i}(\mathcal{U}) \oplus \beta$ is prime was seen in 1.1.50. Conversely if $\mathfrak{I}$ is prime then it follows from 1.1.48 that $\mathfrak{i}\mathfrak{f}(\mathfrak{I}) \oplus c(\mathfrak{I}) \subseteq \mathfrak{I}$ and since the other inclusion always holds it follows that $\mathfrak{i}\mathfrak{f}(\mathfrak{I}) \oplus c(\mathfrak{I}) = \mathfrak{I}$. The result now follows again from 1.1.50. □

1.1.52 Proposition *If $\mathfrak{I}$ is a prime functional ideal and $\mathfrak{H}$ is a finer functional ideal then $\mathfrak{H}$ is prime too and moreover there exists $\alpha \geq 0$ such that $\mathfrak{H} = \mathfrak{I} \oplus \alpha$.*

Proof This follows from 1.1.49 and 1.1.51. □

1.1.53 Proposition *If $\mathfrak{I}$ is a proper functional ideal then the set of all finer prime functional ideals has minimal elements.*

Proof If $(\mathfrak{I}_j)_j$ is a chain of prime functional ideals finer than $\mathfrak{I}$ then it follows from the chain condition that $\bigcap_j \mathfrak{I}_j$ too is a prime functional ideal finer than $\mathfrak{I}$. The result now follows from a standard application of Zorn's lemma. □

In the sequel we will denote the set of all functional ideals on X as $\mathfrak{F}(X)$ and the set of all prime functional ideals as $\mathfrak{P}(X)$.

We denote the collection of all prime functional ideals finer that $\mathfrak{I}$ by $\mathfrak{P}(\mathfrak{I})$ and the subcollection of *minimal prime functional ideals* finer than $\mathfrak{I}$ by $\mathfrak{P}_m(\mathfrak{I})$. An important characterization of this latter set is given by the following result.

1.1.54 Theorem *Given a proper functional ideal $\mathfrak{I}$ the following properties are equivalent.*

1. $\mathfrak{H} \in \mathfrak{P}_m(\mathfrak{I})$.
2. *There exists an $\alpha \geq c(\mathfrak{I})$ and an ultrafilter $\mathcal{U}$ finer than $\mathfrak{f}_\alpha(\mathfrak{I})$ such that $\mathfrak{H} = \mathfrak{I} \vee \mathfrak{i}(\mathcal{U})$.*

Moreover, if $\mathcal{U}$ is an ultrafilter which is finer than some $\mathfrak{f}_\alpha(\mathfrak{I})$ then

$$c(\mathfrak{I} \vee \mathfrak{i}(\mathcal{U})) = \inf\{\gamma \mid \mathfrak{f}_\gamma(\mathfrak{I}) \subseteq \mathcal{U}\}.$$

Proof 2 $\Rightarrow$ 1. Take $\alpha \geq c(\mathfrak{I})$ and $\mathcal{U}$ an ultrafilter such that $\mathcal{U} \supset \mathfrak{f}_\alpha(\mathfrak{I})$. First note that $\mathfrak{I} \vee \mathfrak{i}(\mathcal{U})$ is proper. Indeed, if $\mu \in \mathfrak{I}$, $U \in \mathcal{U}$, $\omega < \infty$ then it is easily verified that

$$\inf_{x \in X} (\mu \vee (\theta_U \wedge \omega))(x) \leq \alpha.$$

Hence it follows from 1.1.50 and 1.1.52 that $\mathfrak{I} \vee \mathfrak{i}(\mathcal{U})$ is a prime functional ideal finer than $\mathfrak{I}$.

Now let $\mathfrak{M}$ be a prime functional ideal such that $\mathfrak{I} \subseteq \mathfrak{M} \subseteq \mathfrak{I} \vee \mathfrak{i}(\mathcal{U})$, and suppose there exists $\mu \in \mathfrak{I}$, $U \in \mathcal{U}$ and $c(\mathfrak{I} \vee \mathfrak{i}(\mathcal{U})) \vee \sup \mu < \omega < \infty$ such that $\mu \vee (\theta_U \wedge \omega) \notin \mathfrak{M}$. Since

$$\mu = (\mu \vee (\theta_U \wedge \omega)) \wedge (\mu \vee (\theta_{X \setminus U} \wedge \omega))$$

and since $\mathfrak{M}$ is prime, it follows that $\mu \vee (\theta_{X \setminus U} \wedge \omega) \in \mathfrak{M} \subseteq \mathfrak{I} \vee \mathfrak{i}(\mathcal{U})$. Hence it follows that

$$\omega = (\mu \vee (\theta_U \wedge \omega)) \vee (\mu \vee (\theta_{X \setminus U} \wedge \omega)) \in \mathfrak{I} \vee \mathfrak{i}(\mathcal{U})$$

which is impossible by the choice of ω.

Consequently $\mathfrak{M} = \mathfrak{I} \vee \mathfrak{i}(\mathcal{U})$ and thus $\mathfrak{I} \vee \mathfrak{i}(\mathcal{U}) \in \mathfrak{P}_m(\mathfrak{I})$.

1 $\Rightarrow$ 2. Take $\mathfrak{H} \in \mathfrak{P}_m(\mathfrak{I})$. Then it follows that $c(\mathfrak{I}) \leq c(\mathfrak{H})$. Put

$$\mathcal{U} := \mathfrak{f}(\mathfrak{H}) = \mathfrak{f}_{c(\mathfrak{H})}(\mathfrak{H}) \supset \mathfrak{f}_{c(\mathfrak{H})}(\mathfrak{I})$$

and consider $\mathfrak{I} \vee \mathfrak{i}(\mathcal{U})$. From the first part we know that this is a minimal prime functional ideal finer than $\mathfrak{I}$. To prove that $\mathfrak{H}$ and $\mathfrak{I} \vee \mathfrak{i}(\mathcal{U})$ are equal it is thus sufficient to prove any inclusion between the two. Since $\mathfrak{P}_m(\mathfrak{H}) = \{\mathfrak{H}\}$ we necessarily have $\mathfrak{H} \vee \mathfrak{i}(\mathcal{U}) = \mathfrak{H}$ and consequently it follows that $\mathfrak{I} \vee \mathfrak{i}(\mathcal{U}) \subseteq \mathfrak{H} \vee \mathfrak{i}(\mathcal{U}) = \mathfrak{H}$.

To show the final claim, note that if $\mathfrak{f}_\gamma(\mathfrak{I}) \subseteq \mathcal{U}$ then it follows that for any $\mu \in \mathfrak{I}$ and $U \in \mathcal{U}$ we have $\inf \mu \vee \theta_U \leq \gamma$ and conversely if $\sup_{\mu \in \mathfrak{I}} \sup_{U \in \mathcal{U}} \inf \mu \vee \theta_U < \gamma$ then $\mathfrak{f}_\gamma(\mathfrak{I}) \subseteq \mathcal{U}$ and we are finished. □

1.1.55 Theorem *If $\mathfrak{I}$ is a proper functional ideal then*

$$\mathfrak{I} = \bigcap\{\mathfrak{H} \mid \mathfrak{H} \in \mathfrak{P}_m(\mathfrak{I})\}.$$

Proof It is well known (see e.g. Bourbaki (1965)) that

$$\mathfrak{I} = \bigcap \{\mathfrak{H} \mid \mathfrak{H} \in \mathfrak{P}(\mathfrak{I})\}$$

from which the result immediately follows. □

The following is a functional ideal and prime functional ideal counterpart to 1.1.4 for filters and ultrafilters.

1.1.56 Proposition *If $\mathfrak{I}$ is a proper functional ideal, and for each minimal prime functional ideal $\mathfrak{K} \in \mathfrak{P}_m(\mathfrak{I})$ we have a function $\rho(\mathfrak{K}) \in \mathfrak{K}$ then for any $\alpha \in [c(\mathfrak{I}), \infty[$ there exists a finite set $\mathfrak{P}_\alpha \subseteq \mathfrak{P}_m(\mathfrak{I})$ such that*

$$\inf_{\mathfrak{K}\in\mathfrak{P}_\alpha} \rho(\mathfrak{K}) \in \mathfrak{I} \vee \mathrm{i}\mathfrak{f}_\alpha(\mathfrak{I}).$$

Proof We put $\mathsf{U}(\mathfrak{I})$ the set of all ultrafilters finer than $\mathfrak{f}_\alpha(\mathfrak{I})$ for some $\alpha \in [c(\mathfrak{I}), \infty[$. Then it follows from 1.1.54 that there exists

$$(\rho_1, \rho_2) : \mathsf{U}(\mathfrak{I}) \longrightarrow \bigcup_{\mathscr{U}\in\mathsf{U}(\mathfrak{I})} \mathfrak{I} \times \mathscr{U}$$

with $\rho_2(\mathscr{U}) \in \mathscr{U}$, such that if $\mathfrak{K} = \mathfrak{I} \vee \mathrm{i}(\mathscr{U}_\mathfrak{K})$ then

$$\rho(\mathfrak{K}) \le \rho_1(\mathscr{U}_\mathfrak{K}) \vee (\theta_{\rho_2(\mathscr{U}_\mathfrak{K})} \wedge \omega_{\mathscr{U}_\mathfrak{K}})$$

for some $\omega_{\mathscr{U}_\mathfrak{K}}$ finite. Then it follows from 1.1.4 that there exists a finite set $\mathsf{U}_\alpha \subseteq \mathsf{U}(\mathfrak{f}_\alpha(\mathfrak{I}))$ such that $\bigcup_{\mathscr{U}\in\mathsf{U}_\alpha} \rho_2(\mathscr{U}) \in \mathfrak{f}_\alpha(\mathfrak{I})$. Put $\mathfrak{P}_\alpha := \{\mathfrak{I} \vee \mathrm{i}(\mathscr{U}) \mid \mathscr{U} \in \mathsf{U}_\alpha\}$, then it follows that if we put $F := \cup_{\mathscr{U}\in\mathsf{U}_\alpha} \rho_2(\mathscr{U})$ then

$$\begin{aligned}\inf_{\mathfrak{K}\in\mathfrak{P}_\alpha} \rho(\mathfrak{K}) &\le \inf_{\mathscr{U}\in\mathsf{U}_\alpha} \rho_1(\mathscr{U}) \vee (\theta_{\rho_2(\mathscr{U})} \wedge \omega_\mathscr{U}) \\ &\le \sup_{\mathscr{U}\in\mathsf{U}_\alpha} \rho_1(\mathscr{U}) \vee (\theta_F \wedge (\sup_{\mathscr{U}\in\mathsf{U}_\alpha} \omega_\mathscr{U})).\end{aligned}$$

Hence $\inf_{\mathfrak{K}\in\mathfrak{P}_\alpha} \rho(\mathfrak{K}) \in \mathfrak{I} \vee \mathrm{i}\mathfrak{f}_\alpha(\mathfrak{I})$. □

In a similar way as for filters we require a *diagonal operation*. For this we need the following definition:

$$l : \mathbb{P}_b^X \longrightarrow \mathbb{P}_b^{\mathfrak{F}(X)} : \mu \mapsto [\mathfrak{I} \mapsto \inf\{\alpha \in \mathbb{P} \mid \mu \in \mathfrak{I} \oplus \alpha\}].$$

It follows from the following proposition that l is well-defined. Furthermore, by (I2) the infimum in the definition is actually a minimum so that for any $\mu \in \mathbb{P}_b^X$ and $\mathfrak{I} \in \mathfrak{F}(X)$ we have $\mu \in \mathfrak{I} \oplus l(\mu)(\mathfrak{I})$.

1.1.57 Proposition *The following properties hold.*

1. *For any* $\mu, \nu \in \mathbb{P}_b^X$ *and* $\mathfrak{I} \in \mathfrak{F}(X)$*:* $l(\mu \vee \nu)(\mathfrak{I}) = l(\mu)(\mathfrak{I}) \vee l(\nu)(\mathfrak{I})$.
2. *For any* $\mu, \nu \in \mathbb{P}_b^X$ *and* $\mathfrak{I} \in \mathfrak{P}(X)$*:* $l(\mu \wedge \nu)(\mathfrak{I}) = l(\mu)(\mathfrak{I}) \wedge l(\nu)(\mathfrak{I})$.
3. *For any* $\mathfrak{I} \in \mathfrak{F}(X)$*, if* θ *is constant then* $l\theta(\mathfrak{I}) = \theta \ominus c(\mathfrak{I})$ *and in particular* $l\theta \leq \theta$.
4. *For any* $\mathfrak{I} \in \mathfrak{F}(X)$ *and for any* $\mu \in \mathbb{P}_b^X$*, if* θ *is constant then*

$$l(\mu + \theta)(\mathfrak{I}) \vee \theta = l(\mu)(\mathfrak{I}) + \theta \text{ and } l(\mu \ominus \theta)(\mathfrak{I}) = l(\mu)(\mathfrak{I}) \ominus \theta,$$

and in particular $l(\mu + \theta) \leq l(\mu) + \theta$.
5. $l\mu$ *is an extension of* μ *in the sense that* $l\mu(\mathrm{i}(\dot{x})) = \mu(x)$ *for any* $x \in X$.
6. *For any* $\mathfrak{I} \in \mathfrak{F}(X)$ *and* $\mu \in \mathbb{P}_b^X$*:* $l\mu(\mathfrak{I}) = 0$ *if and only if* $\mu \in \mathfrak{I}$ *and in particular* $l\mu(\mathfrak{Z}_X) = 0$.

Proof All properties follow by straightforward verification and hence we leave this to the reader. □

Given sets J and X, a map $\mathfrak{s} : J \longrightarrow \mathfrak{F}(X)$ and a functional ideal $\mathfrak{I} \in \mathfrak{F}(J)$ then we define the *diagonal functional ideal of* $\mathfrak{s}$ *with respect to* $\mathfrak{I}$ as

$$\Sigma\mathfrak{s}(\mathfrak{I}) := \{\mu \in \mathbb{P}_b^X \mid l(\mu) \in \mathfrak{s}(\mathfrak{I})\}.$$

It follows from 1.1.57 that $\Sigma\mathfrak{s}(\mathfrak{I})$ is a well-defined functional ideal. We refer to $(\mathfrak{s}(j))_{j \in J}$ as being a *selection of functional ideals*.

We now give a useful alternative characterization of $\Sigma\mathfrak{s}(\mathfrak{I})$.

1.1.58 Theorem *If* X *and* J *are sets,* $\mathfrak{s} : J \longrightarrow \mathfrak{F}(X)$ *and* $\mathfrak{I} \in \mathfrak{F}(J)$ *then*

$$\Sigma\mathfrak{s}(\mathfrak{I}) = \bigcup_{\nu \in \mathfrak{I}} \bigcap_{j \in J} \mathfrak{s}(j) \oplus \nu(j).$$

Proof Suppose that $\mu \in \bigcup_{\nu \in \mathfrak{I}} \bigcap_{j \in J} \mathfrak{s}(j) \oplus \nu(j)$ then there exists $\nu \in \mathfrak{I}$ such that for all $j \in J$

$$l(\mu)(\mathfrak{s}(j)) \leq \nu(j)$$

which proves that $l(\mu) \circ \mathfrak{s} \in \mathfrak{I}$ and thus that $l(\mu) \in \mathfrak{s}(\mathfrak{I})$. This shows that $\mu \in \Sigma\mathfrak{s}(\mathfrak{I})$.

Conversely, let μ be such that $l(\mu) \in \mathfrak{s}(\mathfrak{I})$ and thus $l(\mu) \circ \mathfrak{s} \in \mathfrak{I}$. Since $\mu \in \bigcap_{j \in J} \mathfrak{s}(j) \oplus l(\mu)(\mathfrak{s}(j))$ this proves the other inclusion and we are finished. □

1.1.59 Proposition *If* X *and* J *are sets,* $\mathfrak{s} : J \longrightarrow \mathfrak{P}(X)$ *and* $\mathfrak{I} \in \mathfrak{P}(J)$ *then* $\Sigma\mathfrak{s}(\mathfrak{I}) \in \mathfrak{P}(X)$.

Proof This follows from the second property in 1.1.57. □

1.1.60 Proposition *If $\mathfrak{s} : J \longrightarrow \mathfrak{F}(X)$ and $\mathfrak{I} \in \mathfrak{F}(J)$ then*

$$\Sigma\mathfrak{s}(\mathfrak{I}) \subseteq \bigvee_{F \in \mathfrak{f}(\mathfrak{I})} \bigcap_{j \in F} \mathfrak{s}(j) \oplus c(\mathfrak{I}),$$

and if moreover $\mathfrak{I} = \mathfrak{i}(\mathscr{F}) \oplus \beta$ for some filter $\mathscr{F}$ and $\beta < \infty$ then we have an equality.

Proof Consider a function $\inf_{j \in J} \mu_j \oplus \mu(j)$ in $\Sigma\mathfrak{s}(\mathfrak{I})$, i.e. such that $\mu \in \mathfrak{I}$ and $\mu_j \in \mathfrak{s}(j)$ for all $j \in J$, then since for any $\varepsilon > 0$ we have

$$\inf_{j \in J} \mu_j \oplus \mu(j) \leq \left(\inf_{\{\mu < c(\mathfrak{I}) + \varepsilon\}} \mu_j \oplus c(\mathfrak{I})\right) + \varepsilon$$

this, by (I2), proves the inclusion.

In order to show equality, provided $\mathfrak{I} = \mathfrak{i}(\mathscr{F}) \oplus \beta$, consider a function of type $\inf_{j \in F} \mu_j \oplus \beta$ where $F \in \mathscr{F}$ and $\mu_j \in \mathfrak{s}(j)$ for each $j \in F$. Since this function is bounded we can choose $\omega < \infty$ which dominates it. We now put

$$\nu := (\theta_F + \beta) \wedge \omega,$$

then clearly $\nu \in \mathfrak{I}$. Further, consider a new choice of functions in the functional ideals $\mathfrak{s}(j)$, $j \in J$

$$\nu_j := \begin{cases} \mu_j & j \in F \\ 0 & j \notin F \end{cases}$$

then it follows that

$$\inf_{j \in F} \mu_j \oplus \beta = \inf_{j \in J} \nu_j \oplus \nu(j)$$

which proves the remaining inclusion. □

By 1.1.51 we have the following corollary.

1.1.61 Corollary *If $\mathfrak{s} : J \longrightarrow \mathfrak{F}(X)$ and $\mathfrak{I} \in \mathfrak{P}(J)$ then*

$$\Sigma\mathfrak{s}(\mathfrak{I}) = \bigvee_{F \in \mathfrak{f}(\mathfrak{I})} \bigcap_{j \in F} \mathfrak{s}(j) \oplus c(\mathfrak{I}).$$

The inclusion in 1.1.60 is, in general, strict.

1.1.62 Example Let $X := \mathbb{R}$ (with usual metric and topology), let $\mathfrak{I}$ be generated by $\{d(0, \cdot) \wedge \omega \mid \omega < \infty\}$ and take as selection $\mathfrak{s} : \mathbb{R} \longrightarrow \mathfrak{P}(\mathbb{R})$ where $\mathfrak{s}(z)$ is generated by $\{\theta_{\{z\}} \wedge \omega \mid \omega < \infty\}$. Then $\Sigma\mathfrak{s}(\mathfrak{I}) = \mathfrak{I}$ whereas $\bigvee_{F \in \mathfrak{f}(\mathfrak{I})} \bigcap_{z \in F} \mathfrak{s}(z) \oplus c(\mathfrak{I}) = \mathfrak{i}(\mathscr{V}(0))$ where $\mathscr{V}(0)$ stands for the usual Euclidean neighbourhood filter of 0.

1.1.63 Proposition *If* $\mathfrak{s} : J \longrightarrow \mathfrak{F}(X)$ *and* $\mathfrak{I} \in \mathfrak{P}(J)$ *then*

$$c(\Sigma\mathfrak{s}(\mathfrak{I})) = \sup_{F\in\mathfrak{f}(\mathfrak{I})} \inf_{j\in F} c(\mathfrak{s}(j)) + c(\mathfrak{I}) = \inf_{F\in\mathfrak{f}(\mathfrak{I})} \sup_{j\in F} c(\mathfrak{s}(j)) + c(\mathfrak{I}).$$

Proof The first equality follows from

$$\begin{aligned} c(\Sigma\mathfrak{s}(\mathfrak{I})) &= \sup_{F\in\mathfrak{f}(\mathfrak{I})} \sup_{\phi\in\prod_{j\in F}\mathfrak{s}(j)} \inf_{x\in X} \inf_{j\in F} \varphi(j)(x) \oplus c(\mathfrak{I}) \\ &= \sup_{F\in\mathfrak{f}(\mathfrak{I})} \inf_{j\in F} \sup_{\mu\in\mathfrak{s}(j)} \inf_{x\in X} \mu(x) \oplus c(\mathfrak{I}) \\ &= \sup_{F\in\mathfrak{f}(\mathfrak{I})} \inf_{j\in F} c(\mathfrak{s}(j)) + c(\mathfrak{I}). \end{aligned}$$

The second equality is a consequence of 1.1.5. □

The reason why we have to consider the improper functional ideal in our considerations is because we require the diagonal operation in our axioms for what we will call functional ideal convergence (see 1.1.65) and even if all functional ideals involved are proper (and prime) the diagonal may nevertheless be improper.

1.1.64 Example Take $X := \mathbb{R}$, let $\mathscr{U}$ be an ultrafilter which converges to ∞ (in $\overline{\mathbb{R}}$), put $\mathfrak{I} := \mathfrak{i}(\mathscr{U})$ and for each $z \in \mathbb{R}$ let $\mathfrak{s}(z) := \mathfrak{i}(\dot{z}) \oplus |z|$. Then $c(\Sigma\mathfrak{s}(\mathfrak{I})) = \sup_{U\in\mathscr{U}} \inf_{z\in U} c(\mathfrak{s}(z)) = \infty$ and hence $\Sigma\mathfrak{s}(\mathfrak{I}) = \mathfrak{Z}_{\mathbb{R}}$.

Given a set X we now define a notion of convergence for functional ideals which we will refer to as the *functional ideal convergence*.

1.1.65 Definition (Functional ideal convergence) A relation $\rightarrowtail \subseteq \mathfrak{F}(X) \times X$ is called a *functional ideal convergence* if it satisfies the following properties.

(F1) For every $x \in X : \mathfrak{i}(\dot{x}) \rightarrowtail x$.
(F2) If $(\mathfrak{I}_j)_{j\in J}$ is a family of functional ideals then $\bigcap_{j\in J} \mathfrak{I}_j \rightarrowtail x$ if and only if $\mathfrak{I}_j \rightarrowtail x$ for every $j \in J$.
(F3) If $\mathfrak{s} : X \longrightarrow \mathfrak{F}(X)$ is a selection of functional ideals such that $\mathfrak{s}(z) \rightarrowtail z$ for all $z \in X$ and $\mathfrak{I}$ is a functional ideal such that $\mathfrak{I} \rightarrowtail x$ then $\Sigma\mathfrak{s}(\mathfrak{I}) \rightarrowtail x$.

1.1.66 Proposition *Given a functional ideal convergence* $\rightarrowtail$ *on* X*, for any proper functional ideal* $\mathfrak{I} \in \mathfrak{F}(X)$ *and* $x \in X$*, the following properties are equivalent.*

1. $\mathfrak{I} \rightarrowtail x$.
2. $\forall \mathfrak{H} \in \mathfrak{P}_m(\mathfrak{I}) : \mathfrak{H} \rightarrowtail x$.
3. $\forall \mathfrak{H} \in \mathfrak{P}(\mathfrak{I}) : \mathfrak{H} \rightarrowtail x$.

Proof This follows from 1.1.55 and (F2). □

As was the case for limit operators, here too we are able to prove a useful alternative characterization which entails a weakening of (F2) and a strengthening of (F3).

1.1.67 Theorem *A relation* $\rightarrowtail \subseteq \mathfrak{F}(X) \times X$ *satisfying* (F1) *is a functional ideal convergence if and only if it satisfies the properties.*

(F2w) *For any* $\mathfrak{K} \subseteq \mathfrak{I}$ *and any* $x \in X$: $\mathfrak{K} \rightarrowtail x \Rightarrow \mathfrak{I} \rightarrowtail x$.
(F) *For any set* J, *for any* $\psi : J \longrightarrow X$, *for any* $\mathfrak{s} : J \longrightarrow \mathfrak{F}(X)$, *for any* $\mathfrak{I} \in \mathfrak{F}(J)$ *and for any* $x \in X$

$$(\forall j \in J : \mathfrak{s}(j) \rightarrowtail \psi(j) \text{ and } \psi(\mathfrak{I}) \rightarrowtail x) \Rightarrow \Sigma\mathfrak{s}(\mathfrak{I}) \rightarrowtail x.$$

Proof To show that (F2) is fulfilled, let $\mathfrak{I}_j$, $j \in J$ be a family of functional ideals such that $\mathfrak{I}_j \rightarrowtail x$ for every $j \in J$. Define $\mathfrak{s}$, ψ and $\mathfrak{I}$ as follows

$$\mathfrak{s} : J \longrightarrow \mathfrak{F}(X) : j \mapsto \mathfrak{I}_j,$$
$$\psi : J \longrightarrow X : j \mapsto x,$$
$$\mathfrak{I} := \dot{0} = \{0\}.$$

Then it follows that $\Sigma\mathfrak{s}(\mathfrak{I}) = \bigcap_{j\in J} \mathfrak{I}_j$, $\psi(\mathfrak{I}) = \mathfrak{i}(\dot{x}) \rightarrowtail x$, which, from (F), shows that $\bigcap_{j\in J} \mathfrak{I}_j \rightarrowtail x$. The other implication follows from (F2w).

That (F) implies (F3) is trivial.

Conversely, suppose that J, $\mathfrak{s}$, ψ and $\mathfrak{I}$ are as in (F) and put

$$\mathfrak{t} : X \longrightarrow \mathfrak{F}(X) : x \mapsto \begin{cases} \mathfrak{i}(\dot{x}) & x \notin \psi(J), \\ \bigcap_{j\in\psi^{-1}(x)} \mathfrak{s}(j) & x \in \psi(J). \end{cases}$$

Then it follows that for any $\mu \in \mathfrak{I}$

$$\begin{aligned} \bigcap_{x\in X} \mathfrak{t}(x) \oplus \psi(\mu)(x) &\subseteq \bigcap_{x\in\psi(J)} \bigcap_{j\in\psi^{-1}(x)} \mathfrak{s}(j) \oplus \psi(\mu)(x) \\ &= \bigcap_{x\in\psi(J)} \bigcap_{j\in\psi^{-1}(x)} \mathfrak{s}(j) \oplus (\inf_{k\in\psi^{-1}(x)} \mu(k)) \\ &\subseteq \bigcap_{x\in\psi(J)} \bigcap_{j\in\psi^{-1}(x)} \mathfrak{s}(j) \oplus \mu(j) \\ &= \bigcap_{j\in J} \mathfrak{s}(j) \oplus \mu(j) \end{aligned}$$

which shows that $\Sigma\mathfrak{t}(\psi(\mathfrak{I})) \subseteq \Sigma\mathfrak{s}(\mathfrak{I})$. Now it follows by (F1) that $\mathfrak{t}(z) \rightarrowtail z$ for $z \notin \psi(J)$ and by (F2) that for $z \in \psi(J)$ also

$$\mathfrak{t}(z) = \bigcap_{j\in\psi^{-1}(z)} \mathfrak{s}(j) \rightarrowtail z.$$

Since moreover $\psi(\mathfrak{I}) \rightarrowtail x$ it follows from (F3) that $\Sigma\mathfrak{t}(\psi(\mathfrak{I})) \rightarrowtail x$ and hence again by (F2) that $\Sigma\mathfrak{s}(\mathfrak{I}) \rightarrowtail x$. □

1.2 The Objects

In spite of the fact that all the concepts which were defined in the foregoing section are both conceptually and technically very different from each other, they are all equivalent. In this section we will prove that one type of structure unambiguously determines a unique structure of each of the other types, and we will also give several precise formulas for going from one structure to another. A structure derived from another one by such a transition will be referred to as an *associated structure*.

Since we have many different types of structures there are of course very many possible transitions. Although we will describe a fair number of these, we will not attempt to describe them all. Rather, we will restrict ourselves to those which are needed to prove the equivalence and to those which are interesting and natural. Each theorem giving such a transition will be marked with an indication of its precise nature.

First, we will give the transitions as indicated in Fig. 1.1, from which it will result that all ten structures are equivalent with each other.

Second, we will give explicit non-circuitous formulas for a number of transitions as indicated in Fig. 1.2. In both these diagrams, the numbers next to the arrows refer to the theorems where the transition formulas are proved. The number above (right) is for the arrow left-right (up-down).

At the end of the book (see pages 431–434) in the appendix, for easy reference, we recall all the formulas which have been proved in both diagrams.

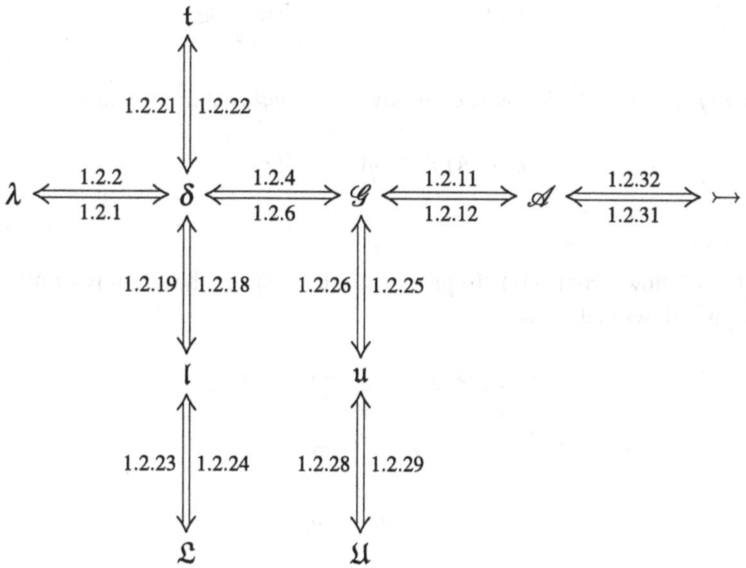

Fig. 1.1 Fundamental equivalences and transitions between the structures

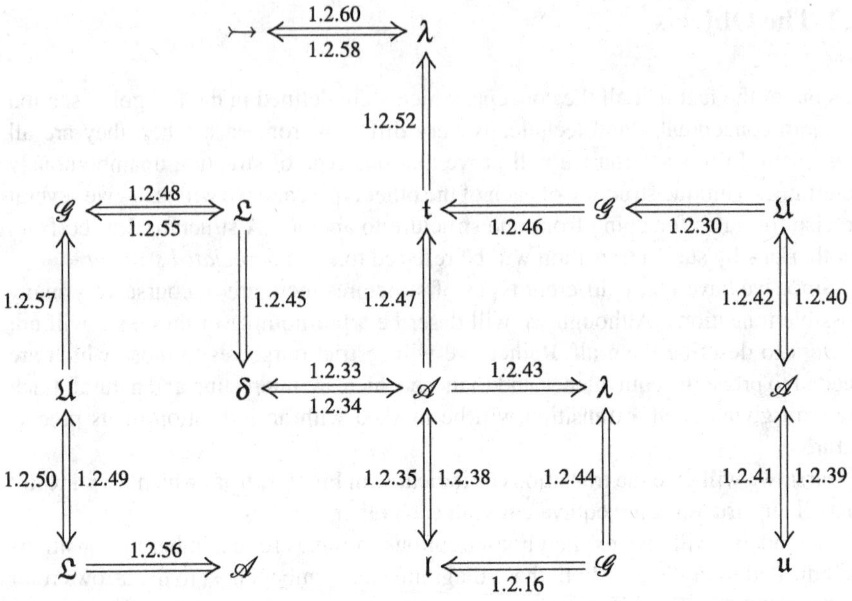

Fig. 1.2 Further transitions between the structures

1.2.1 Theorem **(D $\rightrightarrows$ L)** *If $\delta : X \times 2^X \longrightarrow \mathbb{P}$ is a distance on X, then the function*

$$\lambda : \mathsf{F}(X) \longrightarrow \mathbb{P}^X : \mathscr{F} \mapsto \sup_{U \in \sec \mathscr{F}} \delta_U$$

is a limit operator on X. Moreover, for any $x \in X$ and $A \in 2^X$, we have

$$\delta(x, A) = \inf_{\mathscr{U} \in \mathsf{U}(A)} \lambda \mathscr{U}(x).$$

Proof (L1) follows from (D1). To prove (L2), let $(\mathscr{F}_j)_{j \in J}$ be a family of filters on X. Then it follows that we have

$$\begin{aligned}
\lambda(\bigcap_{j \in J} \mathscr{F}_j) &= \sup_{U \in \sec \bigcap_{j \in J} \mathscr{F}_j} \delta_U \\
&= \sup_{U \in \bigcup_{j \in J} \sec \mathscr{F}_j} \delta_U \\
&= \sup_{j \in J} \sup_{U \in \sec \mathscr{F}_j} \delta_U \\
&= \sup_{j \in J} \lambda(\mathscr{F}_j).
\end{aligned}$$

To prove (L3), let $\mathscr{F} \in \mathsf{F}(X)$ and let $(\sigma(y))_{y\in X}$ be a selection of filters on X. Next let

$$\varepsilon := \sup_{y\in X} \lambda(\sigma(y))(y).$$

First suppose that all filters involved are ultrafilters. For any $D \in \Sigma\sigma(\mathscr{F})$, there exists an $F \in \mathscr{F}$ such that, for all $y \in F$, $D \in \sigma(y)$. Consequently,

$$\delta(y, D) \leq \lambda(\sigma(y))(y) \leq \varepsilon.$$

This proves that $D^{(\varepsilon)} \in \mathscr{F}$ and then it follows from (D4) that

$$\delta_D \leq \delta_{D^{(\varepsilon)}} + \varepsilon \leq \lambda\mathscr{F} + \varepsilon.$$

By the arbitrariness of $D \in \Sigma\sigma(\mathscr{F})$ it follows that

$$\lambda\Sigma\sigma(\mathscr{F}) = \sup_{D\in\Sigma\sigma(\mathscr{F})} \delta_D \leq \lambda\mathscr{F} + \varepsilon.$$

Second we consider the case where all filters involved are arbitrary. For each selection $\rho \in \prod_{y\in X} \mathsf{U}(\sigma(y))$, let

$$\varepsilon_\rho := \sup_{y\in X} \lambda(\rho(y))(y).$$

A straightforward verification shows that

$$\varepsilon = \sup_{\rho\in\prod_{y\in X}\mathsf{U}(\sigma(y))} \varepsilon_\rho.$$

From the result for ultrafilters it follows that, for any $\rho \in \prod_{y\in X} \mathsf{U}(\sigma(y))$ and $\mathscr{U} \in \mathsf{U}(\mathscr{F})$,

$$\lambda\Sigma\rho(\mathscr{U}) \leq \lambda\mathscr{U} + \varepsilon_\rho.$$

It then follows from (L2) and 1.1.3 that

$$\begin{aligned}
\lambda\Sigma\sigma(\mathscr{F}) &= \sup_{\rho\in\prod_{y\in X}\mathsf{U}(\sigma(y))} \sup_{\mathscr{U}\in\mathsf{U}(\mathscr{F})} \lambda\Sigma\rho(\mathscr{U}) \\
&\leq \sup_{\rho\in\prod_{y\in X}\mathsf{U}(\sigma(y))} \sup_{\mathscr{U}\in\mathsf{U}(\mathscr{F})} (\lambda\mathscr{U} + \varepsilon_\rho) \\
&= \lambda\mathscr{F} + \varepsilon.
\end{aligned}$$

To prove the final claim of the theorem, first note that one inequality is clear. To prove the other one, let $x \in X$ and $A \in 2^X$. It follows from the definition of λ, and upon applying complete distributivity, that

$$\inf_{\mathcal{U} \in \mathsf{U}(A)} \lambda \mathcal{U} = \sup_{\zeta \in \prod_{\mathcal{U} \in \mathsf{U}(A)} \mathcal{U}} \inf_{\mathcal{U} \in \mathsf{U}(A)} \delta_{\zeta(\mathcal{U})}.$$

Now, by 1.1.4, for each $\zeta \in \prod_{\mathcal{U} \in \mathsf{U}(A)} \mathcal{U}$, we can find a finite subset $\mathsf{U}_\zeta \subseteq \mathsf{U}(A)$ such that $A \subseteq \bigcup_{\mathcal{U} \in \mathsf{U}_\zeta} \zeta(\mathcal{U})$. Consequently it follows from (D3) that

$$\inf_{\mathcal{U} \in \mathsf{U}(A)} \lambda \mathcal{U} \leq \sup_{\zeta \in \prod_{\mathcal{U} \in \mathsf{U}(A)} \mathcal{U}} \delta_{\left(\bigcup_{\mathcal{U} \in \mathsf{U}_\zeta} \zeta(\mathcal{U})\right)} \leq \delta_A,$$

which proves the remaining inequality. □

1.2.2 Theorem **(L $\rightrightarrows$ D)** *If* $\lambda : \mathsf{F}(X) \longrightarrow \mathbb{P}^X$ *is a limit operator on* X*, then the function*

$$\delta : X \times 2^X \longrightarrow \mathbb{P} : (x, A) \mapsto \inf_{\mathcal{U} \in \mathsf{U}(A)} \lambda \mathcal{U}(x)$$

is a distance on X*. Moreover, for any* $\mathcal{F} \in \mathsf{F}(X)$ *and* $x \in X$*, we have*

$$\lambda \mathcal{F}(x) = \sup_{U \in \sec \mathcal{F}} \delta(x, U).$$

Proof (D1) follows from (L1), (D2) is trivial, and (D3) follows from the fact that, for any $A, B \in 2^X$, we have $\mathsf{U}(\text{stack} A \cup B) = \mathsf{U}(\text{stack} A) \cup \mathsf{U}(\text{stack} B)$. We will now prove the final claim of the theorem since we will require this in the proof of (D4). Let λ' be defined as

$$\lambda' : \mathsf{F}(X) \longrightarrow \mathbb{P}^X : \mathcal{F} \mapsto \sup_{U \in \sec \mathcal{F}} \delta_U.$$

Let $\mathcal{U} \in \mathsf{U}(X)$. Then first we have

$$\lambda' \mathcal{U} = \sup_{U \in \mathcal{U}} \delta_U = \sup_{U \in \mathcal{U}} \inf_{\mathcal{W} \in \mathsf{U}(U)} \lambda \mathcal{W} \leq \lambda \mathcal{U}.$$

Second, by complete distributivity, we have

$$\begin{aligned} \lambda' \mathcal{U} &= \sup_{U \in \mathcal{U}} \inf_{\mathcal{W} \in \mathsf{U}(U)} \lambda \mathcal{W} \\ &= \inf_{\theta \in \prod_{U \in \mathcal{U}} \mathsf{U}(U)} \sup_{U \in \mathcal{U}} \lambda(\theta(U)). \end{aligned}$$

Furthermore, for any $\theta \in \prod_{U \in \mathscr{U}} \mathsf{U}(U)$ and any $U \in \mathscr{U}$, we have $U \in \theta(U)$ and thus $\bigcap_{U \in \mathscr{U}} \theta(U) \subseteq \mathscr{U}$. From (L2) it then follows that

$$\lambda \mathscr{U} \leq \lambda(\bigcap_{U \in \mathscr{U}} \theta(U)).$$

Since this holds for all $\theta \in \prod_{U \in \mathscr{U}} \mathsf{U}(U)$, it follows that

$$\lambda \mathscr{U} \leq \lambda' \mathscr{U}.$$

Consequently λ and λ' coincide on ultrafilters. From the definition of λ' and the fact that λ fulfils (L2), this, however, suffices to conclude that $\lambda = \lambda'$.

In order now to prove (D4), let $A \in 2^X$, $\varepsilon \in \mathbb{R}^+$, and $\mathscr{W} \in \mathsf{U}(A^{(\varepsilon)})$
Claim: $\forall y \in A^{(\varepsilon)}, \exists \sigma(y) \in \mathsf{U}(A)$ such that $\lambda(\sigma(y))(y) \leq \varepsilon$.
Indeed, if not then, for some $y \in A^{(\varepsilon)}$ and for all $\mathscr{U} \in \mathsf{U}(A)$, we have

$$\varepsilon < \lambda \mathscr{U}(y) = \lambda' \mathscr{U}(y) = \sup_{U \in \mathscr{U}} \delta(y, U).$$

Consequently, for all $\mathscr{U} \in \mathsf{U}(A)$, we can find $U_{\mathscr{U}} \in \mathscr{U}$ such that $\varepsilon < \delta(y, U_{\mathscr{U}})$. By 1.1.4 we can then find $\mathscr{U}_1, \ldots, \mathscr{U}_n \in \mathsf{U}(A)$ such that $A \subseteq \bigcup_{i=1}^{n} U_{\mathscr{U}_i}$. It then follows from (D3) that

$$\varepsilon < \inf_{i=1}^{n} \delta(y, U_{\mathscr{U}_i}) = \delta(y, \bigcup_{i=1}^{n} U_{\mathscr{U}_i}) \leq \delta(y, A),$$

which is in contradiction with the choice of y. This proves our claim. Consequently, for all $y \in A^{(\varepsilon)}$, we can fix some $\sigma(y) \in \mathsf{U}(A)$ such that $\lambda(\sigma(y))(y) \leq \varepsilon$. For $y \notin A^{(\varepsilon)}$ we let $\sigma(y) := \dot{y}$. Next we let

$$\varepsilon' := \sup_{y \in X} \lambda(\sigma(y))(y).$$

Since $A \in \bigcap_{y \in A^{(\varepsilon)}} \sigma(y)$ it follows that $A \in \Sigma\sigma(\mathscr{W})$ and consequently, by 1.1.3, $\Sigma\sigma(\mathscr{W}) \in \mathsf{U}(A)$. From the definition of δ and making use of (L3) we then obtain, for all $x \in X$,

$$\begin{aligned} \delta(x, A) &\leq \lambda(\Sigma\sigma(\mathscr{W}))(x) \\ &\leq \lambda \mathscr{W}(x) + \varepsilon' \\ &\leq \lambda \mathscr{W}(x) + \varepsilon. \end{aligned}$$

Since this holds for all $\mathscr{W} \in \mathsf{U}(A^{(\varepsilon)})$, it follows that

$$\delta(x, A) \leq \delta(x, A^{(\varepsilon)}) + \varepsilon. \qquad \square$$

The foregoing two theorems show that distances and limit operators are equivalent concepts. Together with approach systems and gauges they form the most important and useful structures.

1.2.3 Corollary *For any* $A \subseteq X$ *and* $x \in X$: $\delta(x, A) = \inf_{\mathscr{F} \in \mathsf{F}(A)} \lambda\mathscr{F}(x)$.

1.2.4 Theorem (D $\rightrightarrows$ G) *If* $\delta : X \times 2^X \longrightarrow \mathbb{P}$ *is a distance on* X*, then*

$$\mathscr{G} := \left\{ d \in q\mathsf{Met}(X) \mid \forall A \subseteq X, \forall x \in X : \inf_{a \in A} d(x, a) \leq \delta(x, A) \right\}$$

is a gauge on X.

Proof Let $\mathscr{G}_0 \in 2^{(\mathscr{G})}$, $x \in X$, and $A \subseteq X$. Then, by complete distributivity, we have that $\inf_{a \in A} \sup_{d \in \mathscr{G}_0} d(x, a) = \sup_{\varphi \in \mathscr{G}_0^A} \inf_{a \in A} \varphi(a)(x, a)$. If we fix $\varphi \in \mathscr{G}_0^A$, then it follows that

$$\begin{aligned}
\inf_{a \in A} \varphi(a)(x, a) &= \inf_{d \in \mathscr{G}_0} \inf_{a \in \varphi^{-1}(d)} d(x, a) \\
&\leq \inf_{d \in \mathscr{G}_0} \delta(x, \varphi^{-1}(d)) \\
&= \delta(x, A).
\end{aligned}$$

Hence it follows that $\mathscr{G}$ is closed under the formation of finite suprema. The remaining properties of an ideal are trivial.

If $d \in q\mathsf{Met}(X)$ is such that, for all $x \in X, \varepsilon > 0$, and $\omega < \infty$, there exists $d' \in \mathscr{G}$ such that $d(x, \cdot) \wedge \omega \leq d'(x, \cdot) + \varepsilon$ then it follows at once that, for any $\varepsilon > 0$ and $\omega < \infty$,

$$\begin{aligned}
\inf_{a \in A} d(x, a) \wedge \omega &\leq \inf_{a \in A} d'(x, a) + \varepsilon \\
&\leq \delta(x, A) + \varepsilon.
\end{aligned}$$

By the arbitrariness of ε and ω this proves that $d \in \mathscr{G}$. Hence $\mathscr{G}$ is locally saturated. $\square$

In order to be able to work with the gauge associated with a distance, as given by the foregoing theorem, it will be useful to have a specified set of quasi-metrics in that gauge. The following proposition provides us with such a collection.

1.2.5 Proposition *If* $\delta : X \times 2^X \longrightarrow \mathbb{P}$ *is a distance on* X*, then, for any* $\zeta \in \mathbb{R}^+$ *and* $Z \subseteq X$*, the function*

$$d_Z^\zeta : X \times X \longrightarrow \mathbb{P} : (x, y) \mapsto (\delta(x, Z) \wedge \zeta) \ominus (\delta(y, Z) \wedge \zeta)$$

is a quasi-metric in the gauge associated with δ.

Proof That d_Z^ζ is a quasi-metric is easily seen. Let $x \in X$ and $A \subseteq X$. Then it follows from (D4) that

$$\begin{aligned}
\inf_{a \in A} d_Z^\zeta(x, a) &= \inf_{a \in A} (\delta(x, Z) \wedge \zeta) \ominus (\delta(a, Z) \wedge \zeta) \\
&\le (\delta(x, Z) \wedge \zeta) \ominus (\sup_{a \in A} \delta(a, Z) \wedge \zeta) \\
&\le ((\delta(x, A) + \sup_{a \in A} \delta(a, Z)) \wedge \zeta) \ominus (\sup_{a \in A} \delta(a, Z) \wedge \zeta) \\
&\le (\delta(x, A) + \sup_{a \in A} \delta(a, Z) \wedge \zeta) \ominus (\sup_{a \in A} \delta(a, Z) \wedge \zeta) \\
&= \delta(x, A).
\end{aligned}$$

By the characterization given in 1.2.4 this proves that d_Z^ζ is indeed a member of the gauge associated with δ. □

1.2.6 Theorem **(G $\rightrightarrows$ D)** *If $\mathcal{H} \subseteq \mathsf{qMet}(X)$ is a gauge basis on X, then the function*

$$\delta : X \times 2^X \longrightarrow \mathbb{P} : (x, A) \mapsto \sup_{d \in \mathcal{H}} \inf_{a \in A} d(x, a)$$

is a distance on X.

Proof Verification of (D1) and (D2) is straightforward. To show (D3), note that one inequality is obvious from the formula. To prove the other one, let $\delta(x, A) \wedge \delta(x, B) > \alpha$, then there exist $d_1, d_2 \in \mathcal{H}$ such that

$$\inf_{a \in A} d_1(x, a) > \alpha \text{ and } \inf_{b \in B} d_2(x, b) > \alpha.$$

Now take $\varepsilon > 0$ and $\omega < \infty$ then by local directedness there exists $d \in \mathcal{H}$ such that

$$\begin{aligned}
\sup_{e \in \mathcal{H}} \inf_{c \in A \cup B} e(x, c) + \varepsilon &\ge \inf_{c \in A \cup B} d(x, c) + \varepsilon \\
&\ge (\inf_{a \in A} d_1 \vee d_2(x, a)) \wedge (\inf_{b \in B} d_1 \vee d_2(x, b)) \wedge \omega \\
&\ge \alpha \wedge \omega
\end{aligned}$$

which by the arbitrariness of ε and ω show the other inequality. To show (D4) let $x \in X$, $A \subseteq X$, and $\varepsilon > 0$ be fixed. Then, for any $b \in A^{(\varepsilon)}$, $d \in \mathcal{H}$, and $\theta > 0$, there exists $a_d \in A$ such that $d(b, a_d) \le \varepsilon + \theta$. Consequently,

$$\begin{aligned} d(x, a_d) &\le d(x, b) + d(b, a_d) \\ &\le d(x, b) + \varepsilon + \theta, \end{aligned}$$

which proves that $\inf_{a \in A} d(x, a) \le \inf_{b \in A^{(\varepsilon)}} d(x, b) + \varepsilon + \theta$. Since this holds for all $d \in \mathscr{H}$, it follows that $\delta(x, A) \le \delta(x, A^{(\varepsilon)}) + \varepsilon$. □

1.2.7 Theorem *If δ is a distance on X and $\mathscr{G}$ is the associated gauge, then for all $x \in X$ and $A \subseteq X$ we have*

$$\delta(x, A) = \sup_{d \in \mathscr{G}} \inf_{a \in A} d(x, a).$$

Proof Let $x \in X$ and $A \subseteq X$. It follows at once from 1.2.4 that $\sup_{d \in \mathscr{G}} \inf_{a \in A} d(x, a) \le \delta(x, A)$. Making use of 1.2.5, we further obtain

$$\begin{aligned} \sup_{d \in \mathscr{G}} \inf_{a \in A} d(x, a) &\ge \sup_{\zeta \in \mathbb{R}^+} \sup_{Z \subseteq X} \inf_{a \in A} d_Z^{\zeta}(x, a) \\ &\ge \sup_{\zeta \in \mathbb{R}^+} \inf_{a \in A} d_A^{\zeta}(x, a) \\ &= \sup_{\zeta \in \mathbb{R}^+} \inf_{a \in A} (\delta(x, A) \wedge \zeta) \ominus (\delta(a, A) \wedge \zeta) \\ &= \sup_{\zeta \in \mathbb{R}^+} \delta(x, A) \wedge \zeta = \delta(x, A), \end{aligned}$$

which proves the other inequality. □

1.2.8 Corollary *If δ is a distance on X, then for any $x \in X$ and $A \subseteq X$*

$$\delta(x, A) = \sup_{\zeta \in \mathbb{R}^+} \sup_{Z \subseteq X} \inf_{a \in A} d_Z^{\zeta}(x, a).$$

In spite of the fact that it is large enough to generate the distance, the collection of all quasi-metrics d_Z^{ζ} as introduced in 1.2.5, is not a basis for the gauge associated with δ since it is not locally directed. In 1.2.55 we will discover an enlarged collection which is a basis for the gauge. See also the remarks after 1.2.12.

1.2.9 Proposition *If δ is a distance on X and $\mathscr{D} \subseteq q\mathsf{Met}(X)$ is locally directed such that, for all $x \in X$ and $A \subseteq X$, we have $\delta(x, A) = \sup_{d \in \mathscr{D}} \inf_{a \in A} d(x, a)$, then $\mathscr{D}$ is a basis for the gauge $\mathscr{G}$ associated with δ.*

Proof It follows from 1.2.4 that $\mathscr{D} \subseteq \mathscr{G}$. Suppose that there exists $d_0 \in \mathscr{G} \setminus \widehat{\mathscr{D}}$. Then d_0 is not locally dominated by $\mathscr{D}$ and hence there exist $x \in X$, $\varepsilon > 0$, and $\omega < \infty$ such that, for all $d \in \mathscr{D}$,

$$d_0(x, \cdot) \wedge \omega \nleq d(x, \cdot) + 2\varepsilon.$$

For each $\mathscr{D}_0 \subseteq \mathscr{D}$ finite, let $A(\mathscr{D}_0) := \{y \in X \mid d_0(x, y) \wedge \omega > \sup_{d \in \mathscr{D}_0} d(x, y) + \varepsilon\}$. By local directedness, given $\mathscr{D}_0 \subseteq \mathscr{D}$ finite we can find $e \in \mathscr{D}$ such that

$$\sup_{d \in \mathscr{D}_0} d(x, y) \wedge \omega \leq e(x, y) + \varepsilon.$$

It then follows that

$$A(\mathscr{D}_0) \supset \{y \in X \mid d_0(x, y) \wedge \omega > e(x, y) + 2\varepsilon\}$$

where the latter set, by supposition, is nonempty. Hence the collection $\{A(\mathscr{D}_0) \mid \mathscr{D}_0 \in 2^{(\mathscr{D})}\}$ is a basis for a filter. Then making use of 1.2.7, we obtain

$$\begin{aligned}
\sup_{\mathscr{D}_0 \in 2^{(\mathscr{D})}} \delta(x, A(\mathscr{D}_0)) \wedge \omega = & \sup_{\mathscr{D}_0 \in 2^{(\mathscr{D})}} \sup_{e \in \mathscr{D}} \inf_{y \in A(\mathscr{D}_0)} e(x, y) \wedge \omega \\
\leq & \sup_{\mathscr{D}_0 \in 2^{(\mathscr{D})}} \sup_{e \in \mathscr{D}} \inf_{y \in A(\mathscr{D}_0 \cup \{e\})} (\sup_{d \in \mathscr{D}_0} d \vee e)(x, y) \wedge \omega \\
= & \sup_{\mathscr{D}_0 \in 2^{(\mathscr{D})}} \inf_{y \in A(\mathscr{D}_0)} \sup_{d \in \mathscr{D}_0} d(x, y) \wedge \omega \\
\leq & \sup_{\mathscr{D}_0 \in 2^{(\mathscr{D})}} \inf_{y \in A(\mathscr{D}_0)} d_0(x, y) \wedge \omega - \varepsilon \\
\leq & \sup_{\mathscr{D}_0 \in 2^{(\mathscr{D})}} \sup_{e \in \mathscr{G}} \inf_{y \in A(\mathscr{D}_0)} e(x, y) \wedge \omega - \varepsilon \\
= & \sup_{\mathscr{D}_0 \in 2^{(\mathscr{D})}} \delta(x, A(\mathscr{D}_0)) \wedge \omega - \varepsilon,
\end{aligned}$$

which is a contradiction. □

1.2.10 Theorem *If $\mathscr{G} \subseteq q\mathsf{Met}(X)$ is a gauge on X and δ is the associated distance, then*

$$\mathscr{G} = \left\{d \in q\mathsf{Met}(X) \mid \forall A \subseteq X, \forall x \in X : \inf_{a \in A} d(x, a) \leq \delta(x, A)\right\}.$$

Proof Since $\mathscr{G}$ is locally directed it follows from 1.2.9 that $\mathscr{G}$ is a basis for the gauge $\left\{d \in q\mathsf{Met}(X) \mid \forall A \subseteq X, \forall x \in X : \inf_{a \in A} d(x, a) \leq \delta(x, A)\right\}$ associated with δ. However it is obviously also a basis for itself and hence the result follows. □

The combined results of 1.2.4, 1.2.6, 1.2.7, and 1.2.10 prove that distances and gauges are equivalent systems.

1.2.11 Theorem **(G $\rightrightarrows$ A)** *If $\mathscr{H} \subseteq q\mathsf{Met}(X)$ is locally directed, then $(\mathscr{B}(x))_{x \in X}$, where, for each $x \in X$,*

$$\mathscr{B}(x) := \{d(x, \cdot) \mid d \in \mathscr{H}\}$$

is an approach basis on X. Moreover, if $\mathcal{G}$ *denotes the gauge generated by* $\mathcal{H}$ *and* $(\mathcal{A}(x))_{x\in X}$ *denotes the approach system generated by* $(\mathcal{B}(x))_{x\in X}$*, then*

$$\mathcal{G} = \{d \in q\mathsf{Met}(X) \mid \forall x \in X : d(x, \cdot) \in \mathcal{A}(x)\}.$$

Proof That, for each $x \in X$, $\mathcal{B}(x)$ is an ideal basis follows from the fact that $\mathcal{H}$ is locally directed, and that it satisfies (B1) and (B2) follows from the fact that the members of $\mathcal{G}$ are quasi-metrics. To prove the final claim of the theorem it suffices to note that $d \in \mathcal{G}$ is equivalent respectively to

$$\begin{aligned}
&\forall x \in X, \forall \varepsilon > 0, \forall \omega < \infty, \exists d' \in \mathcal{H} : d(x, \cdot) \wedge \omega \le d'(x, \cdot) + \varepsilon \\
\Leftrightarrow &\forall x \in X, \forall \varepsilon > 0, \forall \omega < \infty, \exists \varphi \in \mathcal{B}(x) : d(x, \cdot) \wedge \omega \le \varphi + \varepsilon \\
\Leftrightarrow &\forall x \in X : d(x, \cdot) \in \mathcal{A}(x).
\end{aligned}$$

□

1.2.12 Theorem **(A $\rightrightarrows$ G)** *If* $(\mathcal{A}(x))_{x\in X}$ *is an approach system on X, then*

$$\mathcal{G} := \{d \in q\mathsf{Met}(X) \mid \forall x \in X : d(x, \cdot) \in \mathcal{A}(x)\}$$

is a gauge on X.

Proof That $\mathcal{G}$ is an ideal in $q\mathsf{Met}(X)$ follows from the fact that, for each $x \in X$, $\mathcal{A}(x)$ is an ideal. That $\mathcal{G}$ is locally saturated follows from the fact that, for each $x \in X$, $\mathcal{A}(x)$ is saturated. □

In order to conclude our proof that gauges and approach systems are equivalent concepts we need some supplementary results. In particular, to be able to work with the gauge associated with an approach system, as given by the foregoing theorem, we again require a specified set of quasi-metrics in that gauge. The following proposition provides us with such a collection.

1.2.13 Proposition *If* $(\mathcal{B}(x))_{x\in X}$ *is an approach basis on X, then, for any* $\zeta \in \mathbb{R}^+$ *and* $Z \subseteq X$*, the function*

$$d_Z^\zeta : X \times X \longrightarrow \mathbb{P}$$

where for any $x, y \in X$

$$d_Z^\zeta(x, y) := \sup_{\varphi\in\mathcal{B}(x)} \inf_{z\in Z} \varphi(z) \wedge \zeta \ominus \sup_{\psi\in\mathcal{B}(y)} \inf_{z\in Z} \psi(z) \wedge \zeta$$

is a quasi-metric in the gauge associated with $\mathcal{A}$*.*

Proof That d_Z^ζ is a quasi-metric is easily seen. Let $x \in X$ and $\varepsilon > 0$ be fixed. Choose $\varphi_0 \in \mathcal{B}(x)$ such that

$$\sup_{\varphi\in\mathcal{B}(x)} \inf_{z\in Z} \varphi(z) \wedge \zeta \le \inf_{z\in Z} \varphi_0(z) \wedge \zeta + \varepsilon$$

and then choose a family $(\psi_u)_{u\in X} \in \prod_{u\in X} \mathscr{B}(u)$ such that, for all $u, z \in X$,

$$\varphi_0(z) \wedge \zeta \leq \psi_x(u) + \psi_u(z) + \varepsilon.$$

Then it follows that, for all $y \in X$,

$$\begin{aligned} d_Z^{\zeta}(x, y) &= (\sup_{\varphi\in\mathscr{B}(x)} \inf_{z\in Z} \varphi(z) \wedge \zeta) \ominus (\sup_{\psi\in\mathscr{B}(y)} \inf_{z\in Z} \psi(z) \wedge \zeta) \\ &\leq (\inf_{z\in Z} \varphi_0(z) \wedge \zeta + \varepsilon) \ominus (\inf_{z\in Z} \psi_y(z) \wedge \zeta) \\ &\leq (\inf_{z\in Z} (\psi_x(y) + \psi_y(z) + \varepsilon) \wedge \zeta + \varepsilon) \ominus (\inf_{z\in Z} \psi_y(z) \wedge \zeta) \\ &\leq (\inf_{z\in Z} (\psi_x(y) + \psi_y(z) \wedge \zeta + \varepsilon) + \varepsilon) \ominus (\inf_{z\in Z} \psi_y(z) \wedge \zeta) \\ &\leq (\psi_x(y) + 2\varepsilon + \inf_{z\in Z} \psi_y(z) \wedge \zeta) \ominus (\inf_{z\in Z} \psi_y(z) \wedge \zeta) \\ &= \psi_x(y) + 2\varepsilon, \end{aligned}$$

which by (A2) proves that, for all $x \in X$, $d_Z^{\zeta}(x, \cdot) \in \mathscr{B}(x)$. Hence d_Z^{ζ} is indeed a member of the gauge associated with $\mathscr{B}$. □

The similarity between the quasi-metrics of the foregoing proposition and those of 1.2.5 is not just a coincidence. Actually it will follow from 1.2.34 that they are the same. Hence, even if it is somewhat prematurely, we are justified in using the same notation.

1.2.14 Proposition *If $(\mathscr{A}(x))_{x\in X}$ is an approach system on X and $\mathscr{G}$ is the associated gauge, then for all $x \in X$ and $A \subseteq X$ we have*

$$\sup_{\varphi\in\mathscr{A}(x)} \inf_{a\in A} \varphi(a) = \sup_{d\in\mathscr{G}} \inf_{a\in A} d(x, a).$$

Proof One inequality is an immediate consequence of the definition of $\mathscr{G}$. Making use of 1.2.13 we further obtain

$$\begin{aligned} \sup_{d\in\mathscr{G}} \inf_{a\in A} d(x, a) &\geq \sup_{\zeta\in\mathbb{R}^+} \sup_{Z\subseteq X} \inf_{a\in A} d_Z^{\zeta}(x, a) \\ &= \sup_{\zeta\in\mathbb{R}^+} \sup_{Z\subseteq X} \inf_{a\in A} ((\sup_{\varphi\in\mathscr{A}(x)} \inf_{z\in Z} \varphi(z) \wedge \zeta) \ominus (\sup_{\psi\in\mathscr{A}(a)} \inf_{z\in Z} \psi(z) \wedge \zeta)) \\ &\geq \sup_{\zeta\in\mathbb{R}^+} \inf_{a\in A} ((\sup_{\varphi\in\mathscr{A}(x)} \inf_{z\in A} \varphi(z) \wedge \zeta) \ominus (\sup_{\psi\in\mathscr{A}(a)} \inf_{z\in A} \psi(z) \wedge \zeta)) \\ &= \sup_{\zeta\in\mathbb{R}^+} \inf_{a\in A} \sup_{\varphi\in\mathscr{A}(x)} \inf_{z\in A} \varphi(z) \wedge \zeta \\ &= \sup_{\varphi\in\mathscr{A}(x)} \inf_{z\in A} \varphi(z), \end{aligned}$$

which proves the other inequality. □

1.2.15 Theorem *If $(\mathcal{A}(x))_{x\in X}$ is an approach system on X and $\mathcal{G}$ is the associated gauge, then, for all $x \in X$, we have that $\mathcal{A}(x) = \widehat{\mathcal{B}(x)}$, where*

$$\mathcal{B}(x) = \{d(x,\cdot) \mid d \in \mathcal{G}\}.$$

Proof It is immediately clear that, for all $x \in X$, $\widehat{\mathcal{B}(x)} \subseteq \mathcal{A}(x)$. To show that the converse inclusion also holds we make use of 1.2.14. Suppose that, for some $x \in X$, there exists $\psi \in \mathcal{A}(x)\setminus \widehat{\mathcal{B}(x)}$. Hence ψ is not dominated by $\mathcal{B}(x)$, which implies that we can find $\varepsilon > 0$ and $\omega < \infty$ such that, for all $d \in \mathcal{G}$,

$$\psi \wedge \omega \nleq d(x,\cdot) + \varepsilon.$$

For each $d \in \mathcal{G}$, let $A(d) := \{y \in X \mid \psi(y) \wedge \omega > d(x,y) + \varepsilon\}$. Then it is clear that, for all $d, e \in \mathcal{G}$, $A(d) \cap A(e) = A(d \vee e) \neq \emptyset$. Hence we obtain

$$\begin{aligned}
\sup_{d\in\mathcal{G}} \sup_{e\in\mathcal{G}} \inf_{y\in A(d)} e(x,y) \wedge \omega &\leq \sup_{d\in\mathcal{G}} \sup_{e\in\mathcal{G}} \inf_{y\in A(d\vee e)} (d \vee e)(x,y) \wedge \omega \\
&= \sup_{d\in\mathcal{G}} \inf_{y\in A(d)} d(x,y) \wedge \omega \\
&\leq \sup_{d\in\mathcal{G}} \inf_{y\in A(d)} (\psi(y) \wedge \omega - \varepsilon) \wedge \omega \\
&\leq \sup_{d\in\mathcal{G}} \inf_{y\in A(d)} \psi(y) \wedge \omega - \varepsilon \\
&\leq \sup_{d\in\mathcal{G}} \sup_{\varphi\in\mathcal{A}(x)} \inf_{y\in A(d)} \varphi(y) \wedge \omega - \varepsilon,
\end{aligned}$$

which, by 1.2.14, is a contradiction. □

The combined results of 1.2.11, 1.2.12, and 1.2.15 show that gauges and approach systems are equivalent systems.

Now we come to the relationship between distances and lower hull operators where we involve the associated gauge.

1.2.16 Theorem (G ⇉ LH) *If $\mathcal{G} \subseteq q\mathsf{Met}(X)$ is a gauge on X then the function $\mathfrak{l} : \mathbb{P}^X \longrightarrow \mathbb{P}^X$, defined by*

$$\mathfrak{l}(\mu)(x) := \sup_{d\in\mathcal{G}} \inf_{y\in X} (\mu(y) + d(x,y))$$

is a lower hull operator on X.

Proof That $\mathfrak{l}$ fulfils (LH1), (LH2), and (LH4) is immediately seen. To prove (LH3), let $\mu \in \mathbb{P}^X$. That $\mathfrak{l}(\mathfrak{l}(\mu)) \leq \mathfrak{l}(\mu)$ follows from (LH1). To show the converse inequality let $x \in X$, then it follows that

$$
\begin{aligned}
\mathfrak{l}(\mathfrak{l}(\mu))(x) &= \sup_{d \in \mathcal{G}} \inf_{y \in X} (\mathfrak{l}(\mu)(y) + d(x, y)) \\
&= \sup_{d \in \mathcal{G}} \inf_{y \in X} (\sup_{d' \in \mathcal{G}} \inf_{z \in X} (\mu(z) + d'(y, z)) + d(x, y)) \\
&\geq \sup_{d \in \mathcal{G}} \inf_{y \in X} (\inf_{z \in X} (\mu(z) + d(y, z)) + d(x, y)) \\
&\geq \sup_{d \in \mathcal{G}} \inf_{z \in X} (\mu(z) + d(x, z)) = \mathfrak{l}(\mu)(x). \qquad \square
\end{aligned}
$$

The foregoing formula also holds with a basis for the gauge rather than the entire gauge, see 1.2.37.

1.2.17 Proposition *If $\delta : X \times 2^X \longrightarrow \mathbb{P}$ is a distance on X, and $\mathfrak{l}$ is the lower hull operator derived from the gauge associated with δ, then for all $x \in X$ and $A \subseteq X$ we have*

$$\delta(x, A) = \mathfrak{l}(\theta_A)(x).$$

Proof This is an immediate consequence of 1.2.7 and 1.2.16. □

1.2.18 Theorem (D $\rightrightarrows$ LH) *If $\delta : X \times 2^X \longrightarrow \mathbb{P}$ is a distance on X, and $\mathfrak{l}$ is the lower hull operator derived from the gauge associated with δ, then for any $\mu \in \mathbb{P}^X$ we have*

$$\mathfrak{l}(\mu) = \sup_{\omega < \infty} \sup_{\varepsilon > 0} (\inf_{i=1}^{n(\varepsilon,\omega)} (m_i^{\varepsilon,\omega} + \delta_{M_i^{\varepsilon,\omega}})$$

where for each $\omega < \infty$, $(\inf_{i=1}^{n(\varepsilon,\omega)} (m_i^{\varepsilon,\omega} + \theta_{M_i^{\varepsilon,\omega}}))_{\varepsilon > 0}$ is a development of $\mu \wedge \omega$.

Proof This follows from 1.1.31 and 1.2.17. □

1.2.19 Theorem (LH $\rightrightarrows$ D) *If $\mathfrak{l} : \mathbb{P}^X \longrightarrow \mathbb{P}^X$ is a lower hull operator on X, then the function*

$$\delta : X \times 2^X \longrightarrow \mathbb{P} : (x, A) \mapsto \mathfrak{l}(\theta_A)(x)$$

is a distance on X.

Proof Properties (D1), (D2), and (D3) are immediate. To prove (D4) first remark that, for any $A \subseteq X$ and $\varepsilon \in \mathbb{R}^+$, it follows from the definition of $A^{(\varepsilon)}$ that

$$\mathfrak{l}(\theta_A) \leq \theta_{A^{(\varepsilon)}} + \varepsilon,$$

and consequently, it follows that

$$\delta_A = \mathfrak{l}(\theta_A) = \mathfrak{l}(\mathfrak{l}(\theta_A)) \leq \mathfrak{l}(\theta_{A^{(\varepsilon)}} + \varepsilon) = \mathfrak{l}(\theta_{A^{(\varepsilon)}}) + \varepsilon = \delta_{A^{(\varepsilon)}} + \varepsilon. \qquad \square$$

1.2.20 Theorem *If $\mathfrak{l} : \mathbb{P}^X \longrightarrow \mathbb{P}^X$ is a lower hull operator on X, δ is the distance associated with $\mathfrak{l}$, and $\mathcal{G}$ is the gauge associated with δ, then for all $\mu \in \mathbb{P}^X$ and $x \in X$*

$$\mathfrak{l}(\mu)(x) = \sup_{d\in\mathcal{G}} \inf_{y\in X} (\mu(y) + d(x, y)).$$

Proof If we define the function $\mathfrak{l}' : \mathbb{P}^X \longrightarrow \mathbb{P}^X$ by

$$\mathfrak{l}'(\mu)(x) := \sup_{d\in\mathcal{G}} \inf_{y\in X} (\mu(y) + d(x, y)),$$

then it follows from 1.2.16 that $\mathfrak{l}'$ is a lower hull operator. By 1.1.30 we know that lower hull operators are completely determined by their restriction to $Ind(X)$. Hence, to prove the theorem, it suffices to note that by 1.2.7, for any $x \in X$ and $A \subseteq X$,

$$\begin{aligned}\mathfrak{l}'(\theta_A)(x) &= \sup_{d\in\mathcal{G}} \inf_{y\in X} (\theta_A(y) + d(x, y))\\ &= \sup_{d\in\mathcal{G}} \inf_{y\in A} d(x, y)\\ &= \delta(x, A) = \mathfrak{l}(\theta_A)(x).\end{aligned}$$ □

Taking into account that we already know that distances and gauges are equivalent systems, the combined results of 1.2.16, 1.2.18, 1.2.19, and 1.2.20 show that distances and lower hull operators are equivalent systems.

1.2.21 Theorem (D $\rightrightarrows$ T) *If $\delta : X \times 2^X \longrightarrow \mathbb{P}$ is a distance on X, then the family $(\mathfrak{t}_\varepsilon)_{\varepsilon\in\mathbb{R}^+}$, where $\mathfrak{t}_\varepsilon : 2^X \longrightarrow 2^X$, defined by*

$$\mathfrak{t}_\varepsilon(A) := A^{(\varepsilon)},$$

is a tower on X. Moreover, for any $x \in X$ and $A \subseteq X$, we have

$$\delta(x, A) = \inf \left\{\varepsilon \in \mathbb{R}^+ \mid x \in \mathfrak{t}_\varepsilon(A)\right\}.$$

Proof (T1), (T2), and (T3) follow easily from (D1), (D2), and (D3). To prove (T4) let $A \subseteq X$, $\varepsilon,\gamma \in \mathbb{R}^+$, and $x \in \mathfrak{t}_\varepsilon(\mathfrak{t}_\gamma(A))$. This implies that $\delta(x, A^{(\gamma)}) \leq \varepsilon$ and thus, by (D4), that $\delta(x, A) \leq \varepsilon + \gamma$, i.e. $x \in \mathfrak{t}_{\varepsilon+\gamma}(A)$. (T5) and the last claim of the theorem follow at once from the definitions. □

1.2.22 Theorem (T $\rightrightarrows$ D) *If $(\mathfrak{t}_\varepsilon)_{\varepsilon\in\mathbb{R}^+}$ is a tower on X, then the function*

$$\delta : X \times 2^X \longrightarrow \mathbb{P},$$

defined by

$$\delta(x, A) := \inf \left\{\varepsilon \in \mathbb{R}^+ \mid x \in \mathfrak{t}_\varepsilon(A)\right\}$$

is a distance on X. Moreover, for all $\varepsilon \in \mathbb{R}^+$ and $A \subseteq X$ we have that

$$\mathfrak{t}_\varepsilon(A) = A^{(\varepsilon)}.$$

Proof (D1), (D2) and (D3) follow easily from (T1), (T2), and (T3). To prove (D4) we first prove the last claim of the theorem. Let $x \in X$, $A \subseteq X$, and $\varepsilon \in \mathbb{R}^+$. Then

$$\begin{aligned} x \in A^{(\varepsilon)} &\Rightarrow \inf\left\{\alpha \in \mathbb{R}^+ \mid x \in \mathfrak{t}_\alpha(A)\right\} \leq \varepsilon \\ &\Rightarrow \forall \alpha > \varepsilon : x \in \mathfrak{t}_\alpha(A) \\ &\Rightarrow x \in \bigcap_{\alpha>\varepsilon} \mathfrak{t}_\alpha(A) = \mathfrak{t}_\varepsilon(A). \end{aligned}$$

Conversely we have

$$x \in \mathfrak{t}_\varepsilon(A) \Rightarrow \delta(x, A) = \inf\left\{\alpha \in \mathbb{R}^+ \mid x \in \mathfrak{t}_\alpha(A)\right\} \leq \varepsilon \Rightarrow x \in A^{(\varepsilon)}.$$

(D4) now follows from the observation that if $\delta(x, A^{(\varepsilon)}) < \alpha$ then $x \in \mathfrak{t}_\alpha(A^{(\varepsilon)}) = \mathfrak{t}_\alpha(\mathfrak{t}_\varepsilon(A)) \subseteq \mathfrak{t}_{\alpha+\varepsilon}(A)$ and thus $\delta(x, A) \leq \alpha + \varepsilon$. □

The foregoing results 1.2.21 and 1.2.22 show that distances and towers are equivalent structures.

1.2.23 Theorem **(LR ⇉ LH)** *If $\mathfrak{L}$ is a lower regular function frame on X, then the function $\mathfrak{l} : \mathbb{P}^X \longrightarrow \mathbb{P}^X$, defined by*

$$\mathfrak{l}(\mu) := \sup\{\nu \in \mathfrak{L} \mid \nu \leq \mu\},$$

is a lower hull operator on X. Moreover,

$$\mathfrak{L} = \left\{\mu \in \mathbb{P}^X \mid \mathfrak{l}(\mu) = \mu\right\}.$$

Proof This is perfectly similar to the analogous result in topological spaces concerning the relationship between closed sets and closure operator, with the exception of (LH4). Let $\mu \in \mathbb{P}^X$ and let $\alpha \in \mathbb{P}$. Then

$$\begin{aligned} \mathfrak{l}(\mu + \alpha) &= \sup\left\{\xi \in \mathfrak{L} \mid \xi \leq \mu + \alpha\right\} \\ &= \sup\left\{\xi \in \mathfrak{L} \mid \alpha \leq \xi \leq \mu + \alpha\right\} \\ &= \sup\left\{\xi - \alpha \mid \xi \in \mathfrak{L}, \alpha \leq \xi, \xi - \alpha \leq \mu\right\} + \alpha \\ &= \sup\left\{\rho \in \mathfrak{L} \mid \rho \leq \mu\right\} + \alpha \\ &= \mathfrak{l}(\mu) + \alpha. \end{aligned}$$ □

1.2.24 Theorem **(LH ⇉ LR)** *If $\mathfrak{l} : \mathbb{P}^X \longrightarrow \mathbb{P}^X$ is a lower hull operator on X, then*

$$\mathfrak{L} := \left\{\mu \in \mathbb{P}^X \mid \mathfrak{l}(\mu) = \mu\right\}$$

is a lower regular function frame on X. Moreover, for all $\mu \in \mathbb{P}^X$ *we have that*

$$\mathfrak{l}(\mu) = \sup\{\nu \in \mathfrak{L} \mid \nu \leq \mu\}.$$

Proof Again, this is perfectly similar to the analogous result in topological spaces concerning the relationship between closed sets and closure operator, with the exception of (LR3) and (LR4). Let $\mu \in \mathscr{R}$. If $\alpha \in \mathbb{P}$, then

$$\mathfrak{l}(\mu + \alpha) = \mathfrak{l}(\mu) + \alpha = \mu + \alpha,$$

and hence $\mu + \alpha \in \mathfrak{L}$. If $\alpha \in [0, \inf \mu]$, then it follows from 1.1.27 that $\mathfrak{l}(\mu - \alpha) = \mu - \alpha$ and hence again $\mu - \alpha \in \mathfrak{L}$. □

The foregoing results 1.2.23 and 1.2.24 show that lower hulls and lower regular function frames are equivalent structures.

1.2.25 Theorem **(G ⇉ UH)** *If* $\mathscr{G}$ *is a gauge on X then the function* $\mathfrak{u} : \mathbb{P}_b^X \longrightarrow \mathbb{P}_b^X$, *defined by*

$$\mathfrak{u}(\mu)(x) := \inf_{d \in \mathscr{G}} \sup_{y \in X} (\mu(y) - d(x, y))$$

is an upper hull operator on X and moreover

$$\mathscr{G} = \{d \in q\mathsf{Met}(X) \mid \forall x \in X, \forall \omega < \infty : \mathfrak{u}(d(x, \cdot) \wedge \omega)(x) = 0\}.$$

Proof That $\mathfrak{u}$ fulfils (UH0), (UH1), (UH2) and (UH4) is immediate. To show (UH3) it suffices to calculate

$$\begin{aligned}
\mathfrak{u}(\mathfrak{u}(\mu))(x) &= \inf_{d \in \mathscr{G}} \sup_{y \in X} (\inf_{e \in \mathscr{G}} \sup_{z \in X} (\mu(z) - e(y, z)) - d(x, y)) \\
&\leq \inf_{d \in \mathscr{G}} \sup_{y \in X} \sup_{z \in X} (\mu(z) - d(y, z) - d(x, y)) \\
&\leq \inf_{d \in \mathscr{G}} \sup_{y \in X} \sup_{z \in X} (\mu(z) - d(x, z)) \\
&= \inf_{d \in \mathscr{G}} \sup_{z \in X} (\mu(z) - d(x, z)) = \mathfrak{u}(\mu)(x).
\end{aligned}$$

For the second claim, first let $d \in \mathscr{G}$ then it follows that for any $\omega < \infty$ and $y \in X$

$$\begin{aligned}
\mathfrak{u}(d(x, \cdot) \wedge \omega)(x) &= \inf_{e \in \mathscr{G}} \sup_{z \in X} (d(x, z) \wedge \omega - e(x, z)) \\
&\leq \sup_{z \in X} (d(x, z) \wedge \omega - d(x, z)) = 0
\end{aligned}$$

which proves one inclusion. To prove the other inclusion, if $d \in q\mathsf{Met}(X)$ is such that for all $x \in X$ and $\omega < \infty$, $\mathfrak{u}(d(x, \cdot) \wedge \omega)(x) = 0$ then it follows that

$$\forall x \in X, \forall \omega < \infty : \inf_{e \in \mathcal{G}} \sup_{z \in X}(d(x,z) \wedge \omega - e(x,z)) = 0$$

which implies that

$$\forall x \in X, \forall \omega < \infty, \forall \varepsilon > 0, \exists e \in \mathcal{G} : d(x,\cdot) \wedge \omega - e(x,\cdot) \leq \varepsilon$$

which in turn implies that $d \in \mathcal{G}$. □

In what follows, given any bounded function $\nu \in \mathbb{P}_b^X$ we define the quasi-metric d^ν as

$$d^\nu : X \times X \longrightarrow \mathbb{R} : (x,y) \mapsto \nu(y) \ominus \nu(x).$$

1.2.26 Theorem **(UH $\rightrightarrows$ G)** *If* u *is an upper hull operator on* X *then*

$$\mathcal{G} := \{d \in q\mathsf{Met}(X) \mid \forall \omega < \infty, \forall x \in X : \mathrm{u}(d(x,\cdot) \wedge \omega)(x) = 0\}$$

is a gauge and moreover, for any μ *bounded,* $d^{\mathrm{u}(\mu)} \in \mathcal{G}_b$ *and for any* $x \in X$ *we have*

$$\mathrm{u}(\mu)(x) = \inf_{d \in \mathcal{G}} \sup_{y \in X}(\mu(y) - d(x,y)).$$

Proof The first claim follows immediately from (UH0), (UH2) and (UH4). The second claim follows from (UH3) since

$$\mathrm{u}(d^{\mathrm{u}(\mu)}(x,\cdot))(x) = \mathrm{u}(\mathrm{u}(\mu) \ominus \mathrm{u}(\mu)(x))(x) = \mathrm{u}(\mathrm{u}(\mu))(x) \ominus \mathrm{u}(\mu)(x) = 0.$$

For the last claim finally

$$\begin{aligned}\inf_{d \in \mathcal{G}} \sup_{y \in X}(\mu(y) - d(x,y)) &\leq \sup_{y \in X}(\mu(y) - \mathrm{u}(\mu)(y) \ominus \mathrm{u}(\mu)(x)) \\ &\leq \sup_{y \in X}(\mu(y) - \mathrm{u}(\mu)(y) + \mathrm{u}(\mu)(x)) \\ &\leq \mathrm{u}(\mu)(x)\end{aligned}$$

and conversely, if $\inf_{d \in \mathcal{G}} \sup_{y \in X}(\mu(y) - d(x,y)) < \alpha$ then there exists $e \in \mathcal{G}$ such that for all $y \in X$, $\mu(y) - e(x,y) \leq \alpha$ and thus $\mu \ominus \alpha \leq e(x,\cdot)$. Consequently

$$\mathrm{u}(\mu)(x) \ominus \alpha = \mathrm{u}(\mu \ominus \alpha)(x) \leq \mathrm{u}(e(x,\cdot))(x) = 0$$

which implies that $\mathrm{u}(\mu)(x) \leq \alpha$. □

The foregoing results 1.2.25 and 1.2.26 show that gauges and upper hulls are equivalent structures.

1.2.27 Proposition *If* u *is an upper hull operator on* X*,* $\mathcal{G}$ *is the associated gauge and* d *is a quasi-metric then the following properties are equivalent.*

1. $d \in \mathcal{G}$.
2. $\forall x \in X, \forall \omega < \infty : \mathfrak{u}(d(x,\cdot) \wedge \omega)(x) = 0$.
3. $\forall x \in X, \forall \omega < \infty : \mathfrak{u}(d(x,\cdot) \wedge \omega) = d(x,\cdot) \wedge \omega$.

Proof That the first and second property are equivalent was shown in 1.2.26 and that the third property implies the second one is evident. To show that the first property implies the third one it suffices to note that for any $x, y \in X$ and any $\omega < \infty$

$$\begin{aligned}\mathfrak{u}(d(x,\cdot) \wedge \omega)(y) &= \inf_{e \in \mathcal{G}} \sup_{z \in X} (d(x,z) \wedge \omega - e(y,z)) \\ &\leq \sup_{z \in X} (d(x,z) \wedge \omega - d(y,z) \wedge \omega) \\ &\leq d(x,y) \wedge \omega.\end{aligned}$$

□

1.2.28 Theorem **(UR $\rightrightarrows$ UH)** *If $\mathfrak{U}$ is an upper regular function frame then*

$$\mathfrak{u} : \mathbb{P}_b^X \longrightarrow \mathbb{P}_b^X : \mu \mapsto \inf\{\nu \in \mathfrak{U} \mid \mu \leq \nu\}$$

is an upper hull operator and moreover $\mathfrak{U} = \{\mu \mid \mathfrak{u}(\mu) = \mu\}$.

Proof This is analogous to 1.2.23 and we leave this to the reader. □

1.2.29 Theorem **(UH $\rightrightarrows$ UR)** *If $\mathfrak{u} : \mathbb{P}_b^X \longrightarrow \mathbb{P}_b^X$ is an upper hull operator then $\mathfrak{U} = \{\mu \mid \mathfrak{u}(\mu) = \mu\}$ is an upper regular function frame and moreover for any $\mu \in \mathbb{P}_b^X$, $\mathfrak{u}(\mu) = \inf\{\nu \in \mathfrak{U} \mid \mu \leq \nu\}$.*

Proof This is analogous to 1.2.24 and we leave this to the reader. □

The foregoing results 1.2.28 and 1.2.29 show that upper hulls and upper regular function frames are equivalent structures.

1.2.30 Corollary **(UR $\Rightarrow$ G)** *If $\mathfrak{U}$ is an upper regular function frame and $\mathcal{G}$ is the associated gauge then for any quasi-metric d we have that $d \in \mathcal{G}$ if and only if for all $x \in X$ and $\omega < \infty$, $d(x,\cdot) \wedge \omega \in \mathfrak{U}$.*

1.2.31 Theorem **(A $\rightrightarrows$ F)** *Given an approach system $\mathcal{A}$ on X the relation*

$$\mathfrak{I} \rightarrowtail x \Leftrightarrow \mathcal{A}_b(x) \subseteq \mathfrak{I}$$

is a functional ideal convergence on X. Moreover, for any $x \in X$

$$\mathcal{A}_b(x) = \bigcap\{\mathfrak{I} \mid \mathfrak{I} \rightarrowtail x\}.$$

Proof (F1) and (F2) are evident from the definition. To prove (F3) take a selection of functional ideals $\mathfrak{s} : X \longrightarrow \mathfrak{F}(X)$ such that $\mathfrak{s}(z) \rightarrowtail z$ for all $z \in X$ and a functional ideal $\mathfrak{I}$ such that $\mathfrak{I} \rightarrowtail x$. For each $\mu \in \mathcal{A}_b(x)$ and $\varepsilon > 0$ there is a family $(\mu_z)_{z \in X} \in \prod_{z \in X} \mathcal{A}_b(z)$ such that for all $y, z \in X$ we have

$$\mu(z) \le \mu_x(y) + \mu_y(z) + \varepsilon.$$

We now define

$$\nu_\varepsilon := \inf_{z \in X} (\mu_z + \mu_x(z)).$$

Since for each $z \in X$ we have $\mathscr{A}_b(z) \subseteq \mathfrak{s}(z)$ the function $\mu_z + \mu_x(z)$ is an element of $\mathfrak{s}(z) \oplus \mu_x(z)$ and hence

$$\nu_\varepsilon \in \bigcap_{z \in X} \mathfrak{s}(z) \oplus \mu_x(z).$$

Further, since $\mu_x \in \mathscr{A}_b(x) \subseteq \mathfrak{J}$ it follows that

$$\nu_\varepsilon \in \bigcup_{\varphi \in \mathfrak{J}} \bigcap_{z \in X} \mathfrak{s}(z) \oplus \varphi(z) = \Sigma\mathfrak{s}(\mathfrak{J}).$$

Finally, since $\mu \le \nu_\varepsilon + \varepsilon$, and we can find such ν_ε for each $\varepsilon > 0$ it follows that $\mu \in \Sigma\mathfrak{s}(\mathfrak{J})$ and hence $\Sigma\mathfrak{s}(\mathfrak{J}) \rightarrowtail x$. The final claim is evident. □

1.2.32 Theorem **(F $\rightrightarrows$ A)** *Given a functional ideal convergence* $\rightarrowtail$ *the collection of functional ideals*

$$\mathscr{A}_b(x) := \bigcap \{\mathfrak{J} \mid \mathfrak{J} \rightarrowtail x\}, \quad x \in X,$$

defines a bounded approach system on X. *Moreover, for any* $\mathfrak{J} \in \mathfrak{F}(X)$ *and* $x \in X$, $\mathfrak{J} \rightarrowtail x$ *if and only if* $\mathscr{A}_b(x) \subseteq \mathfrak{J}$.

Proof Because $\mathfrak{i}(\dot{x}) \rightarrowtail x$ it follows that $\mu(x) = 0$ for each $\mu \in \mathscr{A}_b(x)$. Now take $\mu \in \mathscr{A}_b(x)$ and $\varepsilon > 0$ and define

$$\mathfrak{s} : X \longrightarrow \mathfrak{F}(X) : z \mapsto \mathscr{A}_b(z).$$

Then it follows from (F3) that $\mathscr{A}_b(x) \subseteq \Sigma\mathfrak{s}(\mathscr{A}_b(x))$. Hence there is $\varphi \in \mathscr{A}_b(x)$ and for each $z \in X$ a $\varphi_z \in \mathscr{A}_b(z)$ such that

$$\mu \le \inf_{z \in X} \varphi_z + \varphi(z) + \varepsilon.$$

If we now define

$$\mu_x := \varphi \vee \varphi_x \text{ and } \mu_z := \varphi_z \text{ for each } z \neq x$$

then it follows that for all $y, z \in X$ we have $\mu(y) \le \mu_x(z) + \mu_z(y) + \varepsilon$. Hence $(\mathscr{A}_b(x))_{x \in X}$ is a bounded approach system.

The final claim is evident. □

The foregoing results 1.2.31 and 1.2.32 show that approach systems and functional ideal convergence are equivalent structures.

We have now seen the basic transitions in the diagram of Fig. 1.1 at the beginning of this section, which prove that the ten different types of structures are equivalent with each other. For the transition from distances to lower hull operators we went via gauges. Often, however, it will be necessary to have a non-circuitous transition from one structure to another. In the rest of this section we will derive some such non-circuitous transitions, some of which we will require in the sequel.

1.2.33 Theorem **(D $\rightrightarrows$ A)** *If $\delta : X \times 2^X \longrightarrow \mathbb{P}$ is a distance on X, then the associated approach system $(\mathcal{A}(x))_{x\in X}$ is given by*

$$\mathcal{A}(x) = \left\{\varphi \in \mathbb{P}^X | \forall A \subseteq X : \inf_{a\in A} \varphi(a) \leq \delta(x, A)\right\}.$$

Proof It follows from 1.2.4 and 1.2.11 that the approach system $(\mathcal{A}(x))_{x\in X}$ associated with δ is generated by the basis $(\mathcal{B}(x))_{x\in X}$ where for each $x \in X$

$$\mathcal{B}(x) = \left\{d(x, \cdot)|d \in q\mathsf{Met}(X), \forall A \subseteq X : \inf_{a\in A} d(x, a) \leq \delta(x, A)\right\}.$$

For all $x \in X$, let us put

$$\mathcal{C}(x) := \left\{\varphi \in \mathbb{P}^X \mid \forall A \subseteq X : \inf_{a\in A} \varphi(a) \leq \delta(x, A)\right\}.$$

Then it is immediately clear that, for all $x \in X$, $\mathcal{A}(x) \subseteq \mathcal{C}(x)$. Let $\mathcal{G}$ stand for the gauge associated with δ and suppose that there exists $\psi \in \mathcal{C}(x)\setminus \mathcal{A}(x)$. Then it follows that ψ is not dominated by $\mathcal{B}(x)$ and hence there exist $\varepsilon > 0$ and $\omega < \infty$ such that, for all $d \in \mathcal{G}$,

$$\psi \wedge \omega \nleq d(x, \cdot) + \varepsilon.$$

For each $d \in \mathcal{G}$, let $A(d) := \{y \in X \mid \psi(y) \wedge \omega > d(x, y) + \varepsilon\}$. Then it is clear that, for all $d, e \in \mathcal{G}$, $A(d) \cap A(e) = A(d \vee e) \neq \emptyset$. Hence, from 1.2.7, we obtain

$$\begin{aligned}
\sup_{d\in\mathcal{G}} \delta(x, A(d)) \wedge \omega &= \sup_{d\in\mathcal{G}} \sup_{e\in\mathcal{G}} \inf_{y\in A(d)} e(x, y) \wedge \omega \\
&\leq \sup_{d\in\mathcal{G}} \sup_{e\in\mathcal{G}} \inf_{y\in A(d)} (d \vee e)(x, y) \wedge \omega \\
&= \sup_{d\in\mathcal{G}} \inf_{y\in A(d)} d(x, y) \wedge \omega \\
&\leq \sup_{d\in\mathcal{G}} \inf_{y\in A(d)} (\psi(y) \wedge \omega - \varepsilon) \wedge \omega \\
&\leq \sup_{d\in\mathcal{G}} \inf_{y\in A(d)} \psi(y) \wedge \omega - \varepsilon
\end{aligned}$$

$$\leq \sup_{d\in\mathcal{G}} \sup_{\varphi\in\mathcal{A}(x)} \inf_{y\in A(d)} \varphi(y) \wedge \omega - \varepsilon$$

$$\leq \sup_{d\in\mathcal{G}} \delta(x, A(d)) \wedge \omega - \varepsilon,$$

which is a contradiction. □

1.2.34 Theorem **(A ⇉ D)** *If* $(\mathcal{A}(x))_{x\in X}$ *is an approach system on* X*, then the associated distance is given by*

$$\delta(x, A) = \sup_{\varphi\in\mathcal{A}(x)} \inf_{a\in A} \varphi(a).$$

Proof This follows from 1.2.6 and 1.2.14. □

1.2.35 Theorem **(LH ⇉ A)** *If* $\mathfrak{l} : \mathbb{P}^X \longrightarrow \mathbb{P}^X$ *is a lower hull operator on* X*, then the associated approach system* $(\mathcal{A}(x))_{x\in X}$ *is given by*

$$\mathcal{A}(x) = \left\{ \varphi \in \mathbb{P}^X \,|\, \forall\mu \in \mathbb{P}^X : \inf_{y\in X} (\mu + \varphi)(y) \leq \mathfrak{l}(\mu)(x) \right\}.$$

Proof This follows from 1.2.33 and the fact that, by 1.1.31, if φ fulfils the property $\inf_{a\in A} \varphi(a) \leq \mathfrak{l}(\theta_A)(x)$, for all $A \subseteq X$, then it also fulfils the property $\inf_{y\in X} (\mu+\varphi)(y) \leq \mathfrak{l}(\mu)(x)$, for all $\mu \in \mathbb{P}^X$, as can easily be verified. □

We often prove and use formulas involving the complete gauge and the complete approach system. The following two propositions show that in most cases we can safely replace a gauge or approach system by a basis.

1.2.36 Proposition *If* $(\mathcal{A}(x))_{x\in X}$ *is an approach system on* X *with basis* $(\mathcal{B}(x))_{x\in X}$*, then the following formulas hold.*

1. *For any* $x \in X$ *and* $A \subseteq X$*:* $\sup_{\varphi\in\mathcal{A}(x)} \inf_{a\in A} \varphi(a) = \sup_{\varphi\in\mathcal{B}(x)} \inf_{a\in A} \varphi(a).$
2. *For any* $x \in X$ *and* $\mu \in \mathbb{P}^X$*:* $\sup_{\varphi\in\mathcal{A}(x)} \inf_{y\in X} (\mu+\varphi)(y) = \sup_{\varphi\in\mathcal{B}(x)} \inf_{y\in X} (\mu+\varphi)(y).$
3. *For any* $x \in X$ *and* $\mu \in \mathbb{P}^X_b$*:* $\inf_{\varphi\in\mathcal{A}(x)} \sup_{y\in X} (\mu-\varphi)(y) = \inf_{\varphi\in\mathcal{B}(x)} \sup_{y\in X} (\mu-\varphi)(y).$
4. *For any* $x \in X$ *and* $\mathcal{F} \in \mathsf{F}(X)$*:* $\sup_{\varphi\in\mathcal{A}(x)} \inf_{F\in\mathcal{F}} \sup_{y\in F} \varphi(y) = \sup_{\varphi\in\mathcal{B}(x)} \inf_{F\in\mathcal{F}} \sup_{y\in F} \varphi(y).$
5. *For any* $x \in X$ *and* $\mathcal{F} \in \mathsf{F}(X)$*:* $\sup_{\varphi\in\mathcal{A}(x)} \sup_{F\in\mathcal{F}} \inf_{y\in F} \varphi(y) = \sup_{\varphi\in\mathcal{B}(x)} \sup_{F\in\mathcal{F}} \inf_{y\in F} \varphi(y).$

Proof We only give the proof of the first and third formula, the other ones are similar.

For the first formula, one inequality is trivial. To prove the other inequality let $\varphi \in \mathcal{A}(x)$ be fixed. Then it follows from the definition of a basis for an approach system that there exists a family $(\varphi_\varepsilon^\omega)_{\varepsilon>0,\omega<\infty}$ in $\mathcal{B}(x)$ which dominates φ. From this it follows that, for all $\omega < \infty$,

$$
\begin{aligned}
\inf_{a\in A} \varphi(a) \wedge \omega &\le \inf_{\varepsilon>0} \inf_{a\in A} \varphi_\varepsilon^\omega(a) + \varepsilon \\
&\le \inf_{\varepsilon>0} \sup_{\psi\in\mathcal{B}(x)} \inf_{a\in A} \psi(a) + \varepsilon \\
&= \sup_{\psi\in\mathcal{B}(x)} \inf_{a\in A} \psi(a).
\end{aligned}
$$

For the third formula, again one inequality is trivial. To prove the other one let $\varphi \in \mathcal{A}(x)$ be fixed. Then, again, from the definition of a basis for an approach system it follows that we can find a family $(\varphi_\varepsilon^\omega)_{\varepsilon>0,\omega<\infty}$ in $\mathcal{B}(x)$ which dominates φ. From this it follows that, for all $\varepsilon > 0$ and $\omega < \infty$

$$\sup_{y\in X}(\mu(y) - \varphi_\varepsilon^\omega(y)) - \varepsilon \le \sup_{y\in X}(\mu(y) - \varphi(y) \wedge \omega)$$

from which it follows that

$$\inf_{\psi\in\mathcal{B}(x)} \sup_{y\in X}(\mu-\psi)(y) \le \inf_{\omega<\infty} \sup_{y\in X}(\mu(y)-\varphi(y)\wedge\omega) \le \sup_{y\in X}(\mu(y)-\varphi(y)). \qquad \square$$

A similar result holds for gauges and gauge bases.

1.2.37 Proposition *If $\mathcal{G}$ is a gauge on X with basis $\mathcal{H}$, then the following formulas hold.*

1. *For any $x \in X$ and $A \subseteq X$:* $\sup_{d\in\mathcal{G}} \inf_{a\in A} d(x,a) = \sup_{d\in\mathcal{H}} \inf_{a\in A} d(x,a)$.
2. *For any $x \in X$ and $\mu \in \mathbb{P}^X$:* $\sup_{d\in\mathcal{G}} \inf_{y\in X} (\mu(y) + d(x,y)) = \sup_{d\in\mathcal{H}} \inf_{y\in X} (\mu(y) + d(x,y))$.
3. *For any $x \in X$ and $\mu \in \mathbb{P}_b^X$:* $\inf_{d\in\mathcal{G}} \sup_{y\in X}(\mu(y) - d(x,y)) = \inf_{d\in\mathcal{H}} \sup_{y\in X}(\mu(y) - d(x,y))$.
4. *For any $x \in X$ and $\mathcal{F} \in \mathsf{F}(X)$:* $\sup_{d\in\mathcal{G}} \inf_{F\in\mathcal{F}} \sup_{y\in F} d(x,y) = \sup_{d\in\mathcal{H}} \inf_{F\in\mathcal{F}} \sup_{y\in F} d(x,y)$.
5. *For any $x \in X$ and $\mathcal{F} \in \mathsf{F}(X)$:* $\sup_{d\in\mathcal{G}} \sup_{F\in\mathcal{F}} \inf_{y\in F} d(x,y) = \sup_{d\in\mathcal{H}} \sup_{F\in\mathcal{F}} \inf_{y\in F} d(x,y)$.

Proof This is precisely the same as the proof of 1.2.36. $\square$

Note that, especially for the third formula in the foregoing two propositions it will often be advantageous to work with bounded bases for the approach systems and the gauges.

1.2.38 Theorem (A $\rightrightarrows$ LH) *If $(\mathcal{A}(x))_{x\in X}$ is an approach system on X, then the associated lower hull operator is given by*

$$\mathfrak{l}(\mu)(x) = \sup_{\varphi\in\mathcal{A}(x)} \inf_{y\in X} (\mu + \varphi)(y).$$

Proof This follows from 1.2.15 and 1.2.16. □

1.2.39 Theorem (A ⇉ UH) *If* $(\mathscr{A}(x))_{x\in X}$ *is an approach system on* X*, then the associated upper hull operator is given by*

$$\mathfrak{u}(\mu)(x) = \inf_{\varphi\in\mathscr{A}(x)} \sup_{y\in X}(\mu - \varphi)(y).$$

Proof This follows from 1.2.15 and 1.2.25. □

It follows from 1.2.36 that in both foregoing results a basis for the approach system can be used rather than the entire system.

1.2.40 Theorem (UR ⇉ A) *If* $\mathfrak{U}$ *is an upper regular function frame then for any* $x \in X$, $\{\mu \in \mathfrak{U} \mid \mu(x) = 0\}$ *is a bounded basis for the associated approach system in* x*, more precisely for every* $\varphi \in \mathscr{A}_b(x)$ *there exists* $\mu \in \mathscr{A}_b(x) \cap \mathfrak{U}$ *such that* $\varphi \leq \mu$.

Proof If $\varphi \in \mathscr{A}_b(x)$ then for any $\varepsilon > 0$ there exists $d^\varepsilon \in \mathscr{G}_b$ such that $\varphi \leq d^\varepsilon(x,\cdot) + \varepsilon$ from which, by 1.2.27 it follows that $\mathfrak{u}(\varphi) \leq d^\varepsilon(x,\cdot) + \varepsilon$ and hence $\mathfrak{u}(\varphi) \in \mathscr{A}(x)$. Consequently $\varphi \in \mathscr{A}_b(x)$ if and only if $\mathfrak{u}(\varphi) \in \mathscr{A}_b(x)$. Since moreover $\varphi \leq \mathfrak{u}(\varphi)$ it follows that $\{\mathfrak{u}(\varphi) \mid \varphi \in \mathscr{A}_b(x)\} = \mathscr{A}_b(x) \cap \mathfrak{U}$ is a basis. On the other hand if $\mu \in \mathfrak{U}$ and $\mu(x) = 0$ then it follows from the fact that $\mathfrak{u}(\mu) = \mu$ that $\mu \in \mathscr{A}_b(x)$. Hence $\{\mu \in \mathfrak{U} \mid \mu(x) = 0\} = \mathscr{A}_b(x) \cap \mathfrak{U}$. □

1.2.41 Corollary (UH ⇉ A) *If* $\mathfrak{u}$ *is an upper hull operator then for any* $x \in X$, $\{\mathfrak{u}(\varphi) \mid \mathfrak{u}(\varphi)(x) = 0\}$ *is a bounded basis for the associated approach system in* x *and*

$$\mathscr{A}_b(x) = \{\varphi \in \mathbb{P}_b^X \mid \mathfrak{u}(\varphi)(x) = 0\}.$$

1.2.42 Theorem (A ⇉ UR) *If* $(\mathscr{A}(x))_{x\in X}$ *is an approach system on* X *and* $\mathfrak{U}$ *is the associated upper regular function frame then for any* $\mu \in \mathbb{P}_b^X$ *we have that* $\mu \in \mathfrak{U}$ *if and only if* $\mu \ominus \mu(x) \in \mathscr{A}(x)$ *for all* $x \in X$.

Proof To show sufficiency take $x \in X$ then

$$\begin{aligned}\mathfrak{u}(\mu)(x) &= \inf_{\varphi\in\mathscr{A}(x)} \sup_{y\in X}(\mu(y) - \varphi(y))\\ &\leq \sup_{y\in X}(\mu(y) - \mu(y) \ominus \mu(x)) \leq \mu(x).\end{aligned}$$

Necessity follows from 1.2.39 since, given $x \in X$, we now have

$$\mu(x) = \inf_{\varphi\in\mathscr{A}(x)} \sup_{y\in X}(\mu - \varphi)(y)$$

and hence for every $\varepsilon > 0$ there exists $\varphi \in \mathscr{A}(x)$ such that $\mu(x) + \varepsilon \geq \mu - \varphi$. □

1.2.43 Theorem (A $\rightrightarrows$ L) *If $(\mathcal{A}(x))_{x\in X}$ is an approach system on X, then the associated limit operator is given by*

$$\lambda\mathcal{F}(x) = \sup_{\varphi\in\mathcal{A}(x)} \inf_{F\in\mathcal{F}} \sup_{y\in F} \varphi(y).$$

Proof Let δ be the distance associated with $(\mathcal{A}(x))_{x\in X}$. Applying complete distributivity, 1.2.1, 1.2.34, and 1.1.5 we obtain

$$\begin{aligned}
\lambda\mathcal{F}(x) &= \sup_{\mathcal{U}\in \mathsf{U}(\mathcal{F})} \sup_{U\in\mathcal{U}} \delta(x, U)\\
&= \sup_{\mathcal{U}\in \mathsf{U}(\mathcal{F})} \sup_{U\in\mathcal{U}} \sup_{\varphi\in\mathcal{A}(x)} \inf_{y\in U} \varphi(y)\\
&= \sup_{\mathcal{U}\in \mathsf{U}(\mathcal{F})} \sup_{\varphi\in\mathcal{A}(x)} \inf_{U\in\mathcal{U}} \sup_{y\in U} \varphi(y)\\
&= \sup_{\varphi\in\mathcal{A}(x)} \inf_{(A_\mathcal{U})_{\mathcal{U}\in\mathsf{U}(\mathcal{F})}\in\prod_{\mathcal{U}\in\mathsf{U}(\mathcal{F})}\mathcal{U}} \sup_{\mathcal{U}\in\mathsf{U}(\mathcal{F})} \sup_{y\in A_\mathcal{U}} \varphi(y)\\
&= \sup_{\varphi\in\mathcal{A}(x)} \inf_{(A_\mathcal{U})_{\mathcal{U}\in\mathsf{U}(\mathcal{F})}\in\prod_{\mathcal{U}\in\mathsf{U}(\mathcal{F})}\mathcal{U}} \sup_{y\in\bigcup_{\mathcal{U}\in\mathsf{U}(\mathcal{F})} A_\mathcal{U}} \varphi(y)\\
&= \sup_{\varphi\in\mathcal{A}(x)} \inf_{F\in\mathcal{F}} \sup_{y\in F} \varphi(y).
\end{aligned}$$

□

1.2.44 Theorem (G $\rightrightarrows$ L) *If $\mathcal{G} \subseteq q\mathsf{Met}(X)$ is a gauge on X, then the associated limit operator is given by*

$$\lambda\mathcal{F}(x) = \sup_{d\in\mathcal{G}} \inf_{F\in\mathcal{F}} \sup_{y\in F} d(x, y).$$

Proof This follows from 1.2.15 and 1.2.43. □

Again, it follows from 1.2.36 that in both foregoing results a basis for the approach system can be used rather than the entire system.

1.2.45 Theorem (LR $\rightrightarrows$ D) *If $\mathfrak{B}$ is a basis for the lower regular function frame on X then the associated distance is given by*

$$\delta(x, A) = \sup\{\rho(x) \mid \rho \in \mathfrak{B}, \rho_{|A} = 0\}.$$

Proof Since $\delta_A = \mathfrak{l}(\theta_A)$ we have $\delta_A \in \mathfrak{L}$ and if $\rho \in \mathfrak{L}$ such that $\rho_{|A} = 0$ then $\rho = \mathfrak{l}(\rho) \le \mathfrak{l}(\theta_A) = \delta_A$. □

1.2.46 Theorem (G $\rightrightarrows$ T) *If $\mathcal{G}$ is a gauge on X then the associated tower is determined by the neighbourhood filters*

$$\mathcal{V}_\varepsilon(x) = \text{stack}\{B_d(x, \gamma) \mid d \in \mathcal{G}, \varepsilon < \gamma\} \quad x \in X, \varepsilon \in \mathbb{R}^+.$$

Proof This follows from 1.2.6 and 1.2.21. □

1.2.47 Corollary (A $\rightrightarrows$ T) *If $(\mathscr{A}(x))_{x\in X}$ is an approach system on X then the associated tower is determined by the neighbourhood filters*

$$\mathscr{V}_\varepsilon(x) = \{\{\varphi < \gamma\} \mid \varphi \in \mathscr{A}(x), \varepsilon < \gamma\} \quad x \in X, \varepsilon \in \mathbb{R}^+.$$

1.2.48 Theorem (G $\rightrightarrows$ LR) *If $\mathscr{G}$ is a gauge on X then*

$$\mathfrak{B} := \{\alpha \ominus d(x,\cdot) \mid \alpha \in \mathbb{R}^+, d \in \mathscr{G}, x \in X\}$$

is a basis for the associated lower regular function frame $\mathfrak{L}$.

Proof First of all, it is easily verified that $\mathfrak{l}(\alpha \ominus d(x,\cdot)) \geq \alpha \ominus d(x,\cdot)$ for any $\alpha \in \mathbb{R}^+$ and $d \in \mathscr{G}$, from which it follows that $\mathfrak{B} \subseteq \mathfrak{L}$. It suffices of course now to look at a bounded function μ in $\mathfrak{L}$. Since for any $x \in X$ we have that $\mu(x) = \sup_{d\in\mathscr{G}} \inf_{y\in X}(\mu(y) + d(x,y))$ it follows that for all $\varepsilon > 0$ we can find $d_\varepsilon \in \mathscr{G}$ such that

$$(\mu(x) \ominus \varepsilon) \ominus d_\varepsilon(x,\cdot) \leq \mu$$

which proves our claim. □

1.2.49 Theorem (UR $\rightrightarrows$ LR) *If $\mu \in \mathfrak{U}$ then $\alpha \ominus \mu \in \mathfrak{L}$ for any α and moreover if $\mathfrak{B}$ is a bounded basis for the upper regular function frame, then $\{\alpha \ominus \mu \mid \mu \in \mathfrak{B}, \sup \mu < \alpha < \infty\}$ is a basis for the lower regular function frame.*

Proof First, making use of 1.2.16 and 1.2.25 it is easily verified that if $\mu \in \mathfrak{U}$ then $\alpha \ominus \mu \in \mathfrak{L}$ for any α. Second for any given $\mu \in \mathfrak{L}$, $\sup \mu < \alpha < \infty$, $x \in X$ and $\varepsilon > 0$, if $\beta \in \mathfrak{B}$ is such that $\beta \leq \alpha \ominus \mu$ and $\alpha \ominus \mu(x) \leq \beta(x) + \varepsilon$ then it follows that $\mu \leq \alpha \ominus \beta$ and $\alpha \ominus \beta(x) - \varepsilon \leq \mu(x)$. □

1.2.50 Theorem (LR $\rightrightarrows$ UR) *If $\mu \in \mathfrak{L}$ then $\alpha \ominus \mu \in \mathfrak{U}$ for any α and moreover if $\mathfrak{B}$ is a bounded basis for the lower regular function frame, then $\{\alpha \ominus \mu \mid \mu \in \mathfrak{B}, \sup \mu < \alpha < \infty\}$ is a basis for the upper regular function frame.*

Proof This is analogous to 1.2.49 and we leave this to the reader. □

1.2.51 Proposition *If $\mathfrak{l}$ and $\mathfrak{u}$ are associated lower and upper hull operators on X, $A \subseteq X$ and $\omega < \infty$ then*

$$\mathfrak{u}(\theta_A^\omega) + \mathfrak{l}(\theta_{X\setminus A}^\omega) = \omega$$

or put differently

$$\mathfrak{u}(\theta_A^\omega) = \omega - \omega \wedge \delta_{X\setminus A}.$$

Proof Note that $\omega \ominus \mathfrak{u}(\theta_A^\omega) \in \mathfrak{L}$ by 1.2.49, and hence $\omega \ominus \mathfrak{u}(\theta_A^\omega) = \mathfrak{l}(\omega \ominus \mathfrak{u}(\theta_A^\omega))$. Now it suffices to expand both the upper hull and the lower hull on the righthand-side, using 1.2.16 and 1.2.25. □

1.2.52 Theorem **(T$\rightrightarrows$L)** *If* $\mathfrak{t}$ *is a tower with neighbourhood filters* $(\mathscr{V}_\varepsilon(x))_{x\in X,\varepsilon\in\mathbb{R}^+}$ *and convergence* $(\overset{\varepsilon}{\longrightarrow})_{\varepsilon\in\mathbb{R}^+}$ *then the associated limit operator* λ *is determined by*

$$\lambda\mathscr{F}(x) \le \varepsilon \iff \mathscr{V}_\varepsilon(x) \subseteq \mathscr{F} \iff \mathscr{F} \overset{\varepsilon}{\longrightarrow} x.$$

Proof This follows from 1.2.43. □

1.2.53 Proposition *If* λ *is a limit operator and* $(\mathscr{V}_\varepsilon(x))_{x\in X,\varepsilon\in\mathbb{R}^+}$ *is the associated neighbourhood tower then for any* $x \in X$ *and* $\varepsilon \in \mathbb{R}^+$ *we have* $\lambda\mathscr{V}_\varepsilon(x)(x) \le \varepsilon$ *and for any* $\varepsilon \in \mathbb{R}^+$ *and* $\gamma \in [\lambda\mathscr{V}_\varepsilon(x)(x), \varepsilon]$ *we have* $\mathscr{V}_\gamma(x) = \mathscr{V}_\varepsilon(x)$.

Proof The first claim follows at once from the definitions. To prove the second claim it is sufficient to prove that if $\gamma := \lambda\mathscr{V}_\varepsilon(x)(x) < \varepsilon$, then $\mathscr{V}_\gamma(x) = \mathscr{V}_\varepsilon(x)$. Take $\varphi \in \mathscr{A}(x)$ and $\gamma < \eta \le \varepsilon$. Using 1.2.43 it then follows from the hypothesis that there exists $\psi \in \mathscr{A}(x)$ and $\theta > \varepsilon$ such that $\{\psi < \theta\} \subseteq \{\varphi < \eta\}$. Hence $\mathscr{V}_\gamma(x) \subseteq \mathscr{V}_\varepsilon(x)$, and since $\gamma < \varepsilon$ we have $\mathscr{V}_\gamma(x) = \mathscr{V}_\varepsilon(x)$. □

1.2.54 Proposition *If* λ *is a limit operator and* $(\mathscr{V}_\varepsilon(x))_{x\in X,\varepsilon\in\mathbb{R}^+}$ *is the associated neighbourhood tower then for any* $x \in X$ *and* $\varepsilon \in \mathbb{R}^+$ *we have*

$$\lambda\mathscr{V}_\varepsilon(x)(x) = \min\{\gamma \mid \mathscr{V}_\gamma(x) = \mathscr{V}_\varepsilon(x)\}.$$

Proof It follows from 1.2.53 that $\lambda\mathscr{V}_\varepsilon(x)(x) = \inf\{\gamma \mid \mathscr{V}_\gamma(x) = \mathscr{V}_\varepsilon(x)\}$ and that the infimum is a minimum follows at once from (T2n). □

In the following, for $\mu \in \mathbb{P}_b^X$, d_μ is defined by

$$d_\mu(x, y) := \mu(x) \ominus \mu(y)$$

and d^μ (see also 1.2.26) is defined by

$$d^\mu(x, y) := \mu(y) \ominus \mu(x).$$

1.2.55 Theorem **(LR $\rightrightarrows$ G)** *If* $\mathfrak{B}$ *is a basis for a lower regular function frame consisting of bounded functions then* $\mathscr{H} := \{d_\mu \mid \mu \in \mathfrak{B}\}$ *is a basis for the associated gauge.*

Proof If $\mu \in \mathfrak{B}$, $x \in X$ and $A \subseteq X$ then

$$\begin{aligned}\inf_{a\in A} d_\mu(x, a) &= \mu(x) \ominus \sup_{a\in A} \mu(a)\\ &= \mathfrak{l}(\mu \ominus \sup_{a\in A} \mu(a))(x)\\ &\le \mathfrak{l}(\theta_A)(x) = \delta(x, A)\end{aligned}$$

which proves that d_μ is an element of the gauge.

Let $\mathfrak{B}_0 \in 2^{(\mathfrak{B})}$ and $x \in X$. For every $\nu \in \mathfrak{B}_0$ we put

$$\zeta_\nu := \sup_{\xi \in \mathfrak{B}_0} \xi(x) - \nu(x).$$

Then for any $\varepsilon > 0$ there exists $\mu \in \mathfrak{B}$ such that

$$\mu \leq \inf_{\nu \in \mathfrak{B}_0} (\nu + \zeta_\nu) \text{ and } \inf_{\nu \in \mathfrak{B}_0} (\nu + \zeta_\nu)(x) \leq \mu(x) + \varepsilon$$

and it then follows that for any $y \in X$

$$\begin{aligned}
\sup_{\nu \in \mathfrak{B}_0} (\nu(x) \ominus \nu(y)) &= \sup_{\nu \in \mathfrak{B}_0} ((\nu(x) + \zeta_\nu) \ominus (\nu(y) + \zeta_\nu)) \\
&= \sup_{\nu \in \mathfrak{B}_0} ((\sup_{\xi \in \mathfrak{B}_0} \xi(x)) \ominus (\nu(y) + \zeta_\nu)) \\
&= \sup_{\xi \in \mathfrak{B}_0} \xi(x) \ominus (\inf_{\nu \in \mathfrak{B}_0} (\nu(y) + \zeta_\nu)) \\
&= (\inf_{\nu \in \mathfrak{B}_0} (\nu(x) + \zeta_\nu)) \ominus (\inf_{\nu \in \mathfrak{B}_0} (\nu(y) + \zeta_\nu)) \\
&\leq (\mu(x) + \varepsilon) \ominus \mu(y) \\
&\leq \mu(x) \ominus \mu(y) + \varepsilon
\end{aligned}$$

which proves that $\mathscr{H}$ is locally directed.

Further, if $x \in X$ and $A \subseteq X$ then with $L(A) := \{\mu \in \mathfrak{B} \mid \mu|_A = 0\}$ it follows from 1.2.45 that

$$\begin{aligned}
\sup_{\mu \in \mathfrak{B}} \inf_{a \in A} d_\mu(x, a) &\geq \sup_{\mu \in L(A)} \inf_{a \in A} d_\mu(x, a) \\
&= \sup_{\mu \in L(A)} \inf_{a \in A} \mu(x) = \delta(x, A).
\end{aligned}$$

Hence by 1.2.9 it follows that $\mathscr{H}$ is indeed a basis for the gauge. □

1.2.56 Corollary **(LR $\rightrightarrows$ A)** *If $\mathfrak{B}$ is a basis for a lower regular function frame consisting of bounded functions then $\mathscr{A}(x) := \{\mu(x) \ominus \mu \mid \mu \in \mathfrak{B}\}$ is a basis for the associated approach system at $x \in X$.*

1.2.57 Theorem **(UR $\rightrightarrows$ G)** *If $\mathfrak{B}$ is a basis for an upper regular function frame consisting of bounded functions then $\mathscr{H} := \{d^\mu \mid \mu \in \mathfrak{B}\}$ is a basis for the associated gauge.*

Proof If $\mu \in \mathfrak{B}$ and $x \in X$ then

$$\mathrm{u}(d^{\mu}(x,\cdot))(x) = \inf_{e\in\mathscr{G}_b} \sup_{y\in X}(\mu(y) \ominus \mu(x) - e(x,y)) = 0$$

since μ is upper regular, and hence by 1.2.27, $d^{\mu} \in \mathscr{G}$.

Analogously to the proof of 1.2.55 one verifies that $\mathscr{H}$ is locally directed and that the property required to apply 1.2.9 is satisfied. □

The following results completely elucidate the relationship between the two ways of viewing convergence in approach spaces, with limit operators of filters on the one hand and with convergence of functional ideals on the other hand.

1.2.58 Theorem **($\mathbf{L} \rightrightarrows \mathbf{F}$)** *Given associated functional ideal convergence and limit operator, for any proper functional ideal $\mathfrak{I}$ and $x \in X$ the following properties are equivalent.*

1. $\mathfrak{I} \rightarrowtail x$.
2. $\forall \alpha \in [c(\mathfrak{I}), \infty[: \lambda \mathfrak{f}_{\alpha}(\mathfrak{I})(x) \leq \alpha$.

Proof To show that the first property implies the second, let $\alpha \in [c(\mathfrak{I}), \infty[$ then

$$\begin{aligned}\lambda \mathfrak{f}_{\alpha}\mathfrak{I}(x) &\leq \lambda \mathfrak{f}_{\alpha}(\mathscr{A}_b(x))(x)\\ &= \sup_{\mu\in\mathscr{A}_b(x)} \inf_{\alpha<\beta} \inf_{\varphi\in\mathscr{A}_b(x)} \sup_{y\in\{\varphi<\beta\}} \mu(y)\\ &\leq \sup_{\mu\in\mathscr{A}_b(x)} \inf_{\alpha<\beta} \inf_{\substack{\varphi\in\mathscr{A}_b(x)\\ \mu\leq\varphi}} \sup_{y\in\{\varphi<\beta\}} \mu(y)\\ &\leq \alpha.\end{aligned}$$

To see that the second property implies the first, suppose that $\mathfrak{I} \not\rightarrowtail x$. Then, by definition there exist $\varphi \in \mathscr{A}_b(x)$ and $\varepsilon > 0$ such that for all $\mu \in \mathfrak{I} : F_{\mu} := \{\varphi > \mu + \varepsilon\} \neq \emptyset$. It is easily seen that the collection of sets F_{μ}, $\mu \in \mathfrak{I}$ generates a filter, say $\mathscr{F}$. Now consider the functional ideal $\mathfrak{H} := \mathrm{i}(\mathscr{F}) \vee \mathfrak{I}$ which is generated by all functions

$$\nu_{\mu}^{\gamma} := (\theta_{F_{\mu}} \wedge \gamma) \vee \mu \text{ where } \mu \in \mathfrak{I}, \gamma \geq 0.$$

Now note that $\mathfrak{I} \subseteq \mathfrak{H}$ and thus $\alpha := c(\mathfrak{H}) \geq c(\mathfrak{I})$ and $\mathfrak{f}_{\alpha}(\mathfrak{I}) \subseteq \mathfrak{f}_{\alpha}(\mathfrak{H})$. It is easily seen that $\mathfrak{f}_{\alpha}(\mathfrak{H})$ is generated by the sets $F_{\mu} \cap \{\mu < \beta\}$ for $\mu \in \mathfrak{I}$ and $\beta > \alpha$. Now it follows that

$$\begin{aligned}\lambda \mathfrak{f}_{\alpha}\mathfrak{I}(x) &\geq \lambda \mathfrak{f}_{\alpha}\mathfrak{H}(x)\\ &= \sup_{\psi\in\mathscr{A}_b(x)} \inf_{\mu\in\mathfrak{I}} \inf_{\alpha<\beta} \sup_{y\in F_{\mu}\cap\{\mu<\beta\}} \psi(y)\\ &\geq \inf_{\mu\in\mathfrak{I}} \inf_{\alpha<\beta} \sup_{y\in F_{\mu}\cap\{\mu<\beta\}} \varphi(y)\end{aligned}$$

$$\begin{aligned}
&\geq \sup_{\mu \in \mathfrak{I}} \sup_{\alpha < \beta} \inf_{y \in F_\mu \cap \{\mu < \beta\}} \varphi(y) \\
&\geq \sup_{\mu \in \mathfrak{I}} \inf_{y \in F_\mu} \mu(y) + \varepsilon \\
&= \sup_{\mu \in \mathfrak{I}} \inf_{y \in X} \theta_{F_\mu} \vee \mu(y) + \varepsilon \\
&\geq \sup_{\mu \in \mathfrak{I}} \sup_{\gamma \geq 0} \inf_{y \in X} (\theta_{F_\mu} \wedge \gamma) \vee \mu(y) + \varepsilon = \alpha + \varepsilon.
\end{aligned}$$
□

1.2.59 Corollary *Given associated functional ideal convergence and limit operator, for any proper prime functional ideal $\mathfrak{I}$ and $x \in X$ the following properties are equivalent.*

1. $\mathfrak{I} \rightarrowtail x$.
2. $\lambda\mathfrak{f}(\mathfrak{I})(x) \leq c(\mathfrak{I})$.

Proof This follows from 1.1.50 and 1.2.58. □

1.2.60 Theorem **(F $\rightrightarrows$ L)** *Given associated functional ideal convergence and limit operator, for any filter $\mathscr{F} \in \mathsf{F}(X)$ and $x \in X$ we have*

$$\lambda\mathscr{F}(x) = \inf\{\alpha \mid \mathfrak{i}(\mathscr{F}) \oplus \alpha \rightarrowtail x\}.$$

Proof This follows from 1.2.58 since

$$\begin{aligned}
\inf\{\alpha \mid \mathfrak{i}(\mathscr{F}) \oplus \alpha \rightarrowtail x\} &= \inf\{\alpha \mid \forall\beta \in [\alpha, \infty[: \lambda\mathfrak{f}_\beta(\mathfrak{i}(\mathscr{F}) \oplus \alpha)(x) \leq \beta\} \\
&= \inf\{\alpha \mid \forall\beta \in [\alpha, \infty[: \lambda\mathscr{F}(x) \leq \beta\} \\
&= \lambda\mathscr{F}(x).
\end{aligned}$$
□

We have now seen all the transitions indicated in the diagrams of Figs. 1.1 and 1.2 at the beginning of this section, for going from one basic structure to another and which we will require in this work. Clearly there are a number of fundamental transitions which have great conceptual interest and technical elegance, whereas some are no more than the composition of other transitions. Again this is not unlike the situation in topology, where defining, for example, the open sets from the neighbourhoods goes in a direct way, whereas defining them in terms of the closure operator goes most naturally via the closed sets.

1.2.61 Definition A pair $(X, \mathfrak{S})$, where $\mathfrak{S}$ is a distance, a limit operator, an approach system, a gauge, a tower, a lower or upper hull operator, a lower or upper regular function frame or a functional ideal convergence is called an *approach space*.

As we have just seen, an approach space can hence be determined by giving any of these equivalent structures and in most of the examples which we will study later on, at least one of them will present itself as the basic one from which the others can be derived. If no confusion is possible, we will always denote a distance by δ and the associated limit operator, approach system, gauge, tower, upper and lower hull operator, upper and lower regular function frame and functional ideal convergence respectively by λ, $\mathscr{A} := (\mathscr{A}(x))_{x\in X}$, $\mathscr{G}$, $\mathfrak{t} := (\mathfrak{t}_\varepsilon)_{\varepsilon\in\mathbb{R}^+}$, $\mathfrak{u}$, $\mathfrak{l}$, $\mathfrak{U}$, $\mathfrak{L}$ and $\rightarrowtail$. It will generally be understood that these structures are associated with one another. If we are dealing with different approach spaces at the same time, we will make clear which structures are associated with each other, by using some appropriate indices or accents.

In general we will also denote an approach space simply by its underlying set and only if required for clarity or unambiguity will we write it as a pair consisting of the underlying set and one of its defining structures. Some spaces may get special notational treatment like those in the next example. When an approach space is determined by a gauge basis (or by an approach basis), then we will usually refer to the other structures of the approach space as being *generated* by that basis. We use both notations $A^{(\varepsilon)}$ and $\mathfrak{t}_\varepsilon(A)$ for $\{x \in X \mid \delta(x, A) \leq \varepsilon\}$. In all that follows we will freely use the different structures defining approach spaces as well as the transitions among them. In the appendix, for easy reference, we give some tables containing the most important transition formulas.

Once we have developed more machinery, we will encounter many individual examples as well as special classes of natural approach spaces. Some particular examples which we can easily introduce at this stage and which we will require quite frequently in the sequel are the following. Note that the examples in 1.2.62 and 1.2.63 are genuine approach spaces in the sense that they are neither metric nor topological (see following chapter).

1.2.62 Example (Approach structures on $\mathbb{P}$)

It will be useful to have at our disposal the following quasi-metrics. We denote by $d_\mathbb{E}$ the metric defined by

$$d_\mathbb{E} : \mathbb{P} \times \mathbb{P} \longrightarrow \mathbb{P} : (x, y) \mapsto |x - y|$$

and by $d_\mathbb{P}$ the quasi-metric defined by

$$d_\mathbb{P} : \mathbb{P} \times \mathbb{P} \longrightarrow \mathbb{P} : (x, y) \mapsto x \ominus y.$$

Note that both are straightforward extensions of the well-known quasi-metric and metric on $[0, \infty[$. Restrictions will be denoted by the same symbols as it will always be clear from the context whether we are considering them on $\mathbb{P}$ or on a subspace. For the topologies generated by $d_\mathbb{E}$ and by $d_\mathbb{P}$ on $\mathbb{P}$, the point ∞ is isolated.

1. Define

$$\delta_\mathbb{E} : \mathbb{P} \times 2^\mathbb{P} \longrightarrow \mathbb{P}$$

by

$$\delta_{\mathbb{E}}(x, A) := \begin{cases} 0 & x = \infty, \ A \text{ unbounded}, \\ \infty & x = \infty, \ A \text{ bounded}, \\ \inf_{a \in A} |x - a| & x \in \mathbb{R}^+. \end{cases}$$

It is easily verified that this function is a distance on $\mathbb{P}$. This distance actually extends the Euclidean metric $d_{\mathbb{E}}$ to $\mathbb{P}$. The associated approach system is the following:

$$\mathscr{A}_{\mathbb{E}}(x) := \begin{cases} \{\varphi \in \mathbb{P}^{\mathbb{P}} \mid \varphi \leq d_{\mathbb{E}}(x, \cdot)\} & x \in \mathbb{R}^+, \\ \widehat{\{\theta_{]a,\infty]} \mid a < \infty\}} & x = \infty. \end{cases}$$

Note that this approach system is of a purely metric nature in the finite points and of a purely topological nature at ∞.

If $\mathbb{P}$ is equipped with this structure then we will denote it $\mathbb{P}_{\mathbb{E}}$, i.e. we put

$$\mathbb{P}_{\mathbb{E}} := (\mathbb{P}, \delta_{\mathbb{E}}).$$

The Euclidean topology on $[-\infty, \infty]$ or any of its subspaces will also be denoted by $\mathscr{T}_{\mathbb{E}}$. We note that the extension of the Euclidean metric on $\mathbb{P}$ (i.e. $d_{\mathbb{E}}$) and a similar extension on $[-\infty, \infty]$ does not generate the Euclidean topology since for the underlying topology of this metric (and also for the underlying topology of the quasi-metric $d_{\mathbb{P}}$) the points $-\infty$ and ∞ are isolated.

2. An analogous example is obtained as follows. Define

$$\delta_{\mathbb{P}} : \mathbb{P} \times 2^{\mathbb{P}} \longrightarrow \mathbb{P}$$

by

$$\delta_{\mathbb{P}}(x, A) := \begin{cases} x \ominus \sup A & A \neq \emptyset, \\ \infty & A = \emptyset. \end{cases}$$

Again this function is a distance on $\mathbb{P}$ which is an extension of the quasi-metric $d_{\mathbb{P}}$ on $\mathbb{R}^+$ to $\mathbb{P}$.

The associated approach system is the following:

$$\mathscr{A}_{\mathbb{P}}(x) := \begin{cases} \{\varphi \in \mathbb{P}^{\mathbb{P}} \mid \varphi \leq d_{\mathbb{P}}(x, \cdot)\} & x \in \mathbb{R}^+, \\ \widehat{\{\theta_{]a,\infty]} \mid a \in \mathbb{R}^+\}} & x = \infty. \end{cases}$$

Here too the approach system is of a purely quasi-metric nature in the finite points and of a purely topological nature at ∞.

The associated limit operator is determined as follows. Given any filter $\mathscr{F}$ on $\mathbb{P}$, let

$$l(\mathcal{F}) := \inf_{U \in \sec \mathcal{F}} \sup U.$$

Then for all $x \in \mathbb{P}$

$$\lambda_{\mathbb{P}}\mathcal{F}(x) = x \ominus l(\mathcal{F}).$$

Notice for example that for any filter $\mathcal{F}$, either $\lambda_{\mathbb{P}}\mathcal{F}(\infty) = 0$ or $\lambda_{\mathbb{P}}\mathcal{F}(\infty) = \infty$. The first case will occur if $l(\mathcal{F}) = \infty$, i.e. if every member of $\sec\mathcal{F}$ is unbounded. A gauge basis is given by $\{d_\alpha \mid \alpha \in \mathbb{R}^+\}$ where

$$d_\alpha(x, y) = x \wedge \alpha \ominus y \wedge \alpha.$$

We denote the approach space thus defined simply by

$$\mathbb{P} := (\mathbb{P}, \delta_{\mathbb{P}}).$$

It will always be clear from the context whether we mean the space $\mathbb{P}$ or the set $\mathbb{P}$ and if we consider any other structure on $\mathbb{P}$ besides $\delta_{\mathbb{P}}$ then this will be mentioned explicitly as in the case of the first example.

We will leave it as an exercise for the reader to determine the other structures associated with these distances.

1.2.63 Example (Filter approach spaces) Let X be an arbitrary set, $\mathcal{F} \in \mathsf{F}(X)$ a fixed filter and $f \in \mathbb{P}^X$ a fixed arbitrary function. We now define

$$\lambda_{(\mathcal{F},f)} : \mathsf{F}(X) \longrightarrow \mathbb{P}^X$$

by

$$\lambda_{(\mathcal{F},f)}\mathcal{G}(x) := \begin{cases} f(x) & \mathcal{F} \cap \text{stack}x \subseteq \mathcal{G}, \mathcal{G} \neq \text{stack}x, \\ \infty & \mathcal{F} \cap \text{stack}x \nsubseteq \mathcal{G}, \\ 0 & \mathcal{G} = \text{stack}x. \end{cases}$$

Then $(X, \lambda_{(\mathcal{F},f)})$ is an approach space. Indeed, (L1) follows at once from the definition. The verification of (L2) is straightforward by considering cases. In order to verify (L3) let $(\sigma(y))_{y\in X}$ be a selection of filters on X and first let $\mathcal{U} \in \mathsf{U}(X)$ be an ultrafilter on X. Given $x \in X$ the only case where

$$\lambda_{(\mathcal{F},f)}\mathcal{U}(x) + \sup_{y\in X} \lambda_{(\mathcal{F},f)}\sigma(y)(y)$$

is not necessarily equal to ∞ is when

$$\mathcal{F} \cap \text{stack}x \subseteq \mathcal{U} \text{ and } \mathcal{F} \cap \text{stack}y \subseteq \sigma(y) \text{ for every } y \in X.$$

Under these two assumptions we consider the following two cases. Either $\mathcal{U} =$ stack$x$ and then $\Sigma\sigma(\mathcal{U}) = \sigma(x)$ from which (L3) follows at once, or $\mathcal{U} \neq$ stack$x$ and then we have $\mathcal{F} \subseteq \mathcal{U}$ and

$$\mathcal{F} \subseteq \bigvee_{U\in\mathcal{U}} \bigcap_{y\in U} (\mathcal{F} \cap \text{stack}y) \subseteq \Sigma\sigma(\mathcal{U})$$

and in that case we have

$$\begin{aligned}\lambda_{(\mathcal{F},f)}\Sigma\sigma(\mathcal{U})(x) &\leq \lambda_{(\mathcal{F},f)}\mathcal{F}(x)\\ &= \lambda_{(\mathcal{F},f)}\mathcal{U}(x)\\ &\leq \lambda_{(\mathcal{F},f)}\mathcal{U}(x) + \sup_{y\in X} \lambda_{(\mathcal{F},f)}\sigma(y)(y).\end{aligned}$$

Second we consider the case of an $\mathcal{H} \in \mathsf{F}(X)$ and then using (L2) we have

$$\begin{aligned}\lambda_{(\mathcal{F},f)}\Sigma\sigma(\mathcal{H}) &= \sup_{\mathcal{U}\in\mathsf{U}(\mathcal{H})} \lambda_{(\mathcal{F},f)}\Sigma\sigma(\mathcal{U})\\ &\leq \sup_{\mathcal{U}\in\mathsf{U}(\mathcal{H})} \lambda_{(\mathcal{F},f)}\mathcal{U} + \sup_{y\in X} \lambda_{(\mathcal{F},f)}\sigma(y)(y)\\ &= \lambda_{(\mathcal{F},f)}\mathcal{H} + \sup_{y\in X} \lambda_{(\mathcal{F},f)}\sigma(y)(y)\end{aligned}$$

which proves (L3) in the general case and we are finished.

The final structural concept which we will introduce is that of an *adherence operator*. Adherence operators constitute the logical counterpart to limit operators in the same way as the notion of an adherence point of a filter is the counterpart of the notion of limit point of a filter. We will see many instances where it is advantageous to use this concept.

1.2.64 Definition Let X be an approach space. We define what we will call the *adherence operator* (associated with δ and with all of the other defining structures) as the function

$$\alpha : \mathsf{F}(X) \longrightarrow \mathbb{P}^X,$$

determined by

$$\alpha\mathcal{F}(x) := \sup_{F\in\mathcal{F}} \delta(x, F).$$

The interpretation of this operator is analogous to that of the limit operator. In each point $x \in X$, the value $\alpha\mathcal{F}(x)$ indicates how far the point x is away from being an adherence point of the filter $\mathcal{F}$.

1.2.65 Proposition *For any pair of filters $\mathcal{F}$ and $\mathcal{G}$ we have*

$$\mathcal{F} \subseteq \mathcal{G} \Rightarrow \alpha\mathcal{F} \leq \alpha\mathcal{G} \leq \lambda\mathcal{G} \leq \lambda\mathcal{F}.$$

Proof This follows from 1.2.1 and the definition of the adherence operator. □

1.2.66 Proposition *For any ultrafilter $\mathcal{U}$ we have*

$$\lambda\mathcal{U} = \alpha\mathcal{U}.$$

Proof If $\mathcal{U}$ is an ultrafilter, then $\sec\mathcal{U} = \mathcal{U}$ and, hence, the result follows from 1.2.1 and the definition of adherence operator. □

The foregoing result generalizes the well-known fact that for ultrafilters limit points and adherence points are the same.

1.2.67 Proposition *For any filter $\mathcal{F}$ and $x \in X$ the following properties hold.*

1. $\lambda\mathcal{F}(x) = \sup_{\mathcal{U} \in \mathsf{U}(\mathcal{F})} \lambda\mathcal{U}(x) = \sup_{\mathcal{U} \in \mathsf{U}(\mathcal{F})} \alpha\mathcal{U}(x).$
2. $\alpha\mathcal{F}(x) = \inf_{\mathcal{U} \in \mathsf{U}(\mathcal{F})} \lambda\mathcal{U}(x) = \inf_{\mathcal{U} \in \mathsf{U}(\mathcal{F})} \alpha\mathcal{U}(x)$ *and moreover the infimum is actually a minimum.*

Proof The first formula follows from 1.2.1 and 1.2.66.

For the second formula, that $\alpha\mathcal{F}(x) \leq \inf_{\mathcal{U} \in \mathsf{U}(\mathcal{F})} \lambda\mathcal{U}(x)$ is evident.

Suppose now that, for all $\mathcal{U} \in \mathsf{U}(\mathcal{F})$, $\alpha\mathcal{F}(x) < \lambda\mathcal{U}(x)$. Then it follows from 1.2.66 and the definition of adherence operator that, for each $\mathcal{U} \in \mathsf{U}(\mathcal{F})$, there exists $U_{\mathcal{U}} \in \mathcal{U}$ such that $\alpha\mathcal{F}(x) < \delta(x, U_{\mathcal{U}})$. It now follows from 1.1.4 that there exists a finite subset $\mathsf{U}_0 \subseteq \mathsf{U}(\mathcal{F})$ such that $\bigcup_{\mathcal{U} \in \mathsf{U}_0} U_{\mathcal{U}} \in \mathcal{F}$ and consequently, by definition of the adherence operator,

$$\begin{aligned}\alpha\mathcal{F}(x) &\geq \delta(x, \bigcup_{\mathcal{U} \in \mathsf{U}_0} U_{\mathcal{U}}) \\ &= \min_{\mathcal{U} \in \mathsf{U}_0} \delta(x, U_{\mathcal{U}}) \\ &> \alpha\mathcal{F}(x),\end{aligned}$$

which is a contradiction. □

The foregoing results of course generalize the facts that in a topological space a filter converges to a point if and only if all ultrafilters which are finer converge to that point and that a filter adheres to a point if and only if there exists an ultrafilter which is finer and which converges to that point. Considering as filter the principal filter generated by a set $A \subseteq X$ it also allows for the following immediate consequence.

1.2.68 Corollary *For any $A \subseteq X$ and $x \in X$ there exists an ultrafilter $\mathcal{U} \in \mathsf{U}(A)$ such that $\delta(x, A) = \lambda\mathcal{U}(x)$.*

The following characterizations of the adherence operator in terms of approach systems and gauges will also be of use.

1.2.69 Proposition *For any filter $\mathcal{F}$ and $x \in X$ we have*

$$\begin{aligned}\alpha\mathcal{F}(x) &= \sup_{\varphi\in\mathcal{A}(x)} \sup_{F\in\mathcal{F}} \inf_{y\in F} \varphi(y)\\ &= \sup_{d\in\mathcal{G}} \sup_{F\in\mathcal{F}} \inf_{y\in F} d(x, y).\end{aligned}$$

Proof This follows from 1.2.34, 1.2.6, and the definition of adherence operator. □

1.2.70 Proposition *For any filter $\mathcal{F}$ and $x, y \in X$ we have*

$$\lambda\mathcal{F}(x) \leq \delta(x, \{y\}) + \lambda\mathcal{F}(y)$$

and

$$\alpha\mathcal{F}(x) \leq \delta(x, \{y\}) + \alpha\mathcal{F}(y).$$

Proof Let $x, y \in X$, then, by 1.2.44, we have

$$\begin{aligned}\lambda\mathcal{F}(x) &= \sup_{d\in\mathcal{G}} \inf_{F\in\mathcal{F}} \sup_{z\in F} d(x, z)\\ &\leq \sup_{d\in\mathcal{G}} \inf_{F\in\mathcal{F}} \sup_{z\in F}(d(x, y) + d(y, z))\\ &\leq \sup_{d\in\mathcal{G}} d(x, y) + \sup_{d\in\mathcal{G}} \inf_{F\in\mathcal{F}} \sup_{z\in F} d(y, z)\\ &= \delta(x, \{y\}) + \lambda\mathcal{F}(y)\end{aligned}$$

which proves the first inequality. By 1.2.67 the second inequality is an immediate consequence of the first one. □

1.3 The Morphisms: Contractions

The morphisms which are naturally associated with the structures defined in the foregoing sections can most elegantly be defined in terms of distances, but we will immediately show that they can equally well be characterized by means of any of the other structures.

1.3.1 Definition A function $f : X \longrightarrow X'$ between approach spaces is called a *contraction* if for all $A \subseteq X$

$$\delta'_{f(A)} \circ f \le \delta_A$$

or explicitly, if for all $x \in X$ and $A \subseteq X$

$$\delta'(f(x), f(A)) \le \delta(x, A).$$

We will require the following lemma.

1.3.2 Lemma *If $f : X \longrightarrow X'$ is a function and $\mathscr{F}$ is a filter on X then for any $\mathscr{W} \in \mathsf{U}(f(\mathscr{F}))$ there exists $\mathscr{U} \in \mathsf{U}(\mathscr{F})$ such that $f(\mathscr{U}) = \mathscr{W}$.*

Proof Suppose that for all $\mathscr{U} \in \mathsf{U}(\mathscr{F})$ we have a set $U_{\mathscr{U}} \in \mathscr{U}$ such that $f(U_{\mathscr{U}}) \notin \mathscr{W}$, then it follows from 1.1.4 that there exist $U_{\mathscr{U}_1} \in \mathscr{U}_1, \ldots, U_{\mathscr{U}_n} \in \mathscr{U}_n$ such that $\bigcup_{i=1}^n U_{\mathscr{U}_i} \in \mathscr{F}$. However, in that case there would exist $k \in \{1, \ldots, n\}$ such that $f(U_{\mathscr{U}_k}) \in f(\mathscr{F}) \subseteq \mathscr{W}$ which is a contradiction. □

We will use these facts for instance in some of the proofs of the following result.

1.3.3 Theorem *For a function $f : X \longrightarrow X'$ between approach spaces the following properties are equivalent.*

1. *f is a contraction.*
2. $\forall \mathscr{F} \in \mathsf{F}(X) : \lambda'(f(\mathscr{F})) \circ f \le \lambda\mathscr{F}$.
3. $\forall \mathscr{F} \in \mathsf{U}(X) : \lambda'(f(\mathscr{F})) \circ f \le \lambda\mathscr{F}$.
4. $\forall x \in X, \forall \varphi' \in \mathscr{A}'(f(x)) : \varphi' \circ f \in \mathscr{A}(x)$.
5. $\forall d' \in \mathscr{G}' : d' \circ (f \times f) \in \mathscr{G}$.
6. $\forall A \subseteq X, \forall \varepsilon \in \mathbb{R}^+ : f(\mathfrak{t}_\varepsilon(A)) \subseteq \mathfrak{t}'_\varepsilon(f(A))$.
7. $\forall x \in X, \forall \varepsilon \in \mathbb{R}^+, \forall V' \in \mathscr{V}'_\varepsilon(f(x)) : f^{-1}(V') \in \mathscr{V}_\varepsilon(x)$.
8. $\forall \mu \in \mathbb{P}^X : \mathfrak{l}'(f(\mu)) \le f(\mathfrak{l}(\mu))$.
9. $\forall \nu \in \mathfrak{L}' : \nu \circ f \in \mathfrak{L}$.
10. $\forall \nu \in \mathfrak{U}' : \nu \circ f \in \mathfrak{U}$.
11. $\forall \mathfrak{J} \in \mathfrak{F}(X), \forall x \in X : \mathfrak{J} \rightarrowtail x \Rightarrow f(\mathfrak{J}) \rightarrowtail f(x)$.
12. $\forall \mathfrak{J} \in \mathfrak{P}(X), \forall x \in X : \mathfrak{J} \rightarrowtail x \Rightarrow f(\mathfrak{J}) \rightarrowtail f(x)$.

In 4 instead of $\mathscr{A}'(f(x))$ a basis is sufficient, in 5 instead of $\mathscr{G}'$ also a basis is sufficient, in 8 instead of $\mathfrak{L}'$ a subbasis is sufficient and in 9 instead of $\mathfrak{U}'$ also a subbasis is sufficient.

Proof 1 ⇒ 2. This follows from 1.2.1 together with the fact that $\sec f(\mathscr{F}) \subseteq f(\sec\mathscr{F})$ as can easily be seen making use of 1.3.2.

2 ⇒ 3. This is evident.

3 ⇒ 1. This follows from 1.2.2 together with the fact that $f(\mathsf{U}(A) \subseteq \mathsf{U}(f(A))$.

6 ⇒ 1. For any $x \in X$ and $A \subseteq X$, we have $x \in A^{(\delta(x,A))}$, and thus $f(x) \in f(A)^{(\delta(x,A))'}$ which precisely means that $\delta'(f(x), f(A)) \le \delta(x, A)$.

1 ⇒ 4. Suppose on the contrary that there exist $x_0 \in X$, $\varphi'_0 \in \mathscr{A}'(f(x_0))$, $\varepsilon_0 > 0$, and $\omega_0 < \infty$ such that, for all $\varphi \in \mathscr{A}(x_0)$,

$$A(\varphi) := \{x \in X \mid \varphi'_0(f(x)) \wedge \omega_0 > \varphi(x) + \varepsilon_0\}$$

is nonempty. Then we obtain

$$\begin{aligned}
\sup_{\varphi\in\mathscr{A}(x_0)} \delta(x_0, A(\varphi)) \wedge \omega_0 &= \sup_{\varphi\in\mathscr{A}(x_0)} \sup_{\psi\in\mathscr{A}(x_0)} \inf_{x\in A(\varphi)} \psi(x) \wedge \omega_0 \\
&\leq \sup_{\varphi\in\mathscr{A}(x_0)} \sup_{\psi\in\mathscr{A}(x_0)} \inf_{x\in A(\varphi\vee\psi)} (\varphi \vee \psi(x)) \wedge \omega_0 \\
&= \sup_{\varphi\in\mathscr{A}(x_0)} \inf_{x\in A(\varphi)} \varphi(x) \wedge \omega_0 \\
&\leq \sup_{\varphi\in\mathscr{A}(x_0)} \inf_{x\in A(\varphi)} \varphi_0'(f(x)) \wedge \omega_0 - \varepsilon_0 \\
&\leq \sup_{\varphi\in\mathscr{A}(x_0)} \sup_{\xi\in\mathscr{A}'(f(x_0))} \inf_{y\in f(A(\varphi))} \xi(y) \wedge \omega_0 - \varepsilon_0 \\
&= \sup_{\varphi\in\mathscr{A}(x_0)} \delta'(f(x_0), f(A(\varphi))) \wedge \omega_0 - \varepsilon_0,
\end{aligned}$$

which is impossible.

4 $\Rightarrow$ 5. Let $d' \in \mathscr{G}'$. Then it follows that, for all $x \in X$, $d'(f(x), \cdot) \in \mathscr{A}'(f(x))$ and hence $d'(f(x), \cdot) \circ f \in \mathscr{A}(x)$. Consequently $d' \circ (f \times f) \in \mathscr{G}$.

5 $\Rightarrow$ 8. Let $\mu \in \mathbb{P}^X$ and $x' \in X'$. If $x' \notin f(X)$, then there is nothing to prove. Suppose therefore that $x' \in f(X)$ and fix $x \in f^{-1}(x')$. Then we obtain

$$\begin{aligned}
\mathfrak{l}'(f(\mu))(x') &= \sup_{d'\in\mathscr{G}'} \inf_{u'\in X'} \Big(\inf_{u\in f^{-1}(u')} (\mu(u) + d'(x', u'))\Big) \\
&\leq \sup_{d'\in\mathscr{G}'} \inf_{v\in X} \Big(\inf_{u\in f^{-1}(f(v))} (\mu(u) + d'(x', f(v)))\Big) \\
&\leq \sup_{d\in\mathscr{G}} \inf_{v\in X} (\mu(v) + d(x, v)) \\
&= \mathfrak{l}(\mu)(x),
\end{aligned}$$

which from the arbitrariness of $x \in f^{-1}(x')$ proves our claim.

6 $\Leftrightarrow$ 7. This is a well-known result in pretopological spaces.

8 $\Rightarrow$ 6. Let $x \in \mathfrak{t}_\varepsilon(A)$. Then $\mathfrak{l}(\theta_A)(x) = \delta_A(x) \leq \varepsilon$. Consequently it follows that

$$\begin{aligned}
\delta'_{f(A)}(f(x)) &= \mathfrak{l}'(\theta_{f(A)})(f(x)) \\
&\leq f(\mathfrak{l}(\theta_A))(f(x)) \\
&= \inf_{z\in f^{-1}(f(x))} \mathfrak{l}(\theta_A)(z) \\
&\leq \mathfrak{l}(\theta_A)(x) \leq \varepsilon,
\end{aligned}$$

which proves the assertion.

8 $\Rightarrow$ 9. Let $\nu \in \mathfrak{L}'$. Then

$$\nu = \mathfrak{l}'(\nu) \leq \mathfrak{l}'(f(\nu \circ f)) \leq f(\mathfrak{l}(\nu \circ f))$$

and it follows that $\nu \circ f \leq \mathfrak{l}(\nu \circ f)$, i.e. $\nu \circ f \in \mathfrak{L}$.

$9 \Rightarrow 8$. Let $\mu \in \mathbb{P}^X$. Then

$$(\mathfrak{l}'(f(\mu))) \circ f \leq f(\mu) \circ f \leq \mu$$

and since $(\mathfrak{l}'(f(\mu))) \circ f \in \mathfrak{L}$, it follows that also $(\mathfrak{l}'(f(\mu))) \circ f \leq \mathfrak{l}(\mu)$ and thus $\mathfrak{l}'(f(\mu)) \leq f(\mathfrak{l}(\mu))$.

$4 \Rightarrow 10$. Let $\mu \in \mathfrak{U}'$ and $x \in X$ then it follows that $\mu \ominus \mu(f(x)) \in \mathscr{A}'(f(x))$ and hence $\mu \circ f \ominus \mu \circ f(x) \in \mathscr{A}(x)$ which by 1.2.42 shows that $\mu \circ f \in \mathfrak{U}$.

$10 \Rightarrow 4$. Let $\mu \in \mathfrak{U}'$ such that $\mu(f(x)) = 0$ then it follows that $\mu \circ f \in \mathfrak{U}$ and hence by 1.2.40 that $\mu \circ f \in \mathscr{A}(x)$.

$1 \Rightarrow 11$. If f is a contraction and $\mathfrak{I} \rightarrowtail x$ then it suffices to apply 1.2.58 to see that also $f(\mathfrak{I}) \rightarrowtail f(x)$.

$11 \Rightarrow 1$. It suffices to apply 1.2.60 in order to see that f is a contraction.

$1 \Leftrightarrow 12$. This is perfectly analogous to the equivalence of 1 and 11 using 1.2.59 instead of 1.2.58. □

Note that in the foregoing theorem items 6 and 7 are of course nothing else but two equivalent ways to say that f is continuous if X and X' are each equipped with the same level tower structures.

Some interesting examples of contractions are given in the following propositions, where the space $\mathbb{P}$ plays an important role.

1.3.4 Proposition *For any $A \subseteq X$, the distance functional*

$$\delta_A : (X, \delta) \longrightarrow \mathbb{P} : x \mapsto \delta(x, A)$$

is a contraction.

Proof Let $x \in X$ and $B \subseteq X$. If $B = \emptyset$, then

$$\delta_{\mathbb{P}}(\delta_A(x), \delta_A(B)) = \delta(x, B) = \infty.$$

Otherwise, since

$$\delta_{\mathbb{P}}(\delta_A(x), \delta_A(B)) = \delta(x, A) \ominus (\sup_{b \in B} \delta(b, A)),$$

the result is an immediate consequence of 1.1.2. □

1.3.5 Proposition *For any function $\xi \in \mathbb{P}^X$, the following properties are equivalent.*

1. *$\xi \in \mathfrak{L}$, i.e. ξ is lower regular.*
2. *$\xi : X \longrightarrow \mathbb{P}$ is a contraction.*

Proof Both the first and the second property can be expressed as local properties of ξ. By 1.2.24 and 1.2.38, $\xi \in \mathfrak{L}$ if and only if

$$\forall x \in \{\xi < \infty\}, \forall \varepsilon > 0, \exists \varphi \in \mathscr{A}(x), \forall y \in X : \xi(x) \leq \xi(y) + \varphi(y) + \varepsilon$$
$$\text{and} \tag{$*$}$$
$$\forall x \in \{\xi = \infty\}, \forall \omega < \infty, \exists \varphi \in \mathscr{A}(x), \forall y \in X : \omega \leq \xi(y) + \varphi(y).$$

From 1.3.3 and the description of the approach system of $\mathbb{P}$ it follows that $\xi : X \longrightarrow \mathbb{P}$ is a contraction if and only if

$$\forall x \in \{\xi < \infty\}, \forall \varepsilon > 0, \forall \omega < \infty, \exists \varphi \in \mathscr{A}(x), \forall y \in \{\xi < \xi(x)\} :$$
$$(\xi(x) - \xi(y)) \wedge \omega \leq \varphi(y) + \varepsilon$$
$$\text{and} \tag{$**$}$$
$$\forall x \in \{\xi = \infty\}, \forall \varepsilon > 0, \forall \omega < \infty, \forall \sigma < \infty, \exists \varphi \in \mathscr{A}(x) :$$
$$(\theta_{]\sigma,\infty]} \circ \xi) \wedge \omega \leq \varphi + \varepsilon.$$

First, let us consider the case $x \in \{\xi < \infty\}$. Then in the first condition of $(**)$ it clearly suffices to consider $\omega = \xi(x)$ and thus this condition is equivalent with the first condition of $(*)$. Next, let us consider the case where $\xi(x) = \infty$. Clearly, in the second condition of $(**)$ it suffices to consider $\omega = \sigma$ and then it is easily seen that this condition is equivalent with

$$\forall x \in \{\xi = \infty\}, \forall \varepsilon > 0, \forall \omega < \infty, \exists \varphi \in \mathscr{A}(x), \forall y \in \{\xi \leq \omega\} : \omega \leq \varphi(y) + \varepsilon,$$

which, taking ω and 2ω for ε and ω, is further seen to be equivalent with

$$\forall x \in \{\xi = \infty\}, \forall \omega < \infty, \exists \varphi \in \mathscr{A}(x), \forall y \in \{\xi \leq \omega\} : \omega \leq \varphi(y).$$

This condition clearly implies the second condition of $(*)$. Conversely the second condition of $(*)$ implies that

$$\forall x \in \{\xi = \infty\}, \forall \omega < \infty, \exists \varphi \in \mathscr{A}(x), \forall y \in X : 2\omega \leq \xi(y) + \varphi(y),$$

which clearly implies the above equivalent form of the second condition of $(**)$. □

1.3.6 Corollary *For any $A \subseteq X$, we have that $\delta_A \in \mathfrak{L}$, i.e. δ_A is lower regular.*

This result shows that indeed the collection of quasi-metrics d_μ from 1.2.55 is an extension of the collection d_Z^ζ from 1.2.5.

The following consequence follows at once from the foregoing and from 1.2.1.

1.3.7 Corollary *For any $\mathscr{F} \in \mathsf{F}(X)$, we have that $\lambda\mathscr{F} \in \mathfrak{L}$ and $\alpha\mathscr{F} \in \mathfrak{L}$, i.e. $\lambda\mathscr{F}$ and $\alpha\mathscr{F}$ are lower regular.*

1.3.8 Definition Approach spaces form the objects and contractions form the morphisms of a category which we denote App.

In the literature the notion of a topological category was introduced by Herrlich (1974b). Although the definition underwent some changes through time, in this book, we will adhere to the original definition. A *topological category* (over Set) is a category $\mathscr{C}$ which is concrete over Set, (precisely, such that there is a forgetful functor $U : \mathscr{C} \longrightarrow$ Set but we will always leave reference to this functor out of the picture) and which, simply put, satisfies the following conditions:

1. *Terminal separator property*: There is only one structure on a singleton set.
2. *Small-fibered*: There is only a set of structures on any given set.
3. *Topological*: Initial and (as a consequence) final structures exist.

For details we refer to Herrlich (1974b).

1.3.9 Theorem App *is a topological category. In particular, given approach spaces $(X_j)_{j\in J}$, consider the source*

$$(f_j : X \longrightarrow X_j)_{j\in J}$$

in App *and suppose that, for each $j \in J$, $(\mathscr{B}_j(x))_{x\in X_j}$ is a basis for the approach system in X_j. Then a basis for the initial approach system on X is given by*

$$\mathscr{B}(x) := \left\{ \sup_{j\in K} \xi_j \circ f_j | K \in 2^{(J)}, \forall j \in K : \xi_j \in \mathscr{B}_j(f_j(x)) \right\}.$$

Proof Clearly App is concrete, small-fibered, and fulfils the terminal separator property. The main property we have to show is the existence of initial structures in App. Let X be a set and let J be a class. For each $j \in J$, let (X_j, δ_j) be an approach space, let $(\mathscr{B}_j(x))_{x\in X_j}$ be a basis for the approach system $(\mathscr{A}_j(x))_{x\in X_j}$ in X_j, and let $f_j : X \longrightarrow X_j$ be a function. For each $x \in X$, let $\mathscr{B}(x)$ be defined as above. It is easily seen that $\mathscr{B}(x)$ is a basis for an ideal and that it fulfils (B1). To prove that $(\mathscr{B}(x))_{x\in X}$ fulfils (B2), let $\sup_{j\in K} \xi_j \circ f_j \in \mathscr{B}(x)$ and let $\varepsilon > 0$ and $\omega < \infty$. For each $j \in K$, there is a family $(\xi^j_z)_{z\in X_j} \in \prod_{z\in X_j} \mathscr{B}_j(z)$ such that, for all $z, y \in X_j$,

$$\xi^j_{f_j(x)}(z) + \xi^j_z(y) + \varepsilon \geq \xi_j(y) \wedge \omega.$$

For each $t \in X$, let

$$\nu_t := \sup_{j\in K} \xi^j_{f_j(t)} \circ f_j.$$

Then $\nu_t \in \mathscr{B}(t)$ and, for any $t, s \in X$, we have

$$\begin{aligned}
(\sup_{j\in K} \xi_j \circ f_j)(s) \wedge \omega &= \sup_{j\in K} \xi_j(f_j(s)) \wedge \omega \\
&\leq \sup_{j\in K}(\xi^j_{f_j(x)}(f_j(t)) + \xi^j_{f_j(t)}(f_j(s)) + \varepsilon) \\
&\leq \nu_x(t) + \nu_t(s) + \varepsilon.
\end{aligned}$$

This proves that $(\mathscr{A}(x))_{x\in X} := (\widehat{\mathscr{B}(x)})_{x\in X}$ is indeed an approach system on X. By construction, all the functions $f_j, j \in J$, are contractions. To prove that $(\mathscr{A}(x))_{x\in X}$ is an initial structure, let $(Z, (\mathscr{C}(z))_{z\in Z})$ be an approach space determined by its approach system and let

$$g : Z \longrightarrow X$$

be such that all compositions

$$f_j \circ g : (Z, (\mathscr{C}(z))_{z\in Z}) \longrightarrow (X_j, (\mathscr{A}_j)_{x\in X_j})$$

are contractions. Further, let $z \in Z$ and let $\sup_{j\in K} \xi_j \circ f_j \in \mathscr{A}(g(z))$. Since

$$(\sup_{j\in K} \xi_j \circ f_j) \circ g = \sup_{j\in K} \xi_j \circ (f_j \circ g)$$

and since, for all $j \in K$,

$$\xi_j \circ (f_j \circ g) \in \mathscr{C}(z),$$

it follows from 1.3.3 that g is a contraction. □

1.3.10 Example (Finest and coarsest structures) In any topological category, on any set there are discrete and indiscrete structures, where a structure is called discrete if any function defined on a set with that structure is a morphism and indiscrete if any function to a set with that structure is a morphism. Given a set X the discrete (or finest) approach structure on it is determined by any (and all) of the following structures:

1. Distance: $\delta : X \times 2^X \longrightarrow \mathbb{P}$ where, for all $x \in X$ and $A \subseteq X$,

$$\delta(x, A) = \begin{cases} 0 & x \in A, \\ \infty & x \notin A. \end{cases}$$

2. Limit: $\lambda : \mathsf{F}(X) \longrightarrow \mathbb{P}^X$ where, for all $x \in X$ and $\mathscr{F} \in \mathsf{F}(X)$,

$$\lambda\mathscr{F} = \begin{cases} \theta_{\{x\}} & \mathscr{F} = \dot{x}, \\ \infty & \mathscr{F} \neq \dot{x}. \end{cases}$$

3. Approach system: $(\mathscr{A}(x))_{x\in X}$ where, for all $x \in X$,

$$\mathscr{A}(x) = \left\{\varphi \in \mathbb{P}^X \mid \varphi(x) = 0\right\}.$$

4. Gauge: $\mathscr{G} = q\mathsf{Met}(X)$.
5. Tower: $(\mathfrak{t}_\varepsilon)_{\varepsilon\in\mathbb{R}^+}$ where, for all $\varepsilon \in \mathbb{R}^+$ and $A \subseteq X$, $\mathfrak{t}_\varepsilon(A) = A$.
6. Lower hull operator: $\mathfrak{l} : \mathbb{P}^X \longrightarrow \mathbb{P}^X$ where, for all $\mu \in \mathbb{P}^X$, $\mathfrak{l}(\mu) = \mu$.
7. Lower regular function frame: $\mathfrak{L} = \mathbb{P}^X$.

8. Upper hull operator: $\mathfrak{u} : \mathbb{P}_b^X \longrightarrow \mathbb{P}_b^X$ where, for all $\mu \in \mathbb{P}_b^X$, $\mathfrak{u}(\mu) = \mu$.
9. Upper regular function frame: $\mathfrak{U} = \mathbb{P}_b^X$.
10. Functional ideal convergence: If $\mathfrak{I} = \mathfrak{i}(\dot{x}) \oplus \alpha$ for some $x \in X$ and $\alpha \in]0, \infty[$ then $\mathfrak{I} \rightarrowtail x$ and otherwise $\mathfrak{I}$ does not converge to any point. The improper functional ideal converges to all points.

Analogously the indiscrete (or trivial, or coarsest) approach structure on X is determined by any (and all) of the following structures:

1. Distance: $\delta : X \times 2^X \longrightarrow \mathbb{P}$ where, for all $x \in X$ and $A \subseteq X$,

$$\delta(x, A) = \begin{cases} 0 & A \neq \emptyset, \\ \infty & A = \emptyset. \end{cases}$$

2. Limit: $\lambda : \mathsf{F}(X) \longrightarrow \mathbb{P}^X$ where, for all $\mathscr{F} \in \mathsf{F}(X)$, $\lambda\mathscr{F} = 0$.
3. Approach system: $(\mathscr{A}(x))_{x \in X}$ where, for all $x \in X$, $\mathscr{A}(x) = \{0\}$.
4. Gauge: $\mathscr{G} = \{0\}$.
5. Tower: $(\mathfrak{t}_\varepsilon)_{\varepsilon \in \mathbb{R}^+}$ where, for all $\varepsilon \in \mathbb{R}^+$ *and* $A \subseteq X$,

$$\mathfrak{t}_\varepsilon(A) = \begin{cases} X & A \neq \emptyset, \\ \emptyset & A = \emptyset. \end{cases}$$

6. Lower hull operator: $\mathfrak{l} : \mathbb{P}^X \longrightarrow \mathbb{P}^X$ where, for all $\mu \in \mathbb{P}^X$, $\mathfrak{l}(\mu) = \inf \mu$.
7. Lower regular function frame: $\mathfrak{L} = \{\varphi \in \mathbb{P}^X | \varphi \text{ constant}\}$.
8. Upper hull operator: $\mathfrak{u} : \mathbb{P}_b^X \longrightarrow \mathbb{P}_b^X$ where, for all $\mu \in \mathbb{P}_b^X$, $\mathfrak{u}(\mu) = \sup \mu$.
9. Upper regular function frame: $\mathfrak{U} = \{\varphi \in \mathbb{P}_b^X | \varphi \text{ constant}\}$.
10. Functional ideal convergence: For all $\mathfrak{I} \in \mathfrak{F}(X)$ and $x \in X$: $\mathfrak{I} \rightarrowtail x$.

Easy and useful characterizations of initial and final structures in App for all the different defining concepts of approach spaces do not exist. However, there are a few important constructions which we have to explain, namely initial structures by means of gauges, distances, limit operators and functional ideal convergence. Furthermore, for completeness and since they are so simple, we will also give the descriptions of both initial and final structures by means of lower and upper regular function frames.

1.3.11 Theorem *Given approach spaces $(X_j)_{j \in J}$, consider the source*

$$(f_j : X \longrightarrow X_j)_{j \in J}$$

in App. *If, for each $j \in J$, $\mathscr{H}_j$ is a basis for the gauge of X_j, then a basis for the initial gauge on X is given by*

$$\mathscr{H} := \left\{ \sup_{j \in K} d_j \circ (f_j \times f_j) | K \in 2^{(J)}, \forall j \in K : d_j \in \mathscr{H}_j \right\}.$$

Proof This goes along the same lines as 1.3.9. □

1.3.12 Theorem *Given approach spaces $(X_j)_{j\in J}$, consider the source*

$$(f_j : X \longrightarrow X_j)_{j\in J}$$

in App. *If for each $j \in J$, λ_j is the limit operator on X_j then the initial limit operator on X is given by*

$$\lambda\mathscr{F} = \sup_{j\in J} \lambda_j(f_j(\mathscr{F})) \circ f_j.$$

Proof It suffices to restrict our attention to ultrafilters since the general case then immediately follows upon applying 1.2.67. We denote by λ the initial limit operator on X, by $\mathscr{A}$ the approach system associated with λ, and by $\mathscr{A}_j$ the approach system associated with λ_j, for each $j \in J$.

Now let $\mathscr{U}$ be an ultrafilter on X and let $x \in X$ be arbitrary. We know from 1.2.66 and 1.2.69 that

$$\begin{aligned}\lambda\mathscr{U}(x) &= \sup_{\varphi\in\mathscr{A}(x)} \sup_{U\in\mathscr{U}} \inf_{y\in U} \varphi(y)\\ &= \sup_{J_0\in 2^{(J)}} \sup_{\varphi\in\prod_{j\in J_0}\mathscr{A}_j(f_j(x))} \sup_{U\in\mathscr{U}} \inf_{y\in U} \sup_{j\in J_0} \varphi_j \circ f_j(y).\end{aligned}$$

Now let $J_0 \in 2^{(J)}$, $\varphi \in \prod_{j\in J_0} \mathscr{A}_j(f_j(x))$, and $U \in \mathscr{U}$ be fixed. Then it follows that

$$\begin{aligned}\inf_{y\in U} \sup_{j\in J_0} \varphi_j \circ f_j(y) &= \sup_{t\in J_0^U} \inf_{y\in U} \varphi_{t(y)} \circ f_{t(y)}(y)\\ &= \sup_{t\in J_0^U} \inf_{j\in J_0} \inf_{y\in t^{-1}(j)} \varphi_j \circ f_j(y).\end{aligned} \qquad (*)$$

Since $\mathscr{U}$ is an ultrafilter, it follows that, for each $t \in J_0^U$, there exists $j_t \in J_0$ such that $t^{-1}(j_t) \in \mathscr{U}$ and consequently we have

$$\begin{aligned}\sup_{t\in J_0^U} \inf_{j\in J_0} \inf_{y\in t^{-1}(j)} \varphi_j \circ f_j(y) &\le \sup_{t\in J_0^U} \inf_{y\in t^{-1}(j_t)} \varphi_{j_t} \circ f_{j_t}(y)\\ &\le \sup_{t\in J_0^U} \sup_{W\in\mathscr{U}} \inf_{y\in W} \varphi_{j_t} \circ f_{j_t}(y)\\ &= \sup_{W\in\mathscr{U}} \sup_{j\in J_0} \inf_{y\in W} \varphi_j \circ f_j(y).\end{aligned} \qquad (**)$$

Because this inequality holds for all choices of $U \in \mathscr{U}$ it follows from $(*)$ and $(**)$ that

$$\sup_{U\in\mathscr{U}} \inf_{y\in U} \sup_{j\in J_0} \varphi_j \circ f_j(y) \le \sup_{W\in\mathscr{U}} \sup_{j\in J_0} \inf_{y\in W} \varphi_j \circ f_j(y).$$

Since, as is immediately clear, the other inequality also holds we can now incorporate this in our previous calculation for $\lambda\mathscr{U}(x)$ in order to obtain

$$\begin{aligned}
\lambda\mathscr{U}(x) &= \sup_{J_0\in 2^{(J)}} \sup_{\varphi\in\prod_{j\in J_0}\mathscr{A}_j(f_j(x))} \sup_{U\in\mathscr{U}} \inf_{y\in U} \sup_{j\in J_0} \varphi_j\circ f_j(y)\\
&= \sup_{J_0\in 2^{(J)}} \sup_{\varphi\in\prod_{j\in J_0}\mathscr{A}_j(f_j(x))} \sup_{U\in\mathscr{U}} \sup_{j\in J_0} \inf_{y\in U} \varphi_j\circ f_j(y)\\
&= \sup_{J_0\in 2^{(J)}} \sup_{\varphi\in\prod_{j\in J_0}\mathscr{A}_j(f_j(x))} \sup_{j\in J_0} \sup_{U\in\mathscr{U}} \inf_{y\in f_j(U)} \varphi_j(y)\\
&= \sup_{j\in J} \sup_{\varphi\in\mathscr{A}_j(f_j(x))} \sup_{U\in\mathscr{U}} \inf_{y\in f_j(U)} \varphi_j(y)\\
&= \sup_{j\in J} \lambda_j(f_j\mathscr{U})(f_j(x)),
\end{aligned}$$

which proves our claim. □

1.3.13 Theorem *Given approach spaces $(X_j)_{j\in J}$, consider the sink*

$$(f_j : X_j \longrightarrow X)_{j\in J}$$

in App. *If for each $j\in J$, $\mathfrak{L}_j$ is the lower regular function frame on X_j then the final lower regular function frame on X is given by*

$$\mathfrak{L} := \{\mu\in\mathbb{P}^X \mid \forall j\in J : \mu\circ f_j\in\mathfrak{L}_j\}.$$

Proof It is straightforward to verify that $\mathfrak{L}$ as defined above is indeed a regular function frame and by construction it is of course the finest one turning all the maps f_j, $j\in J$, into contractions. □

In the following theorem, given a set of functions $\mathscr{B}\subseteq\mathbb{P}^X$, we denote by $\mathscr{B}^\wedge$ the set of all infima of finite sets of members of $\mathscr{B}$ and by $\mathscr{B}^\vee$ the set of all suprema of arbitrary families of members of $\mathscr{B}$.

1.3.14 Theorem *Given approach spaces $(X_j)_{j\in J}$ consider the source*

$$(f_j : X \longrightarrow X_j)_{j\in J}$$

in App. *If for each $j\in J$, $\mathfrak{L}_j$ is the lower regular function frame on X_j then the initial lower regular function frame on X is given by*

$$\mathfrak{L} := \{\mu\circ f_j \mid j\in J, \mu\in\mathfrak{L}_j\}^{\wedge\vee}.$$

Proof Since translations commute with infima, suprema, and composition, it is at once clear by construction that $\mathfrak{L}$ as defined above is indeed a regular function frame and that it is the coarsest one turning all the maps f_j, $j \in J$, into contractions. □

1.3.15 Theorem *Given approach spaces $(X_j)_{j\in J}$, consider the sink*

$$(f_j : X_j \longrightarrow X)_{j\in J}$$

in App. *If for each $j \in J$, $\mathfrak{U}_j$ is the upper regular function frame on X_j then the final upper regular function frame on X is given by*

$$\mathfrak{U} := \{\mu \in \mathbb{P}_b^X \mid \forall j \in J : \mu \circ f_j \in \mathfrak{U}_j\}.$$

Proof This is analogous to 1.3.13 and we leave this to the reader. □

In the following theorem, given a set of functions $\mathscr{B} \subseteq \mathbb{P}^X$, we denote by $\mathscr{B}^\vee$ the set of all suprema of finite sets of members of $\mathscr{B}$ and by $\mathscr{B}^\wedge$ the set of all infima of arbitrary nonempty families of members of $\mathscr{B}$.

1.3.16 Theorem *Given approach spaces $(X_j)_{j\in J}$ consider the source*

$$(f_j : X \longrightarrow X_j)_{j\in J}$$

in App. *If for each $j \in J$, $\mathfrak{U}_j$ is the upper regular function frame on X_j then the initial upper regular function frame on X is given by*

$$\mathfrak{U} := \{\mu \circ f_j \mid j \in J, \mu \in \mathfrak{U}_j\}^{\vee\wedge}.$$

Proof This is analogous to 1.3.14 and we leave this to the reader. □

Given a subset $A \subseteq X$ we denote by $\mathsf{P}(A)$ the set of finite covers of A by means of subsets of A, and similarly by $\mathsf{R}(A)$ the set of finite partitions of A.

1.3.17 Theorem *Given approach spaces $(X_j)_{j\in J}$, consider the source*

$$(f_j : X \longrightarrow X_j)_{j\in J}$$

in App. *If for each $j \in J$, δ_j is the distance on X_j, then the initial distance is given by*

$$\begin{aligned}\delta(x, A) &:= \sup_{\mathscr{P}\in\mathsf{P}(A)} \min_{P\in\mathscr{P}} \sup_{j\in J} \delta_j(f_j(x), f_j(P))\\ &= \sup_{\mathscr{P}\in\mathsf{R}(A)} \min_{P\in\mathscr{P}} \sup_{j\in J} \delta_j(f_j(x), f_j(P)).\end{aligned}$$

Proof We only prove the case of finite covers, the case for partitions easily follows from this. If δ_{in} is the initial distance on X then it immediately follows from (D3) that $\delta \leq \delta_{in}$.

To prove the remaining inequality suppose that $0 < \alpha < \delta_{in}(x, A)$ then according to 1.3.11 this means that

$$\alpha < \sup_{K \in 2^{(J)}} \sup_{(d_k)_{k\in K} \in \prod_{k\in K} \mathcal{G}_k} \inf_{a\in A} \sup_{k\in K} d_k(f_k(x), f_k(a)).$$

Consequently there exist $K \subseteq J$ finite, and $d_k \in \mathcal{G}_k, k \in K$ such that for all $a \in A$ there exists a $k \in K$ such that $d_k(f_k(x), f_k(a)) \geq \alpha$. This means that

$$A \cap (\bigcap_{k\in K} f_k^{-1}(B_{d_k}(f_k(x), \alpha))) = \emptyset.$$

Put

$$B_k := A \cap f_k^{-1}(X_k \setminus B_{d_k}(f_k(x), \alpha))$$

then $A = \bigcup_{k\in K} B_k$ and it follows that for any $k \in K$

$$\begin{aligned}\sup_{j\in J} \delta_j(f_j(x), f_j(B_k)) \geq \delta_k(f_k(x), f_k(B_k))& \\ &= \sup_{d\in\mathcal{G}_k} \inf_{a\in B_k} d(f_k(x), f_k(a)) \\ &\geq \inf_{a\in B_k} d_k(f_k(x), f_k(a)) \\ &\geq \alpha.\end{aligned}$$

Consequently

$$\alpha \leq \sup_{\mathcal{P}\in \mathsf{P}(A)} \min_{P\in\mathcal{P}} \sup_{j\in J} \delta_j(f_j(x), f_j(P))$$

and we are finished. □

1.3.18 Theorem *Given approach spaces* $(X_j)_{j\in J}$, *consider the source*

$$(f_j : X \longrightarrow X_j)_{j\in J}$$

in App. *Then the initial functional ideal convergence is determined by*

$$\mathfrak{I} \rightarrowtail x \Leftrightarrow \forall j \in J : f_j(\mathfrak{I}) \rightarrowtail f_j(x).$$

Proof This follows from 1.3.12, 1.2.58 and 1.2.60. □

We end this section by showing that App is simply generated, i.e. it has an *initially dense object*.

1.3.19 Theorem $\mathbb{P}$ *is initially dense in* App. *More precisely, for any approach space X, both the sources*

$$(\delta_A : X \longrightarrow \mathbb{P})_{A \in 2^X} \text{ and } (\xi : X \longrightarrow \mathbb{P})_{\xi \in \mathfrak{L}}$$

are initial.

Proof From 1.3.4 we already know that if δ_{in} stands for the initial distance, then $\delta_{in} \leq \delta$. Conversely it follows from 1.3.12 that, for any $\mathcal{U} \in \mathsf{U}(X)$,

$$\begin{aligned}\lambda_{in}\mathcal{U}(x) &= \sup_{A \in 2^X} \lambda_{\mathbb{P}}(\delta_A(\mathcal{U}))(\delta_A(x)) \\ &= \sup_{A \in 2^X} \sup_{U \in \mathcal{U}} \delta_{\mathbb{P}}(\delta_A(x), \delta_A(U)) \\ &\geq \sup_{U \in \mathcal{U}} \delta_{\mathbb{P}}(\delta_U(x), \delta_U(U)) \\ &\geq \sup_{U \in \mathcal{U}} \delta_U(x) \\ &= \lambda\mathcal{U}(x),\end{aligned}$$

which proves our claim. That the second source too is initial follows from the fact that the first one is and from 1.3.5. □

Alternative interesting initially dense objects as compared to $\mathbb{P}$ where found by Claes (2009). We will come back to this in Sect. 2.4 when we have seen metric approach spaces (see 2.4.13).

1.4 Closed and Open Expansions and Proper Contractions

The morphisms in our category being contractions (i.e. essentially non-expansive maps) it is normal to consider also their counterpart, namely expansions. However whereas there is only one reasonable way in which to express that a function is contractive, there are two natural ways to express that a function is expansive, and since in the topological case these coincide with respectively open and closed maps (see Chap. 2) we will define these two different concepts and refer to them as open and closed expansions respectively. Note that as is usual, we give these definitions for arbitrary functions, not only for contractions. Only in the case that they are also contractions are we dealing with morphisms in App which will be the case for what we will call proper contractions.

Finally, in this section we only give the basic definitions and characterizations since we will come back to these concepts more in depth in the section on morphism indices in Chap. 3.

1.4.1 Definition A function $f : X \longrightarrow X'$ between approach spaces is called *closed expansive* or a *closed expansion* if for all $A \subseteq X$

$$f(\delta_A) \leq \delta'_{f(A)}$$

or explicitly, if for all for all $y \in X'$ and $A \subseteq X$

$$\inf_{x \in f^{-1}(y)} \delta(x, A) \leq \delta'(y, f(A)).$$

1.4.2 Theorem *For a function $f : X \longrightarrow X'$ between approach spaces the following properties are equivalent.*

1. *f is a closed expansion.*
2. $\forall \mu \in \mathbb{P}^X : f(\mathfrak{l}(\mu)) \leq \mathfrak{l}'(f(\mu))$.
3. $\forall \mathscr{F} \in \mathsf{F}(X) : \sup_{T \in \sec \mathscr{F}} f(\delta_T) \leq \lambda'(f(\mathscr{F}))$.
4. $\forall \mathscr{U} \in \mathsf{U}(X) : \sup_{T \in \mathscr{U}} f(\delta_T) \leq \lambda'(f(\mathscr{U}))$.
5. $\forall \mu \in \mathfrak{L} : f(\mu) \in \mathfrak{L}'$.
6. $\forall A \subseteq X, \forall \alpha \in \mathbb{R}^+ : f(A)^{(\alpha)} \subseteq \bigcap_{\varepsilon > 0} f(A^{(\alpha+\varepsilon)})$.

In 5 it suffices to have the property for a basis of $\mathfrak{L}$.

Proof 1 $\Rightarrow$ 2. Let $(A_j)_{j \in J}$ be a finite collection of subsets of X, for any $j \in J$ let $m_j \in \mathbb{P}$, and let $\mu = \inf_{j \in J}(m_j + \theta_{A_j})$. Further fix $x' \in X'$, then

$$f(\mu) = \inf_{j \in J}(m_j + f(\theta_{A_j})) \text{ and } \mathfrak{l}(\mu) = \inf_{j \in J}(m_j + \mathfrak{l}(\theta_{A_j}))$$

and therefore

$$\begin{aligned} f(\mathfrak{l}(\mu))(x') &= \inf_{j \in J} \inf_{x \in f^{-1}(x')} (m_j + \mathfrak{l}(\theta_{A_j})(x)) = \inf_{j \in J}(m_j + f(\mathfrak{l}(\theta_{A_j}))(x')) \\ &\leq \inf_{j \in J}(m_j + \mathfrak{l}'(f(\theta_{A_j})))(x') = \mathfrak{l}'(f(\mu))(x'). \end{aligned}$$

If $\mu \in \mathbb{P}_b^X$, $\varepsilon > 0$ and $\mu_\varepsilon \leq \mu \leq \mu_\varepsilon + \varepsilon$ with $\mu_\varepsilon \in \mathsf{Fin}(X)$, then

$$f(\mu_\varepsilon) \leq f(\mu) \leq f(\mu_\varepsilon) + \varepsilon \text{ and } \mathfrak{l}(\mu_\varepsilon) \leq \mathfrak{l}(\mu) \leq \mathfrak{l}(\mu_\varepsilon) + \varepsilon,$$

and consequently

$$f(\mathfrak{l}(\mu)) \leq \mathfrak{l}'(f(\mu)) + \varepsilon,$$

and the result follows from the arbitrariness of ε.

Finally, if $\mu \in \mathbb{P}^X$ then for any $\alpha \in [0, \infty[$ we obtain

$$\alpha \wedge f(\mathfrak{l}(\mu)) = f(\mathfrak{l}(\alpha \wedge \mu)) \leq \mathfrak{l}'(f(\alpha \wedge \mu)) \leq \mathfrak{l}'(f(\mu)),$$

and the result again follows by arbitrariness of α.

$2 \Rightarrow 1$. This follows immediately choosing $\mu = \theta_A$.

$1 \Rightarrow 3$. Indeed

$$\sup_{T \in \sec \mathcal{F}} f(\delta_T) \leq \sup_{T \in \sec \mathcal{F}} \delta'_{f(T)} = \sup_{W \in \sec f(\mathcal{F})} \delta'_W = \lambda'(f(\mathcal{F})).$$

$3 \Rightarrow 4$. This is evident.

$4 \Rightarrow 1$. Applying 1.2.1 we obtain

$$\delta'_{f(A)}(x') = \inf_{\mathcal{U} \in \mathsf{U}(A)} \lambda' f(\mathcal{U})(x') \geq \inf_{\mathcal{U} \in \mathsf{U}(A)} \sup_{T \in \mathcal{U}} f(\delta_T)(x') \geq f(\delta_A)(x').$$

$2 \Rightarrow 5$. This is immediate since $\mathfrak{l}(\mu) = \mu$ if $\mu \in \mathfrak{L}$.

$5 \Rightarrow 2$. This follows from 1.2.23.

$1 \Leftrightarrow 6$. This follows from 1.2.21 and 1.2.22. □

Note that being closed-expansive does not mean being closed as a map for all the level tower structures as can easily be seen from 6.

1.4.3 Proposition *For a function $f : X \longrightarrow X'$ between approach spaces the following properties are equivalent.*

1. *f is an injective closed expansive contraction.*
2. *f is an embedding (i.e. an initial injection) such that $\delta'_{f(X)} = \theta_{f(X)}$.*

Proof $1 \Rightarrow 2$. If f is an injective closed expansive contraction then it follows immediately from the definitions that for all $x \in X$ and $A \subseteq X$

$$\delta'(f(x), f(A)) = \delta(x, A),$$

and hence f is an embedding. Moreover, $\delta'_{f(X)} = f(\delta_X) = f(\theta_X) = \theta_{f(X)}$.

$2 \Rightarrow 1$. Conversely, if f is an embedding and $\delta'_{f(X)} = \theta_{f(X)}$ then f is obviously injective and a contraction. Let $A \subseteq X$ and $x' \in X'$. If $x' \in f(X)$ and $x \in X$ is the unique point such that $f(x) = x'$ then

$$f(\delta_A)(x') = \delta(x, A) = \delta'(f(x), f(A)) = \delta'_{f(A)}(x'),$$

and if $x' \notin f(X)$ then

$$f(\delta_A)(x') = \infty = \theta_{f(X)}(x') = \delta'_{f(X)}(x') \leq \delta'_{f(A)}(x').$$

Hence $f(\delta_A) \leq \delta'_{f(A)}$ which shows that f is closed expansive. □

1.4.4 Proposition *Let $f : X \longrightarrow X'$ be a closed expansion between approach spaces and let $Z \subseteq X'$ then the restriction $g := f|_{f^{-1}(Z)} : f^{-1}(Z) \longrightarrow Z$ is a closed expansion.*

Proof Let $A \subseteq f^{-1}(Z)$ and $y \in Z$, then

$$g(\delta_A)(y) = f(\delta_A)(y) \leq \delta'_{f(A)}(y) = \delta'_{g(A)}(y).$$ □

1.4.5 Definition A function $f : X \longrightarrow X'$ between approach spaces is called *open expansive* or an *open expansion* if for all $A \subseteq X'$ we have

$$\delta_{f^{-1}(A)} \leq \delta'_A \circ f$$

or explicitly, if for all $x \in X$ and $A \subseteq X'$

$$\delta(x, f^{-1}(A)) \leq \delta'(f(x), A).$$

1.4.6 Theorem *For a function $f : X \longrightarrow X'$ between approach spaces the following properties are equivalent.*

1. *f is an open expansion.*
2. $\forall A \subseteq X', \forall \varepsilon \in \mathbb{R}^+ : f^{-1}(A^{(\varepsilon)}) \subseteq (f^{-1}(A))^{(\varepsilon)}$.
3. $\forall x \in X, \forall \varphi \in \mathcal{A}(x) : f(\varphi) \in \mathcal{A}'(f(x))$.
4. $\forall \nu \in \mathfrak{U} : f(\nu) \in \mathfrak{U}'$.
5. $\forall x \in X, \forall \varepsilon \in \mathbb{R}^+, \forall V \in \mathcal{V}_\varepsilon(x) : f(V) \in \mathcal{V}_\varepsilon(f(x))$.

In 3 it is sufficient to have the property for a basis for $\mathcal{A}(x)$ and in 4 it is sufficient to have the property for a basis of $\mathfrak{U}$.

If f is surjective then these are moreover equivalent to

6. $\forall x \in X, \forall \mathcal{U}' \in \mathsf{U}(X')$ *there exists* $\mathcal{U} \in \mathsf{U}(X)$ *such that* $f(\mathcal{U}) = \mathcal{U}'$ *and* $\lambda\mathcal{U}(x) \leq \lambda'(\mathcal{U}')(f(x))$.
7. $\forall x \in X, \forall \mathcal{U}' \in \mathsf{U}(X') : \inf\{\lambda\mathcal{U}(x) \mid f(\mathcal{U}) = \mathcal{U}'\} \leq \lambda'(\mathcal{U}')(f(x))$.

Proof 1 ⇔ 2 This follows from the relationship between distances and towers.

3 ⇒ 1. Let $A \subseteq X'$ and $x \in X$ then

$$\begin{aligned}\delta'(f(x), A) &= \sup_{\varphi' \in \mathcal{A}'(f(x))} \inf_{b \in A} \varphi'(b)\\ &\geq \sup_{\varphi \in \mathcal{A}(x)} \inf_{b \in A} f(\varphi)(b)\\ &= \sup_{\varphi \in \mathcal{A}(x)} \inf_{z \in f^{-1}(A)} \varphi(z)\\ &= \delta(x, f^{-1}(A)).\end{aligned}$$

1 ⇒ 3. Suppose there exist an $x_0 \in X$, a $\varphi_0 \in \mathcal{A}(x_0)$, and $\varepsilon > 0$, $\omega < \infty$ such that for every $\varphi' \in \mathcal{A}'(f(x_0))$, $f(\varphi_0) \wedge \omega \not\leq \varphi' + \varepsilon$. For every $\varphi' \in \mathcal{A}'(f(x_0))$, define

$$B(\varphi') := \Big\{ y \in X' \mid f(\varphi_0)(y) \wedge \omega > \varphi'(y) + \varepsilon \Big\}.$$

Notice that for every $\varphi', \xi' \in \mathscr{A}'(f(x_0))$, $B(\varphi' \vee \xi') = B(\varphi') \cap B(\xi')$ and $B(\varphi') \neq \emptyset$, consequently we obtain

$$\begin{aligned}
\sup_{\varphi' \in \mathscr{A}'(f(x_0))} \delta'(f(x_0), B(\varphi')) \wedge \omega &= \sup_{\varphi' \in \mathscr{A}'(f(x_0))} \sup_{\xi' \in \mathscr{A}'(f(x_0))} \inf_{y \in B(\varphi')} \xi'(y) \wedge \omega \\
&\leq \sup_{\varphi' \in \mathscr{A}'(f(x_0))} \sup_{\xi' \in \mathscr{A}'(f(x_0))} \inf_{y \in B(\varphi' \vee \xi')} (\varphi' \vee \xi')(y) \wedge \omega \\
&= \sup_{\varphi' \in \mathscr{A}'(f(x_0))} \inf_{y \in B(\varphi')} \varphi'(y) \wedge \omega \\
&\leq \sup_{\varphi' \in \mathscr{A}'(f(x_0))} \inf_{y \in B(\varphi')} \big(f(\varphi_0)(y) \wedge \omega\big) - \varepsilon \\
&\leq \sup_{\varphi' \in \mathscr{A}'(f(x_0))} \sup_{\xi \in \mathscr{A}(x_0)} \inf_{x \in f^{-1}(B(\varphi'))} (\xi(x) \wedge \omega) - \varepsilon \\
&= \sup_{\varphi' \in \mathscr{A}'(f(x_0))} \big(\delta(x_0, f^{-1}(B(\varphi'))) \wedge \omega\big) - \varepsilon,
\end{aligned}$$

which contradicts 1.

3 ⇒ 4. Take $\nu \in \mathfrak{U}$ then it follows from 1.2.42 that for all $x \in X$, $\nu \ominus \nu(x) \in \mathscr{A}(x)$ and consequently that $f(\nu) \ominus f(\nu)(f(x)) \in \mathscr{A}'(f(x))$. Since for $y \notin f(X)$ we have that $f(\nu) \ominus f(\nu)(y) = 0 \in \mathscr{A}'(y)$ this proves that $f(\nu) \in \mathfrak{U}'$.

4 ⇒ 3. This follows from 1.2.40.

2 ⇔ 5. This follows from the relation between the various descriptions of towers (see remarks after 1.1.25).

Now let us suppose that f is surjective.

1 ⇒ 6. Let $\mathscr{U}' \in \mathsf{U}(X')$ and $x \in X$. Suppose now that for all $\mathscr{U} \in \mathsf{U}(X)$ with $f(\mathscr{U}) = \mathscr{U}'$ we have $\lambda'(\mathscr{U}')(f(x)) < \lambda\mathscr{U}(x)$. Then for each such $\mathscr{U}$ there exists $U_{\mathscr{U}} \in \mathscr{U}$ such that

$$\lambda'(\mathscr{U}')(f(x)) < \delta(x, U_{\mathscr{U}})$$

and by 1.1.4 we can find $U_{\mathscr{U}_1}, \ldots, U_{\mathscr{U}_n}$ such that $\cup_{i=1}^n U_{\mathscr{U}_i} \in f^{-1}(\mathscr{U}')$. Choose $W \in \mathscr{U}'$ such that $f^{-1}(W) \subseteq \cup_{i=1}^n U_{\mathscr{U}_i}$. We then have

$$\begin{aligned}
\min_{i=1}^{n} \delta(x, U_{\mathscr{U}_i}) &= \delta(x, \cup_{i=1}^n U_{\mathscr{U}_i}) \\
&\leq \delta(x, f^{-1}(W)) \\
&\leq \delta'(f(x), W) \\
&\leq \lambda'\mathscr{U}'(f(x))
\end{aligned}$$

which is a contradiction.

6 ⇒ 7. This is evident.

7 ⇒ 1. Let $A' \subseteq X'$ and $x \in X$. Then we find

$$\begin{aligned}\delta(x, f^{-1}(A')) &= \inf_{\mathscr{U} \in \mathsf{U}(f^{-1}(A'))} \lambda\mathscr{U}(x)\\ &\leq \inf_{\mathscr{U}' \in \mathsf{U}(A')} \lambda'\mathscr{U}'(f(x))\\ &= \delta'(f(x), A').\end{aligned}$$ □

Note that items 2 and 5 in the foregoing result are equivalent ways of saying that f is open as a map between each pair of corresponding level tower structures.

1.4.7 Proposition *Let* $(X_i)_{i \in I}$ *be a family of approach spaces then for every* $k \in I$*, the projection* $\mathrm{pr}_k : \prod_{i \in I} X_i \longrightarrow X_k$ *is an open expansion.*

Proof Choose $k \in I$, a finite subset $J \subseteq I$ and $\varphi_j \in \mathscr{A}_j(x_j)$ for $j \in J$. If $y \in X_k$ then we define $z^0 = (z_i^0)_{i \in I}$ as

$$z_i^0 := \begin{cases} y & i = k \\ x_i & i \in J \setminus \{k\} \\ \text{arbitrary} & \text{elsewhere} \end{cases}$$

then, if $k \in J$

$$\begin{aligned}\mathrm{pr}_k(\sup_{j \in J} \varphi_j \circ \mathrm{pr}_j)(y) &= \inf_{z_k^0 = y} \sup_{j \in J} \varphi_j(z_j^0)\\ &= \varphi_k(y) \vee \inf_{z_k^0 = y} \sup_{j \in J \setminus \{k\}} \varphi_j(z_j^0)\\ &\leq \varphi_k(y) \vee \sup_{j \in J \setminus \{k\}} \varphi_j(x_j)\\ &= \varphi_k(y)\end{aligned}$$

and if $k \notin J$ then

$$\mathrm{pr}_k\left(\sup_{j \in J} \varphi_j \circ \mathrm{pr}_j\right)(y) \leq \sup_{j \in J} \varphi_j(x_j) = 0.$$

Hence we have $\mathrm{pr}_k\left(\sup_{j \in J} \varphi_j \circ \mathrm{pr}_j\right) \leq \varphi_k$ and by 1.4.6 we are finished. □

1.4.8 Proposition *Let* $f : X \longrightarrow X'$ *be an open expansion between approach spaces and let* $Z \subseteq X'$ *then the restriction* $g := f|_{f^{-1}(Z)} : f^{-1}(Z) \longrightarrow Z$ *is an open expansion.*

Proof This is analogous to 1.4.4 and we leave this to the reader. □

For a category with a given factorization structure $(\mathscr{E}, \mathscr{M})$ (see e.g. Herrlich and Strecker 1973), satisfying properties (F0–F2) below, (Clementino et al. 2003), Clementino, Giuli and Tholen formulate three further axioms which a class $\mathscr{F}$ of morphisms has to fulfil for it to be a viable class of closed morphisms in that category. We recall the axioms (F0–F2):

(F0) $\mathcal{M}$ is a class of monomorphisms and $\mathcal{E}$ is a class of epimorphisms and both are closed under composition with isomorphisms.
(F1) Every morphism f decomposes as $f = m \circ e$ with $m \in \mathcal{M}$ and $e \in \mathcal{E}$.
(F2) Every $e \in \mathcal{E}$ is orthogonal to every $m \in \mathcal{M}$, that is, given any morphisms u and v such that $m \circ u = v \circ e$ there exists a unique morphism w making the following diagram commutative

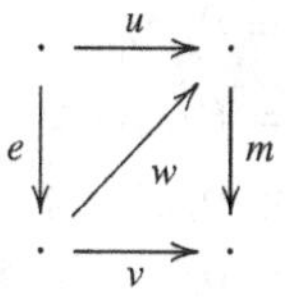

The axioms which $\mathcal{F}$ has to fulfil are:

(F3) $\mathcal{F}$ contains all isomorphisms and is closed under composition.
(F4) $\mathcal{F} \cap \mathcal{M}$ is stable under pullbacks.
(F5) Whenever $g \circ f \in \mathcal{F}$ and $f \in \mathcal{E}$ then $g \in \mathcal{F}$.

Note that in the abstract categorical setting these axioms necessarily are formulated in terms of morphisms, whereas our notion of closed expansiveness also makes sense for an arbitrary function, hence the supplementary (and required) condition in 3 of 1.4.9.

There are many different factorization structures $(\mathcal{E}, \mathcal{M})$ on a topological category, but we will consider the most usual one where $\mathcal{E}$ are the epimorphisms (i.e. the surjective contractions) and $\mathcal{M}$ are the extremal monomorphisms (i.e. the embeddings). As in any topological category, this factorization structure satisfies the aforementioned conditions (F0–F2). For now, in this section, we let $\mathcal{F} :=$ the class of all closed expansive contractions.

That isomorphisms are closed expansive and that closed expansive contractions are stable under composition is evident from the definition. This implies that $\mathcal{F}$ satisfies axiom (F3).

We will now point out that $\mathcal{F}$ also satisfies the remaining two axioms with regard to the given factorization structure.

1.4.9 Proposition *The following properties hold.*

1. *$\mathcal{F} \cap \mathcal{M}$ is stable under pullbacks.*
2. *If $g \circ f$ is closed expansive and f is a surjective contraction then g is closed expansive.*
3. *If g is a contraction, $g \circ f \in \mathcal{F}$ and $f \in \mathcal{E}$ then $g \in \mathcal{F}$.*

Proof 1. Consider the pullback diagram

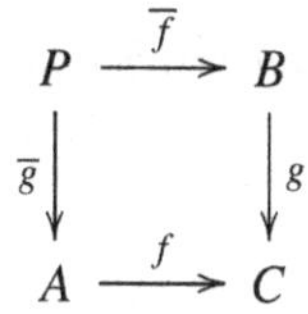

where f is a closed embedding. We mention that, as in all topological categories, we can take $P = \{(a, b) \in A \times B \mid f(a) = g(b)\}$ where $A \times B$ carries the product structure and P the subspace structure and where $\overline{f}$ and $\overline{g}$ are the restrictions of the projections. Since $\mathscr{M}$ is stable under pullbacks we already obtain that $\overline{f}$ is an embedding. Hence it remains to show that $\delta_{\overline{f}(P)} = \theta_{\overline{f}(P)}$. Since g is a contraction we have that

$$\begin{aligned}\theta_{\overline{f}(P)} &= \theta_{g^{-1}(f(A))}\\ &= \theta_{f(A)} \circ g\\ &= \delta_{f(A)} \circ g\\ &\leq \delta_{gg^{-1}f(A)} \circ g\\ &\leq \delta_{g^{-1}(f(A))} = \delta_{\overline{f}(P)}.\end{aligned}$$

Since the other inequality always holds this proves that $\overline{f}$ is in $\mathscr{F} \cap \mathscr{M}$.

2 and 3. Let $f : X \longrightarrow Y$, $g : Y \longrightarrow Z$ be as stated and let $B \subseteq Y$. Then we have

$$\begin{aligned}g(\delta_B) &= g \circ f(\delta_B \circ f)\\ &\leq g \circ f(\delta_{f^{-1}(B)})\\ &\leq \delta_{gff^{-1}(B)} = \delta_{g(B)},\end{aligned}$$

which shows that g is closed expansive. □

By the foregoing result, (F4) and (F5) are fulfilled. From Clementino et al. (2003) we adopt the following definition of proper morphism.

1.4.10 Definition A contraction $f : X \longrightarrow Y$ is called a *proper contraction* if it belongs stably to $\mathscr{F}$, i.e. whenever

$$\begin{array}{ccc} W & \xrightarrow{\overline{f}} & Z \\ {\scriptstyle \overline{g}}\downarrow & & \downarrow{\scriptstyle g} \\ X & \xrightarrow{f} & Y \end{array}$$

is a pullback diagram, then $\overline{f} \in \mathscr{F}$.

In view of the following property, verifying properness of a contraction can be done by a simple criterion mentioned in Clementino et al. (2003).

1.4.11 Proposition *$\mathscr{F}$ is stable under restrictions.*

Proof This follows from 1.4.4. □

The following results are immediate consequences of the general results proved in Clementino et al. (2003).

1.4.12 Proposition *A contraction $f : X \longrightarrow X'$ is proper if and only if for each approach space Z the map $f \times 1_Z : X \times Z \longrightarrow X' \times Z$ is closed expansive.*

Put $\mathscr{F}^*$ for the class of proper contractions. From the definition, it is immediately clear that $\mathscr{F}^* \subseteq \mathscr{F}$.

1.4.13 Proposition *The class of proper contractions fulfils the following stability properties.*

1. *$\mathscr{F}^*$ is stable under composition.*
2. *$\mathscr{F}^*$ contains all closed embeddings.*
3. *$\mathscr{F}^*$ is the largest pullback-stable subclass of $\mathscr{F}$.*
4. *If $g \circ f \in \mathscr{F}^*$ and g is an injective contraction, then $f \in \mathscr{F}^*$.*

1.5 Comments

1. **Asymptotic radius and center**

The idea of a limit operator exists in the literature, although it was never formalized in a setting as approach spaces like here or used as characterizing concept of a structure of any kind. In approximation theory we find the notion of *asymptotic radius* and *asymptotic center* (see e.g. Benavides and Lorenzo 2004; Kirk 1990; Lim 1980; Edelstein 1972, 1974). These concepts were introduced to be able to work with non-convergent sequences and to be able to gauge to what extent they are non-convergent. See more details in the comments on the second chapter.

2. **An interior-counterpart to distance**

The following gives a "partial" counterpart to the concept of a distance derived from a lower hull operator. Suppose given an upper hull operator $\mathfrak{u}$ on X as in 1.2.19. For any $\omega \in \mathbb{R}^+$ we define

$$\iota^\omega : X \times 2^X \longrightarrow \mathbb{P} : (x, A) \mapsto \mathfrak{u}(\theta_A^\omega)(x).$$

Then the reader can verify that ι^ω satisfies the following properties.

1. $\forall x \in X, \forall A \subseteq X : \iota^\omega(x, A) = 0 \Rightarrow x \in A.$
2. $\forall x \in X, \iota^\omega(x, X) = 0.$
3. $\forall x \in X, \forall A, B \in 2^X : \iota^\omega(x, A \cap B) = \max\{\iota^\omega(x, A), \iota^\omega(x, B)\}.$
4. $\forall x \in X, \forall A \subseteq X, \forall \varepsilon < \omega : \iota^\omega(x, A_{(\varepsilon)}) \leq \iota^\omega(x, A) + \varepsilon$ where

$$A_{(\varepsilon)} := \{y \in X \mid \iota^\omega(y, A) < \omega - \varepsilon\}.$$

3. **Local contractivity**

The concept of a contraction between approach spaces, just as the topological notion of continuous map, has a local version of contractivity in a point. It suffices to fix the point x in definition 1.3.1. Results similar to those which can be obtained for local continuity can then be proved.

4. **Open expansion in the literature**

The same concept as our notion of an open expansion appears in the work of Ioffe on subdifferential calculus (Ioffe 1981, 1990, 2000), although there it is on the one hand slightly more general since it deals with multivalued maps and allows for Lipschitz constants in the inequalities but on the other hand is more restricted since it deals with metric spaces. However, it is the same intuition which lies at the basis of both concepts.

5. **Properness**

In this chapter we defined properness linked to contractivity. This is also the way it is done in topology where a proper map always implies continuity. However, this is not necessary, just as with open and closed expansions we can isolate the properness-part without requiring contractivity. The reason we did not do this in the present chapter is to show the link with the work of Clementino et al. (2003). We will come back to these concepts in the chapter on index analysis where we will isolate the concept of properness from contractivity.

6. **Approach spaces in other theories**

Approach spaces appear as the penultimate example of two other general theories in categorical topology.

The first concerns *metrically generated theories* as introduced by Colebunders and Lowen (2005). The theory of approach spaces is embedded in metrically generated theories via the description with gauges. Metrically generated theories were further studied in Claes (2009) concerning initially dense objects, Colebunders et al. (2006) concerning function spaces and one-point extensions, Colebunders et al. (2007) where completeness in a symmetric setup is treated, Colebunders and Lowen (2009) where the embedding of bornological spaces in metrically generated theories is considered, Colebunders et al. (2012) where local metrically generated theories are introduced, Claes et al. (2007) concerning co-wellpoweredness, Colebunders and Gerlo (2007) where firm reflections are treated and Colebunders and Vandersmissen (2010) where completeness aspects in a non-symmetric setup are handled. More information can also be found in the PhD theses of Gerlo (2007), Vandersmissen (2008) and Van Geenhoven (2010).

The second is the theory of *lax algebras* as studied extensively in the book Monoidal Topology (Hofmann et al. 2014). We come back to this in the last chapter on categorical aspects where we will especially be looking at the way approach spaces can be described as a category of lax algebras (see Clementino and Hofmann 2003; Clementino et al. 2004).

Further, approach spaces also appear in the theory of approach frames. This theory was first suggested by Banaschewski in some lectures at UCT in Cape Town. This finally resulted in a series of papers by Banaschewski et al. (2006, 2007, 2012) and the PhD thesis of Van Olmen (2005). An approach frame is a frame L with top $\top$ and bottom $\bot$ equipped with two families of unary operations, addition and subtraction of $\alpha \in \mathbb{P}$, denoted respectively $A_\alpha : L \longrightarrow L$ and $S_\alpha : L \longrightarrow L$ which satisfy all identities valid for ordinary addition and subtraction by α, the frame operations in $\mathbb{P}$ and the implications $A_\alpha \bot = \bot \Rightarrow \alpha = 0$ and $S_\alpha \top = \top \Rightarrow \alpha \neq \infty$. Morphisms between approach frames are frame homomorphisms which commute with additions and subtractions. Approach spaces then are "embedded" into the theory of approach frames via their lower regular function frames.

7. **Links to κ-metrizability**

Several notions generalizing metrizability and involving a type of "distance functions" have been introduced in the literature. The one which is most closely related to our concept of approach structure is the concept of a κ-metric as introduced by Shchepin (1976a). See also Shchepin (1976b, 1980), Isiwata (1985, 1987, 1988), and Suzuki et al. (1989). A Tychonoff space X is called κ-metrizable if there exists a function $\rho : X \times RC(X) \longrightarrow \mathbb{R}$, where $RC(X)$ stands for the collection of regularly closed sets in X, fulfilling the following properties.

1. $\rho(x, F) = 0$ if and only if $x \in F$.
2. $F \subseteq F' \Rightarrow \rho(\cdot, F') \leq \rho(\cdot, F)$.
3. For all $F \in RC(X)$: $\rho(\cdot, F)$ is continuous.
4. $\rho(\cdot, \mathrm{cl}(\mathrm{int}(\bigcup_{j \in J} F_j))) = \inf_{j \in J} \rho(\cdot, F_j)$ for any collection $(F_j)_j$ in $RC(X)$ which is increasing and totally ordered.

Such a function is called a κ-metric for X. The class of κ-metrizable topological spaces contains all metrizable topological spaces and is closed under the formation of products.

Analogous ideas also appeared in the work of Borges (1966), Nagata (1992), and Naimpally and Pareek (2014) where the concept of an annihilator was used. An annihilator basically is more general than a κ-metric, only fulfils the first and third properties above, and can be defined on $X \times C(X)$ where $C(X)$ is an arbitrary collection of closed sets. Much work in this area was aimed at finding supplementary conditions on a κ-metric or an annihilator to insure metrizability of the given space, see e.g. the paper of Suzuki, Tamano and Tanaka (1989).

8. **Towers in the literature**

Towers are a useful concept to define or characterize various extensions of classical categories. They were used by Brock and Kent (1997a) to define the category of so-called limit-tower spaces which turns out to be isomorphic to the category CAp which we study in Sect. 12.2. They were also used by Nauwelaerts (2000) to study certain categorical hulls in approach theory. Further they were used by Zhang (2001) to construct tower extensions of general topological categories and by Herrlich and Zhang in their study of categorical properties of probabilistic convergence spaces Herrlich and Zhang (1998).

Chapter 2
Topological and Metric Approach Spaces

Well, I use the metric system. It's the only way to get really exact numbers.

(Catherynne M. Valente, in The Girl Who Fell Beneath Fairyland and Led the Revels There)

As every mathematician knows, nothing is more fruitful than these obscure analogies, these indistinct reflections of one theory into another, these furtive caresses, these inexplicable disagreements ...

(André Weil)

Both topological and metric spaces can be viewed as special types of approach spaces. More precisely, both the categories of topological spaces and continuous maps, Top, and of (quasi)-metric spaces and non-expansive maps, (q)Met, can actually be embedded as full and isomorphism-closed subcategories of App. In this chapter we will see various characterizations of topological and of (quasi-)metric spaces as approach spaces and we will see exactly how Top and qMet (respectively Met) are embedded in App. For Top the embedding will turn out to be both concretely reflective and concretely coreflective. For both Met and qMet the embedding will turn out to be concretely coreflective but not reflective. We will demonstrate that it is precisely the failure of Met and qMet to be embedded reflectively in App which makes the theory of approach spaces particularly interesting in any situation in mathematics where initial structures of (quasi-)metric or (quasi-)metrizable topological spaces occur.

2.1 Topological Approach Spaces

As far as notation is concerned, from now on, whenever we say that X is a topological space, $\mathscr{T}$ will stand for the collection of open sets. Structures derived from $\mathscr{T}$, such as the associated closure operator, will be denoted, for example, by $\mathrm{cl}_{\mathscr{T}}$. If no confusion can arise we may also drop reference to $\mathscr{T}$ altogether. We put usc (respectively lsc) for upper semicontinuous (respectively lower semicontinuous) and $(\mathscr{V}_{\mathscr{T}}(x))_x$ or shortly $(\mathscr{V}(x))_x$ for the neighbourhood filters of a topological space. Given a quasi-metric d,

R. Lowen, *Index Analysis*, Springer Monographs in Mathematics,
DOI 10.1007/978-1-4471-6485-2_2

we let $\mathscr{T}_d$ stand for the underlying topology. Given a filter $\mathscr{F}$ in a topological space, the set of adherence points of $\mathscr{F}$ is denoted by adh$\mathscr{F}$ and the set of limit points by lim $\mathscr{F}$. That a filter $\mathscr{F}$ converges to a point x is written as $\mathscr{F} \longrightarrow x$ and that it adheres to x is written as $\mathscr{F} \rightsquigarrow x$.

Given a topological space $(X, \mathscr{T})$ we associate with it a natural approach space in the following way. Let

$$\delta_{\mathscr{T}} : X \times 2^X \longrightarrow \mathbb{P} : (x, A) \mapsto \begin{cases} 0 & x \in \mathrm{cl}_{\mathscr{T}}(A), \\ \infty & x \notin \mathrm{cl}_{\mathscr{T}}(A). \end{cases}$$

2.1.1 Proposition *If $(X, \mathscr{T})$ is a topological space, then the function*

$$\delta_{\mathscr{T}} : X \times 2^X \longrightarrow \mathbb{P}$$

is a distance on X and the associated structures are given as follows.

1. *For any filter $\mathscr{F}$ on X: $\alpha_{\mathscr{T}}\mathscr{F} = \theta_{\mathrm{adh}\,\mathscr{F}}$ and $\lambda_{\mathscr{T}}\mathscr{F} = \theta_{\lim \mathscr{F}}$.*
2. *For any $x \in X$: $\mathscr{A}_{\mathscr{T}}(x) := \{\varphi \in \mathbb{P}^X \mid \varphi(x) = 0, \varphi$ usc in $x\}$ and a basis is given by $\mathscr{B}_{\mathscr{T}}(x) := \{\theta_V \mid V \in \mathscr{V}_{\mathscr{T}}(x)\}$.*
3. *$\mathscr{G}_{\mathscr{T}} := \{d \in q\mathsf{Met}(X) \mid \mathscr{T}_d \subseteq \mathscr{T}\}$.*
4. *The tower is given by the family $(\mathfrak{t}_\varepsilon^{\mathscr{T}})_{\varepsilon \in \mathbb{R}^+}$ where for each $\varepsilon \in \mathbb{R}^+$, $\mathfrak{t}_\varepsilon^{\mathscr{T}}$ coincides with $\mathrm{cl}_{\mathscr{T}}$.*
5. *For any $\mu \in \mathbb{P}^X$ and $x \in X$: $\mathfrak{l}_{\mathscr{T}}(\mu)(x) := \sup_{V \in \mathscr{V}_{\mathscr{T}}(x)} \inf_{y \in V} \mu(y)$ i.e. $\mathfrak{l}_{\mathscr{T}}(\mu)$ is the largest lower semicontinuous function smaller than μ.*
6. *$\mathfrak{L}_{\mathscr{T}} := \{\mu \in \mathbb{P}^X \mid \mu$ lsc$\}$.*
7. *For any $\mu \in \mathbb{P}_b^X$ and $x \in X$: $\mathfrak{u}_{\mathscr{T}}(\mu)(x) := \inf_{V \in \mathscr{V}_{\mathscr{T}}(x)} \sup_{y \in V} \mu(y)$ i.e. $\mathfrak{u}_{\mathscr{T}}(\mu)$ is the smallest upper semicontinuous function larger than μ.*
8. *$\mathfrak{U}_{\mathscr{T}} := \{\mu \in \mathbb{P}_b^X \mid \mu$ usc$\}$.*
9. *For any functional ideal $\mathfrak{I}$ on X: $\mathfrak{I} \rightarrowtail x$ if and only $\mathfrak{f}_\alpha(\mathfrak{I}) \longrightarrow x$ for all $\alpha \in [c(\mathfrak{I}), \infty[$.*

Proof (D1) and (D2) are immediate. (D3) follows from the fact that, for any $A, B \subseteq X$, we have $\mathrm{cl}_{\mathscr{T}}(A \cup B) = \mathrm{cl}_{\mathscr{T}}(A) \cup \mathrm{cl}_{\mathscr{T}}(B)$. (D4) follows from the fact that, for all $\varepsilon < \infty$, $A^{(\varepsilon)} = \mathrm{cl}_{\mathscr{T}}(A)$ and $A^{(\infty)} = X$.

1. We give the proof for the adherence operator; the one for the limit operator is precisely the same. It follows from 1.2.64 that,

$$\alpha_{\mathscr{T}}\mathscr{F} = \sup_{F \in \mathscr{F}} (\delta_{\mathscr{T}})_F = \sup_{F \in \mathscr{F}} \theta_{\mathrm{cl}_{\mathscr{T}}(F)} = \theta_{\bigcap_{F \in \mathscr{F}} \mathrm{cl}_{\mathscr{T}}(F)} = \theta_{\mathrm{adh}\,\mathscr{F}}.$$

2. From 1.2.33 it follows that, for all $x \in X$,

$$\mathscr{A}_{\mathscr{T}}(x) = \left\{ \varphi \in \mathbb{P}^X \mid \forall A \subseteq X : x \in \mathrm{cl}_{\mathscr{T}}(A) \Rightarrow \inf_{a \in A} \varphi(a) = 0 \right\}.$$

If $\varphi \in \mathscr{A}_{\mathscr{T}}(x)$, then, since $x \in \mathrm{cl}_{\mathscr{T}}(\{x\})$, it follows that $\varphi(x) = 0$. Now let $\alpha > 0$. If $x \in \mathrm{cl}_{\mathscr{T}}(\{\varphi \geq \alpha\})$, then

$$\inf_{a \in \{\varphi \geq \alpha\}} \varphi(a) = 0,$$

which is absurd. Hence, for all $\alpha > 0$, $x \in \mathrm{int}_{\mathscr{T}}(\{\varphi < \alpha\})$, which proves that φ is upper semicontinuous in x. The converse follows from the fact that if φ is upper semicontinuous in x and $\varphi(x) = 0$, then, for any $\alpha > 0$, $\{\varphi < \alpha\}$ is a neighbourhood of x. That $\mathscr{B}_{\mathscr{T}}(x)$ is a basis for $\mathscr{A}_{\mathscr{T}}(x)$ follows easily from the definitions.

3. From 1.2.4 we obtain that

$$\mathscr{G} = \left\{ d \in q\mathsf{Met}(X) \mid x \in \mathrm{cl}_{\mathscr{T}}(A) \Rightarrow \delta_d(x, A) = 0 \right\}$$

from which the result immediately follows.

4. For any $\varepsilon \in \mathbb{R}^+$ and $A \subseteq X$, we have

$$\begin{aligned} \mathrm{t}_{\varepsilon}^{\mathscr{T}}(A) &= \left\{ x \in X \mid \delta_{\mathscr{T}}(x, A) \leq \varepsilon \right\} \\ &= \left\{ x \in X \mid \delta_{\mathscr{T}}(x, A) = 0 \right\} \\ &= \mathrm{cl}_{\mathscr{T}}(A). \end{aligned}$$

5. The formula follows from 1.2.38. The alternative description is well known and can be found for instance in Bourbaki (1960).

6. This follows from 1.2.24.

7. This is analogous to 5.

8. This follows from 1.2.29.

9. This follows from 1.2.58. □

An approach space of type $(X, \delta_{\mathscr{T}})$ for some topology $\mathscr{T}$ on X will be called a *topological approach space*. Analogously all associated structures will be referred to as being topological and will be denoted in a similar way with a subscript or superscript referring to the original topology. Note that in particular 5 and 7 are the well-known lower semicontinuous and upper semicontinuous regularization of functions, see e.g. Bourbaki (1960).

The next result gives an internal characterization of these spaces.

2.1.2 Proposition *An approach space (X, δ) is topological if and only if any of the following equivalent properties holds.*

1. $\delta(X \times 2^X) \subseteq \{0, \infty\}$.
2. *For any filter $\mathscr{F} \in \mathsf{F}(X)$ we have that $\lambda\mathscr{F}(X) \subseteq \{0, \infty\}$.*
3. *For any lower regular function μ we have that also $\theta_{\{\mu=0\}}$ is lower regular.*
4. *For any lower regular function μ and for any $\varepsilon \in \mathbb{R}^+$ also $\theta_{\{\mu \leq \varepsilon\}}$ is lower regular.*

5. *There exists a subbasis $\mathcal{M}$ for the lower regular function frame such that for all $\mu \in \mathcal{M}$ also $\theta_{\{\mu=0\}}$ is lower regular.*
6. *For any upper regular function ν and for any $\alpha, \beta \in]0, \infty[$ also $\theta_{\{\nu<\alpha\}} \wedge \beta$ is upper regular.*
7. *There exists a subbasis $\mathcal{M}$ for the upper regular function frame such that for all $\nu \in \mathcal{M}$ and for any $\alpha, \beta \in]0, \infty[$ also $\theta_{\{\nu<\alpha\}} \wedge \beta$ is upper regular.*
8. *For any $d \in \mathcal{G}$ and $\alpha \in \mathbb{R}^+$ also $\alpha d \in \mathcal{G}$.*

Proof The only-if part of 1 follows from the definition of a topological approach space. To show the if part it suffices to note that if, for all $A \subseteq X$, we put $\mathrm{cl}(A) := \{x \in X \mid \delta(x, A) = 0\}$, then cl is a topological closure operator and δ is the associated distance. Characterization 2 is an immediate consequence. To prove 3 it suffices to look at the functions $\mu = \delta_A$ for $A \subseteq X$ and 4 is clearly equivalent to 3 because of the translation-invariance of $\mathfrak{L}$. Claims 5 to 7 follow by analogous reasoning. That 8 is necessary follows from 2.1.1 and that it is sufficient follows from 1.2.6 and the first claim. □

2.2 Embedding Top in App

In the foregoing section we have seen that a topological space can be viewed as a special type of approach space. That, moreover, Top is concretely embedded in App is a consequence of the fact that given topological spaces $(X, \mathcal{T})$ and $(X', \mathcal{T}')$ a function $f : X \longrightarrow X'$ will be continuous as a map between the topological spaces if and only if it is a contraction as a map between the associated approach spaces, as follows at once e.g. from the observation that if $A \subseteq X$, then

$$f(\mathrm{cl}_{\mathcal{T}} A) \subseteq \mathrm{cl}_{\mathcal{T}'}(f(A)) \Leftrightarrow \forall \varepsilon \in \mathbb{R}^+ : f(A^{(\varepsilon)}) \subseteq (f(A))^{(\varepsilon)'},$$

which by 1.3.3 proves our claim. Hence the concrete functor from Top to App which takes $(X, \mathcal{T})$ to $(X, \delta_{\mathcal{T}})$ is a full embedding of Top into App.

We will now prove that Top is actually very nicely embedded in App. In contrast to most known topological categories which do not have subcategories which are simultaneously reflectively and coreflectively embedded, such as for instance Top itself, we will prove that Top is simultaneously concretely reflectively and concretely coreflectively embedded in App. This is what, in 12.1 we call a *stable subcategory*.

The fact that the embeddings, reflections and coreflections are concrete implies in particular that the reflection and coreflection morphisms are carried by the identity map of the underlying set. Hence, throughout this work, when referring to a reflection or coreflection, we will only mention the objects and never the reflection or coreflection morphisms.

We recall that two important aspects of the fact that Top is reflectively embedded in App are: (1) for each approach structure, there exists a finest coarser topological structure on the same underlying set and (2) initial structures of topological spaces are the same whether they are taken in Top or in App. We refer to the seminal work of Herrlich on these matters (Herrlich 1968, 1983).

2.2.1 Proposition *For any approach space X, the operation defined by*

$$\mathrm{cl}(A) := \{x \in X \mid \delta(x, A) < \infty\}$$

is a pretopological closure operator.

Proof This follows from (D1), (D2), and (D3). □

We recall that the category PrTop of pretopological spaces and continuous maps is a topological category in which Top is reflectively embedded. A *pretopological space* is a set equipped with a closure operation which satisfies all the usual axioms with the possible exception of idempotency. For more information on PrTop we refer the reader to Choquet (1947) and Colebunders (1989).

2.2.2 Theorem Top *is embedded as a concretely reflective subcategory of* App. *For any approach space (X, δ), its* Top*-reflection is determined by the distance δ^{tr} associated with the topological reflection of the pretopological closure operator* cl.

Proof The topological reflection of this closure operator is obtained by a standard transfinite process which produces a topological closure operator, cl_{tr}. To verify that $1_X : (X, \delta) \longrightarrow (X, \delta^{tr})$ is a contraction it suffices to note that if $\delta(x, A) < \infty$, for some $x \in X$ and $A \subseteq X$, then $\delta^{tr}(x, A) = 0$. Now suppose that $(Y, \mathcal{T})$ is a topological space and that

$$f : (X, \delta) \longrightarrow (Y, \delta_{\mathcal{T}})$$

is a contraction. Then, for any $x \in X$ and $A \subseteq X$, we have

$$\begin{aligned} x \in \mathrm{cl}(A) &\Rightarrow \delta(x, A) < \infty \\ &\Rightarrow \delta_{\mathcal{T}}(f(x), f(A)) < \infty \\ &\Rightarrow f(x) \in \mathrm{cl}_{\mathcal{T}}(f(A)). \end{aligned}$$

Hence $f : (X, \mathrm{cl}) \longrightarrow (Y, \mathrm{cl}_{\mathcal{T}})$ is continuous as a function between pretopological spaces. It then follows that also $f : (X, \mathrm{cl}_{tr}) \longrightarrow (Y, \mathrm{cl}_{\mathcal{T}})$ is continuous as a function between topological spaces, which in turn means that $f : (X, \delta^{tr}) \longrightarrow (Y, \delta_{\mathcal{T}})$ is a contraction. □

2.2.3 Corollary Top *is closed under the formation of limits and initial structures in* App. *In particular, a product in* App *of a family of topological approach spaces is a topological approach space and, likewise, a subspace in* App *of a topological approach space is a topological approach space.*

Although, as the previous results show, it is important to know from a structural point of view that Top is reflectively embedded in App, we will not often have recourse to considering the Top-reflection of an approach space. It is easily seen that if the distance is finite, then the Top-reflection is indiscrete. This shows that it will usually not be a very interesting topology to consider. The situation becomes totally different for the dual property, coreflectivity, as we will now see. We also recall that two important aspects of the fact that Top is coreflectively embedded in App are: (1) for each approach structure there exists a coarsest finer topological structure on the same underlying set and (2) final structures of topological spaces are the same whether they are taken in Top or in App.

2.2.4 Theorem Top *is embedded as a concretely coreflective subcategory of* App. *For any approach space* (X, δ), *its* Top*-coreflection is determined by the distance* δ^{tc} *associated with the topological closure operator given by*

$$\mathrm{cl}_\delta(A) := \left\{x \in X \mid \delta(x, A) = 0\right\}.$$

Proof It is easily verified that cl_δ is indeed a topological closure operator and that $1_X : (X, \delta^{tc}) \longrightarrow (X, \delta)$ is a contraction. Now suppose that $(Y, \mathscr{T})$ is a topological space and that

$$f : (Y, \delta_{\mathscr{T}}) \longrightarrow (X, \delta)$$

is a contraction. Then, for any $x \in Y$ and $A \subseteq Y$ such that $x \in \mathrm{cl}_{\mathscr{T}}(A)$, we have

$$\delta(f(x), f(A)) \leq \delta_{\mathscr{T}}(x, A) = 0,$$

which proves that

$$f : (Y, \delta_{\mathscr{T}}) \longrightarrow (X, \delta^{tc})$$

is also a contraction. □

2.2.5 Corollary Top *is closed under the formation of colimits and final structures in* App. *In particular, a coproduct in* App *of a family of topological approach spaces is a topological approach space and, likewise, a quotient in* App *of a topological approach space is a topological approach space.*

In the sequel, in order not to overload notation and terminology, we will call the topological spaces associated with the topological reflection and coreflection also simply the topological reflection and coreflection of a given approach space. In other words, unless required for technical reasons, we do not differentiate between a topological space and the associated approach space.

At the end of this chapter we will discuss the importance of the Top-coreflection of an approach space. Anticipating this, we will now describe this coreflection by means of the most important other basic structures. Given an approach space (X, δ), we will denote the topology underlying the topological coreflection by $\mathscr{T}_\delta$, or with any other index referring to the original structure.

2.2.6 Proposition *If X is an approach space with $(\mathscr{B}(x))_{x\in X}$ a basis for the approach system and $\mathscr{H}$ a basis for the gauge, then the following properties hold.*

1. *Convergence in the topological coreflection is characterized by*

$$\mathscr{F} \to x \Leftrightarrow \lambda\mathscr{F}(x) = 0 \text{ and } \mathscr{F} \rightsquigarrow x \Leftrightarrow \alpha\mathscr{F}(x) = 0.$$

2. *The neighbourhoods in the topological coreflection are characterized by*

$$\begin{aligned}\mathscr{V}(x) &= \left\{V \in 2^X \mid \exists\varepsilon > 0, \exists\varphi \in \mathscr{B}(x) : \{\varphi < \varepsilon\} \subseteq V\right\}\\ &= \left\{V \in 2^X \mid \exists\varepsilon > 0, \exists d \in \mathscr{H} : B_d(x,\varepsilon) \subseteq V\right\}.\end{aligned}$$

Proof 1. This follows from

$$\begin{aligned}\mathscr{F} \to x \Leftrightarrow x &\in \bigcap_{A\in\sec(\mathscr{F})} \mathrm{cl}_\delta(A)\\ &\Leftrightarrow \sup_{A\in\sec(\mathscr{F})} \delta(x, A) = 0\\ &\Leftrightarrow \lambda\mathscr{F}(x) = 0,\end{aligned}$$

and analogously for the adherence.

2. For the approach system this follows from

$$\begin{aligned}V \in \mathscr{V}(x) &\Leftrightarrow x \notin \mathrm{cl}_\delta(X \setminus V)\\ &\Leftrightarrow \exists\varepsilon > 0 : \sup_{\varphi\in\mathscr{B}(x)} \inf_{y\in X\setminus V} \varphi(y) > \varepsilon\\ &\Leftrightarrow \exists\varepsilon > 0, \exists\varphi \in \mathscr{B}(x) : \{\varphi < \varepsilon\} \subseteq V\end{aligned}$$

and for the gauge it is entirely similar. □

2.2.7 Example We refer to the two examples which we considered in 1.2.62.

The distance of the first example is $\delta_\mathbb{E} : \mathbb{P}\times 2^\mathbb{P} \longrightarrow \mathbb{P}$. The topological coreflection of this space is $(\mathbb{P}, \mathscr{T}_\mathbb{E})$, where $\mathscr{T}_E$ is the topology of the Alexandroff compactification of $[0, \infty[$ with the usual (Euclidean) topology.

The distance of the second example is $\delta_\mathbb{P} : \mathbb{P} \times 2^\mathbb{P} \longrightarrow \mathbb{P}$. The topological coreflection of this space is $(\mathbb{P}, \mathscr{T}_\mathbb{P})$, where

$$\mathscr{T}_\mathbb{P} := \{]a, \infty] | a \in \mathbb{P}\} \cup \{\mathbb{P}\}.$$

2.2.8 Proposition *If X is an approach space, then any lower regular function $\xi \in \mathfrak{L}$ considered as a map $\xi : (X, \mathscr{T}_\delta) \longrightarrow (\mathbb{P}, \mathscr{T}_\mathbb{P})$ is continuous, in particular the distance functionals, and limits and adherences of filters are continuous maps.*

Proof This follows from 1.3.5, 2.2.4, and 2.2.7. □

If, in the foregoing result, we replace the topology $\mathscr{T}_{\mathbb{P}}$ by the Euclidean topology $\mathscr{T}_{\mathbb{E}}$, then all maps are lower semicontinuous.

In the foregoing chapter we introduced three types of maps, namely, open and closed expansions and proper contractions. The following result tells us that the terminology for closed and open was appropriately chosen.

2.2.9 Proposition (Top) *If X and X' are topological spaces and $f : X \longrightarrow X'$ is a map then the following properties hold.*

1. *f is closed as a map between the topological spaces if and only if it is a closed expansion between the associated approach spaces.*
2. *f is open as a map between the topological spaces if and only if it is an open expansion between the associated approach spaces.*

Proof (1) This follows from 2.1.1 (6), 1.4.2 (5) and the observation that if f is closed and μ is lsc then $f(\mu)$ is lsc.

(2) Analogously, this follows from 2.1.1 (8), 1.4.6 (5) and the observation that if f is open and μ is usc then $f(\mu)$ is usc. □

We will treat the case of proper contractions later when we have a more appropriate characterization at our disposal (see 4.3.30).

2.3 (Quasi-)Metric Approach Spaces

Given a quasi-metric space (X, d), we associate with it a natural approach space by defining in the usual way the function

$$\delta_d : X \times 2^X \longrightarrow \mathbb{P} : (x, A) \mapsto \inf_{a \in A} d(x, a).$$

2.3.1 Proposition *If (X, d) is a quasi-metric space, then the function*

$$\delta_d : X \times 2^X \longrightarrow \mathbb{P}$$

is a distance on X and the associated structures are given as follows.

1. *For all $\mathscr{F} \in \mathsf{F}(X)$ and $x \in X$:*

   a. $\alpha_d \mathscr{F}(x) = \sup_{F \in \mathscr{F}} \inf_{y \in F} d(x, y) = \sup_{F \in \mathscr{F}} \delta_d(x, F)$,

   b. $\lambda_d \mathscr{F}(x) = \inf_{F \in \mathscr{F}} \sup_{y \in F} d(x, y)$.

2. *For any $x \in X$: $\mathscr{A}_d(x) := \{\varphi \in \mathbb{P}^X \mid \varphi \leq d(x, \cdot)\}$ and a basis is given by the singleton $\{d(x, \cdot)\}$.*

3. $\mathscr{G}_d := \{e \in q\mathsf{Met}(X) \mid e \leq d\}$ *and a basis is given by the singleton* $\{d\}$.
4. *The tower is given by the family* $(\mathfrak{t}^d_\varepsilon)_{\varepsilon \in \mathbb{R}^+}$ *where*

$$\mathfrak{t}^d_\varepsilon : 2^X \longrightarrow 2^X : A \mapsto \{x \in X \mid \delta_d(x, A) \leq \varepsilon\}.$$

5. *For any* $\mu \in \mathbb{P}^X$ *and* $x \in X$: $\mathfrak{l}_d(\mu)(x) := \inf_{y \in X}(\mu(y) + d(x, y))$ *i.e.* $\mathfrak{l}_d(\mu)$ *is the largest non-expansive map smaller than* μ.
6. $\mathfrak{L}_d := \{\mu \in \mathbb{P}^X \mid \mu$ *non-expansive*$\}$.
7. *For any* $\mu \in \mathbb{P}^X_b$ *and* $x \in X$: $\mathfrak{u}_d(\mu)(x) := \sup_{y \in X}(\mu(y) - d(x, y))$ *i.e.* $\mathfrak{u}_d(\mu)$ *is the smallest non-expansive map larger than* μ.
8. $\mathfrak{U}_d := \{\mu \in \mathbb{P}^X_b \mid \mu$ *non-expansive*$\} = \mathfrak{L}_d \cap \mathbb{P}^X_b$.
9. *For any functional ideal* $\mathfrak{I}$ *on* X: $\mathfrak{I} \rightarrowtail x$ *if and only if* $d(x, \cdot) \wedge \omega \in \mathfrak{I}$ *for each* $\omega \in \mathbb{R}^+$.

Proof That δ_d is a distance is merely the special case of 1.2.6 where we take for the gauge basis $\mathscr{H} := \{d\}$.

1. This is an immediate consequence of the definition of the adherence operator 1.2.64 and of 1.2.44.

2. By 1.2.33 we have that, for all $x \in X$,

$$\mathscr{A}_d(x) = \left\{\varphi \in \mathbb{P}^X \mid \forall A \in 2^X : \inf_{a \in A} \varphi(a) \leq \inf_{a \in A} d(x, a)\right\}.$$

Clearly, $\inf_{a \in A} \varphi(a) \leq \inf_{a \in A} d(x, a)$ holds, for all $A \subseteq X$, if and only if $\varphi(a) \leq d(x, a)$ holds, for all $a \in X$.

3. Referring to 1.2.4 instead of to 1.2.33, this is precisely the same as the proof of the foregoing result.

4. This follows from 1.2.21.

5. This follows from 1.2.16.

6. This follows from 1.2.24.

7. This follows from 1.2.25.

8. This follows from 1.2.29 and 6.

9. This follows from 1.2.31 and 2. □

The above expression for δ_d is of course well known and notationally often no distinction is made between d and δ_d. We emphasize, however, that for our purposes it is important to use different notations for a quasi-metric and for the distance derived from it in the sense of the foregoing definition. In the first place, the two functions have different domains and in the second place, they determine categorically different structures. We have seen that Top is simultaneously reflectively and coreflectively embedded in App, and hence for topological approach spaces, it makes no difference whether we perform constructions, such as the making of limits, colimits, initial, and final structures, in Top or in App. However, for quasi-metric approach spaces, as we will see later in this chapter when we study the precise way in which $q\mathsf{Met}$ is embedded in App, it does make a difference whether we make initial structures of quasi-metric approach spaces in qMet or in App. Hence it is important to make the distinction.

For a sequence $(x_n)_n$ we denote the generated filter by $\langle (x_n)_n \rangle$.

2.3.2 Corollary *If (X, d) is a quasi-metric space, $(x_n)_n$ is a sequence in X and $x \in X$, then the following formulas hold.*

1. $\alpha_d \langle (x_n)_n \rangle (x) = \liminf_{n\to\infty} d(x, x_n)$.
2. $\lambda_d \langle (x_n)_n \rangle (x) = \limsup_{n\to\infty} d(x, x_n)$.

An approach space of type (X, δ_d), for some quasi-metric d on X, will be called a *(quasi-)metric approach space*. Analogously all associated structures will be referred to as being (quasi-)metric.

2.3.3 Proposition *An approach space is quasi-metric if and only if it has one, and hence all of the following equivalent properties.*

1. *For all $x \in X$ and $\mathscr{A} \subseteq 2^X$, we have $\delta(x, \cup\mathscr{A}) = \inf_{A\in\mathscr{A}} \delta(x, A)$.*
2. *For all $x \in X$ and $A \subseteq X$, we have $\delta(x, A) = \inf_{a\in A} \delta(x, \{a\})$.*
3. *The lower regular function frame is closed under the formation of arbitrary infima.*
4. *The upper regular function frame is closed under the formation of arbitrary bounded suprema.*
5. *The gauge is a principal gauge generated by a unique function, which necessarily is a quasi-metric.*

Proof The equivalence of the first and second property with being quasi-metric follows at once from the definition. That a quasi-metric space fulfils the third property follows from the description of the lower regular function frame in 2.3.1. Conversely it follows from 3 that given $x \in X$ and $A \subseteq X$ the function

$$\eta : X \longrightarrow \mathbb{P} : y \mapsto \inf_{a\in A} \delta(y, \{a\})$$

belongs to $\mathfrak{L}$ and vanishes on A. Consequently it follows from 1.2.45 that

$$\begin{aligned}\delta(x, A) &= \sup\{\mu(x) \mid \mu \in \mathfrak{L}, \mu_{|A} = 0\}\\ &\geq \eta(x)\\ &= \inf_{a\in A} \delta(x, \{a\}).\end{aligned}$$

Since the other inequality always holds this proves 3. The third and fourth properties are obviously equivalent. Property 5 finally is evident. □

2.3.4 Proposition *An approach space is metric if and only if for all $A, B \in 2^X$*

$$\inf_{a\in A} \delta(a, B) = \inf_{b\in B} \delta(b, A).$$

Proof This follows from the foregoing result and the symmetry of metrics. □

The above results clarify our terminology of local distances. In a quasi-metric approach space the basic local distance at a point x is simply the quasi-metric "localized" at x.

The pretopological closures $\mathfrak{t}_{\varepsilon}^{d}$, $\varepsilon \in \mathbb{R}^+$, are referred to, in Beer and Luchetti (1993), as (ε)-enlargement operators.

Making use of the results of this section it is possible to formulate some of the transitions in a more concise way.

2.3.5 Proposition *The following formulas for various transitions hold.*

1. *The transition from distances to gauges:* $\mathcal{G} = \{d \in q\mathsf{Met}(X) \mid \delta_d \leq \delta\}$.
2. *The transition from gauges to distances:* $\delta = \sup_{d \in \mathcal{G}} \delta_d$.
3. *The transition from gauges to lower hull operators:* $\mathfrak{l} = \sup_{d \in \mathcal{G}} \mathfrak{l}_d$.
4. *The transition from gauges to upper hull operators:* $\mathfrak{u} = \inf_{d \in \mathcal{G}} \mathfrak{u}_d$.
5. *The transition from gauges to limit operators:* $\lambda = \sup_{d \in \mathcal{G}} \lambda_d$.
6. *The transition from gauges to adherence operators:* $\alpha = \sup_{d \in \mathcal{G}} \alpha_d$.

Proof This follows from 2.3.1 and all the respective transition formulas proved in the first chapter. □

2.3.6 Proposition *If* (X, d) *is a quasi-metric space,* $\mathcal{F}$ *is a filter on* X *and* $x \in X$, *then the following properties hold.*

1. $\mathcal{F} \rightsquigarrow x$ *in* $(X, \mathcal{T}_d)$ *if and only if* $\alpha_d \mathcal{F}(x) = 0$.
2. $\mathcal{F} \to x$ *in* $(X, \mathcal{T}_d)$ *if and only if* $\lambda_d \mathcal{F}(x) = 0$.

Proof This follows from 2.3.1. □

2.4 Embedding qMet in App

In 2.3 we have seen that quasi-metric spaces can be viewed as special types of approach spaces. That qMet is concretely embedded in App is a consequence of the fact that given quasi-metric spaces (X, d) and (X', d') a function $f : X \longrightarrow X'$ will be nonexpansive between the quasi-metric spaces if and only if it is a contraction between the associated approach spaces. Hence the concrete functor from qMet to App which takes (X, d) to (X, δ_d) is a full embedding of qMet into App.

In 2.2 we were able to show both concrete reflectivity and coreflectivity of the embedding of Top in App. For Met and qMet only concrete coreflectivity of the embedding holds. However, as we will see later, it is precisely the fact that neither Met nor qMet is embedded reflectively in App which makes the theory of approach spaces especially interesting.

2.4.1 Theorem *q*Met *is embedded as a concretely coreflective subcategory of* App. *For any approach space* (X, δ)*, its q*Met*-coreflection is determined by the distance* δ^{qm} *associated with the quasi-metric*

$$d_\delta : X \times X \longrightarrow \mathbb{P} : (x, y) \mapsto \delta(x, \{y\}).$$

Proof To show that $1_X : (X, \delta^{qm}) \longrightarrow (X, \delta)$ is a contraction let $x \in X$ and let $A \subseteq X$. Then we have

$$\delta(x, A) \leq \inf_{a \in A} \delta(x, \{a\}) = \delta^{qm}(x, A).$$

Now suppose that (Y, d) is a quasi-metric space and that

$$f : (Y, \delta_d) \longrightarrow (X, \delta)$$

is a contraction. Then, for any $x \in Y$ and $A \subseteq Y$, we have

$$\begin{aligned} \delta^{qm}(f(x), f(A)) &= \inf_{a \in A} \delta(f(x), \{f(a)\}) \\ &\leq \inf_{a \in A} \delta_d(x, \{a\}) \\ &= \delta_d(x, A), \end{aligned}$$

which proves that

$$f : (Y, \delta_d) \longrightarrow (X, \delta^{qm})$$

is also a contraction. □

2.4.2 Corollary *q*Met *is closed under the formation of colimits and final structures in* App. *In particular, a coproduct in* App *of a family of quasi-metric approach spaces is a quasi-metric approach space and, likewise, a quotient in* App *of a quasi-metric approach space is a quasi-metric approdch space.*

In the following result we describe the *q*Met-coreflection of an approach space by means of approach systems and gauges. This result is the counterpart of 2.2.6.

2.4.3 Proposition *If* (X, δ) *is an approach space with* $(\mathcal{B}(x))_{x \in X}$ *a basis for the approach system and* $\mathcal{H}$ *a basis for the gauge, then for any* $x, y \in X$: $d_\delta(x, y) = \sup_{\varphi \in \mathcal{B}(x)} \varphi(y) = \sup_{d \in \mathcal{H}} d(x, y)$.

Proof The first equality follows from 1.2.34 and 2.4.1 while the second one follows from 1.2.6 and 2.4.1. □

Given a quasi-metric d on a set X we call d^-, defined by

$$d^-(x, y) := d(y, x)$$

the *adjoint quasi-metric*. Further we put $d^* := d \vee d^-$.

2.4.4 Theorem Met *is embedded as a concretely coreflective subcategory of* App. *For any approach space* (X, δ), *its* Met*-coreflection is determined by the distance* δ^m *associated with the metric*

$$d_\delta^* : X \times X \longrightarrow \mathbb{P} : (x, y) \mapsto d_\delta(x, y) \vee d_\delta^-(x, y).$$

Proof This is analogous to 2.4.1 and we leave this to the reader. □

2.4.5 Corollary Met *is closed under the formation of colimits and final structures in* App. *In particular, a coproduct in* App *of a family of metric approach spaces is a metric approach space and, likewise, a quotient in* App *of a metric approach space is a metric approach space.*

The description of the Met-coreflection of an approach space by means of a basis for the approach system or a basis for the gauge is easily deduced from 2.4.3 and 2.4.4. For instance, if $\mathscr{H}$ is a basis for the gauge associated with δ, then $d_\delta^*(x, y) = \sup_{d \in \mathscr{H}} d(x, y) \vee d(y, x)$, for all $x, y \in X$.

2.4.6 Example Again we refer to the examples which we considered in 1.2.62.

The distance of the first example is $\delta_\mathbb{E} : \mathbb{P} \times 2^\mathbb{P} \longrightarrow \mathbb{P}$, and both the qMet-coreflection and the Met-coreflection of this space are given by $(\mathbb{P}, d_\mathbb{E})$ where $d_\mathbb{E}$ is the "Euclidean" metric on $\mathbb{P}$, i.e. for all $x, y \in \mathbb{P}$

$$d_\mathbb{E}(x, y) := |x - y|.$$

The distance of the second example is $\delta_\mathbb{P} : \mathbb{P} \times 2^\mathbb{P} \longrightarrow \mathbb{P}$, and the qMet-coreflection of this space is $(\mathbb{P}, d_\mathbb{P})$, where for all $x, y \in \mathbb{P}$

$$d_\mathbb{P}(x, y) := (x - y) \vee 0$$

and the Met-coreflection is $(\mathbb{P}, d_\mathbb{E})$.

2.4.7 Proposition *If* (X, δ) *is an approach space, then for any* $\xi \in \mathfrak{L}$,

$$\xi : (X, d_\delta) \longrightarrow (\mathbb{P}, d_\mathbb{P})$$

is a nonexpansive map. In particular distance functionals, limits and adherences are nonexpansive maps.

Proof This follows from 2.4.1, and 2.4.6. □

From 2.3.1 it follows at once that, given a quasi-metric space (X, d), $\xi \in \mathfrak{L}_d$ if and only if the function $\xi : (X, d) \longrightarrow (\mathbb{P}, d_{\mathbb{P}})$ is nonexpansive, which means that for quasi-metric approach spaces the condition given in 2.4.7 is both necessary and sufficient. This also implies that, for any $\mu \in \mathbb{P}^X$, the lower hull $\mathfrak{l}_d(\mu)$, as given in 2.3.1, can be described as the largest nonexpansive function smaller than μ. This alternative characterization is well known in the classical situation (i.e. when considering only real-valued functions), and can for instance be found in Singer (1986).

2.4.8 Proposition *If (X, d) is a quasi-metric space, then the following properties hold.*

1. *The* Top*-coreflection of (X, δ_d) is $(X, \delta_{\mathcal{T}_d})$, where $\mathcal{T}_d$ is the topology generated by d.*
2. *The* Met*-coreflection of (X, δ_d) is (X, δ_{d^*}).*

Proof 1. This follows from the second property in 2.2.6. This result indeed implies that the Top-coreflection of (X, δ_d) has as neighbourhood system

$$\mathcal{V}(x) = \left\{V \in 2^X \mid \exists \varepsilon > 0 : B_d(x, \varepsilon) \subseteq V\right\}.$$

2. This follows from 2.4.4. □

2.4.9 Proposition *(q*Met*) If X and X' are quasi-metric approach spaces and $f : X \longrightarrow X'$ is a map, then the following properties hold.*

1. *f is open expansive if and only if for all $x \in X$ and $y \in X'$:*

$$\inf_{z \in f^{-1}(y)} d(x, z) \leq d'(f(x), y).$$

2. *f is closed expansive if and only if for all $x \in X$ and $y \in X'$:*

$$\inf_{z \in f^{-1}(y)} d(z, x) \leq d'(y, f(x)).$$

Proof This follow from 1.4.5 and the definition of δ_d for a quasi-metric d. □

2.4.10 Corollary (Met) *If X and X' are metric approach spaces and $f : X \longrightarrow X'$ is a map, then f is closed-expansive if and only if it is open-expansive.*

2.4.11 Example 1. The situation with closed- and open-expansiveness is quite different in the metric case when compared to the topological case. For instance, whereas a projection $\mathbb{R}^2 \longrightarrow \mathbb{R}$ is open but not closed in the topological sense when both spaces are equipped with their usual Euclidean topologies, it is both closed-expansive and open-expansive when both spaces are equipped with their usual Euclidean metrics.

2. Let $X := \{a, b_1, b_2\}$ and $X' := \{c_1, c_2\}$ with quasi-metrics defined by

$$d(b_1, a) = d(b_1, b_2) = d(b_2, b_1) = 1, d(b_2, a) = 2, d(a, b_1) = d(a, b_2) = 0$$

and

$$d(c_1, c_2) = 1, d(c_2, c_1) = 0$$

and let $f : X \longrightarrow X'$ be the function defined as $f(a) = c_2, f(b_1) = f(b_2) = c_1$. Then it is easily verified that f is closed-expansive but not open-expansive. Replacing both quasi-metrics by their adjoints (see 3.1, $d^-(x, y) := d(y, x)$) gives an example of a function which is open-expansive but not closed-expansive.

2.4.12 Example 1. If $(X, \delta_{\mathcal{T}})$ is a topological approach space then the qMet-coreflection is given by (X, δ_{d_1}), where d_1 is the quasi-metric

$$d_1(x, y) := \begin{cases} 0 & x \in \mathrm{cl}_{\mathcal{T}}\{y\}, \\ \infty & x \notin \mathrm{cl}_{\mathcal{T}}\{y\}, \end{cases}$$

and the Met-coreflection is given by (X, δ_{d_0}), where d_0 is the metric

$$d_0(x, y) := \begin{cases} 0 & x \in \mathrm{cl}_{\mathcal{T}}\{y\} \text{ and } y \in \mathrm{cl}_{\mathcal{T}}\{x\}, \\ \infty & x \notin \mathrm{cl}_{\mathcal{T}}\{y\} \text{ or } y \notin \mathrm{cl}_{\mathcal{T}}\{x\}. \end{cases}$$

Hence we can deduce that $(X, \mathcal{T})$ is T_1 if and only if d_1 is a separated metric and that it is T_0 if and only if d_0 is a separated metric.

2. An object in App is at the same time topological and quasi-metric if and only if it is a finitely generated topological space. (Recall that a topological space is said to be finitely generated if the closure is entirely determined by the closures of the singletons in the sense that a point will be in the closure of a set if and only if it is in the closure of a point of the set.)

3. An object in App is at the same time topological and metric if and only if it is a coproduct of indiscrete topological spaces.

4. An object in App is at the same time topological and separated metric if and only if it is discrete.

We now return to the initially dense objects which where found by Claes in (2009). In that paper the research was performed in the setting of metrically generated theories, and the results are far more general that what we require. Hence we will give short proofs restricted to our case (see also Colebunders et al. 2011). We consider the same underlying set as $\mathbb{P}$ but now equipped with the following distances

$$\mathbb{P} \times 2^{\mathbb{P}} \longrightarrow \mathbb{P} : (x, A) \mapsto \inf_{a \in A} a \ominus x,$$

and

$$\mathbb{P} \times 2^{\mathbb{P}} \longrightarrow \mathbb{P} : (x, A) \mapsto \inf_{a \in A} x \ominus a.$$

Note that both distances are quasi-metric, precisely, the first distance is nothing else but $\delta_{d_{\mathbb{P}}^-}$ and the second distance is $\delta_{d_{\mathbb{P}}}$.

2.4.13 Theorem *Both* $(\mathbb{P}, \delta_{d_{\mathbb{P}}^-})$ *and* $(\mathbb{P}, \delta_{d_{\mathbb{P}}})$ *are initially dense objects in* App.

Proof Since we already know one initially dense object, namely $(\mathbb{P}, \delta_{\mathbb{P}})$, it suffices to show that we can obtain that object via initial sources from either of the two objects above. We recall (see 1.2.62) that a gauge basis for $(\mathbb{P}, \delta_{\mathbb{P}})$ is given by the family $\{d_\alpha \mid \alpha \in \mathbb{R}^+\}$ where $d_\alpha(x, y) = (x \wedge \alpha) \ominus (y \wedge \alpha)$.

For the first we consider the following source:

$$(f_\alpha : (\mathbb{P}, \delta_{\mathbb{P}}) \longrightarrow (\mathbb{P}, \delta_{d_{\mathbb{P}}^-}) : x \mapsto \alpha \ominus x)_{\alpha \in \mathbb{R}^+}$$

then the equality $f_\alpha(y) \ominus f_\alpha(x) = d_\alpha(x, y)$ holds because if $x \leq y$ and $\alpha < y < x$ both sides are zero, if $y < x$ and $y \leq \alpha \leq x$ both sides are equal to $\alpha - y$ and if $y < x \leq \alpha$ both sides are equal to $x - y$. Hence for any $\alpha \in \mathbb{R}^+$ we have $d_{\mathbb{P}}^- \circ (f_\alpha \times f_\alpha) = d_\alpha$ which shows that this first source is initial.

For the second we consider the source:

$$(g_\alpha : (\mathbb{P}, \delta_{\mathbb{P}}) \longrightarrow (\mathbb{P}, \delta_{d_{\mathbb{P}}}) : x \mapsto x \wedge \alpha)_{\alpha \in \mathbb{R}^+}$$

then here too it follows that for any $\alpha \in \mathbb{R}^+$ we have $d_{\mathbb{P}} \circ (g_\alpha \times g_\alpha) = d_\alpha$, which shows that this source too is initial. □

2.4.14 Corollary App *is the epireflective hull of* qMet *in* App.

2.4.15 Theorem *If* (X, δ) *is an approach space and we put* $J := \mathbb{R}^+ \times 2^X$, *then*

$$((X, \delta) \longrightarrow (\mathbb{P}, \delta_{d_{\mathbb{P}}}) : x \mapsto \delta(x, A) \wedge \alpha)_{(\alpha, A) \in J}$$

and

$$((X, \delta) \longrightarrow (\mathbb{P}, \delta_{d_{\mathbb{P}}^-}) : x \mapsto \alpha \ominus \delta(x, A))_{(\alpha, A) \in J}$$

are initial sources.

Proof This follows from the combination of the initial sources in 1.3.19 and in 2.4.13. □

We know that Top consists precisely of subspaces of products (in Top) of quasi-metrizable topological spaces, Herrlich (1968). This is strengthened in the approach case.

In the following theorem we use the notation of 1.2.5. This means that, given an approach space (X, δ), $Z \subseteq X$, and $\zeta \in \mathbb{R}^+$, we consider the quasi-metric d_Z^ζ which, for all $x, y \in X$, is given by

$$d_Z^\zeta(x, y) = (\delta(x, Z) \wedge \zeta) \ominus (\delta(y, Z) \wedge \zeta).$$

We will prove the following result by a straightforward calculation of the distances involved.

2.4.16 Theorem *If X is an approach space and we put $J := \mathbb{R}^+ \times 2^X$, then*

$$\psi : (X, \delta) \longrightarrow (X^J, \prod_{(\zeta,Z)\in J} \delta_{d_Z^\zeta}) : x \mapsto (x_{(\zeta,Z)} := x)_{(\zeta,Z)\in J}$$

is an embedding.

Proof Let us put $\delta^* := \prod_{(\zeta,Z)\in J} \delta_{d_Z^\zeta}$, the product distance on X^J. Let $x \in X$ and $A \subseteq X$. Then, making use of 1.3.11, on the one hand we have

$$\begin{aligned}
\delta^*(\psi(x), \psi(A)) &= \sup_{\mathscr{Z}\in 2^{(2^X)}} \sup_{\zeta\in\mathbb{R}^+} \inf_{a\in A} \sup_{Z\in\mathscr{Z}} d_Z^\zeta(x, a) \\
&= \sup_{\mathscr{Z}\in 2^{(2^X)}} \sup_{\zeta\in\mathbb{R}^+} \sup_{\varphi\in\mathscr{Z}^A} \inf_{a\in A} d_{\varphi(a)}^\zeta(x, a) \\
&= \sup_{\mathscr{Z}\in 2^{(2^X)}} \sup_{\zeta\in\mathbb{R}^+} \sup_{\varphi\in\mathscr{Z}^A} \inf_{Z\in\mathscr{Z}} \inf_{a\in\varphi^{-1}(Z)} d_Z^\zeta(x, a) \\
&= \sup_{\mathscr{Z}\in 2^{(2^X)}} \sup_{\zeta\in\mathbb{R}^+} \sup_{\varphi\in\mathscr{Z}^A} \inf_{Z\in\mathscr{Z}} (\delta(x, Z) \wedge \zeta) \ominus (\sup_{a\in\varphi^{-1}(Z)} \delta(a, Z) \wedge \zeta) \\
&\leq \sup_{\mathscr{Z}\in 2^{(2^X)}} \sup_{\zeta\in\mathbb{R}^+} \sup_{\varphi\in\mathscr{Z}^A} \inf_{Z\in\mathscr{Z}} \delta(x, \varphi^{-1}(Z)) \\
&= \sup_{\mathscr{Z}\in 2^{(2^X)}} \sup_{\zeta\in\mathbb{R}^+} \sup_{\varphi\in\mathscr{Z}^A} \delta(x, A) = \delta(x, A).
\end{aligned}$$

On the other hand it follows from 1.2.8 that we have

$$\begin{aligned}
\delta(x, A) &= \sup_{Z\in 2^X} \sup_{\zeta\in\mathbb{R}^+} \inf_{a\in A} d_Z^\zeta(x, a) \\
&\leq \sup_{\mathscr{Z}\in 2^{(2^X)}} \sup_{\zeta\in\mathbb{R}^+} \inf_{a\in A} \sup_{Z\in\mathscr{Z}} d_Z^\zeta(x, a) \\
&= \delta^*(\psi(x), \psi(A)).
\end{aligned}$$

This proves that (X, δ) is indeed embedded in (X^J, δ^*). □

The foregoing results are important for the sequel. A fundamental relationship among the different types of structures which we are considering in this work is that of a topology generated by a metric. As we argued in the introduction, it is the failure of this relation to be well behaved with respect to products in particular and initial structures in general which is one of the motivations for considering approach spaces.

What the foregoing results tell us is that the operation of taking the topology underlying a (quasi-)metric is recaptured in App as a canonical functor, namely the

Top-coreflector restricted to qMet. In the case of a quasi-metric space the Top-coreflector gives us precisely the underlying topological space. It is natural therefore to extend this interpretation to the whole of App and given an arbitrary approach space (X, δ), we will speak of (X, δ^{tc}) or $(X, \mathcal{T}_\delta)$ (the Top-coreflection of (X, δ)) as the *underlying topological approach space* and of the topology $\mathcal{T}_\delta$ as the *topology underlying* δ or the topology *generated by* δ. The situation is clarified in the following commutative diagram.

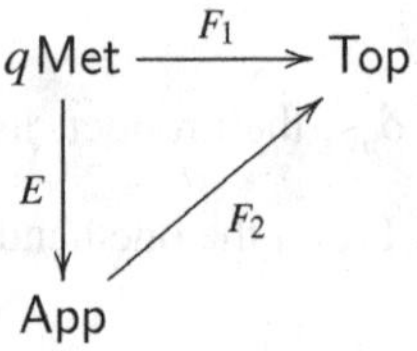

The functor E is the embedding of qMet in App, F_1 is the forgetful functor associating with each quasi-metric space its underlying topological space, and F_2 is the Top-coreflector. The diagram commutes and F_2 thus is an extension of F_1.

Although it is a fundamental aspect of the theory of approach spaces that qMet and, especially also, Met are not epireflectively embedded in App, for the restricted case of subspaces we do have the following result.

2.4.17 Theorem qMet *and* Met *are closed under the formation of subspaces in* App.

Proof This follows from the definitions. □

Referring to the foregoing diagram we can now further point out that the problem of the non-(quasi-)metrizability, in general, of initial topologies of (quasi-)metric topological spaces gets completely resolved in the setting of approach spaces. Consider the source in Top

$$(f_i : X \longrightarrow (X_i, \mathcal{T}_{d_i}))_{i \in I}$$

where d_i is a (quasi-)metric on X_i for each $i \in I$. The initial topology of this source is in general not (quasi-)metrizable. However, it is sufficient to embed the (quasi-) metric spaces (X_i, d_i) in App, there to consider the source

$$(f_i : X \longrightarrow (X_i, \delta_{d_i}))_{i \in I}$$

and then to take the initial approach structure on X and finally to apply the Top-coreflection to this initial structure. As coreflections preserve initial structures, that topological coreflection will be exactly the initial topology, which is generated, not by a (quasi-)metric but by an approach structure.

In the following diagram we give an overview of the categorical situation.

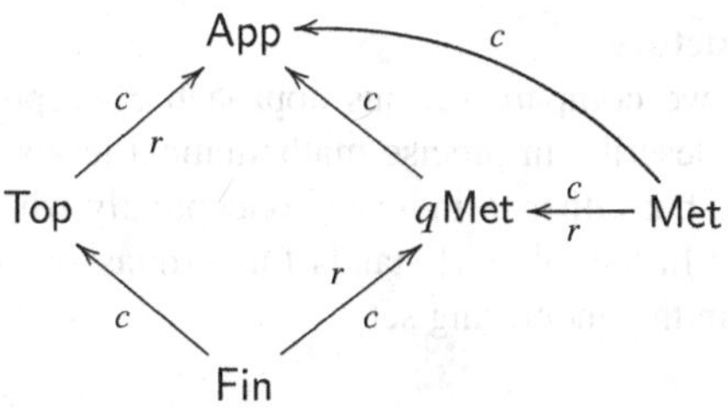

Both qMet and Met are concretely coreflectively embedded in App. Initial structures can be taken in App rather than in either qMet or Met (both of which do indeed have initial structures, neither being compatible with the initial structures of the underlying topologies) and then the coreflection to Top can be applied. The concept of a metric simply is too restricted, what is required to resolve the incompatibility is precisely the category of approach spaces.

Now further note that in the diagram Fin stands for the category of finitely generated topological spaces. This category is not only a coreflective subcategory of Top but also of qMet. Given a finitely generated space it suffices to define a quasi-metric by

$$d(x, y) := \begin{cases} 0 & x \in \mathrm{cl}\{y\}, \\ \infty & x \notin \mathrm{cl}\{y\}. \end{cases}$$

Thus, within App it turns out that Fin is precisely the intersection of Top and qMet. In a certain sense, with regard to Top, Fin therefore plays the role that qMet plays with regard to App. A metric distance from x to A is completely determined by the distances between x and the points of A, and analogously for a finitely generated topology, whether x is in the closure of A or not, is entirely determined by whether x is in the closure of any of the points of A. Referring to the diagram in the introduction we see that App fills in the place of the first question mark (Fig. 2.1).

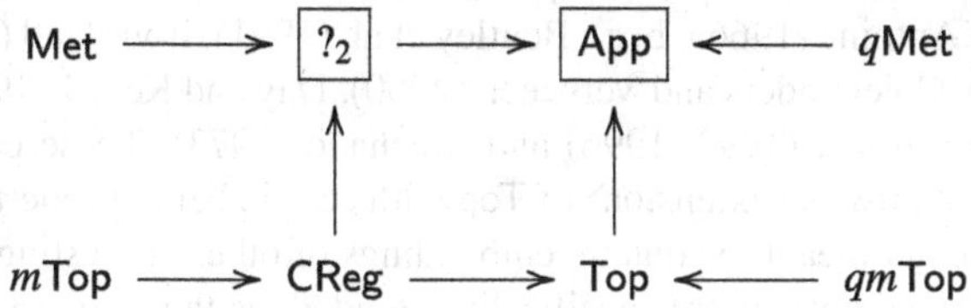

Fig. 2.1 The categorical position of App with regard to Top and (q)Met

2.5 Comments

1. **Comparison of structures**

In the table below we compare various approach concepts in Top and qMet. The purpose is not to describe in precise mathematical terms what these different approach concepts are, but rather to indicate conceptually what are the basic ideas behind these concepts. In the table, d stands for a (quasi-)metric, $\mathcal{F}$ stands for a filter, and x is a point in the underlying set.

Concept in App	**Basic** Top-**analog**	**Basic** qMet-**analog**
Distance	Closure operator	Point-set distance
Adherence of $\mathcal{F}$	Adherence points of $\mathcal{F}$	$\liminf d(\mathcal{F}, \cdot)$
Limit of $\mathcal{F}$	Limit points of $\mathcal{F}$	$\limsup d(\mathcal{F}, \cdot)$
Approach system	Neighbourhoods	Localized quasi-metrics $d(x, \cdot)$
Gauge	Quasi-metrics determining coarser topologies	Quasi-metrics smaller than d
Tower	Closure operator	Enlargement operators
Lower hull operator	Lower semicontinuous regularization	Nonexpansive regularization
Lower regular function frame	Lower semicontinuous $\mathbb{P}$-valued functions	Nonexpansive $\mathbb{P}$-valued functions
Upper hull operator	Upper semicontinuous regularization	Nonexpansive regularization
Upper regular function frame	Upper semicontinuous $\mathbb{P}$-valued functions	Nonexpansive $\mathbb{P}$-valued functions
Contraction	Continuous map	Nonexpansive map

2. **Supercategories of** Top

There are many topological categories wherein Top is embedded as a full subcategory in a more or less nice way. Some of these categories are intended for their better categorical properties such as the category PrTop of pretopological spaces which is extensional (see Herrlich 1987, 1988a, b) and the category PsTop of pseudotopological spaces which is a quasi-topos or topological universe (see Herrlich et al. 1991). We also refer to Antoine (1966a, b, c), Bentley et al. (1991), Bourdaud (1975, 1976), Choquet (1947), Colebunders and Verbeeck (2000), Day and Kelley (1970), Lowen-Colebunders and Sonck (1993, 1996) and Machado (1973). These categories are basically smallest possible extensions of Top with certain better properties, and they do not, and were not meant to contain, embeddings of other interesting categories.

Some categories however are specifically intended, as the category of approach spaces, to merge two familiar and somewhat related categories in one supercategory.

Another typical such example is the category of *nearness spaces* as introduced by Herrlich (1974a). Nearness spaces constitute a supercategory of the categories Unif of uniform spaces and R_0Top of R_0-topological spaces. For further information we refer to Bentley et al. (1998), Császár (1963) (for related concepts), Herrlich (1974b), Herrlich et al. (1991) and Hušek (1964a, b).

A historical overview can be found in "Handbook of the History of General Topology" Volume 3 (edit: Aull and Lowen 2001) in the articles "Supercategories of Top and the inevitable emergence of topological constructs" by Colebunders and Lowen and "The historical development of uniform, proximal and nearness concepts in topology" by Bentley, Herrlich and Hušek.

3. **Limit operators and approximation theory**

The formula given in 2.3.2 for the limit operator of a sequence in a metric space is well known in approximation theory. It was introduced in this field in 1972 by Edelstein in (1972) and was later, in 1980, generalized for nets by Lim (1980) (see also Amir et al. 1982; Benyamini 1985; Lami Dozo 1981; Liu 2001). The setting there was mainly restricted to bounded sequences or nets in closed convex subsets of a Banach space E and the main interest was to find a point where the limit operator would be minimal. Such a point is called an *asymptotic center*, i.e. a point x where

$$\lambda_d \langle (x_n)_n \rangle (x) = \min\{\lambda_d \langle (x_n)_n \rangle (y) \mid y \in E\}$$

and the value of the limit operator at such an asymptotic center is called the *asymptotic radius*, i.e. $\inf_{x \in X} \lambda_d \langle (x_n)_n \rangle (x)$. Since an asymptotic center, if it exists, need not be unique, the term asymptotic center is also used for the set of all asymptotic centers in the sense of the first definition, i.e.

$$\{x \mid \lambda_d \langle (x_n)_n \rangle (x) = \min_{y \in E} \lambda_d \langle (x_n)_n \rangle (y)\}.$$

As an example, consider the real line $\mathbb{R}$ with the usual Euclidean metric and topology, and consider the sequences $(x_n)_n$ and $(y_n)_n$ where

$$x_n := \begin{cases} \varepsilon & n \text{ even} \\ -\varepsilon & n \text{ odd,} \end{cases} \quad \text{and } y_n := \begin{cases} n & n \text{ even} \\ -n & n \text{ odd.} \end{cases}$$

Neither of these sequences converges. The first one, however, has two main convergent subsequences, and from the point of view of numerical analysis or approximation theory, for a "sufficiently small" ε, the sequence itself might actually be considered "sufficiently" convergent, e.g. to 0. The second sequence on the other hand has no convergent subsequences, and could not even remotely be considered to be "approximately convergent" to any point of $\mathbb{R}$.

A more striking example is obtained as follows. Let $\varphi : \mathbb{R} \longrightarrow \,]-\varepsilon, \varepsilon[$ be a homeomorphism, and let $(r_n)_n$ be an enumeration of the rationals. The sequence $(r_n)_n$ is not remotely "approximately convergent" to any point of $\mathbb{R}$. The sequence $(\varphi(r_n))_n$ on the other hand, for a "sufficiently small" ε, might again be considered "sufficiently" convergent.

By means of the topology of $\mathbb{R}$, not only can we not detect the different behaviour of these sequences, we must conclude that they are "identical". In order to "see" the difference we require the metric and the concepts of limit and adherence operator.

Chapter 3
Approach Invariants

Indeed, in mathematics it seems as though the invariants are the most beautiful and elegant constructs. Invariants are unchanging by definition, and descriptive by nature, but most importantly they are the twisting trunks of magnificent trees from which all of mathematics can blossom forth.

(Matthew Strauss)

In this chapter we will study some natural invariants, or properties, of approach spaces. In view of the fact that approach spaces generalize at the same time topological spaces and metric spaces, there will be properties which are more of a topological nature while others are more of a metric nature. In the general setting of topological categories several concepts, inspired by topological properties, have been defined by Marny (1979), Preuss (1987) and Clementino et al. (2003). We will not systematically try to generalize all possible concepts in topological and metric spaces to the setting of approach spaces but rather concentrate ourselves on those concepts which demonstrate some interesting aspects, appear to live naturally in the realm of approach theory or which are required, in particular, for the applications later on.

Uniformity and symmetry refer to a concept which is at the same time topological and metric in nature. From the topological point of view it is the generalization of complete regularity and from the metric point of view it relates to the difference between quasi-metrics and metrics.

Weak adjointness is a concept which is formulated in terms of the quasi-metrics in the gauge of a space and is basically a pure approach concept which does not have an immediate counterpart in topology but which is known in the theory in quasi-metric spaces.

Some lower separation properties, namely T_0, T_1 and T_2 are shown to be purely topological in nature. Regularity however will turn out to be a very interesting concept with various characterizations and allowing also for an extension theorem.

There are various countability properties related respectively to approach systems, lower or upper regular function frames and gauges. The one referring to approach systems is the logical counterpart of first countability in topology and the one referring

R. Lowen, *Index Analysis*, Springer Monographs in Mathematics,
DOI 10.1007/978-1-4471-6485-2_3

to lower or upper regular function frames is the counterpart of second countability in topology. The one referring to gauges is a pure approach concept.

The last section in this chapter deals with the notion of completeness which clearly is a metric-like concept.

Some evidently important concepts are missing from this list for reasons which will become clear in the chapter on index analysis.

3.1 Uniformity and Symmetry

The epireflective hull of Met in App consists of all subspaces of products of metric spaces. We will show that it is precisely this epireflective hull which fills in the position of the second question mark in the diagram of the introduction. Actually, there is a good reason why this is plausible at this point, namely the fact that Creg (the full subcategory of Top with objects all completely regular spaces) consists precisely of subspaces of products (in Top) of metrizable topological spaces.

We first recall products of metric spaces in App (see also 2.4.16).

3.1.1 Theorem *If $(X_j, d_j)_{j\in J}$ is an arbitrary family of metric spaces, then the product of the family $(X_j, \delta_{d_j})_{j\in J}$ (in* App*) is the approach space*

$$(\prod_{j\in J} X_j, \prod_{j\in J} \delta_{d_j})$$

where $\prod_{j\in J} \delta_{d_j}$ is the distance generated by the gauge basis

$$\mathscr{H} = \left\{ \sup_{k\in K} d_k \circ (\mathrm{pr}_k \times \mathrm{pr}_k) \mid K \subseteq J \text{ finite} \right\}.$$

In particular, if we put $X := \prod_{j\in J} X_j$ and $\delta := \prod_{j\in J} \delta_{d_j}$ then this distance is given by

$$\delta(x, A) = \sup_{K\in 2^{(J)}} \inf_{a\in A} \sup_{k\in K} d_k(\mathrm{pr}_k(x), \mathrm{pr}_k(a)).$$

Proof This follows from the description of initial structures in 1.3.11. □

It is a fundamental aspect of the theory of approach spaces that this product, in general, is not a metric space. More precisely, in general, there does not exist a metric d on X such that $\delta = \delta_d$, except for finite products, as the following result shows.

3.1.2 Proposition *If $(X_j, d_j)_{j\in J}$ is a finite family of metric spaces, then the product of the family $(X_j, \delta_{d_j})_{j\in J}$ (in* App*) is a metric space.*

Proof Using the same notations as in the foregoing result, it suffices to note that, for any $x \in X$ and $A \subseteq X$, we have

$$\begin{aligned}\delta(x, A) &= \sup_{K \in 2^{(J)}} \inf_{a \in A} \sup_{k \in K} d_k(\mathrm{pr}_k(x), \mathrm{pr}_k(a)) \\ &= \inf_{a \in A} \sup_{j \in J} d_j(\mathrm{pr}_j(x), \mathrm{pr}_j(a)) \\ &= \delta_d(x, A),\end{aligned}$$

where the metric d is defined by $d := \sup_{j \in J} d_j \circ (\mathrm{pr}_j \times \mathrm{pr}_j)$. □

We now turn to subspaces of products of metric spaces in App. The internal characterizations which follow will be important for our further considerations. Before giving these characterizations, however, we need to introduce some more concepts and terminology.

3.1.3 Definition A gauge is called a *symmetric gauge* if it has a basis consisting of metrics. A gauge basis $\mathscr{H}$ consisting only of metrics is called a *symmetric gauge basis*.

3.1.4 Proposition *If X is an approach space with symmetric gauge $\mathscr{G}$, then $\mathscr{G} \cap \mathsf{Met}(X)$ is the largest basis for $\mathscr{G}$ consisting of metrics.*

Proof If $\mathscr{H}$ is a basis for $\mathscr{G}$ consisting only of metrics, then clearly $\mathscr{H} \subseteq \mathscr{G} \cap \mathsf{Met}(X)$ and hence $\mathscr{G} \cap \mathsf{Met}(X)$ too is a basis for $\mathscr{G}$. That it is the largest basis consisting of metrics is evident. □

The foregoing result says that, if $\mathscr{G}$ is a symmetric gauge, then $\mathscr{G} \cap \mathsf{Met}(X)$ is its largest symmetric basis. Given a symmetric gauge basis, it is sometimes interesting to be able to work not with all the members of the gauge generated by it, but only with the metrics in it. Thus for a symmetric gauge basis $\mathscr{H} \subseteq \mathsf{Met}(X)$ we may sometimes consider the set

$$\{d \in \mathsf{Met}(X) \mid \mathscr{H} \text{dominates } d\} = \widehat{\mathscr{H}} \cap \mathsf{Met}(X).$$

We call this set the *symmetric saturation* of $\mathscr{H}$ and we denote it by $\mathscr{H}^s$. Note that a symmetric gauge basis $\mathscr{H}$ can thus be saturated in two ways: the saturation $\widehat{\mathscr{H}}$ taken in $q\mathsf{Met}(X)$ and the symmetric saturation $\mathscr{H}^s = \widehat{\mathscr{H}} \cap \mathsf{Met}(X)$ taken in $\mathsf{Met}(X)$. If $\mathscr{G}$ is a gauge then $\mathscr{G}^s$ simply consists of all metrics in $\mathscr{G}$. Hence if δ is the distance associated with $\mathscr{G}$ then it follows from 1.2.4 that $\mathscr{G}^s = \{d \in \mathsf{Met}(X) \mid \delta_d \leq \delta\}$.

3.1.5 Proposition *If X is an approach space with gauge $\mathscr{G}$, then the following properties are equivalent.*

1. *X is a subspace of a product of metric spaces in* App.
2. *There exists a symmetric gauge basis for $\mathscr{G}$.*

3. $\mathscr{G}^s$ *is a basis for* $\mathscr{G}$.
4. $\mathscr{G}$ *is a symmetric gauge.*

Proof 1 $\Rightarrow$ 2. Let $(X_j, d_j)_{j\in J}$ be an arbitrary family of metric spaces, and suppose that X is a subspace of the product $(\prod_{j\in J} X_j, \prod_{j\in J} \delta_{d_j})$. Then it follows from 3.1.1 that

$$\mathscr{H} := \left\{ \sup_{k\in K} d_k \circ (\mathrm{pr}_k \times \mathrm{pr}_k)_{|X\times X} \mid K \subseteq J \text{ finite} \right\}$$

is a symmetric gauge basis for $\mathscr{G}$.

2 $\Leftrightarrow$ 3. This follows from 3.1.4.

3 $\Rightarrow$ 4. This follows from the definition.

4 $\Rightarrow$ 1. Let $\mathscr{H}$ be a symmetric gauge basis for $\mathscr{G}$ and consider the diagonal injection

$$\psi : (X, \delta) \longrightarrow (X^{\mathscr{H}}, \prod_{d\in\mathscr{H}} \delta_d) : x \mapsto (x_d := x)_{d\in\mathscr{H}}.$$

Let $\delta^* := \prod_{d\in\mathscr{H}} \delta_d$. For any $x \in X$ and $A \subseteq X$, making use of the fact that $\mathscr{H}$ is a basis for $\mathscr{G}$, on the one hand we then immediately have

$$\begin{aligned}\delta^*(\psi(x), \psi(A)) &= \sup_{\mathscr{D}_0\in 2^{(\mathscr{H})}} \inf_{a\in A} \sup_{d\in\mathscr{D}_0} d(x, a)\\ &\geq \sup_{d\in\mathscr{H}} \inf_{a\in A} d(x, a)\\ &= \delta(x, A),\end{aligned}$$

and on the other hand, since $\mathscr{H}$ is locally directed, for any $\omega < \infty$ and $\varepsilon > 0$

$$\begin{aligned}\delta^*(\psi(x), \psi(A)) \wedge \omega &= \sup_{\mathscr{D}_0\in 2^{(\mathscr{H})}} \inf_{a\in A} \sup_{d\in\mathscr{D}_0} d(x, a) \wedge \omega\\ &\leq \sup_{d\in\mathscr{H}} \inf_{a\in A} d(x, a) + \varepsilon\\ &= \delta(x, A) + \varepsilon,\end{aligned}$$

which proves our claim. □

3.1.6 Definition A subspace of a product of metric approach spaces in App is called a *uniform approach space*. The full subcategory consisting of all these spaces, i.e. the epireflective hull of Met in App, will be denoted by UAp. Structures on uniform approach spaces will also be called uniform, i.e. we will speak of a *uniform distance*, a *uniform approach system*, and so on. As before (see 1.2.61), if a uniform approach space is determined by a particular symmetric gauge basis then we will usually refer to the structures of that approach space as being *generated* by this basis.

Referring to the problem explained under the first item in the introduction, the foregoing results show that UAp is a candidate to fill in the position of the second question mark in the diagram. It follows from the body of results of this chapter that it is the right candidate.

We now give an internal characterization of uniform approach spaces, not unlike the characterization of completely regular topological spaces. We denote by $\mathscr{K}(X)$ (respectively $\mathscr{K}^*(X)$) the set of all contractions (respectively bounded contractions) from (X, δ) to $(\mathbb{R}, \delta_{d_{\mathbb{E}}})$.

3.1.7 Theorem *An approach space X is uniform if and only if for all $x \in X$, $A \subseteq X$, $\varepsilon > 0$ and $\omega < \infty$ there exists $f \in \mathscr{K}(X)$ (respectively $f \in \mathscr{K}^*(X)$) such that*

1. $f(x) = 0$.
2. $f|_A + \varepsilon \geq \delta(x, A) \wedge \omega$.

Proof To show the only-if part, let $\mathscr{H}$ be a symmetric gauge basis for (X, δ) consisting of bounded metrics and let $x \in X$, $A \subseteq X$, $\varepsilon > 0$, and $\omega < \infty$ be fixed. Then it follows from 3.1.5 that there exists $d \in \mathscr{H}$ such that

$$\delta_d(x, A) + \varepsilon \geq \delta(x, A) \wedge \omega.$$

Now, if we put $f := d(x, \cdot)$, then it is immediately clear that $f \in \mathscr{K}^*(X)$ and that it fulfils the stated condition.

To show the if part, for each $f \in \mathscr{K}(X)$, put $d_f := d_{\mathbb{E}} \circ (f \times f)$ and consider the collection of metrics

$$\mathscr{H} := \left\{ \sup_{f \in \mathscr{K}_0} d_f \mid \mathscr{K}_0 \in 2^{(\mathscr{K}(X))} \right\}.$$

It follows at once that d_f is a member of the gauge associated with δ for any $f \in \mathscr{K}(X)$. Since $\mathscr{H}$ is a basis for the initial gauge for the source

$$(f : X \longrightarrow (\mathbb{R}, \delta_{d_{\mathbb{E}}}))_{f \in \mathscr{K}^*(X)}$$

it follows that for any $x \in X$ and $A \subseteq X$

$$\delta(x, A) \geq \sup_{\mathscr{K}_0 \in 2^{\mathscr{K}(X)}} \inf_{a \in A} \sup_{f \in \mathscr{K}_0} d_f(x, a) \geq \sup_{f \in \mathscr{K}(X)} \inf_{a \in A} d_f(x, a).$$

On the other hand if $x \in X$, $A \subseteq X$, $\varepsilon > 0$, and $\omega < \infty$ are fixed and $f \in \mathscr{K}(X)$ is chosen according to the condition stated in the theorem, then it follows that

$$\inf_{a \in A} |f(x) - f(a)| + \varepsilon = \inf_{a \in A} f(a) + \varepsilon \geq \delta(x, A) \wedge \omega$$

which proves that $\sup_{f \in \mathscr{K}(X)} \inf_{a \in A} d_f(x, a) \geq \delta(x, A)$. Hence, for all $x \in X$ and $A \subseteq X$, we have that $\delta(x, A) = \sup_{d \in \mathscr{H}} \inf_{a \in A} d(x, a)$ which by 1.2.9, together with the fact that $\mathscr{H}$ is obviously locally directed, implies that $\mathscr{H}$ is a basis for the gauge associated with δ. Since, moreover, all members of $\mathscr{H}$ are metrics, $\mathscr{H}$ is a symmetric gauge and it follows from 3.1.5 that (X, δ) is a uniform approach space. □

Notice that in the foregoing result it suffices to impose the stated condition for closed sets A and for points x not belonging to A, which emphasizes even more the similarity with complete regularity.

We will say that a set $\mathscr{F} \subseteq \mathscr{K}(X)$ *generates* the structure, or is *generating* if

$$(f : (X, \delta) \longrightarrow (\mathbb{R}, \delta_{d_{\mathbb{E}}}))_{f \in \mathscr{F}}$$

is an initial source.

We then obtain the following further characterizations of uniform approach spaces.

3.1.8 Proposition *For an approach space X the following properties are equivalent.*

1. *X is a uniform approach space, i.e. it is an object in* UAp.
2. $\mathscr{K}^*(X)$ *is generating.*
3. $\mathscr{K}(X)$ *is generating.*
4. *There exists a subset* $\mathscr{F} \subseteq \mathscr{K}(X)$ *which is generating.*
5. *There exists a subset* $\mathscr{F} \subseteq \mathscr{K}^*(X)$ *which is generating.*

Proof 1 ⇒ 2. This follows from 3.1.7.

2 ⇒ 3 ⇒ 4. This is evident.

4 ⇒ 5. If $\mathscr{F} \subseteq \mathscr{K}(X)$ is generating, then it is easily seen that also

$$\mathscr{F}^* := \{(f \wedge \omega) \vee (-\omega) \mid f \in \mathscr{F}, \omega < \infty\}$$

is generating.

5 ⇒ 1. If $\mathscr{F} \subseteq \mathscr{K}^*(X)$ is a set of contractions which is generating, then, as before, the set of metrics

$$\mathscr{D}_F := \left\{ \sup_{f \in \mathscr{H}} d_f \mid \mathscr{H} \in 2^{(\mathscr{F})} \right\},$$

is easily seen to be a symmetric gauge basis and hence X is uniform. □

We will now see how UAp is embedded in App and how it is related to other subcategories of App. Note that by definition we already know that UAp is epireflective in App.

3.1.9 Theorem UAp *is a concretely reflective subcategory of* App. *If* (X, δ) *is an approach space then its* UAp*-reflection is determined by the distance* δ^u *generated by the gauge basis* $\mathscr{G}^s$.

Proof Since indiscrete objects are metric it follows immediately that UAp is concretely reflective in App. That $1_X : (X, \delta) \longrightarrow (X, \delta^u)$ is a contraction, by 1.3.3, follows from the fact that $\mathcal{G}^s \subseteq \mathcal{G}$. Now suppose that (X', δ') is a uniform approach space where δ' is generated by the symmetric gauge $\mathcal{G}'$, and let

$$f : (X, \delta) \longrightarrow (X', \delta')$$

be a contraction. Then, for any $d' \in (\mathcal{G}')^s$, we have $d' \circ (f \times f) \in \mathcal{G}^s$. Again, by 1.3.3, this implies that

$$f : (X, \delta^u) \longrightarrow (X', \delta')$$

is a contraction. □

3.1.10 Corollary UAp *is closed under the formation of limits and initial structures in* App. *In particular, a product in* App *of a family of uniform approach spaces is a uniform approach space and, likewise, a subspace in* App *of a uniform approach space is a uniform approach space.*

3.1.11 Proposition *If X is a uniform approach space with symmetric gauge basis $\mathcal{H}$, then the following properties hold.*

1. *The topological coreflection of X is determined by the topology $\mathcal{T}_\delta$ which has as a basis for the neighbourhoods the collections*

$$\{B_d(x, \varepsilon) \mid d \in \mathcal{H}, \varepsilon > 0\}, \quad x \in X$$

   and as such $\mathcal{T}_\delta$ is a completely regular topology, uniformizable by the uniformity generated by the collection $\mathcal{H}$.
2. *The metric coreflection of X is determined by the metric*

$$d_\delta : X \times X \longrightarrow \mathbb{P} : (x, y) \mapsto \sup_{d \in \mathcal{H}} d(x, y).$$

Proof This follows from 2.2.6 and 2.4.4. □

Note that in the foregoing result the symmetry of $\mathcal{H}$ is only needed to conclude that $\mathcal{T}_\delta$ is completely regular and that d_δ is a metric. The expression for the neighbourhood basis of $\mathcal{T}_\delta$ and the formula for d_δ hold for any gauge basis as we have seen in 2.2.6 and 2.4.3 respectively.

By definition, UAp contains all metric approach spaces. The following result shows that it also contains the right topological approach spaces.

3.1.12 Proposition (Top) *An approach space is at the same time uniform and topological if and only if it is of type $(X, \delta_{\mathcal{T}})$, for some completely regular topology $\mathcal{T}$ on X. In particular,* Creg *is embedded as a full simultaneously concretely reflective and concretely coreflective subcategory of* UAp.

Proof The only-if part follows from 3.1.11. To show the if part, let $\mathcal{T}$ be a completely regular topology on X. It is well known that $(X, \mathcal{T})$ can then be embedded in a product of metrizable topological spaces, say

$$\psi : (X, \mathcal{T}) \longrightarrow (\prod_{j \in J} X_j, \prod_{j \in J} \mathcal{T}_j).$$

It then follows from 2.2.2 that

$$\psi : (X, \delta_{\mathcal{T}}) \longrightarrow (\prod_{j \in J} X_j, \prod_{j \in J} \delta_{\mathcal{T}_j})$$

is also an embedding. For each $j \in J$, let d_j be a metric which metrizes $\mathcal{T}_j$. Now, for each $j \in J$ also, consider the function

$$\theta_j : (X_j, \delta_{\mathcal{T}_j}) \longrightarrow (X_j^{\mathbb{N}}, \prod_{n \in \mathbb{N}} \delta_{nd_j}) : x \mapsto (x_n := x)_{n \in \mathbb{N}}$$

and put $\delta^j := \prod_{n \in \mathbb{N}} \delta_{nd_j}$. For any $x \in X_j$ and $A \subseteq X_j$, we have

$$\begin{aligned}
\delta^j(\theta_j(x), \theta_j(A)) &= \sup_{K \in 2^{(\mathbb{N})}} \inf_{a \in A} \sup_{n \in K} nd_j(x, a) \\
&= \sup_{n \in \mathbb{N}} \inf_{a \in A} nd_j(x, a) \\
&= \sup_{n \in \mathbb{N}} n\delta_{d_j}(x, A) \\
&= \begin{cases} 0 & \delta_{d_j}(x, A) = 0, \\ \infty & \delta_{d_j}(x, A) \neq 0. \end{cases}
\end{aligned}$$

Consequently, for all $j \in J$, θ_j is an embedding. Since a product of embeddings is an embedding, this proves that

$$\varphi : (X, \delta_{\mathcal{T}}) \longrightarrow (\prod_{j \in J} X_j^{\mathbb{N}}, \prod_{j \in J} \prod_{n \in \mathbb{N}} \delta_{nd_j}) : x \mapsto (\theta_j \circ \mathrm{pr}_j \circ \psi(x))_{j \in J}$$

is an embedding.

That Creg is concretely reflective in UAp follows at once from the facts that both Top (see 2.2.2) and UAp (see 3.1.9) are reflectively embedded in App. That Creg is concretely coreflectively embedded in UAp follows from the observation that the Top-coreflection of a UAp-object is completely regular (see 3.1.11). □

3.1.13 Proposition (qMet) *A quasi-metric approach space is uniform if and only if it is a metric.*

Proof This follows from 3.1.5. □

It is well known that Unif is a simultaneously concretely reflective and coreflective subcategory of qUnif and in the same way Met is a simultaneously concretely reflective and coreflective subcategory of qMet. However we know that CReg is not a coreflective subcategory of Top and in spite of the characterization of approach spaces with gauges which has great resemblance to both uniform and metric spaces, neither is UAp a coreflective subcategory of App.

3.1.14 Example Given the fact that a uniform approach space has a symmetric gauge basis, one might be inclined to think that if $d \in \mathcal{G}$ then also $d^* \in \mathcal{G}$. However this is absolutely not the case, and it suffices to look at a completely regular topology. Given a topological space $(X, \mathcal{T})$, for any $G \in \mathcal{T}$ the quasi-metric

$$d_G(x, y) := \begin{cases} 0 & (x, y) \notin G \times (X \setminus G), \\ \infty & (x, y) \in G \times (X \setminus G), \end{cases}$$

belongs to the gauge of the associated approach space. The only balls for this quasi-metric are G and X. However if one considers d_G^* then this metric produces as balls G and $X \setminus G$. This means that the topology generated by the metrics of type d_G^* will contain all closed sets for the topology of d. In particular then this topology will be finer than the original topology and moreover have all original closed sets as open sets.

It is nevertheless interesting to see in general precisely what the structure is that one obtains by considering the set $\mathcal{G}^* := \{d^* \mid d \in \mathcal{G}\}$ as a basis for a gauge. We describe it via the limit operator.

3.1.15 Proposition *If X is an approach space with gauge $\mathcal{G}$ and limit operator λ, then the approach space determined by $\mathcal{G}^*$ has as limit operator*

$$\lambda^* = \lambda \vee \lambda_{d_\lambda^-}$$

where d_λ is the quasi-metric coreflection of the original structure.

Proof Let $\mathcal{F}$ be a filter on X and let $x \in X$. Then

$$\begin{aligned} \lambda^* \mathcal{F}(x) &= \sup_{d \in \mathcal{G}} \inf_{H \in \mathcal{F}} \sup_{y \in H} d^*(x, y) \\ &\leq (\sup_{d \in \mathcal{G}} \inf_{H \in \mathcal{F}} \sup_{y \in H} d(x, y)) \vee (\sup_{d \in \mathcal{G}} \inf_{H \in \mathcal{F}} \sup_{y \in H} d(y, x)) \\ &\leq \lambda \mathcal{F}(x) \vee \lambda_{d_\lambda^-} \mathcal{F}(x). \end{aligned}$$

Now for any $z \in X$ and $\omega < \infty$ put

$$d_z^{\omega}(x, y) := (\lambda \dot{z}(x) \wedge \omega) \ominus (\lambda \dot{z}(y) \wedge \omega)$$

then it follows from 1.2.5 that these quasi-metrics belong to $\mathcal{G}$. Hence we have

$$\begin{aligned}
\lambda_{d_{\lambda}^{-}}\mathcal{F}(x) &= \inf_{H\in\mathcal{F}}\sup_{y\in H} d_{\lambda}(y,x)\\
&= \sup_{\omega<\infty}\inf_{H\in\mathcal{F}}\sup_{y\in H} d_{\lambda}(y,x)\wedge\omega\\
&= \sup_{\omega<\infty}\inf_{H\in\mathcal{F}}\sup_{y\in H} d_{x}^{\omega}(y,x)\\
&\leq \sup_{d\in\mathcal{G}}\inf_{H\in\mathcal{F}}\sup_{y\in H} d(y,x)\\
&\leq \lambda^{*}\mathcal{F}(x).
\end{aligned}$$

Since obviously $\lambda \leq \lambda^{*}$ this concludes the argument. □

An interesting relationship between distance and limit operator in uniform approach spaces is the following.

3.1.16 Proposition *If X is a uniform approach space, $\mathcal{F}$ is a filter on X and $x, y \in X$, then*

$$\delta(x,\{y\}) \leq \lambda\mathcal{F}(x) + \lambda\mathcal{F}(y).$$

Proof Let $x, y \in X$, then

$$\begin{aligned}
\lambda\mathcal{F}(x)+\lambda\mathcal{F}(y) &= \sup_{d\in\mathcal{G}^{s}}\inf_{F\in\mathcal{F}}\sup_{z\in F} d(x,z) + \sup_{d\in\mathcal{G}^{s}}\inf_{F\in\mathcal{F}}\sup_{z\in F} d(y,z)\\
&\geq \sup_{d\in\mathcal{G}^{s}}\inf_{F\in\mathcal{F}}\sup_{z\in F} d(x,y)\\
&= \delta(x,\{y\}).
\end{aligned}$$ □

Uniform spaces were originally introduced by Weil (1937), making use of entourages of the diagonal. We will reserve the term *diagonal uniformity* for such a collection of entourages. For our purposes, in view of the link with distances, the most convenient way to introduce uniformities is via collections of metrics. We recall the definitions (see e.g. Gillman and Jerison 1976).

A set $\mathcal{G}$ of metrics is called a *uniform structure* if it fulfils the following properties.

1. $\mathcal{G}$ is closed under the formation of finite suprema,
2. if e is a metric and if, for any $\varepsilon > 0$, there exists $d \in \mathcal{G}$ and $\delta > 0$ such that for all $x, y \in X$: $d(x, y) \leq \delta \Rightarrow e(x, y) \leq \varepsilon$, then $e \in \mathcal{G}$.

The set $\mathcal{G}$ is sometimes also called a *uniform gauge* (see e.g. Dugundji 1967). However, note that we will use this terminology for another concept in the chapter on uniform gauge spaces. In general terms when we speak of the uniformity of a space we mean both the uniform structure and the diagonal uniformity since one completely determines the other.

Given such a uniform structure, the diagonal uniformity $\mathcal{U}(\mathcal{G})$ derived from $\mathcal{G}$ is given by

$$\mathscr{U}(\mathscr{G}) := \left\{U \in 2^{X \times X} \mid \exists d \in \mathscr{G}, \exists \varepsilon > 0 : \{d < \varepsilon\} \subseteq U\right\}.$$

Any collection of metrics determines a uniform structure, even if it fulfils no conditions at all. It suffices to note that any intersection of uniform structures is again a uniform structure and therefore, given a set $\mathscr{D}$ of metrics, there always is a smallest uniform structure $\mathscr{G}_u(\mathscr{D})$ containing $\mathscr{D}$. Thus any collection of metrics $\mathscr{D}$ also determines a diagonal uniformity $\mathscr{U}(\mathscr{G}_u(\mathscr{D}))$, which we denote by $\mathscr{U}(\mathscr{D})$ for short.

Conversely, given a diagonal uniformity $\mathscr{U}$, the uniform structure derived from $\mathscr{U}$ is given by

$$\mathscr{G}_u(\mathscr{U}) := \left\{d \in \mathbb{P}^{X \times X} \mid d \text{ a uniformly continuous metric}\right\},$$

where uniform continuity is understood with respect to the product uniformity $\mathscr{U} \times \mathscr{U}$ on $X \times X$ and the usual uniformity on $\mathbb{P}$.

The topology underlying a diagonal uniformity $\mathscr{U}$ is denoted by $\mathscr{T}(\mathscr{U})$. This assignment generates a natural forgetful functor from Unif to Creg. This functor, moreover, has an adjoint, called the *fine functor*, which associates with each completely regular topological space the uniform space equipped with the finest uniformity compatible with the given topology. If $\mathscr{T}$ is a completely regular topology on X, then we denote the *fine diagonal uniformity* by $\mathscr{U}_{\text{fine}}(\mathscr{T})$.

Given a uniform approach space (X, δ), we have seen in 3.1.4 that there exists a largest collection of metrics generating δ, namely $\mathscr{G}^s$. This collection in turn determines a diagonal uniformity $\mathscr{U}(\mathscr{G}^s)$, which we will denote by $\mathscr{U}(\delta)$ for short.

We now introduce a number of functors which will help us clarify the relationship between the main categories involved.

The first functor associates with a given uniform space the topological approach space determined by the uniform topology. In fact, via the embedding of Top in App, this is nothing else than the concrete forgetful functor which associates the underlying topological space to a given uniform space,

$$D : \mathsf{Unif} \longrightarrow \mathsf{UAp} : (X, \mathscr{U}) \mapsto (X, \delta_{\mathscr{T}(\mathscr{U})}).$$

The second concrete functor associates with a given uniform approach space the uniform space equipped with the fine uniformity generated by the underlying completely regular topology. Restricted to Creg, this is the fine functor,

$$U_1 : \mathsf{UAp} \longrightarrow \mathsf{Unif} : (X, \delta) \mapsto (X, \mathscr{U}_{\text{fine}}(\mathscr{T}_\delta)).$$

The third concrete functor finally associates with a given uniform approach space the uniform space equipped with the uniformity generated by the largest symmetric gauge basis for the given approach space,

$$U_2 : \mathsf{UAp} \longrightarrow \mathsf{Unif} : (X, \delta) \mapsto (X, \mathscr{U}(\delta)).$$

Furthermore, we will denote the coreflector from App to Top by C.

3.1.17 Proposition *All assignments* $\mathsf{Unif} \xrightarrow{D} \mathsf{UAp}$, $\mathsf{UAp} \xrightarrow{U_1} \mathsf{Unif}$ *and* $\mathsf{UAp} \xrightarrow{U_2} \mathsf{Unif}$ *are faithful functors and moreover the diagrams*

$$\begin{array}{ccc} \mathsf{UAp} & \xrightarrow{U_1} & \mathsf{Unif} \\ {\scriptstyle C}\downarrow & \swarrow {\scriptstyle D} & \\ \mathsf{UAp} & & \end{array} \qquad \begin{array}{ccc} \mathsf{UAp} & \xrightarrow{U_2} & \mathsf{Unif} \\ {\scriptstyle C}\downarrow & \swarrow {\scriptstyle D} & \\ \mathsf{UAp} & & \end{array}$$

are commutative.

Proof For the first claim, since D is actually the forgetful functor which associates the underlying topological space to a given uniform space, D is faithful. If U_1 and U_2 are functors, then they are obviously faithful. Let (X, δ), (X', δ') be uniform approach spaces and let $f : (X, \delta) \longrightarrow (X', \delta')$ be a contraction. To show that U_1 is a functor, it suffices to note that, by 2.2.4, $f : (X, \mathcal{T}_\delta) \longrightarrow (X', \mathcal{T}_{\delta'})$ is continuous and consequently

$$f : (X, \mathcal{U}_{\text{fine}}(\mathcal{T}_\delta)) \longrightarrow (X', \mathcal{U}_{\text{fine}}(\mathcal{T}_{\delta'}))$$

is uniformly continuous. To show that U_2 is a functor, let $d' \in \mathcal{G}(\delta')$. Then, for all $x \in X$ and $A \subseteq X$, we have

$$\delta_{d'}(f(x), f(A)) \leq \delta'(f(x), f(A)) \leq \delta(x, A),$$

which proves that $d' \circ (f \times f) \in \mathcal{G}(\delta)$. By the arbitrariness of $d' \in \mathcal{G}(\delta')$ this shows that $f : (X, \mathcal{U}(\delta)) \longrightarrow (X', \mathcal{U}(\delta'))$ is uniformly continuous.

The second claim follows from the definitions of the functors involved. □

3.1.18 Proposition (Top, Met) *If* (X, δ_d) *is a metric approach space, then* $\mathcal{U}(\delta_d)$ *coincides with the uniformity generated by the metric* d *and if* $(X, \delta_{\mathcal{T}})$ *is a topological uniform approach space, then* $\mathcal{U}(\delta_{\mathcal{T}})$ *coincides with the fine uniformity determined by* $\mathcal{T}$.

Proof Both properties follow from the foregoing considerations. □

3.1.19 Proposition *Given the uniform distances* δ *and* δ' *on* X, *the following implications hold.*

$$\delta = \delta' \Rightarrow \mathcal{U}(\delta) = \mathcal{U}(\delta') \Rightarrow \mathcal{T}(\delta) = \mathcal{T}(\delta').$$

Proof This follows from the definitions. □

3.1.20 Example 3.1.19 cannot be improved upon. Different uniform distances can give rise to both the same underlying topologies and the same underlying uniformities. Also it is possible for two different collections of metrics to generate the same distance but different uniformities.

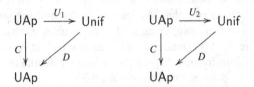

1. Consider the real line $\mathbb{R}$ equipped with two different (not uniformly equivalent) metrics,

$$d_{\mathbb{E}} : \mathbb{R} \times \mathbb{R} \longrightarrow \mathbb{P} : (x, y) \mapsto |x - y|$$

and

$$d_0 : \mathbb{R} \times \mathbb{R} \longrightarrow \mathbb{P} : (x, y) \mapsto |\arctan x - \arctan y| .$$

Then $\mathcal{G}(\delta_{d_{\mathbb{E}}}) \neq \mathcal{G}(\delta_{d_0})$ and thus $\delta_{d_{\mathbb{E}}} \neq \delta_{d_0}$ and $\mathcal{U}(\delta_{d_{\mathbb{E}}}) \neq \mathcal{U}(\delta_{d_0})$. However, it follows from 3.1.11 that $\mathcal{T}_{\delta_{d_{\mathbb{E}}}} = \mathcal{T}_{\delta_{d_0}}$.

2. If (X, δ) is any non-topological uniform approach space then it follows that $(X, 2\delta)$ too is a uniform approach space and in spite of the fact that $\delta \neq 2\delta$, we have $\mathcal{U}(\delta) = \mathcal{U}(2\delta)$.

3. Consider the real line $\mathbb{R}$ and the following two collections of metrics. First, for each real number $a > 0$ define the function

$$f_a : \mathbb{R} \longrightarrow \mathbb{R} : x \mapsto \begin{cases} -a & \text{if } x \leq -a, \\ x & \text{if } -a \leq x \leq a, \\ a & \text{if } a \leq x. \end{cases}$$

Second, consider the metrics $d_a(x, y) := |f_a(x) - f_a(y)|$, and let $\mathcal{D} := \{d_a \mid a > 0\}$ and $\mathcal{D}' := \{d_{\mathbb{E}}\}$, where $d_{\mathbb{E}}$ stands for the Euclidean metric.

Then we have that both $\mathcal{D}$ and $\mathcal{D}'$ are bases for the same symmetric gauge, but $\mathcal{U}(\mathcal{D}) \neq \mathcal{U}(\mathcal{D}')$.

The results of the foregoing sections now allow us to complete the first part of our diagram of categories, namely the local part. In all four of the columns the transition from the first row to the second row consists in forgetting the numerical information and only retaining topological information on a set-theoretical level. In both of the rows the transition from the first column to the second column consists in going from a single metric to "arbitrary" families of metrics and the transition from the second column to the third column consists in going from metrics to quasi-metrics.

Referring to the diagram in the introduction we see that UAp fills in the place of the second question mark (Fig. 3.1).

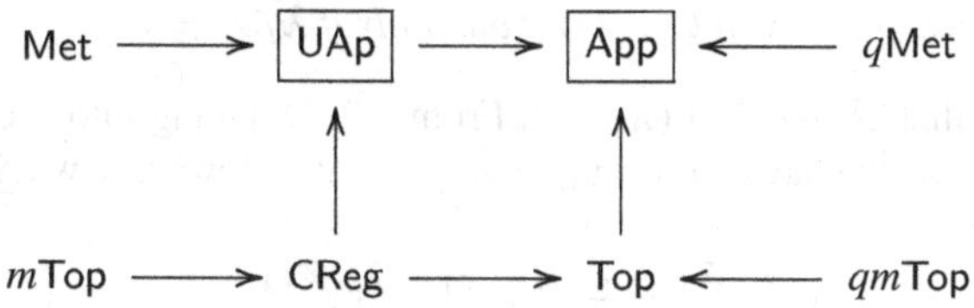

Fig. 3.1 The categorical position of App and UAp with regard to Top and Met

3.2 Weak Adjointness

In general, the topological coreflection together with the metric coreflection of an approach space do not determine the structure of the space. However, under the right conditions in specific subspaces the structure is completely determined by these two coreflections. A condition wherein this happens is weak adjointness which we first study for quasi-metric spaces.

3.2.1 Definition Let (X, d) be a quasi-metric space. We call this space and its quasi-metric *weakly adjoint* if the topology generated by the adjoint quasi-metric is coarser than the topology generated by the original quasi-metric, i.e. $\mathscr{T}_{d^-} \subseteq \mathscr{T}_d$.

3.2.2 Proposition *If (X, d) is a weakly adjoint quasi-metric space, then the underlying topology is metrizable.*

Proof It suffices to note that

$$\mathscr{T}_{d^*} = \mathscr{T}_d \vee \mathscr{T}_{d^-} = \mathscr{T}_d$$

and hence d^* is a metric for the underlying topology. □

3.2.3 Lemma *If (X, d) is a quasi-metric space, $x \in X$ and $B \subseteq X$ then*

$$\sup_{z \in B} d(x, z) = \sup_{z \in cl_{\mathscr{T}_{d^-}} B} d(x, z).$$

Proof Let $\varepsilon > 0$ and for any $z \in \mathrm{cl}_{\mathscr{T}_{d^-}} B$ let $y_z \in B$ be such that $d^-(z, y_z) < \varepsilon$. Then it follows that

$$\begin{aligned}
\sup_{z \in \mathrm{cl}_{\mathscr{T}_{d^-}} B} d(x, z) &\leq \sup_{z \in \mathrm{cl}_{\mathscr{T}_{d^-}} B} (d(x, y_z) + d(y_z, z)) \\
&\leq \sup_{z \in \mathrm{cl}_{\mathscr{T}_{d^-}} B} d(x, y_z) + \varepsilon \\
&\leq \sup_{z \in B} d(x, z) + \varepsilon.
\end{aligned}$$

□

3.2.4 Proposition *If (X, d) is a weakly adjoint quasi-metric space, $\mathscr{F}$ is a filter on X and $x \in X$, then $\mathscr{F} \to x$ in $(X, \mathscr{T}_d)$ if and only if $\lambda_d \mathscr{F} = d(\cdot, x)$.*

Proof Suppose that $\mathscr{F} \to x$ in $(X, \mathscr{T}_d)$. From 1.2.70, taking into account the fact that, for any $y \in X$, we have $\delta_d(y, \{x\}) = d(y, x)$, it follows that we have

$$\begin{aligned}
\lambda_d \mathscr{F}(y) &\leq d(y, x) + \lambda_d \mathscr{F}(x) \\
&= d(y, x),
\end{aligned}$$

which proves one inequality.

On the other hand, using 3.2.3 we find

$$\begin{aligned}\lambda_d\mathscr{F}(y) &= \inf_{F\in\mathscr{F}}\sup_{z\in F} d(y,z)\\ &= \inf_{F\in\mathscr{F}}\sup_{z\in \mathrm{cl}_{\mathscr{T}_{d^-}} F} d(y,z)\\ &\geq \inf_{F\in\mathscr{F}} d(y,x) = \delta_d(y,\{x\})\end{aligned}$$

where the inequality follows from the fact that by weak adjointness $\mathscr{F}\to x$ implies that for all $F\in\mathscr{F}$, $x\in \mathrm{cl}_{\mathscr{T}_{d^-}} F$.

The other implication follows from 2.3.6. □

The foregoing result of course holds in Met but does not necessarily hold in qMet in general, as the following example shows.

3.2.5 Example Consider the real line $\mathbb{R}$ equipped with the quasi-metric $d_{\mathbb{P}}$ defined in 1.2.62 which is clearly not weakly adjoint. Fix points $x, y\in\mathbb{R}$ such that $x<y$. The filter $\dot{y}$ converges to x in $\mathscr{T}_d$. However, from 2.3.1 we have

$$\begin{aligned}\lambda_d(\dot{y})(z) &= \inf_{F\in\dot{y}}\sup_{x\in F} d(z,x)\\ &= d(z,y)\end{aligned}$$

and $d(z,y)\neq d(z,x)$ whenever $x<z$.

We recall that a filter is called *total* if all finer ultrafilters are convergent (not necessarily to the same point). This notion was introduced by Pettis (1969) and extensively used and studied by Vaughan (1976a, b). In regular spaces, if a filter $\mathscr{F}$ is total then it is *adherent–convergent* (Pettis 1969). This means that every open set containing the set of adherence points of $\mathscr{F}$ is a member of $\mathscr{F}$. Since weakly adjoint spaces are metrizable this characterization also holds in them.

3.2.6 Theorem *If (X,d) is a weakly adjoint quasi-metric space, $\mathscr{F}$ is total and $x\in X$, then the following properties hold.*

1. $\alpha_d\mathscr{F}(x) = \inf_{y\in \mathrm{adh}_{\mathscr{T}_d}\mathscr{F}} d(x,y) = \delta_d(x, \mathrm{adh}_{\mathscr{T}_d}\mathscr{F})$.
2. $\lambda_d\mathscr{F}(x) = \sup_{y\in \mathrm{adh}_{\mathscr{T}_d}\mathscr{F}} d(x,y)$.

Proof 1. Suppose that $\delta_d(x,\mathrm{adh}_{\mathscr{T}_d}\mathscr{F})<\varepsilon$. Then there exists $y\in\bigcap_{F\in\mathscr{F}}\mathrm{cl}_{\mathscr{T}_d} F$ for which $d(x,y)<\varepsilon$ and it follows that

$$\begin{aligned}\alpha_d \mathscr{F}(x) &= \sup_{F \in \mathscr{F}} \delta_d(x, F) \\ &= \sup_{F \in \mathscr{F}} \delta_d(x, \mathrm{cl}_{\mathscr{T}_d} F) \\ &\leq d(x, y) < \varepsilon,\end{aligned}$$

which proves that $\alpha_d \mathscr{F}(x) \leq \delta_d(x, \mathrm{cl}_{\mathscr{T}_d} \mathscr{F})$. To prove the other inequality let $\varepsilon > 0$. Since

$$\left\{\delta_{\mathrm{adh}_{\mathscr{T}_d} \mathscr{F}} < \varepsilon\right\} = \bigcup_{y \in \mathrm{adh}_{\mathscr{T}_d} \mathscr{F}} B_{d^-}(y, \varepsilon)$$

it follows by weak adjointness that this set is open and hence, since

$$(\mathrm{adh}_{\mathscr{T}_d} \mathscr{F})^{(\varepsilon)} \supset \left\{\delta_{\mathrm{adh}_{\mathscr{T}_d} \mathscr{F}} < \varepsilon\right\}$$

it follows from the fact that $\mathscr{F}$ is adherent–convergent that

$$(\mathrm{adh}_{\mathscr{T}_d} \mathscr{F})^{(\varepsilon)} \in \mathscr{F}.$$

Consequently

$$\begin{aligned}\delta_d(x, \mathrm{adh}_{\mathscr{T}_d} \mathscr{F}) &\leq \delta_d(x, (\mathrm{adh}_{\mathscr{T}_d} \mathscr{F})^{(\varepsilon)}) + \varepsilon \\ &\leq \alpha_d \mathscr{F}(x) + \varepsilon.\end{aligned}$$

2. Since an ultrafilter finer than $\mathscr{F}$ converges to a point in $\mathrm{adh}_{\mathscr{T}_d} \mathscr{F}$, it follows from the first property, 1.2.66, and 1.2.67 that

$$\begin{aligned}\lambda_d \mathscr{F}(x) &= \sup_{\mathscr{U} \in \mathsf{U}(\mathscr{F})} \lambda_d \mathscr{U}(x) \\ &= \sup_{\mathscr{U} \in \mathsf{U}(\mathscr{F})} \delta_d(x, \mathrm{adh}_{\mathscr{T}_d} \mathscr{U}) \\ &\leq \sup_{y \in \mathrm{adh}_{\mathscr{T}_d} \mathscr{F}} d(x, y).\end{aligned}$$

For the converse inequality, from 3.2.3 it follows that

$$\lambda_d \mathscr{F}(x) = \inf_{F \in \mathscr{F}} \sup_{z \in F} d(x, z) = \inf_{F \in \mathscr{F}} \sup_{z \in \mathrm{cl}_{\mathscr{T}_{d^-}} F} d(x, z).$$

Now if $y \in \mathrm{adh}_{\mathscr{T}_d} \mathscr{F}$ then from weak adjointness it follows that also for all $F \in \mathscr{F}$ we have $y \in \mathrm{cl}_{\mathscr{T}_{d^-}} F$ and hence for any such y

$$\inf_{F \in \mathcal{F}} \sup_{z \in \mathrm{cl}_{\mathcal{T}_{d^-}} F} d(x, z) \geq \inf_{F \in \mathcal{F}} d(x, y) = d(x, y)$$

and thus

$$\lambda_d \mathcal{F}(x) \geq \sup_{y \in \mathrm{adh}_{\mathcal{T}_d} \mathcal{F}} d(x, y). \qquad \square$$

Again it should be pointed out that the foregoing results hold in metric spaces but do not necessarily hold in a non-weakly adjoint quasi-metric space.

3.2.7 Example Consider the set of real numbers $[0, \infty[$ equipped with the quasi-metric $d_{\mathbb{P}}$. Fix a point $x_0 \in]0, \infty[$. For the filter $\mathcal{F} := \dot{x}_0$ we have $\mathrm{adh}_{\mathcal{T}_{d_{\mathbb{P}}}} \mathcal{F} = [0, x_0]$. Consequently it follows that, for all $x \in [0, \infty[$,

$$\sup_{y \in \mathrm{adh}\, \mathcal{T}_{d_{\mathbb{P}}} \mathcal{F}} d_{\mathbb{P}}(x, y) = d_{\mathbb{P}}(x, 0),$$

whereas

$$\lambda_{d_{\mathbb{P}}} \mathcal{F}(x) = d_{\mathbb{P}}(x, x_0).$$

Totality too is a necessary condition in 3.2.6.

3.2.8 Example Consider $\mathbb{R} \setminus \{0\}$ equipped with the usual metric $d_{\mathbb{E}}$ and the usual topology $\mathcal{T}_{\mathbb{E}}$, and consider the sequence $(z_n)_{n \geq 1}$, where

$$z_n := \begin{cases} n & n \text{ even}, \\ \dfrac{1}{n} & n \text{ odd}. \end{cases}$$

Denote by $\mathcal{F}$ the filter generated by $(z_n)_{n \geq 1}$. Then, for all $x \in \mathbb{R} \setminus \{0\}$ the adherence and limit of $\mathcal{F}$ are given respectively by

$$\alpha_{d_{\mathbb{E}}} \mathcal{F}(x) = |x|$$

and

$$\lambda_{d_{\mathbb{E}}} \mathcal{F}(x) = \infty.$$

Now clearly, $\mathcal{F}$ is not total and, actually, $\mathrm{adh}_{\mathcal{T}_{\mathbb{E}}} \mathcal{F} = \emptyset$. Consequently, for any $x \in \mathbb{R} \setminus \{0\}$, the formulas of 3.2.6 give us $\inf_{y \in \mathrm{adh}_{\mathcal{T}_{\mathbb{E}}} \mathcal{F}} d(x, y) = \infty$, which differs from $\alpha_{d_{\mathbb{E}}} \mathcal{F}(x)$ and $\sup_{y \in \mathrm{adh}_{\mathcal{T}_{\mathbb{E}}} \mathcal{F}} d(x, y) = 0$, which differs from $\lambda_{d_{\mathbb{E}}} \mathcal{F}(x)$.

We now generalize the foregoing concepts so as to be meaningful for approach spaces in general.

3.2.9 Definition We call an approach space and its structures *weakly adjoint* if there exists a basis for the gauge consisting of weakly adjoint quasi-metrics. We denote the full subcategory of App with objects all weakly adjoint approach spaces by App_{Wa}.

3.2.10 Theorem *A uniform approach space is weakly adjoint.*

Proof This follows from the definitions. □

The converse does not necessarily hold as the following example shows.

3.2.11 Example The real line equipped with the quasi-metric

$$d(x, y) := \begin{cases} |x - y| & x \leq y, \\ 2|x - y| & y \leq x, \end{cases}$$

is weakly adjoint but not uniform.

Together with uniform approach spaces, weakly adjoint spaces share the following property.

3.2.12 Proposition *If X is a weakly adjoint approach space then the topological coreflection is completely regular.*

Proof This follows from 3.2.2 and the definition. □

3.2.13 Proposition *If (X, δ) is weakly adjoint, $\mathcal{F}$ is a filter and $y \in X$, then $\mathcal{F} \to y$ if and only if $\lambda\mathcal{F} = \delta(\cdot, \{y\})$.*

Proof One implication is evident. To show the other one, note that $\lambda\mathcal{F}(x) \leq \delta(x, \{y\})$ follows at once from 1.2.70.

On the other hand, let $\mathcal{H}$ be a basis for the gauge consisting of weakly adjoint quasi-metrics. Using 3.2.3, as in 3.2.4, we find

$$\begin{aligned} \lambda\mathcal{F}(x) &= \sup_{d \in \mathcal{H}} \inf_{F \in \mathcal{F}} \sup_{z \in F} d(x, z) \\ &= \sup_{d \in \mathcal{H}} \inf_{F \in \mathcal{F}} \sup_{z \in \mathrm{cl}_{\mathcal{T}_{d^-}} F} d(x, z) \\ &\geq \sup_{d \in \mathcal{H}} \inf_{F \in \mathcal{F}} d(x, y) = \delta(x, \{y\}) \end{aligned}$$

where, again, the inequality follows from the fact that by weak adjointness $\mathcal{F} \to y$ implies that for all $F \in \mathcal{F}$, $y \in \mathrm{cl}_{\mathcal{T}_{d^-}} F$. □

We are now in a position to generalize 3.2.6 to the setting of weakly adjoint approach spaces.

3.2.14 Theorem *If (X, δ) is weakly adjoint and $\mathscr{F}$ is total, then the following properties hold.*

1. $\alpha_\delta \mathscr{F}(x) = \delta_{d_\delta}(x, \mathrm{adh}_{\mathscr{T}_\delta} \mathscr{F})$.
2. $\lambda_\delta \mathscr{F}(x) = \sup_{y \in \mathrm{adh}_{\mathscr{T}_\delta} \mathscr{F}} d_\delta(x, y)$.

Proof 1. Let $\mathscr{H}$ be a gauge basis consisting of weakly adjoint quasi-metrics. If $\mathscr{F}$ is total with respect to $\mathscr{T}_\delta$, then obviously it is also total with respect to $\mathscr{T}_d$ for all $d \in \mathscr{H}$. Consequently it follows from 3.2.6 that

$$\begin{aligned}\alpha\mathscr{F}(x) &= \sup_{d \in \mathscr{H}} \alpha_d \mathscr{F}(x)\\ &= \sup_{d \in \mathscr{H}} \delta_d(x, \mathrm{adh}_{\mathscr{T}_d} \mathscr{F})\\ &\leq \sup_{d \in \mathscr{H}} \delta_d(x, \mathrm{adh}_{\mathscr{T}_\delta} \mathscr{F})\\ &\leq \delta_{d_\delta}(x, \mathrm{adh}_{\mathscr{T}_\delta} \mathscr{F}).\end{aligned}$$

To prove the other inequality, first of all note that from 1.2.67 it follows that there exists $\mathscr{U} \in \mathsf{U}(\mathscr{F})$ such that $\alpha\mathscr{F}(x) = \lambda\mathscr{U}(x)$. Now since $\mathscr{F}$ is total, there exists $y \in \mathrm{adh}_{\mathscr{T}_\delta} \mathscr{F}$ such that $\lambda\mathscr{U}(y) = 0$, i.e. such that $\mathscr{U}$ converges to y in $\mathscr{T}_\delta$. Then it follows from 3.2.13 that $\lambda\mathscr{U}(x) = \delta(x, \{y\}) = \delta_{d_\delta}(x, \{y\})$, from which it further follows that

$$\alpha\mathscr{F}(x) = \delta_{d_\delta}(x, \{y\}) \geq \delta_{d_\delta}(x, \mathrm{adh}_{\mathscr{T}_\delta} \mathscr{F}).$$

2. Making use of the fact that each ultrafilter finer than $\mathscr{F}$ converges to some point of $\mathrm{adh}_{\mathscr{T}_\delta} \mathscr{F}$, it follows from the first property, 1.2.66, and 1.2.67 that

$$\begin{aligned}\lambda\mathscr{F}(x) &= \sup_{\mathscr{U} \in \mathsf{U}(\mathscr{F})} \lambda\mathscr{U}(x)\\ &= \sup_{\mathscr{U} \in \mathsf{U}(\mathscr{F})} \delta_{d_\delta}(x, \mathrm{adh}_{\mathscr{T}_\delta} \mathscr{U})\\ &\leq \sup_{y \in \mathrm{adh}_{\mathscr{T}_\delta} \mathscr{F}} d_\delta(x, y).\end{aligned}$$

Conversely, it follows from 3.2.6 that

$$\begin{aligned}\lambda\mathscr{F}(x) &= \sup_{d \in \mathscr{H}} \lambda_d \mathscr{F}(x)\\ &= \sup_{d \in \mathscr{H}} \sup_{y \in \mathrm{adh}_{\mathscr{T}_d} \mathscr{F}} d(x, y)\\ &\geq \sup_{d \in \mathscr{H}} \sup_{y \in \mathrm{adh}_{\mathscr{T}_\delta} \mathscr{F}} d(x, y)\\ &= \sup_{y \in \mathrm{adh}_{\mathscr{T}_\delta} \mathscr{F}} d_\delta(x, y).\end{aligned}$$

□

3.2.15 Corollary *If* (X, δ) *is weakly adjoint,* $A \subseteq X$ *is relatively compact and* $x \in X$*, then* $\delta(x, A) = \delta_{d_\delta}(x, cl_{\mathcal{T}_\delta} A)$.

Proof If A is relatively compact then the filter generated by $\{A\}$ is total and the result follows immediately from 3.2.14. □

The foregoing corollary has an important interpretation. In a weakly adjoint space, since subsets of relatively compact sets are relatively compact, we obtain that the restriction of the approach structure to relatively compact subspaces is completely determined by the topological and metric coreflections.

3.2.16 Theorem App_{Wa} *is concretely reflective in* App.

Proof If X is an approach space with gauge $\mathcal{G}$ then it suffices to take the subset $\mathcal{G}_0 \subseteq \mathcal{G}$ consisting of all weakly adjoint quasi-metrics. This then is a basis for a weakly adjoint gauge, giving us the weakly adjoint approach space, which clearly defines a concrete reflection. □

3.3 Separation

In this section we will see that the lower separation axioms T_0, T_1 and T_2 are of a purely topological nature, in the sense that the most meaningful concept in App that can be given is that the topological coreflection be T_0, T_1 or T_2. Regularity, as we will see, is a different matter.

This is most clearly seen for the T_0 axiom. In Marny (1979) a categorical definition of T_0-objects in the setting of topological categories was given and it was shown there that the resulting subcategory of all T_0-objects is the largest epireflective, not reflective subcategory of the considered topological category. We will now identify the T_0-objects in App. I_2 stands for the two-point indiscrete space.

3.3.1 Definition An approach space X is called T_0 if every contraction $f : I_2 \longrightarrow X$ is constant. We write App_0 for the full subcategory of App consisting of all T_0-objects.

A collection of functions on a set X is said to be *point-separating* if for any two different points $x, y \in X$ there exists a function f in the collection such that $f(x) \neq f(y)$.

3.3.2 Theorem *For an approach space X the following properties are equivalent.*

1. *X is a* T_0*-space.*
2. *X is in the epireflective hull of* $\mathbb{P}$.
3. $x \neq y \Rightarrow ((\exists \varphi \in \mathcal{A}(x) : \varphi(y) > 0)$ *or* $(\exists \varphi \in \mathcal{A}(y) : \varphi(x) > 0))$.
4. $x \neq y \Rightarrow \mathfrak{i}(\dot{x}) \not\to y$ *or* $\mathfrak{i}(\dot{y}) \not\to x$.
5. $\mathfrak{L}$ *is point-separating.*
6. $\mathfrak{U}$ *is point-separating.*

7. $x \neq y \Rightarrow \exists d \in \mathcal{G} : d(x, y) > 0$ *or* $d(y, x) > 0$.
8. $x \neq y \Rightarrow \mathcal{A}(x) \neq \mathcal{A}(y)$.
9. $(X, \mathcal{T}_\delta)$ *is a* T_0*-space.*

Proof This is an easy exercise making use of the various structures defining an approach space and the transitions between them, and hence we leave this to the reader. □

This shows that the T_0-property is in fact a purely topological property.

3.3.3 Corollary App_0 *is an epireflective subcategory of* App.

We recall that a topological category is called *universal* if it is the reflective hull of its T_0-objects (Marny 1979).

3.3.4 Theorem App *is universal, moreover, every epireflector from* App *onto one of its subcategories is either a reflector or the composition of a reflector, followed by the* App_0*-epireflector.*

Proof The universality follows immediately from 1.3.19, because $\mathbb{P}$ is in App_0. The second part is proved in Marny (1979). □

3.3.5 Theorem *For an approach space X, the following properties are equivalent.*

1. $x \neq y \Rightarrow (\exists \varphi \in \mathfrak{L} : \varphi(x) < \varphi(y))$.
2. $x \neq y \Rightarrow (\exists \varphi \in \mathcal{A}(x) : \varphi(y) > 0)$.
3. $x \neq y \Rightarrow \mathcal{A}(x) \not\subseteq \mathcal{A}(y)$.
4. $x \neq y \Rightarrow \mathfrak{i}(\dot{x}) \not\rightarrowtail y$.
5. $(X, \mathcal{T}_\delta)$ *is a* T_1*-space.*

Proof This is proved in the same way as 3.3.2 and we leave this to the reader. □

Comparing this with the way T_1-objects are defined in Top and the characterizations of T_0-objects in App, justifies the following definition.

3.3.6 Definition An approach space is called T_1 if it satisfies any and hence all the equivalent statements from 3.3.5 and we denote App_1 the full subcategory of App with objects all T_1 approach spaces.

3.3.7 Corollary App_1 *is an epireflective subcategory of* App.

3.3.8 Theorem *For an approach space X, the following properties are equivalent.*

1. $x \neq y \Rightarrow c(\mathcal{A}_b(x) \vee \mathcal{A}_b(y)) > 0$.
2. $x \neq y \Rightarrow (\exists \varphi \in \mathcal{A}(x), \exists \psi \in \mathcal{A}(y) : \inf_{s \in X} (\varphi \vee \psi)(s) > 0$.
3. $\mathfrak{I} \in \mathfrak{F}(X), c(\mathfrak{I}) = 0, \mathfrak{I} \rightarrowtail x, \mathfrak{I} \rightarrowtail y \Rightarrow x = y$.
4. $(X, \mathcal{T}_\delta)$ *is Hausdorff.*

Proof This is analogous to 3.3.2 and 3.3.5 and we leave this to the reader. □

Again, a comparison with the classical topological situation and taking into account that the role of neighbourhood filters in topology is here played by approach systems, makes it plausible to define Hausdorff objects in App in the following, again topological, way.

3.3.9 Definition An approach space is called T_2 if it satisfies the equivalent statements from 3.3.8. We define App_2 to be the full subcategory of App with objects all T_2 approach spaces.

3.3.10 Corollary App_2 *is an epireflective subcategory of* App.

We now study the notion of regularity in approach spaces. For $\mathcal{F} \in \mathsf{F}(X)$ and $\gamma \in \mathbb{P}$ we define $\mathcal{F}^{(\gamma)}$ as the filter generated by $\{F^{(\gamma)} \mid F \in \mathcal{F}\}$.

3.3.11 Lemma *Let $\mathcal{F}$ be a filter on X, let $\gamma \in \mathbb{P}$ and let $\mathcal{U} \in \mathsf{U}(\mathcal{F}^{(\gamma)})$ then there exists an ultrafilter $\mathcal{W} \in \mathsf{U}(\mathcal{F})$ such that $\mathcal{W}^{(\gamma)} \subseteq \mathcal{U}$.*

Proof Indeed, if not, then in each $\mathcal{W} \in \mathsf{U}(\mathcal{F})$ we can choose $W_{\mathcal{W}} \in \mathcal{W}$ such that $W_{\mathcal{W}}^{(\gamma)} \notin \mathcal{U}$. Then it follows from 1.1.4 that there exists a finite subset $\mathsf{U}_0 \subseteq \mathsf{U}(\mathcal{F})$ such that $\bigcup_{\mathcal{W} \in \mathsf{U}_0} W_{\mathcal{W}} \in \mathcal{F}$. Now since

$$\bigcup_{\mathcal{W} \in \mathsf{U}_0} W_{\mathcal{W}}^{(\gamma)} = (\bigcup_{\mathcal{W} \in \mathsf{U}_0} W_{\mathcal{W}})^{(\gamma)} \in \mathcal{U}$$

this is a contradiction. □

3.3.12 Theorem *For an approach space X the following properties are equivalent.*

1. *For any $\mathcal{F} \in \mathsf{F}(X)$ and $\gamma \in \mathbb{P}$ we have*

$$\lambda \mathcal{F}^{(\gamma)} \leq \lambda \mathcal{F} + \gamma.$$

2. *For any $\mathcal{W}, \mathcal{U} \in \mathsf{U}(X)$ and $\gamma \in \mathbb{P}$ we have*

$$\mathcal{W}^{(\gamma)} \subseteq \mathcal{U} \Rightarrow \lambda \mathcal{U} \leq \lambda \mathcal{W} + \gamma.$$

Proof 1 ⇒ 2. This is evident.

2 ⇒ 1. Let $\mathcal{F}$ be a filter on X and let $\gamma \in \mathbb{P}$. For every $\mathcal{U} \in \mathsf{U}(\mathcal{F}^{(\gamma)})$ let $\mathcal{W}_{\mathcal{U}} \in \mathsf{U}(\mathcal{F})$ be as guaranteed by 3.3.11, then it follows that

$$\begin{aligned} \lambda \mathcal{F}^{(\gamma)} &= \sup_{\mathcal{U} \in \mathsf{U}(\mathcal{F}^{(\gamma)})} \lambda \mathcal{U} \\ &\leq \sup_{\mathcal{U} \in \mathsf{U}(\mathcal{F}^{(\gamma)})} \lambda \mathcal{W}_{\mathcal{U}} + \gamma \\ &\leq \sup_{\mathcal{W} \in \mathsf{U}(\mathcal{F})} \lambda \mathcal{W} + \gamma \\ &= \lambda \mathcal{F} + \gamma. \end{aligned}$$

□

3.3.13 Definition An approach space is called *regular* if it satisfies the equivalent statements of 3.3.12 and we write App_{Rg} for the full subcategory of App with objects all regular spaces.

3.3.14 Proposition (Top) *A topological approach space is regular if and only if the underlying topology is regular.*

Proof This follows from the definition since a topological space is regular if and only if for any filter $\mathcal{F}$ the filter $\overline{\mathcal{F}}$ generated by the closures of the sets in $\mathcal{F}$ has the same limit points as $\mathcal{F}$. □

3.3.15 Proposition (qMet) *A quasi-metric approach space is regular if and only if it is metric.*

Proof The if-part follows from 3.3.17 since metric spaces are uniform approach spaces. To show the only-if-part let d be a quasi-metric on X, let $x, y \in X$ and let $\varepsilon \in \mathbb{P}$. Then

$$\lambda\dot{x}(y) = d(y, x) \text{ and } \lambda(\dot{x})^{(\varepsilon)}(y) = \sup_{z\in\{x\}^{(\varepsilon)}} d(y, z).$$

Hence it follows from the definition of regularity that $\sup_{z\in\{x\}^{(\varepsilon)}} d(y, z) \leq d(y, x)+\varepsilon$ and thus that

$$d(z, x) \leq \varepsilon \Rightarrow d(y, z) \leq d(y, x) + \varepsilon$$

and letting $y = x$ this shows that $d(z, x) \leq \varepsilon \Rightarrow d(x, z) \leq \varepsilon$ which by the arbitrariness of x, z and ε shows that $d(z, x) = d(x, z)$. □

Note that, where the lower separation axioms in App which we discussed all turn out to be topological in the sense that an approach space is T_i if and only if its topological coreflection is T_i in the classical sense (with $i \in \{0, 1, 2\}$), the regularity condition stated above is of a purely "approach" nature.

3.3.16 Example Consider the real line equipped with the quasi-metric

$$d(x, y) := \begin{cases} |x-y| & x \leq y, \\ 2|x-y| & y \leq x. \end{cases}$$

Then by 3.3.15, $(\mathbb{R}, \delta_d)$ is not regular but $\mathcal{T}_{\delta_d}$ is the usual Euclidean topology.

3.3.17 Theorem *A uniform approach space is regular.*

Proof Let X be a uniform approach space with $\mathcal{H}$ a symmetric gauge basis, let $\mathcal{F}$ be a filter on X and let $\varepsilon > 0$. Then for any $F \in \mathcal{F}$, $z \in F^{(\varepsilon)}$, $d \in \mathcal{H}$ and $\varepsilon' > \varepsilon$ there exists $y_d^z \in F$ such that $d(z, y_d^z) \leq \varepsilon'$. Hence for any $x \in X$ we have

$$\lambda\mathcal{F}^{(\varepsilon)}(x) = \sup_{d\in\mathcal{H}} \inf_{F\in\mathcal{F}} \sup_{z\in F^{(\varepsilon)}} d(x, z)$$

$$
\begin{aligned}
&\leq \sup_{d\in\mathscr{H}} \inf_{F\in\mathscr{F}} \sup_{z\in F^{(\varepsilon)}} (d(x, y_d^z) + d(y_d^z, z))\\
&\leq \sup_{d\in\mathscr{H}} \inf_{F\in\mathscr{F}} \sup_{z\in F^{(\varepsilon)}} d(x, y_d^z) + \varepsilon'\\
&\leq \sup_{d\in\mathscr{H}} \inf_{F\in\mathscr{F}} \sup_{y\in F} d(x, y) + \varepsilon'\\
&= \lambda\mathscr{F}(x) + \varepsilon'
\end{aligned}
$$

which by the arbitrariness of x and $\varepsilon' > \varepsilon$ shows that $\lambda\mathscr{F}^{(\varepsilon)} \leq \lambda\mathscr{F} + \varepsilon$. □

3.3.18 Theorem App_{Rg} *is a concretely reflective subcategory of* App.

Proof Consider an initial source $\left(f_j : (Y, \lambda) \to (X_j, \lambda_j)\right)_{j\in J}$, where each space (X_j, λ_j), $j \in J$ is regular and let $\mathscr{F} \in \mathsf{F}(Y)$ and $\gamma \in \mathbb{P}$. Then by 1.3.12 we have

$$
\begin{aligned}
\lambda\mathscr{F}^{(\gamma)} &= \sup_{j\in J} \lambda_j\left(f_j(\mathscr{F}^{(\gamma)})\right) \circ f_j\\
&\leq \sup_{j\in J} \lambda_j\left(f_j(\mathscr{F})^{(\gamma)}\right) \circ f_j\\
&\leq \sup_{j\in J} \lambda_j\left(f_j(\mathscr{F})\right) \circ f_j + \gamma\\
&= \lambda\mathscr{F} + \gamma.
\end{aligned}
$$

□

We now come to the characterizations of regularity as given by Brock and Kent (1997a, b) to which we alluded earlier.

3.3.19 Proposition *For an approach space X, the following properties are equivalent.*

1. *X is regular.*
2. *For any nonempty set J,* $\psi : J \to X, \sigma : J \to \mathsf{U}(X)$, *and* $\mathscr{U} \in \mathsf{U}(J)$:

$$\lambda\psi\mathscr{U} \leq \lambda\Sigma\sigma(\mathscr{U}) + \inf_{U\in\mathscr{U}} \sup_{j\in U} \lambda\sigma(j)(\psi(j)).$$

3. *For any nonempty set J,* $\psi : J \to X, \sigma : J \to \mathsf{F}(X)$, *and* $\mathscr{F} \in \mathsf{F}(J)$:

$$\lambda\psi\mathscr{F} \leq \lambda\Sigma\sigma(\mathscr{F}) + \inf_{F\in\mathscr{F}} \sup_{j\in F} \lambda\sigma(y)(\psi(y)).$$

Proof 1 ⇒ 2. Let J, ψ, σ, and $\mathscr{U}$ be as stated and put

$$\inf_{U\in\mathscr{U}} \sup_{j\in U} \lambda\sigma(j)(\psi(j)) < \gamma,$$

where $\gamma < \infty$. Then there exists $U_0 \in \mathscr{U}$ such that $\lambda\sigma(j)(\psi(j)) < \gamma$ for all $j \in U_0$. Now take $A \in \Sigma\sigma(\mathscr{U})$ and $U \in \mathscr{U}$. Then there exists $U_1 \in \mathscr{U}$ such that $U_1 \subseteq U_0$

and such that $A \in \bigcap_{j \in U_1} \sigma(j)$. Now if we take $j \in U \cap U_1$ then $\psi(j) \in \psi(U)$ and $\lambda\sigma(j)(\psi(j)) < \gamma$ and since $\sigma(j) \in A$ this implies that $\psi(j) \in A^{(\gamma)}$. Consequently $A^{(\gamma)} \cap \psi(U) \neq \emptyset$ which proves that

$$(\Sigma\sigma(\mathcal{U}))^{(\gamma)} \subseteq \psi(\mathcal{U}).$$

The result now follows from 3.3.12 and the arbitrariness of γ.

2 ⇒ 1. Let $\mathcal{U} \in \mathsf{U}(X)$ and let $\gamma \in \mathbb{P}$. Put

$$J := \{(\mathcal{G}, y) \mid \lambda\mathcal{G}(y) < \gamma\},$$

and put $\sigma := \mathrm{pr}_1$ and $\psi := \mathrm{pr}_2$. Further consider the filter $\mathcal{S}$ on J generated by the sets

$$S_U := \{(\mathcal{G}, y) \in J \mid U \in \mathcal{G}\}.$$

Then it is immediately verified that $\mathcal{U} \subseteq \Sigma\sigma(\mathcal{S})$.

Now let $\mathcal{W} \in \mathsf{U}(\mathcal{U}^{(\gamma)})$. We claim that there exists $\mathcal{R}_{\mathcal{W}} \in \mathsf{U}(\mathcal{S})$ such that $\psi(\mathcal{R}_{\mathcal{W}}) \subseteq \mathcal{W}$. Suppose not, then for each $\mathcal{R} \in \mathsf{U}(\mathcal{S})$ there exists $R_{\mathcal{R}} \in \mathcal{R}$ such that $\psi(R_{\mathcal{R}}) \notin \mathcal{W}$. By 1.1.4 there exists a finite subset $\mathsf{U}_0 \subseteq \mathsf{U}(\mathcal{S})$ such that $\bigcup_{\mathcal{R} \in \mathsf{U}_0} R_{\mathcal{R}} \in \mathcal{S}$. Then it follows that $\bigcup_{\mathcal{R} \in \mathsf{U}_0} \psi(R_{\mathcal{R}}) \in \psi(\mathcal{S}) = \mathcal{U}^{(\gamma)} \subseteq \mathcal{W}$ which by finiteness of U_0 is a contradiction.

Hence, let $\mathcal{R}_{\mathcal{W}}$ be as stated then it follows that $\mathcal{U} \subseteq \Sigma\sigma(\mathcal{S}) \subseteq \Sigma\sigma(\mathcal{R}_{\mathcal{W}})$ and thus

$$\begin{aligned}
\lambda\mathcal{U}^{(\gamma)} &= \sup_{\mathcal{W} \in \mathsf{U}(\mathcal{U}^{(\gamma)})} \lambda\mathcal{W} \\
&\leq \sup_{\mathcal{W} \in \mathsf{U}(\mathcal{U}^{(\gamma)})} \lambda\psi(\mathcal{R}_{\mathcal{W}}) \\
&\leq \sup_{\mathcal{W} \in \mathsf{U}(\mathcal{U}^{(\gamma)})} \Big(\lambda\Sigma\sigma(\mathcal{R}_{\mathcal{W}}) + \inf_{R \in \mathcal{R}_{\mathcal{W}}} \sup_{j \in R} \lambda\sigma(j)(\psi(j))\Big) \\
&\leq \sup_{\mathcal{W} \in \mathsf{U}(\mathcal{U}^{(\gamma)})} (\lambda\mathcal{U} + \gamma) \\
&= \lambda\mathcal{U} + \gamma.
\end{aligned}$$

3 ⇒ 2. This is evident.

1 ⇒ 3. This is perfectly similar to the proof that 1 implies 2, again making use of the appropriate characterization in 3.3.12. □

The following characterizations making use of the gauge of an approach space were first given by Robeys (1992).

3.3.20 Proposition *For an approach space X the following properties are equivalent.*

1. *X is regular.*

2. $\forall x \in X, \forall \varepsilon_2 > \varepsilon_1 > 0, \forall \theta > 0, \forall d \in \mathcal{G}, \exists e \in \mathcal{G}$ *such that*

$$\{e(x,\cdot) < \varepsilon_1\}^{(\theta)} \subseteq \{d(x,\cdot) < \varepsilon_2 + \theta\}.$$

3. $\forall x \in X, \forall A \in 2^X, \forall \varepsilon, \theta > 0:$

$$\{d(x,\cdot) < \varepsilon\}^{(\theta)} \cap A \neq \emptyset, \forall d \in \mathcal{G} \Rightarrow \delta(x, A) \leq \varepsilon + \theta.$$

Proof 1 $\Rightarrow$ 2. If $x \in X$ and $\varepsilon_1 > 0$, consider the filter $\mathcal{F}$ generated by the family $\{\{d(x,\cdot) < \varepsilon_1\} \mid d \in \mathcal{G}\}$. Then $\lambda\mathcal{F}(x) \leq \varepsilon_1$. If $\varepsilon_2 > \varepsilon_1$ and $\theta > 0$ then we have by hypothesis

$$\lambda\mathcal{F}^{(\theta)}(x) \leq \lambda\mathcal{F}(x) + \theta \leq \varepsilon_1 + \theta < \varepsilon_2 + \theta.$$

So if $d \in \mathcal{G}$, then $\{d(x,\cdot) < \varepsilon_2 + \theta\}$ contains a set $\{e(x,\cdot) < \varepsilon_1\}^{(\theta)}$, with $e \in \mathcal{G}$.

2 $\Rightarrow$ 3. This follows from the fact that $\delta(x, A) = \sup_{d\in\mathcal{G}} \inf_{a\in A} d(x, a)$.

3 $\Rightarrow$ 1. Let $\mathcal{F} \in \mathsf{F}(X)$, $\varepsilon > 0$ and $x \in X$ be such that $\lambda\mathcal{F}(x) < \varepsilon$ then for each $\theta > 0$ and $d \in \mathcal{G}$ we have $\{d(x,\cdot) < \varepsilon\}^{(\theta)} \in \mathcal{F}^{(\theta)}$. Hence if $A \in \sec(\mathcal{F}^{(\theta)})$ then $\delta(x, A) \leq \varepsilon + \theta$ by 3 and hence $\lambda\mathcal{F}^{(\theta)} = \sup_{A\in\sec(\mathcal{F}^{(\theta)})} \delta(x, A) \leq \varepsilon + \theta$. □

We end this section with an extension theorem which was first proved in a different way by Jaeger (2014).

We consider the following situation. Let X and Y be approach spaces and consider a map $f : A \to Y$, where $A \subseteq X$. If f is a contraction, we want to find conditions under which we can extend it to a contraction g from $\overline{A}$ or a suitable subset of $\overline{A}$ to Y.

For $x \in X$ and $\varepsilon \in \mathbb{P}$ we introduce the following notations:

$$H_A^\varepsilon(x) := \{\mathcal{F} \in \mathsf{F}(X) \mid A \in \mathcal{F}, \lambda_X(\mathcal{F})(x) \leq \varepsilon\},$$

$$F_A^\varepsilon(x) := \begin{cases} \{y \in Y \mid \forall \mathcal{F} \in H_A^\varepsilon(x) : \lambda_Y(f(\mathcal{F}_{|A}))(y) \leq \varepsilon\} & \text{if } H_A^\varepsilon(x) \neq \emptyset, \\ Y & \text{if } H_A^\varepsilon(x) = \emptyset, \end{cases}$$

where $\mathcal{F}_{|A}$ stands for the restriction of the filter $\mathcal{F}$ to the set A.

Note that for $\varepsilon \leq \delta$, we have $H_A^\varepsilon(x) \subseteq H_A^\delta(x)$ and that $x \in A^{(\varepsilon)}$ if and only if $H_A^\varepsilon(x) \neq \emptyset$.

For a subset $A \subseteq X$, we now define A^* as follows

$$A^* := \Big\{x \in \overline{A} \mid \bigcap_{\alpha\in[0,\infty]} F_A^\alpha(x) \neq \emptyset\Big\}.$$

It is immediately verified that $A \subseteq A^* \subseteq \overline{A}$.

3.3.21 Theorem *Let X and Y be approach spaces where Y is regular. If $A \subseteq X$ and $f : A \to Y$ is a contraction, then there exists a contraction $g : A^* \to Y$ such that $g_{|A} = f$.*

Proof For $x \in A^* \setminus A$ we choose a value $y_x \in \bigcap_{\alpha \in [0,\infty]} F_A^\alpha(x)$ and we define

$$g(x) := \begin{cases} f(x) & \text{if } x \in A, \\ y_x & \text{if } x \in A^* \setminus A. \end{cases}$$

We show that g is a contraction. Let $\mathcal{G} \in \mathsf{F}(A^*)$, let $x_0 \in A^*$ and let $\alpha := \lambda_X \mathcal{G}(x_0)$.

Consider the selection $\sigma : A^* \to \mathsf{F}(A^*)$, where $\sigma(x)$ is such that $A \in \sigma(x)$ and $\lambda_X \sigma(x)(x) = 0$. The diagonal condition (L3) gives us

$$\begin{aligned} \lambda_X \Sigma\sigma(\mathcal{G})(x_0) &\leq \lambda_X \mathcal{G}(x_0) + \sup_{x \in A^*} \lambda_X \sigma(x)(x) \\ &= \lambda_X \mathcal{G}(x_0) \\ &= \alpha. \end{aligned}$$

Since $\Sigma\sigma(\mathcal{G}) = \bigcup_{G \in \mathcal{G}} \bigcap_{x \in G} \sigma(x)$ and $A \in \sigma(x)$, for every $x \in A^*$, we get $A \in \Sigma\sigma(\mathcal{G})$. Hence $\Sigma\sigma(\mathcal{G}) \in H_A^\alpha(x_0)$ and since $\Sigma\sigma(\mathcal{G})$ has a trace on A, we find

$$\lambda_Y f\big((\Sigma\sigma(\mathcal{G}))_{|A}\big)(g(x_0)) \leq \alpha.$$

We now consider the functions $\psi := g$ and

$$\rho : A^* \to \mathsf{F}(Y) : x \mapsto f(\sigma(x)_{|A}).$$

Since Y is regular, by 3.3.19, we find

$$\begin{aligned} \lambda_Y g(\mathcal{G})(g(x_0)) &\leq \lambda_Y \Sigma\rho(\mathcal{G})(g(x_0)) + \inf_{G \in \mathcal{G}} \sup_{x \in G} \lambda_Y f(\sigma(x)_{|A})(g(x)) \\ &= \lambda_Y \Sigma\rho(\mathcal{G})(g(x_0)). \end{aligned}$$

Hence from the fact that

$$\begin{aligned} \Sigma\rho(\mathcal{G}) &= \bigcup_{G \in \mathcal{G}} \bigcap_{x \in G} f(\sigma(x)_{|A}) \\ &= f\Big(\bigcup_{G \in \mathcal{G}} \bigcap_{x \in G} \sigma(x)_{|A}\Big) \\ &= f\big((\Sigma\sigma(\mathcal{G}))_{|A}\big), \end{aligned}$$

it now follows that

$$\lambda_Y g(\mathcal{G})(g(x_0)) \leq \lambda_Y f\big((\Sigma\sigma(\mathcal{G}))_{|A}\big)(g(x_0)) \leq \alpha,$$

which completes the proof. □

If Y is moreover a T_2-space then we see that for each $x \in X$, $F_A^0(x)$ has at most one point. In this case $\bigcap_{\alpha \in [0,\infty]} F_A^\alpha(x)$ has at most one point and the extension g will be unique.

3.3.22 Theorem *Let X and Y be approach spaces where Y is regular and T_2. If $A \subseteq X$ is dense and $f : A \to Y$ is a contraction, then the following properties are equivalent.*

1. *There is a unique contraction $g : X \to Y$ such that $g_{|A} = f$.*
2. *For each $x \in X$: $\bigcap_{\alpha \in [0,\infty]} F_A^\alpha(x) \neq \emptyset$.*

Proof 1 $\Rightarrow$ 2. If f has such an extension g, then for $x \in X$ we have $g(x) \in \bigcap_{\alpha \in [0,\infty]} F_A^\alpha(x)$. Indeed, from the density of A it follows that $H_A^\alpha(x) \neq \emptyset$ for any $\alpha \in [0, \infty]$. For $\mathscr{F} \in H_A^\alpha(x)$ we have that $A \in \mathscr{F}$ and $\lambda_X(\mathscr{F})(x) \leq \alpha$. Hence $\lambda_Y g(\mathscr{F})(g(x)) \leq \alpha$ and it follows that

$$\lambda_Y f(\mathscr{F}_{|A})(g(x)) = \lambda_Y g(\mathscr{F}_{|A})(g(x)) \leq \lambda_Y g(\mathscr{F})(g(x)) \leq \alpha,$$

i.e. $g(x) \in F_A^\alpha(x)$.

2 $\Rightarrow$ 1. If, conversely, $\bigcap_{\alpha \in [0,\infty]} F_A^\alpha(x) \neq \emptyset$ for all $x \in X$, then the existence of an extension $g : X \to Y$ follows from 3.3.21 and uniqueness follows from the fact that Y is T_2. □

3.3.23 Corollary (Top) *If X is a topological space, $A \subseteq X$ is dense and Y is a Hausdorff regular topological space, then a continuous function $f : A \longrightarrow Y$ will have a unique continuous extension to X if and only if for any $x \in X$ the filter $f(\mathscr{V}(x)_{|A})$ converges.*

3.4 Countability

In contrast to the situation in topological spaces, in approach spaces there are at least three defining structures which may be generated by countable collections. The approach system, a regular function frame and the gauge. The first two correspond to first and second countability in the case of topological spaces. The latter on the other hand is a purely approach notion which, as we will see, refers to quasi-metrizability.

3.4.1 Definition An approach space is called *locally countable* if in each point the approach system has a countable basis.

3.4.2 Theorem *An approach space is locally countable if and only if all pretopologies from its tower are first countable.*

Proof The only if-part follows immediately from 1.2.47. To show the if-part, let $\mathscr{B}_\gamma(x)$ be a countable basis for $\mathscr{V}_\gamma(x)$ for each $x \in X$ and $\gamma \in \mathbb{R}^+$. We may of course suppose that all members of $\mathscr{B}_\gamma(x)$ are of type $\{\varphi < \beta\}$ for some $\varphi \in \mathscr{A}(x)$ and $\beta > \gamma$. For each γ put $\mathscr{F}_\gamma(x)$ the (countable) set of functions from $\mathscr{A}(x)$ which appear in $\mathscr{B}_\gamma(x)$. Let $\psi \in \mathscr{A}(x)$ be bounded and let $\varepsilon > \frac{1}{m} > 0$. Let n be such that $\frac{n-1}{m} < \sup \psi \leq \frac{n}{m}$. For each $k = 1, \ldots, n$ take γ_k, β_k and $\varphi_k \in \mathscr{F}_{\frac{k}{m}}(x)$ be such that $\frac{k}{m} < \gamma_k < \frac{k+1}{m}$ and $\frac{k}{m} < \beta_k < \frac{k+1}{m}$ and such that for all k

$$\{\varphi_k < \gamma_k\} \subseteq \{\psi < \beta_k\}.$$

Put $\varphi := \sup_{k=1}^n \varphi_k$ then it follows that $\psi \leq \varphi + 2\varepsilon$. Hence

$$\mathscr{B}(x) := \{\sup_{\varphi \in \mathscr{E}} \varphi \mid \mathscr{E} \subseteq \cup_{q \in \mathbb{Q}} \mathscr{F}_q(x) \text{ finite}\}$$

is a countable basis for $\mathscr{A}(x)$. □

3.4.3 Proposition (Top, qMet) *A topological approach space is locally countable if and only if the underlying topology is first countable and a quasi-metric approach space is always locally countable.*

Proof This follows at once from the foregoing result and the definition and we leave this to the reader. □

3.4.4 Corollary *If an approach space is locally countable then the topological coreflection is first countable.*

The converse of the foregoing corollary does not hold. In order to see this it suffices to take a set X and consider a tower consisting of the discrete topology for $0 \leq \varepsilon < 1$ and a non-first countable topology for $1 \leq \varepsilon$. Then the topological coreflection is discrete but the space is obviously not locally countable.

In the following result we put $\mathsf{S}(A)$ for the set of all sequences in A.

3.4.5 Proposition *If an approach space X is locally countable then for any $A \subseteq X$ and $x \in X$*

$$\delta(x, A) = \inf_{(x_n)_n \in \mathsf{S}(A)} \lambda\langle(x_n)_n\rangle(x)$$

Proof One inequality is trivial, to show the other one let $\delta(x, A)$ be finite and let $\{\varphi_n \mid n \in \mathbb{N}\}$ be a countable increasing basis for $\mathscr{A}(x)$. Then, making use of 1.2.34, it can easily be seen that for any $\varepsilon > 0$ there exists a sequence $(x_n)_n$ in A such that for any $m \leq n$: $\varphi_m(x_n) - \varepsilon \leq \delta(x, A)$. From this, making use of 1.2.43, it immediately follows that $\lambda\langle(x_n)_n\rangle(x) - \varepsilon \leq \delta(x, A)$. □

3.4.6 Proposition *For approach spaces X and X', if X is locally countable and $f : X \longrightarrow X'$ then the following properties are equivalent.*

1. *f is a contraction.*
2. *For any $(x_n)_n$ and x in X: $\lambda'\langle(f(x_n))_n\rangle(f(x)) \leq \lambda\langle(x_n)_n\rangle(x)$.*

Proof This follows from the foregoing result, 1.2.2 and the definition of contraction. □

3.4.7 Proposition *A countable product of locally countable approach spaces is locally countable, in particular a countable product of quasi-metric spaces is locally countable.*

Proof This follows from 1.3.9 and the definition. □

The concept of local countability will be especially useful in the sections dealing with applications in probability theory.

3.4.8 Definition An approach space is called *gauge-countable* if its gauge has a countable basis.

3.4.9 Theorem *An approach space is gauge-countable if and only if it is a subspace of a countable product of quasi-metric spaces.*

Proof The if-part follows from 1.3.11. Conversely, if $\mathscr{B}$ is a countable basis for the gauge $\mathscr{G}$ of X then the following is an embedding

$$(X, \mathscr{G}) \longrightarrow \prod_{d \in \mathscr{B}} (X, d) : x \mapsto (x_d := x)_{d \in \mathscr{B}}.$$

This immediately implies the only if-part. □

3.4.10 Proposition *If an approach space is gauge-countable then its topological coreflection is quasi-metrizable.*

Proof With the usual proof it is easily seen that if $\{d_n \mid n \in \mathbb{N}\}$ is a countable basis for the gauge of X then

$$d := \sup_{n \in \mathbb{N}} \frac{1}{n+1} d_n \wedge 1$$

is a quasi-metric for the topological coreflection. □

3.4.11 Proposition (Top, qMet) *A topological approach space is gauge-countable if and only if the underlying topology is quasi-metrizable and a quasi-metric approach space is always gauge-countable.*

Proof The only if-part follows from 3.4.10. Conversely, if d is any quasi-metric which metrizes the underlying topology then the family $\{nd \mid n \in \mathbb{N}\}$ is a basis for the gauge. To see this, let e be such that $\mathscr{T}_e \subseteq \mathscr{T}_d$. Fix $x \in X$, $\varepsilon > 0$ and $\omega < \infty$. Divide $[0, \omega]$ into a finite number of intervals with length less than ε, $0 < \varepsilon_1 < \varepsilon_2 < \cdots < \varepsilon_m < \omega$. Then for each $k = 1, \ldots, m$ there exists $\delta_k > 0$ such that $B_d(x, \delta_k) \subseteq B_e(x, \varepsilon_k)$. For each $k = 1, \ldots, m$ take n_k such that $\varepsilon_k \leq n_k \delta_k$ and put $n := \max_{k=1}^m n_k$. Then one can verify that $B_{nd}(x, \varepsilon_k) \subseteq B_e(x, \varepsilon_k)$ for each $k = 1, \ldots, m$ from which it follows that $e(x, \cdot) \wedge \omega \leq nd(x, \cdot) + \varepsilon$. The result now follows from 2.1.1.

The second claim is evident. □

3.4.12 Proposition *A countable product of gauge-countable approach spaces is gauge-countable.*

Proof This follows from 3.4.9. □

3.4.13 Definition An approach space is called *regular-countable* if there exists a countable basis $\mathfrak{B}$ for the lower regular function frame.

Note also that from 1.2.49 and 1.2.50 it follows that the lower regular function frame has a countable basis if and only if the upper regular function frame has a countable basis.

3.4.14 Proposition *If an approach space is uniform, gauge-countable and has a separable topological coreflection then it is regular-countable.*

Proof If $\mathscr{B}$ is a countable gauge basis and $Y \subseteq X$ is a countable dense subset then it is easily verified that

$$\mathfrak{B} := \{\alpha \ominus d(x, \cdot) \mid \alpha \in \mathbb{P} \cap \mathbb{Q}, d \in \mathscr{B}, x \in Y\}$$

is a countable basis for the lower regular function frame. □

3.4.15 Proposition (Top) *A topological approach space is regular-countable if and only if the underlying topology is second countable.*

Proof If $\{B_n \mid n \in \mathbb{N}\}$ is a countable basis for the topology then

$$\{r \wedge \theta_{X\setminus B_n} + s \mid r, s \in \mathbb{Q}_+, n \in \mathbb{N}\}$$

is a countable basis for the lower regular function frame. Conversely if a countable basis $\{\mu_n \mid n \in \mathbb{N}\}$ for the lower regular function frame is given then

$$\{\{\mu_n > r\} \mid r \in \mathbb{Q}_+, n \in \mathbb{N}\}$$

is a countable basis for the topology. □

3.4.16 Proposition (Met) *A metric approach space is regular-countable if and only if the underlying topology is second countable.*

Proof The if-part follows from 3.4.14. The only-if-part is analogous to 3.4.15. □

3.4.17 Proposition *If an approach space is regular-countable then it is gauge-countable and if it is gauge-countable then it is locally countable.*

Proof The first claim follows from 1.2.55 and the second from 1.2.11. □

That in the foregoing result neither arrow can be reversed can already be seen from the characterizations for topological approach spaces.

3.5 Completeness

We will show that there exists a natural notion of completeness which, moreover, in the setting of uniform approach spaces, allows for a categorically sound completion theory as we will see in Chap. 6.

Intuitively, a Cauchy filter is a filter which has all the properties of a convergent filter, except that its limit points may be missing. This idea can be nicely captured using the concept of limit operator.

3.5.1 Definition A filter $\mathcal{F}$ in an approach space X is called a *δ-Cauchy filter* or a *Cauchy filter* for short if

$$\inf_{x \in X} \lambda\mathcal{F}(x) = 0.$$

We say that $\mathcal{F}$ is *convergent* (to x) if $\lambda\mathcal{F}(x) = 0$, i.e. if $\mathcal{F}$ converges to x in the topological coreflection.

Hence in any approach space, a convergent filter is a Cauchy filter.

3.5.2 Proposition *If X is a uniform approach space with symmetric gauge basis $\mathcal{D}$, then for a filter $\mathcal{F}$ on X the following properties are equivalent.*

1. *$\mathcal{F}$ is a Cauchy filter.*
2. $\forall \varepsilon > 0, \exists x \in X, \forall d \in \mathcal{D} : B_d(x, \varepsilon) \in \mathcal{F}$.

Proof From 1.2.44 it follows that $\mathcal{F}$ is a Cauchy filter if and only if

$$\inf_{x \in X} \sup_{d \in \mathcal{D}} \inf_{F \in \mathcal{F}} \sup_{y \in F} d(x, y) = 0,$$

from which the result follows immediately. □

If (X, δ) is an arbitrary approach space, then we can consider its Met-coreflection (X, d_δ) and we have a metric notion of Cauchy filter at our disposal. If (X, δ) is a uniform approach space and δ is generated by the symmetric gauge basis $\mathcal{D}$, then we can consider the underlying uniform space $(X, \mathcal{U}(\mathcal{D}))$ and we have a uniform notion of Cauchy filter at our disposal. The following result gives the relationship between these concepts.

3.5.3 Proposition *If X is an approach space, then a d_δ-Cauchy filter is a δ-Cauchy filter, and if X is uniform with $\mathcal{D}$ a symmetric gauge basis, then moreover a δ-Cauchy filter is a $\mathcal{U}(\mathcal{D})$-Cauchy filter.*

Proof That $\mathcal{F}$ is a d_δ-Cauchy filter means that

$$\forall \varepsilon > 0, \exists x \in X : \bigcap_{d \in \mathcal{D}} B_d(x, \varepsilon) \in \mathcal{F}.$$

That $\mathscr{F}$ is a $\mathscr{U}(\mathscr{D})$-Cauchy filter means that it is a d-Cauchy filter, for all $d \in \mathscr{D}$, i.e.

$$\forall \varepsilon > 0, \forall d \in \mathscr{D}, \exists x \in X : B_d(x, \varepsilon) \in \mathscr{F}.$$

Hence, the result follows from 3.5.2. □

3.5.4 Example The implications in the foregoing proposition are strict in general.

1. Let $X := \mathbb{R} \setminus \{0\}$ and let $d_{\mathbb{E}}$ stand for the Euclidean metric on X. If we let $\mathscr{D} := \{\alpha d_{\mathbb{E}} \mid \alpha > 0\}$, then the distance δ generated by $\mathscr{D}$ is topological. The restriction of the Euclidean neighbourhood filter of 0 to X, say $\mathscr{F}$, does not converge in (X, δ) and hence, by 3.5.6, it is not a $\delta_{\mathscr{D}}$-Cauchy filter. However, $\mathscr{U}(\mathscr{D})$ is the usual uniformity on X and hence $\mathscr{F}$ is a $\mathscr{U}(\mathscr{D})$-Cauchy filter.

2. Let $\mathbb{R}$ be equipped with the Euclidean metric and consider the function space $\mathbb{R}^{\mathbb{R}}$ to be equipped with the product distance. For each $\alpha > 0$, consider the function

$$f_\alpha : \mathbb{R} \longrightarrow \mathbb{R} : x \longrightarrow \alpha x,$$

and let Ψ be the filter on $\mathbb{R}^{\mathbb{R}}$ generated by the basis

$$\{\{f_\beta \mid \beta \leq \alpha\} \mid \alpha > 0\}.$$

Since Ψ is convergent for the pointwise topology it follows that Ψ is a δ-Cauchy filter. However, since d_δ is the uniform metric, Ψ is not a d_δ-Cauchy filter.

3.5.5 Proposition *In a uniform approach space the following properties hold.*

1. *If $\mathscr{F}$ is a Cauchy filter, then $\lambda\mathscr{F} = \alpha\mathscr{F}$.*
2. *If $\mathscr{F}$ and $\mathscr{G}$ are Cauchy filters and $\mathscr{F} \subseteq \mathscr{G}$, then $\lambda\mathscr{F} = \lambda\mathscr{G}$.*

Proof 1. Let $\mathscr{D}$ be a symmetric gauge basis which generates δ and let $\mathscr{F}$ be a Cauchy filter. Let $\varepsilon > 0$ and by 3.5.2 choose $z \in X$ such that for all $d \in \mathscr{D}$: $B_d(z, \varepsilon) \in \mathscr{F}$. Then we obtain

$$\begin{aligned}
\lambda\mathscr{F}(x) &= \sup_{d\in\mathscr{D}} \inf_{F\in\mathscr{F}} \sup_{y\in F} d(x, y) \\
&\leq \sup_{d\in\mathscr{D}} \sup_{y\in B_d(z,\varepsilon)} d(x, y) \\
&\leq \sup_{d\in\mathscr{D}} \inf_{y\in B_d(z,\varepsilon)} d(x, y) + 2\varepsilon \\
&\leq \alpha\mathscr{F}(x) + 2\varepsilon
\end{aligned}$$

and the result follows from the arbitrariness of ε and 1.2.65.

2. This follows from 1 since, again applying 1.2.65, we now have

$$\lambda\mathscr{F} = \alpha\mathscr{F} = \alpha\mathscr{G} = \lambda\mathscr{G}.$$

□

3.5.6 Proposition (Top, Met) *For a filter $\mathcal{F} \in \mathsf{F}(X)$ the following properties hold.*

1. *If $(X, \delta_{\mathcal{T}})$ is a topological approach space, then $\mathcal{F}$ is a Cauchy filter if and only if it is convergent in $(X, \mathcal{T})$.*
2. *If (X, δ_d) is a metric approach space, then $\mathcal{F}$ is a Cauchy filter if and only if it is a Cauchy filter in (X, d).*

Proof 1. This follows from the fact that in topological approach spaces the limit operator attains only the values 0 and ∞.
2. This follows from 3.5.2. □

3.5.7 Definition An approach space is called *complete* if every Cauchy filter converges.

3.5.8 Proposition (Top, Met) *A topological approach space is always complete and a metric approach space is complete if and only if it is complete as a metric space.*

Proof This follows from 3.5.6 and 2.3.6. □

3.5.9 Proposition *If X is a uniform approach space with symmetric gauge basis $\mathcal{D}$ and $(X, \mathcal{U}(\mathcal{D}))$ is complete, then X is complete.*

Proof Let $\mathcal{F}$ be a Cauchy filter. It follows from 3.5.3 that $\mathcal{F}$ is a $\mathcal{U}(\mathcal{D})$-Cauchy filter. Since $(X, \mathcal{U}(\mathcal{D}))$ is complete it follows that $\mathcal{F}$ converges. □

The converse of the foregoing result does not hold as shown by the following example.

3.5.10 Example Consider the first example from 3.5.4. Since X is topological it is complete. However, since $\mathcal{U}(\mathcal{D})$ is the usual uniformity $(X, \mathcal{U}(\mathcal{D}))$ is not complete.

3.5.11 Theorem *A uniform approach space (X, δ) is complete if and only if (X, d_δ) is complete.*

Proof Suppose that (X, δ) is complete and that δ is generated by the symmetric gauge basis $\mathcal{D}$. Let $(x_n)_n$ be a Cauchy sequence in (X, d_δ). Then we have that

$$\forall \varepsilon > 0, \exists n_0, \forall p, q \geq n_0, \forall d \in \mathcal{D} : d(x_p, x_q) \leq \varepsilon.$$

This implies that

$$\lambda \langle (x_n)_n \rangle (x_{n_0}) = \sup_{d \in \mathcal{D}} \inf_m \sup_{k \geq m} d(x_{n_0}, x_k) \leq \varepsilon,$$

and hence it follows that $\langle (x_n)_n \rangle$ is a δ-Cauchy filter. Since (X, δ) is complete this implies that there exists $x \in X$ such that $\lambda \langle (x_n)_n \rangle (x) = 0$, i.e.

$$\forall d \in \mathcal{D} : \lim_{n \to \infty} d(x_n, x) = 0.$$

Now let $\varepsilon > 0$. Then from the fact that $(x_n)_n$ is a d_δ-Cauchy sequence it follows that we can find n_0 such that

$$\forall d \in \mathscr{D}, \forall p, q \geq n_0 : d(x_p, x_q) \leq \varepsilon.$$

Taking the limit for $q \to \infty$ it follows that

$$\forall d \in \mathscr{D}, \forall p \geq n_0 : d(x_p, x) \leq \varepsilon,$$

which implies that $(x_n)_n \longrightarrow x$ in (X, d_δ).

Conversely, suppose that (X, d_δ) is complete and let $\mathscr{F}$ be a δ-Cauchy filter. Then

$$\inf_{x \in X} \lambda\mathscr{F}(x) = \inf_{x \in X} \sup_{d \in \mathscr{D}} \inf_{F \in \mathscr{F}} \sup_{y \in F} d(x, y) = 0,$$

which implies that

$$\forall n > 0, \exists x_n \in X, \forall d \in \mathscr{D} : B_d(x_n, \frac{1}{n}) \in \mathscr{F}.$$

Since the intersection of any two of these balls is nonempty this implies that for any $p, q > 0$

$$d_\delta(x_p, x_q) = \sup_{d \in \mathscr{D}} d(x_p, x_q) \leq \frac{1}{p} + \frac{1}{q},$$

from which it follows that $(x_n)_n$ is a d_δ-Cauchy sequence. By the completeness of (X, d_δ) it follows that there exists $x \in X$ such that $(x_n)_n \xrightarrow{d_\delta} x$. It then follows that

$$\begin{aligned} \lambda\mathscr{F}(x) &= \sup_{d \in \mathscr{D}} \inf_{F \in \mathscr{F}} \sup_{y \in F} d(x, y) \\ &\leq \sup_{d \in \mathscr{D}} \inf_{n > 0} \sup_{y \in B_d(x_n, \frac{1}{n})} d(x, y) \\ &\leq \sup_{d \in \mathscr{D}} \inf_{n > 0} (d(x, x_n) + \frac{1}{n}) = 0, \end{aligned}$$

and hence $\mathscr{F}$ converges in (X, δ). □

3.6 Comments

1. **Uniformity in topology**

Besides the original definition by Weil (1937), making use of entourages of the diagonal, there are several other ways in which uniformity in the setting of topology can be characterized. See, for example, the work by Herrlich on nearness structures (Herrlich 1974a) and by Isbell on covering uniformities (Isbell 1964). The characterization which however is most closely related to our setup is the one via collections of metrics (see e.g. Gillman and Jerison 1976). Both the definitions of gauges and of (quasi-)uniform gauges, which we will define in Chap. 5, bear similarity to the definition of uniformities by means of ideals of (quasi-)metrics.

2. **Weak-adjointness in the literature**

Weak-adjointness is a concept which is known in the theory of quasi-uniform spaces. An alternative characterization states that a quasi-metric space is weakly adjoint if and only if for all $x \in X$ the map $d(\cdot, x)$ is upper semicontinuous. See e.g. Fletcher and Lindgren (1982), where 3.2.2 is also mentioned.

3. **Reflectivity of** App_0, App_1 **and** App_2

For any approach space X the quotient defined by the equivalence relation $x \sim y \Leftrightarrow \mathscr{A}(x) = \mathscr{A}(y)$ determines the App_0-reflection.

In order to arrive at the App_1-reflection we first define an approach space to be R_0 if $\mathscr{A}(x) = \bigcap_{y \sim_{R_0} x} \mathscr{A}(y)$ for all x where $y \sim_{R_0} x \Leftrightarrow \delta(x, \{y\}) = 0$. The subcategory consisting of all R_0-spaces is concretely reflective and using transfinite iteration, we can construct an R_0-reflection. The App_1-reflection then is obtained by taking this reflection followed by the T_0-reflection.

In a similar fashion, in order to obtain the App_2-reflection, we call an approach space an R-space if $\mathscr{A}(x) = \bigcap_{y \sim_R x} \mathscr{A}(y)$ for all x where $y \sim_R x$ if there exist $x_1 := x, \ldots, x_n := y$ such that for all $i \in \{1, \ldots, n-1\}$: $c(\mathscr{A}(x_i) \vee \mathscr{A}(x_{i+1})) = 0$. Again the subcategory consisting of all R-spaces is concretely reflective and by a similar procedure as in the T_1-case we construct an R-reflection. This reflection followed by the T_0-reflection is the App_2-reflection.

For details we refer the interested reader to Lowen and Sioen (2003).

4. **Completeness in metrically generated theories**

The notion of completeness in UAp as explained in this chapter is a special case of a more general approach in the setting of metrically generated theories (see Colebunders and Lowen 2005; Colebunders and Vandersmissen 2010). See also the comments on Chap. 6.

5. **Sequential approach spaces**

In an unpublished partial manuscript Gutierres, Hofmann and Van Olmen studied what they called "sequential approach spaces" and "Fréchet" approach spaces. For instance, an approach X space is said to be Fréchet if for any $x \in X$ and $A \subseteq X$

$$\delta(x, A) = \inf_{(x_n)_n \in S(A)} \sup_{k\uparrow} \delta(x, \{x_{k_n} \mid n \in \mathbb{N}\})$$

where $S(A)$ stands for the set of all sequences in A. Sequentiality is defined in a analogous manner, and they show several results similar to what is obtained in the topological case.

Chapter 4
Index Analysis

When you can measure what you are talking about and express it in numbers, you know something about it.

(Lord William Thomson Kelvin)

Finally - and frankly it's a relief to see it - Karl Weierstrass sorted out the muddle in 1850 or thereabouts by taking the phrase "as near as we please" seriously. How near do we please?

(Ian Stewart)

Before we start with the formal definitions and results we would like to reflect on the precise nature of approach spaces and the possibilities for defining natural invariants.

Rather than being restricted to asserting that a space, or a subset of a space, or a function, or any other item defined in terms of the structures at hand, does or does not have a certain property we now have at our disposal a machinery by means of which we can define numerical indices of properties. The smaller the index the better the property is approximated.

This idea is not new. In Kuratowski (1930) the author introduced what he called a measure of noncompactness in the setting of complete metric spaces. Various other measures were also introduced in the literature, especially the so-called Hausdorff measure of noncompactness, and used in a variety of fields, especially in the theory of Banach spaces, functional analysis, fixed point theory and hyperspaces (see the comments at the end of this chapter for detailed references). The purely topological concept of compactness was adapted in a numerical way to be used in the setting of metric spaces in order to measure the deviation an object may have from being compact. However, it is quite clear from the original definition of this measure that it is actually a measure of not being totally bounded, which only in the case that one works with complete sets can be referred to compactness. We will see how these facts, which reveal at the same time similarity and dualism between the topological and metric aspects, completely resolve themselves in the setting of approach spaces.

R. Lowen, *Index Analysis*, Springer Monographs in Mathematics,
DOI 10.1007/978-1-4471-6485-2_4

In this chapter it will become abundantly clear that the systematic use of indices lies at the heart of approach theory as they are built into the basics of the theory. Besides the indices which we will introduce in the present chapter, now that we know how the categories of approach spaces, topological spaces and (quasi-)metric spaces relate to one another, it is also entirely natural to consider:

1. δ distance = index of closure, $\delta(x, A)$ indicates how far x is away from being in the closure of A,
2. λ limit operator = index of convergence, $\lambda\mathscr{F}(x)$ indicates how far x is away from being a limit point of $\mathscr{F}$,
3. α adherence operator = index of adherence, $\alpha\mathscr{F}(x)$ indicates how far x is away from being an adherence point of $\mathscr{F}$.

Since distance and limit operator are defining structures of the theory of approach spaces, the concept of indices is endemic to approach theory.

4.1 Morphism- and Object-Indices

We will formally describe two types of indices which are of paramount importance to the theory: indices for objects and indices for morphisms. Of course other types of indices will appear where needed (or have already been encountered as for example distance, limit and adherence operator mentioned above) but these two are structurally sufficiently important to merit a short formal treatment. Since we will use these considerations not only in the local setting of approach spaces but also in the uniform setting of uniform gauge spaces which we introduce in the following chapter, we present the concepts for an arbitrary topological category.

We suppose that $\mathscr{C}$ is a topological category, concrete over Set and we define $\mathscr{C}_{\mathsf{Set}}$ to be the category with objects the same as those in $\mathscr{C}$ and morphisms, functions between the underlying sets.

4.1.1 Definition An *object-index* is an assignment

$$\chi : \mathrm{Obj}\mathscr{C}_{\mathsf{Set}} = \mathrm{Obj}\mathscr{C} \longrightarrow \mathbb{P}$$

which satisfies the following property.

(OI) If X and Y are isomorphic objects in $\mathscr{C}$ then $\chi(X) = \chi(Y)$.

Functions $f : X \longrightarrow Y$ and $g : U \longrightarrow V$ (i.e. morphisms in $\mathscr{C}_{\mathsf{Set}}$) are said to be *isomorphic* if there exist isomorphisms (in $\mathscr{C}_{\mathsf{Set}}$, i.e. bijections) $h : X \longrightarrow U$ and $k : Y \longrightarrow V$ such that the diagram

$$\begin{array}{ccc} X & \xrightarrow{f} & Y \\ {\scriptstyle h}\downarrow & & \downarrow{\scriptstyle k} \\ U & \xrightarrow[g]{} & V \end{array}$$

is commutative.

This means exactly that f and g are isomorphic objects in the arrow category of $\mathscr{C}_{\mathsf{Set}}$. If we impose the supplementary condition that h and k have to be isomorphisms in $\mathscr{C}$, and not merely in $\mathscr{C}_{\mathsf{Set}}$, i.e. merely bijective functions, we obtain a stronger concept which we will refer to as f and g being *$\mathscr{C}$-isomorphic*.

4.1.2 Definition A *morphism-index* is an assignment

$$\chi : \mathrm{Mor}\mathscr{C}_{\mathsf{Set}} = \mathrm{MorSet} \longrightarrow \mathbb{P}$$

which satisfies the following properties.

(MI1) χ vanishes on identity mappings.
(MI2) For any $f : X \longrightarrow X'$ and $g : X' \longrightarrow X''$ we have $\chi(g \circ f) \leq \chi(f) + \chi(g)$.
(MI3) If f and g are $\mathscr{C}$-isomorphic then $\chi(f) = \chi(g)$.

We make no difference in notation between object- and morphism-indices since it will always be clear from the context which type of index is meant.

Object-indices will gauge to what extent certain properties of objects are satisfied and similarly morphism-indices gauge to what extent certain properties of functions are satisfied.

It should be pointed out that we are of course in the first place thinking of a morphism-index which gauges to what extent functions deviate from being morphisms in $\mathscr{C}$. In the case of App this then means to what extent functions deviate from being contractions. However, in topological categories, and especially in App we also have natural other concepts for functions as we have seen in Sect. 1.4. Hence we will also introduce morphism-indices gauging to what extent functions deviate from being closed-expansive, open-expansive and proper. The above definition is of the right generality to allow for these various indices.

4.1.3 Proposition *Let χ be a morphism-index. If $f : X \longrightarrow Y$ and $g : U \longrightarrow V$ are isomorphic functions then*

$$|\chi(f) - \chi(g)| \leq \max\{\chi(h) + \chi(k^{-1}), \chi(h^{-1}) + \chi(k)\}$$

for any bijections h and k such that $k \circ f = g \circ h$.

Proof This follows from (MI2) since if we consider the diagram

$$\begin{array}{ccc} X & \xrightarrow{f} & Y \\ {\scriptstyle h}\downarrow & & \downarrow{\scriptstyle k} \\ U & \xrightarrow[g]{} & V \end{array}$$

we have

$$\chi(g) \le \chi(h^{-1}) + \chi(f) + \chi(k) \quad \text{and} \quad \chi(f) \le \chi(h) + \chi(g) + \chi(k^{-1}). \qquad \square$$

4.1.4 Definition Let χ be a morphism-index which vanishes on morphisms in the category $\mathscr{C}$. If an object-index χ_0 satisfies the property that there exists a constant c such that for all spaces X and X' with equipotent underlying sets

$$|\chi_0(X) - \chi_0(X')| \le c \inf\{\chi(f) + \chi(f^{-1}) \mid f : X \longrightarrow X' \text{ bijective}\},$$

we say that χ_0 is *(c, χ)-layered*.

The idea behind this is to sharpen condition (OI) by not only asking that an object-index be constant on the equivalence classes of isomorphic objects but that it would also vary less as objects become "more isomorphic", provided of course that the associated morphism-index vanishes on morphisms of the category $\mathscr{C}$, in particular that it is an index that gauges to what extent a function deviates from being a morphism. Hence, in our setting, we will only be interested in the morphism-index of contractivity which we introduce in the following section. In keeping with the philosophy of the theory the relation imposed is a Lipschitz-type inequality.

In order to verify that an object-index is (c, χ)-layered it is sufficient to test it on identity functions in $\mathscr{C}_{\mathsf{Set}}$. In the following we denote the identity function between the underlying sets of objects $U := (X, s)$ and $V := (X, s')$ in $\mathscr{C}$ by 1_{UV}.

4.1.5 Proposition *Let χ be a morphism-index which vanishes on morphisms in the category $\mathscr{C}$. An object-index χ_0 is (c, χ)-layered if and only if for any objects $U := (X, s)$ and $V := (X, s')$ in $\mathscr{C}$, with the same underlying set we have that*

$$|\chi_0(U) - \chi_0(V)| \le c(\chi(1_{UV}) + \chi(1_{VU})).$$

Proof The only-if part is evident. To show the if-part let $f : X \longrightarrow Y$ be a bijective function between $\mathscr{C}$-objects. Consider the diagram

$$\begin{array}{ccc} (X, s_X) & \xrightarrow{f} & (Y, s_Y) \\ {\scriptstyle 1_X}\downarrow & & \downarrow{\scriptstyle g} \\ U & \xrightarrow[1_{UV}]{} & V \end{array}$$

where $U := (X, s_X)$. Further $V := (X, s')$ is obtained by considering the sink $g : (Y, s_Y) \longrightarrow X$ where g, as a function, is equal to f^{-1} and s' stands for the final structure on X. Then both 1_X and g are isomorphisms in $\mathscr{C}$ and it follows that $\chi(f) = \chi(1_{UV})$. Upon interchanging the roles of X and Y we obtain that also $\chi(f^{-1}) = \chi(1_{VU})$ and we are finished. □

Throughout the sequel, if an index is equal to 0 we will often refer to this as a 0-property. For instance, once we have introduced the index of compactness χ_c, when a space satisfies $\chi_c(X) = 0$ then we will say the space is 0-compact. This convention will not be used in case the vanishing of a certain index is equivalent to an earlier defined property. For instance, when we define the index of contractivity in the next section, it will turn out that being 0-contractive means being contractive. In that case we obviously drop the 0.

4.1.6 Definition If two categories $\mathscr{C}$ and $\mathscr{D}$ are equipped with morphism-indices $\chi_{\mathscr{C}}$ and $\chi_{\mathscr{D}}$ and $F : \mathscr{C}_{\mathsf{Set}} \longrightarrow \mathscr{D}_{\mathsf{Set}}$ is a functor then we say that F is a $(\chi_{\mathscr{C}}, \chi_{\mathscr{D}})$*-true functor* if $\chi_{\mathscr{D}}(Ff) \leq \chi_{\mathscr{C}} f$ for any function f.

4.2 $\mathsf{App}_{\mathsf{Set}}$-Morphism-Indices

In this section we will define four morphism-indices related to the concepts of contraction (1.3.1), closed expansion (1.4.1), open expansion (1.4.5) and proper map (1.4.10).

4.2.1 Theorem *If X and X' are approach spaces, $f : X \longrightarrow X'$ is a function and $c \in \mathbb{P}$, then the following properties are equivalent.*

1. $\forall A \subseteq X : \delta'_{f(A)} \circ f \leq \delta_A + c.$
2. $\forall \mathscr{F} \in \mathsf{F}(X) : \lambda'(f(\mathscr{F})) \circ f \leq \lambda \mathscr{F} + c.$
3. $\forall \mathscr{F} \in \mathsf{U}(X) : \lambda'(f(\mathscr{F})) \circ f \leq \lambda \mathscr{F} + c.$
4. $\forall x \in X, \forall \varphi' \in \mathscr{A}'(f(x)) : \varphi' \circ f \ominus c \in \mathscr{A}(x).$
5. $\forall d' \in \mathscr{G}' : d' \circ (f \times f) \ominus c \in \mathscr{G}.$
6. $\forall A \subseteq X, \forall \varepsilon \in \mathbb{R}^+ : f(t_\varepsilon(A)) \subseteq t'_{\varepsilon+c}(f(A)).$
7. $\forall x \in X, \forall \varepsilon \in \mathbb{R}^+, \forall V' \in \mathscr{V}'_{\varepsilon+c}(f(x)) : f^{-1}(V') \in \mathscr{V}_\varepsilon(x).$
8. $\forall \mu \in \mathbb{P}^X : \mathfrak{l}'(f(\mu)) \circ f \leq \mathfrak{l}(\mu) + c.$
9. $\forall \mu' \in \mathfrak{L}' : \mu' \circ f \leq \mathfrak{l}(\mu' \circ f) + c.$
10. $\forall \mu' \in \mathfrak{U}' : \mathfrak{u}(\mu' \circ f) \leq \mu' \circ f + c.$
11. $\forall \mathfrak{I} \in \mathfrak{F}(X) : \mathfrak{I} \rightarrowtail x \Rightarrow f(\mathfrak{I}) \oplus c \rightarrowtail f(x).$
12. $\forall \mathfrak{I} \in \mathfrak{P}(X) : \mathfrak{I} \rightarrowtail x \Rightarrow f(\mathfrak{I}) \oplus c \rightarrowtail f(x).$

In (4) *a basis for $\mathscr{A}'(f(x))$ is sufficient, in* (5) *a basis for $\mathscr{G}'$ is sufficient and in* (8) *and* (9) *subbases for respectively $\mathfrak{L}'$ and $\mathfrak{U}'$ are sufficient.*

Proof This is analogous to 1.3.3 and we leave this to the reader. □

4.2.2 Definition (Index of contractivity) A function f is called *c-contractive* if it has one, and therefore all the properties mentioned in 4.2.1.

We define the *index of contractivity* as

$$\boxed{\chi_c : \mathsf{MorApp}_{\mathsf{Set}} \longrightarrow \mathbb{P} : f \mapsto \min\{c \mid f \text{ is } c\text{-contractive}\}}$$

Note that the minimum actually exists and that this definition implies that for any function $f : X \longrightarrow X'$ between approach spaces and for any subset $A \subseteq X$

$$\delta'_{f(A)} \circ f \leq \delta_A + \chi_c(f).$$

Of course analogous formulas hold derived from the other equivalent statements in 4.2.1, replacing c by $\chi_c(f)$. Note in particular that, by the saturation property we also have

$$\chi_c(f) = \min\{c \mid \forall d' \in \mathcal{G}', \forall \varepsilon > 0, \forall \omega < \infty : \exists d \in \mathcal{G} : d' \circ (f \times f) \wedge \omega \leq d + \varepsilon + c\}$$

and hence $\chi_c(f)$ is also characterized by the claim that for any $d' \in \mathcal{G}'$, $\varepsilon > 0$ and $\omega < \infty$ there exists $d \in \mathcal{G}$ such that

$$d' \circ (f \times f) \wedge \omega \leq d + \varepsilon + \chi_c(f).$$

An analogous reasoning of course holds for the characterization with approach systems.

That this definition indeed provides us with a morphism-index follows from the following two results.

4.2.3 Proposition *If X, Y, U and V are approach spaces and $f : X \longrightarrow Y$ and $g : U \longrightarrow V$ are approach isomorphic then $\chi_c(f) = \chi_c(g)$.*

Proof This is straightforward and we leave this to the reader. □

4.2.4 Proposition *If X, X' and X'' are approach spaces and $f : X \longrightarrow X'$ and $g : X' \longrightarrow X''$ are functions, then the following properties hold.*

1. *f is a contraction if and only if $\chi_c(f) = 0$.*
2. *$\chi_c(g \circ f) \leq \chi_c(f) + \chi_c(g)$.*

Proof This is straightforward and we leave this to the reader. □

4.2.5 Example If $f : \mathbb{R} \longrightarrow \mathbb{R}$ is the characteristic function of $\mathbb{Q}$, then $f \circ f = 1$ and this proves that the inequality in 4.2.4 is strict in general, since for $\mathbb{R}$ considered as the usual topological space, $\chi_c(f) = \infty$ and $\chi_c(1) = 0$.

4.2.6 Proposition *If $f_j : X_j \longrightarrow X'_j$, $j \in J$, is a family of functions between approach spaces and*

$$\otimes_j f_j : \prod_j X_j \longrightarrow \prod_j X'_j : (x_j)_j \mapsto (f_j(x_j))_j$$

then

$$\chi_c(\otimes_j f_j) = \sup_j \chi_c(f_j).$$

Proof This follows from the facts that for any $\mathscr{F} \in \mathsf{F}(\prod_j X_j)$ we have

$$\lambda'(\otimes_j f_j(\mathscr{F})) = \sup_j \lambda'_j(f_j(\mathrm{pr}_j(\mathscr{F}))) \circ \mathrm{pr}_j \text{ and } \lambda\mathscr{F} = \sup_j \lambda_j(\mathrm{pr}_j(\mathscr{F})) \circ \mathrm{pr}_j$$

and that for any k

$$f_k = \mathrm{pr}'_k \circ \otimes_j f_j \circ \mathrm{in}_k$$

where in_k is any canonical embedding of X_k into $\prod_j X_j$. □

4.2.7 Proposition *If $f_j : X \longrightarrow X'_j$, $j \in J$, is a family of functions between approach spaces and*

$$\prod_j f_j : X \longrightarrow \prod_j X'_j : x \mapsto (f_j(x))_j$$

then

$$\chi_c(\prod_j f_j) = \sup_j \chi_c(f_j).$$

Proof This follows from the fact that for any $\mathscr{F} \in \mathsf{F}(X)$ we have

$$\lambda'(\prod_j f_j(\mathscr{F})) = \sup_j \lambda'_j(f_j(\mathscr{F})).$$ □

4.2.8 Proposition *If $f : X \longrightarrow X'$ is a function between approach spaces, $Z \subseteq X$ and we consider the restriction $g := f|_Z : Z \longrightarrow X'$, then $\chi_c(g) \leq \chi_c(f)$.*

Proof This follows from the definitions. □

4.2.9 Theorem *If X and X' are approach spaces, $f : X \longrightarrow X'$ is a function and $c \in \mathbb{P}$, then the following properties are equivalent.*

1. $\forall A \subseteq X : f(\delta_A) \leq \delta'_{f(A)} + c.$
2. $\forall \mu \in \mathbb{P}^X : f(\mathfrak{l}(\mu)) \leq \mathfrak{l}'(f(\mu)) + c.$
3. $\forall \mathscr{F} \in \mathsf{F}(X) : \sup_{T \in \sec \mathscr{F}} f(\delta_T) \leq \lambda'(f(\mathscr{F})) + c.$
4. $\forall \mathscr{U} \in \mathsf{U}(X) : \sup_{T \in \mathscr{U}} f(\delta_T) \leq \lambda'(f(\mathscr{U})) + c.$

5. $\forall \mu \in \mathfrak{L} : f(\mu) \leq \mathfrak{l}'(f(\mu)) + c$.
6. $\forall \mu \in \mathfrak{L}, \exists \mu' \in \mathfrak{L}' : f(\mu) \leq \mu' + c \leq f(\mu) + c$.
7. $\forall A \subseteq X, \forall \alpha \in \mathbb{R}^+ : f(A)^{(\alpha)} \subseteq \bigcap_{\varepsilon>0} f(A^{(\alpha+c+\varepsilon)})$.

In (5) and (6) a subbasis for $\mathfrak{L}$ is sufficient.

Proof This is analogous to 1.4.2 and we leave this to the reader. □

4.2.10 Definition **(Index of closed expansiveness)** A function f is called *c-closed-expansive* if it has one, and therefore all the properties mentioned in 4.2.9.

We define the *index of closed expansiveness* as

$$\boxed{\chi_{ce} : \mathsf{MorApp}_{\mathsf{Set}} \longrightarrow \mathbb{P} : f \mapsto \min\{c \mid f \text{ is } c\text{-closed expansive}\}}$$

Note that here too the minimum is well-defined and that for a function $f : X \longrightarrow X'$ between approach spaces and for any subset $A \subseteq X$

$$f(\delta_A) \leq \delta'_{f(A)} + \chi_{ce}(f).$$

Note also that analogous formulas hold derived from the other equivalent statements in 4.2.9 replacing c by $\chi_{ce}(f)$.

That χ_{ce} is a morphism-index follows from the following results where we also show some relations with the index of contractivity.

4.2.11 Proposition *If X, Y, U and V are approach spaces and $f : X \longrightarrow Y$ and $g : U \longrightarrow V$ are approach isomorphic then $\chi_{ce}(f) = \chi_{ce}(g)$.*

Proof This is straightforward and we leave this to the reader. □

4.2.12 Proposition *If X, X' and X'' are approach spaces and $f : X \longrightarrow X'$ and $g : X' \longrightarrow X''$ are functions, then the following properties hold.*

1. *f is closed expansive if and only if $\chi_{ce}(f) = 0$.*
2. $\chi_{ce}(g \circ f) \leq \chi_{ce}(f) + \chi_{ce}(g)$.
3. *If f is surjective, then*

$$\chi_{ce}(g) \leq \chi_c(f) + \chi_{ce}(g \circ f) \text{ and } \chi_c(g) \leq \chi_{ce}(f) + \chi_c(g \circ f).$$

4. *If g is injective, then*

$$\chi_{ce}(f) \leq \chi_c(g) + \chi_{ce}(g \circ f) \text{ and } \chi_c(f) \leq \chi_{ce}(g) + \chi_c(g \circ f).$$

Proof We only prove the first inequality of 3 to the reader. Let $B \subseteq X'$ then it follows from the surjectivity of f and the definitions of the indices of contractivity and closed expansiveness that

$$\begin{aligned} g(\delta'_B) &= g \circ f(\delta'_B \circ f) \\ &= g \circ f(\delta'_{f(f^{-1}(B))} \circ f) \\ &\leq g \circ f(\delta_{f^{-1}(B)} + \chi_c(f)) \\ &= g \circ f(\delta_{f^{-1}(B)}) + \chi_c(f) \\ &\leq \delta''_{g \circ f(f^{-1}(B))} + \chi_{ce}(g \circ f) + \chi_c(f) \\ &= \delta''_{g(B)} + \chi_{ce}(g \circ f) + \chi_c(f) \end{aligned}$$

which proves that $\chi_{ce}(g) \leq \chi_c(f) + \chi_{ce}(g \circ f)$. □

4.2.13 Corollary (Top) *If X, X' and X'' are topological spaces and $f : X \longrightarrow X'$ and $g : X' \longrightarrow X''$ are functions, then the following properties hold.*

1. *If f is surjective and continuous (respectively closed) and $g \circ f$ is closed (respectively continuous) then g is closed (respectively continuous).*
2. *If g is injective and continuous (respectively closed) and $g \circ f$ is closed (respectively continuous) then f is closed (respectively continuous).*

4.2.14 Proposition *If $f : X \longrightarrow X'$ is a function between approach spaces, $Z \subseteq X'$ and we consider the restriction $g := f|_{f^{-1}(Z)} : f^{-1}(Z) \longrightarrow Z$, then $\chi_{ce}(g) \leq \chi_{ce}(f)$.*

Proof Let $A \subseteq f^{-1}(Z)$ and $y \in Z$, then

$$g(\delta_A)(y) = f(\delta_A)(y) \leq \delta'_{f(A)}(y) + \chi_{ce}(f) = \delta'_{g(A)}(y) + \chi_{ce}(f)$$

which proves the claim. □

4.2.15 Theorem *If X and X' are approach spaces, $f : X \longrightarrow X'$ is a function and $c \in \mathbb{P}$, then the following properties are equivalent.*

1. $\forall A \subseteq X : \delta_{f^{-1}(A)} \leq \delta'_A \circ f + c$.
2. $\forall x \in X, \forall \varphi \in \mathscr{A}(x) : f(\varphi) \ominus c \in \mathscr{A}'(f(x))$.
3. $\forall A \subseteq X', \forall \varepsilon \in \mathbb{R}^+ : f^{-1}(A^{(\varepsilon)}) \subseteq (f^{-1}(A))^{(\varepsilon+c)}$.
4. $\forall \nu \in \mathfrak{U}, \exists \nu' \in \mathfrak{U}' : \nu' \leq f(\nu) + c \leq \nu' + c$.
5. $\forall x \in X, \forall \varepsilon \in \mathbb{R}^+, \forall V \in \mathscr{V}_{\varepsilon+c}(x) : f(V) \in \mathscr{V}_\varepsilon(f(x))$.

In (2) *a basis for $\mathscr{A}'(f(x))$ is sufficient and in* (4) *a subbasis for $\mathfrak{U}$ is sufficient. In the case that f is surjective these are moreover equivalent to the following properties.*

6. *$\forall x \in X, \forall \mathscr{U}' \in \mathsf{U}(X')$ there exists $\mathscr{U} \in \mathsf{U}(X)$ such that $f(\mathscr{U}) = \mathscr{U}'$ and $\lambda\mathscr{U}(x) \leq \lambda'(\mathscr{U}')(f(x)) + c$.*
7. $\forall x \in X, \forall \mathscr{U}' \in \mathsf{U}(X') : \inf\{\lambda\mathscr{U}(x) \mid f(\mathscr{U}) = \mathscr{U}'\} \leq \lambda'(\mathscr{U}')(f(x)) + c$.

Proof This is analogous to 1.4.6 and we leave this to the reader. □

4.2.16 Definition **(Index of open expansiveness)** A function f is called *c-open expansive* if it has one, and therefore all the properties mentioned in 4.2.15.

We define the *index of open expansiveness* as

$$\boxed{\chi_{oe} : \mathsf{MorApp}_{\mathsf{Set}} \longrightarrow \mathbb{P} : f \mapsto \min\{c \mid f \text{ is } c\text{-open expansive}\}}$$

Again note that the minimum is well-defined and that for any function $f : X \longrightarrow X'$ between approach spaces and for any subset $A \subseteq X$

$$\delta_{f^{-1}(A)} \leq \delta'_A \circ f + \chi_{oe}(f).$$

Again note that analogous expressions hold, derived from the other equivalent statements in 4.2.15 replacing c with $\chi_{oe}(f)$. Note in particular that, by the saturation property $\chi_{oe}(f)$ is also characterized by the claim that for any $x \in X$, $\varphi \in \mathscr{A}(x)$, $\varepsilon > 0$ and $\omega < \infty$ there exists $\varphi' \in \mathscr{A}'(f(x))$ such that

$$(f(\varphi) \ominus \chi_{oe}(f)) \wedge \omega \leq \varphi' + \varepsilon.$$

That χ_{oe} is a morphism-index follows from the following results where, again, we show some relations with the index of contractivity.

4.2.17 Proposition *If X, Y, U and V are approach spaces and $f : X \longrightarrow Y$ and $g : U \longrightarrow V$ are approach isomorphic then $\chi_{oe}(f) = \chi_{oe}(g)$.*

Proof This is straightforward and we leave this to the reader. □

4.2.18 Proposition *If X, X' and X'' are approach spaces and $f : X \longrightarrow X'$ and $g : X' \longrightarrow X''$ are functions, then the following properties hold.*

1. *f is open expansive if and only if $\chi_{oe}(f) = 0$.*
2. *$\chi_{oe}(g \circ f) \leq \chi_{oe}(f) + \chi_{oe}(g)$.*
3. *If f is surjective, then*

$$\chi_{oe}(g) \leq \chi_c(f) + \chi_{oe}(g \circ f) \text{ and } \chi_c(g) \leq \chi_{oe}(f) + \chi_c(g \circ f).$$

4. *If g is injective, then*

$$\chi_{oe}(f) \leq \chi_c(g) + \chi_{oe}(g \circ f) \text{ and } \chi_c(f) \leq \chi_{oe}(g) + \chi_c(g \circ f).$$

Proof This is analogous to 4.2.12 and we will only prove the second inequality in 3 leaving the remaining claims to the reader. Let $x' \in X'$ and $A' \subseteq X'$ and take $x \in X$ such that $f(x) = x'$ then

$$\begin{aligned}\delta''_{g(A')} \circ g(x') &= \delta''_{g \circ f(f^{-1}(A'))} \circ g(f(x)) \\ &\leq \delta_{f^{-1}(A')}(x) + \chi_c(g \circ f) \\ &\leq \delta'_{A'}(x') + \chi_{oe}(f) + \chi_c(g \circ f)\end{aligned}$$

which proves that $\chi_c(g) \leq \chi_{oe}(f) + \chi_c(g \circ f)$. □

4.2.19 Corollary (Top) *If X, X' and X'' are topological spaces and $f : X \longrightarrow X'$ and $g : X' \longrightarrow X''$ are functions, then the following properties hold.*

1. *If f is surjective and continuous (respectively open) and $g \circ f$ is open (respectively continuous) then g is open (respectively continuous).*
2. *If g is injective and continuous (respectively open) and $g \circ f$ is open (respectively continuous) then f is open (respectively continuous).*

4.2.20 Proposition *If $f : X \longrightarrow X'$ is a function between approach spaces, $Z \subseteq X'$ and we consider the restriction $g := f|_{f^{-1}(Z)} : f^{-1}(Z) \longrightarrow Z$ then $\chi_{oe}(g) \leq \chi_{oe}(f)$.*

Proof This is analogous to 4.2.14 and we leave this to the reader. □

4.2.21 Proposition (Met) *If X and X' are metric spaces and $f : X \longrightarrow X'$ then $\chi_{ce}(f) = \chi_{oe}(f)$.*

Proof This is analogous to 2.4.9 and 2.4.10 and we leave this to the reader. □

4.2.22 Theorem *If $f : X \longrightarrow X'$ is a function between approach spaces and $c \in \mathbb{P}$, then the following properties are equivalent.*

1. $\forall \mathcal{F} \in \mathsf{F}(X) : f(\alpha\mathcal{F}) \leq \alpha' f(\mathcal{F}) + c$.
2. $\forall \mathcal{U} \in \mathsf{U}(X) : f(\lambda\mathcal{U}) \leq \lambda' f(\mathcal{U}) + c$.
3. $\forall \mathfrak{I} \in \mathfrak{P}(X) : f(\mathfrak{I}) \rightarrowtail y \Rightarrow \forall \varepsilon > 0, \exists x \in f^{-1}(y) : \mathfrak{I} \oplus (c + \varepsilon) \rightarrowtail x$.

Proof This is straightforward and we leave this to the reader. □

4.2.23 Definition (Index of properness) A function f is called *c-proper* if it has any and hence all of the properties in 4.2.22. We define the *index of properness* as

$$\boxed{\chi_p : \mathrm{Mor}\mathsf{App}_{\mathsf{Set}} \longrightarrow \mathbb{P} : f \mapsto \min\{c \mid f \text{ is } c\text{-proper}\}}$$

Again, the minimum is well-defined and for any function $f : X \longrightarrow X'$ and any $\mathcal{F} \in \mathsf{F}(X)$

$$f(\alpha\mathcal{F}) \leq \alpha' f(\mathcal{F}) + \chi_p(f).$$

An analogous formula holds for the other characterizations in 4.2.22.

Note also that, contrary to our treatment of proper contractions in the first chapter, here we do not presuppose that f is a contraction.

That χ_p is a morphism-index follows from the following results.

4.2.24 Proposition *If X, Y, U and V are approach spaces and $f : X \longrightarrow Y$ and $g : U \longrightarrow V$ are approach isomorphic, then $\chi_p(f) = \chi_p(g)$.*

Proof This is straightforward and we leave this to the reader. □

4.2.25 Proposition *If X, X' and X'' are approach spaces and $f : X \longrightarrow X'$ and $g : X' \longrightarrow X''$ are functions, then the following properties hold.*

1. *f is a proper contraction if and only if $\chi_p(f) = \chi_c(f) = 0$.*
2. $\chi_p(g \circ f) \leq \chi_p(f) + \chi_p(g)$.
3. *If f is surjective, then $\chi_p(g) \leq \chi_p(g \circ f) + \chi_c(f)$.*
4. *If g is injective, then $\chi_p(f) \leq \chi_p(g \circ f) + \chi_c(g)$.*
5. $\chi_{ce} \leq \chi_p$.

Proof We leave the verification of 1, 2 and 3 to the reader. To prove 4, let $\mathcal{U} \in \mathsf{U}(X)$ then

$$\begin{aligned} f(\lambda \mathcal{U}) &= g(f(\lambda \mathcal{U})) \circ g \\ &\leq \lambda''(g(f(\mathcal{U}))) \circ g + \chi_p(g \circ f) \\ &\leq \lambda' f(\mathcal{U}) + \chi_c(g) + \chi_p(g \circ f). \end{aligned}$$

To prove 5, let $A \subseteq X$ then

$$\begin{aligned} f(\delta_A) &= \inf_{\mathcal{U} \in \mathsf{U}(A)} f(\lambda \mathcal{U}) \\ &\leq \inf_{\mathcal{U} \in \mathsf{U}(A)} \lambda' f(\mathcal{U}) + \chi_p(f) \\ &= \delta'_{f(A)} + \chi_p(f). \end{aligned}$$

□

4.2.26 Corollary (Top) *If X, X' and X'' are topological spaces and $f : X \longrightarrow X'$ and $g : X' \longrightarrow X''$ are functions, then the following properties hold.*

1. *If f is surjective and continuous and $g \circ f$ is proper then g is proper.*
2. *If g is injective and continuous and $g \circ f$ is proper then f is proper.*

4.2.27 Proposition *If $f : X \longrightarrow X'$ is a function between approach spaces, $Z \subseteq X'$ and we consider the restriction $g := f|_{f^{-1}(Z)} : f^{-1}(Z) \longrightarrow Z$, then $\chi_p(g) \leq \chi_p(f)$.*

Proof This is analogous to 4.2.14 and we leave this to the reader. □

4.2.28 Example Let $\mathbb{R}$ be equipped with the usual Euclidean metric. If $f : \mathbb{R} \longrightarrow \mathbb{R}$ is a contraction (respectively closed expansive, open expansive or proper) then for $g := f + c1_{\mathbb{Q}}$ we have $\chi_c(g) = c$ (respectively $\chi_{ce}(g) = c, \chi_{oe}(g) = c, \chi_p(g) = c$).

4.3 Compactness Indices

A particularly interesting variant of Kuratowski's original measure of noncompactness is the so-called Hausdorff or ball measure of noncompactness, Kuratowski (1930), Banaś and Goebel (1980). It is defined as follows.

Suppose that (X, d) is a metric space and that $A \subseteq X$. Then

$$m_H(A) := \inf\{\varepsilon \in \mathbb{R}_0^+ \mid \exists x_1, \ldots, x_n \in X : A \subseteq \bigcup_{i=1}^{n} B(x_i, \varepsilon)\}$$

is called the *Hausdorff measure of noncompactness*. It differs from the original measure introduced by Kuratowski only slightly.

The value

$$m_K(A) := \inf\{\varepsilon \in \mathbb{R}_0^+ \mid \exists X_1, \ldots, X_n \subseteq X : \max_{i=1}^{n} \operatorname{diam}(X_i) \leq \varepsilon, A \subseteq \bigcup_{i=1}^{n} X_i\}$$

is called the *Kuratowski measure of noncompactness*. It is easily seen that for any $A \subseteq X$ we have $m_H(A) \leq m_K(A) \leq 2m_H(A)$.

A remarkable fact is that both measures are metric dependent, whereas compactness is only dependent on the topology underlying the metric. This rather confusing state of affairs is cleared up if we reconsider the ideas behind these measures in the setting of approach spaces. In this section we show that the Hausdorff measure of noncompactness arises as a very canonical index in the setting of approach spaces. As far as terminology is concerned, and in spite of the fact that this may cause some confusion, we prefer to call this an index of compactness, rather than a measure of noncompactness. If a space has an index of compactness equal to 0 then we will sometimes say it is 0-*compact.* In this chapter we will encounter indices of various types of compactness and each time that such an index is 0 we will use the same convention and speak of e.g. 0-*sequentially compact*, 0-*countably compact*, and so on.

4.3.1 Definition (Index of compactness) Given an approach space X, we define the *index of compactness* of X as

$$\boxed{\chi_c(X) := \sup_{\mathcal{F} \in \mathsf{F}(X)} \inf_{x \in X} \alpha\mathcal{F}(x)}$$

The idea behind this definition is the following: compactness means every filter should have an adherence point and therefore the information given by χ_c is based on the verification for all filters of what their "best" adherence points are. Before continuing we give a number of equivalent expressions of this index, which emphasize its canonicity.

4.3.2 Theorem *For any approach space X, we have*

$$\begin{aligned}\chi_c(X) &= \sup_{\mathcal{U} \in \mathsf{U}(X)} \inf_{x \in X} \alpha\mathcal{U}(x)\\ &= \sup_{\mathcal{U} \in \mathsf{U}(X)} \inf_{x \in X} \lambda\mathcal{U}(x)\end{aligned}$$

$$= \sup_{\varphi \in \prod_{x \in X} \mathscr{B}(x)} \inf_{Y \in 2^{(X)}} \sup_{z \in X} \inf_{x \in Y} \varphi(x)(z)$$

$$= \sup_{\varphi \in \mathscr{H}^X} \inf_{Y \in 2^{(X)}} \sup_{z \in X} \inf_{x \in Y} \varphi(x)(x, z)$$

$$= \inf\{\alpha \mid \forall \mathfrak{J} \in \mathfrak{P}(X), c(\mathfrak{J}) \geq \alpha, \exists x \in X : \mathfrak{J} \rightarrowtail x\}$$

where $(\mathscr{B}(x))_{x \in X}$ is a basis for the approach system and $\mathscr{H}$ is a basis for the gauge.

Proof Let us denote the five expressions of the theorem (in order) by $c_1, \ldots, c_5$. Obviously $c_3 = c_4$. Further it follows from 1.2.66 that $c_1 = c_2$ and it is obvious that we have $\chi_c(X) \geq c_1$. To prove that $c_3 \geq \chi_c(X)$, suppose that $\chi_c(X) > r$. Then by 1.2.43 and 1.2.66 we can find an ultrafilter $\mathscr{U}$ such that

$$\sup_{\psi \in \prod_{x \in X} \mathscr{B}(x)} \inf_{x \in X} \inf_{U \in \mathscr{U}} \sup_{y \in U} \psi(x)(y) = \inf_{x \in X} \sup_{\psi \in \mathscr{B}(x)} \inf_{U \in \mathscr{U}} \sup_{y \in U} \psi(y)$$

$$= \inf_{x \in X} \alpha\mathscr{U}(x) > r.$$

Consequently, there exists $\psi \in \prod_{x \in X} \mathscr{B}(x)$ such that, for all $x \in X$,

$$\inf_{U \in \mathscr{U}} \sup_{y \in U} \psi(x)(y) > r.$$

Now suppose that for some $Y \in 2^{(X)}$ we have $\sup_{z \in X} \inf_{x \in Y} \psi(x)(z) \leq r$. Then the collection $\{\{\psi(x) \leq r\} \mid x \in Y\}$ is a finite cover of X and thus there exists some $x_0 \in Y$ such that $\{\psi(x_0) \leq r\} \in \mathscr{U}$. Then, however, we find that

$$\inf_{U \in \mathscr{U}} \sup_{y \in U} \psi(x_0)(y) \leq \sup_{y \in \{\psi(x_0) \leq r\}} \psi(x_0)(y) \leq r,$$

which is a contradiction. Consequently, for all $Y \in 2^{(X)}$, we have

$$\sup_{z \in X} \inf_{x \in Y} \psi(x)(z) > r,$$

which proves that $c_3 \geq r$. From the arbitrariness of r this shows that $c_3 \geq \chi_c(X)$. To show that $c_1 \geq c_3$ suppose that $c_3 > r$. Then there exists $\varphi_0 \in \prod_{x \in X} \mathscr{B}(x)$ such that, for all $Y \in 2^{(X)}$,

$$\sup_{z \in X} \inf_{x \in Y} \varphi_0(x)(z) > r.$$

Consequently, if, for all $Y \in 2^{(X)}$, we let

$$F_Y := \left\{ \inf_{x \in Y} \varphi_0(x) > r \right\},$$

then it follows that the collection $\{F_Y \mid Y \in 2^{(X)}\}$ is a basis for a filter $\mathscr{F}$ on X. Let $\mathscr{W}$ be an arbitrary ultrafilter finer than $\mathscr{F}$. Then it follows that

$$\begin{aligned} c_1 &= \sup_{\mathscr{U} \in \mathsf{U}(X)} \inf_{x \in X} \alpha\mathscr{U}(x) \\ &\geq \inf_{x \in X} \sup_{\varphi \in \mathscr{B}(x)} \sup_{W \in \mathscr{W}} \inf_{y \in W} \varphi(y) \\ &= \sup_{\varphi \in \prod_{x \in X} \mathscr{B}(x)} \inf_{x \in X} \sup_{W \in \mathscr{W}} \inf_{y \in W} \varphi(x)(y) \\ &\geq \inf_{x \in X} \sup_{W \in \mathscr{W}} \inf_{y \in W} \varphi_0(x)(y) \\ &\geq \inf_{x \in X} \inf_{y \in F_{\{x\}}} \varphi_0(x)(y) \geq r, \end{aligned}$$

which by the arbitrariness of r shows that $c_1 \geq c_3$. That $c_2 = c_5$ finally follows from a straightforward application of 1.2.59 and 1.2.60. □

4.3.3 Example 1. In a metric space $\chi_c = m_H$ and in a non-Archimedean metric space $\chi_c = m_K = m_H$.

2. Take for X the closed unit ball in $\mathbb{R}^2$ equipped with the radial metric

$$d(x, y) := \begin{cases} d_{\mathbb{E}}(x, y) & x, y, 0 \text{ collinear} \\ d_{\mathbb{E}}(x, 0) + d_{\mathbb{E}}(y, 0) & \text{otherwise,} \end{cases}$$

then $\chi_c(X) = 1$ and $m_K(X) = 2$.

3. Let B_E be the closed unit ball of an infinite-dimensional normed space E, then $\chi_c(B_E) = 1$. (See also the chapter on applications in functional analysis and 4.3.28).

4. Let X be a metric space and let $\mathscr{N}$ be the set of all nonempty compact subsets of X, then for any bounded set $B \subseteq X$ we have $\chi_c(B) = d_H(B, \mathscr{N})$ where d_H is the Hausdorff distance (see Malkowsky and Rakočević Preprint).

In the chapter on applications to the theory of hyperspaces, in particular the Vietoris hyperspace, it will be advantageous to have a characterization of the index of compactness in terms of lower regular function frames. This requires some preliminary work.

If $\mathfrak{L}$ is a lower regular function frame on X and $\mathfrak{B} \subseteq \mathfrak{L}$ we write $\mathfrak{I}(\mathfrak{B})$ for the smallest ideal in $\mathfrak{L}$ containing $\mathfrak{B}$. We say that $\mathfrak{I}(\mathfrak{B})$ is *generated* by $\mathfrak{B}$. Clearly

$$\mathfrak{I}(\mathfrak{B}) = \{\mu \in \mathfrak{L} \mid \exists \beta_1, \beta_2, \ldots, \beta_n \in \mathfrak{B} : \mu \leq \beta_1 \vee \beta_2 \vee \cdots \vee \beta_n\}.$$

If for all $\alpha, \beta \in \mathfrak{B}$ there exists $\gamma \in \mathfrak{B}$ such that $\alpha \vee \beta \leq \gamma$, we call $\mathfrak{B}$ an *ideal basis*. In that case $\mathfrak{I}(\mathfrak{B}) = \{\mu \in \mathfrak{L} \mid \exists \beta \in \mathfrak{B} : \mu \leq \beta\}$.

Note that by definition, we are actually allowing for the improper ideal. However we will only be interested in ideals which are very far from being improper. An ideal $\mathfrak{I} \subseteq \mathfrak{L}$ is called *small* if $\inf \mu = 0$ for all $\mu \in \mathfrak{I}$.

We will say that $\mathfrak{B} \subseteq \mathfrak{L}$ has the *finite-sup property* if $\inf \sup_{\mu \in \mathfrak{C}} \mu = 0$ for each finite subset $\mathfrak{C} \subseteq \mathfrak{B}$.

It is clear that an ideal $\mathfrak{I} \subseteq \mathfrak{L}$ is small if and only if it has a basis $\mathfrak{B}$ such that $\inf \mu = 0$ for each $\mu \in \mathfrak{B}$ and if and only if it is generated by a set with the finite-sup property.

4.3.4 Lemma *Every set with the finite-sup property is contained in a maximal set with the finite-sup property and this set is actually a small ideal.*

Proof If $\mathfrak{B} \subseteq \mathfrak{L}$ has the finite-sup property then it is easily verified that

$$\{\mathfrak{P} \mid \mathfrak{B} \subseteq \mathfrak{P} \subseteq \mathfrak{L}, \mathfrak{P} \text{ has the finite-sup property}\}$$

is inductively ordered by inclusion and the result then follows by a standard application of Zorn's lemma. □

4.3.5 Lemma *If $\mathfrak{M} \subseteq \mathfrak{L}$ is a maximal small ideal then it is prime in the sense that if $\rho_1 \wedge \rho_2 \in \mathfrak{M}$, then $\rho_1 \in \mathfrak{M}$ or $\rho_2 \in \mathfrak{M}$.*

Proof If $\rho_k \in \mathfrak{L} \setminus \mathfrak{M}$ for $k \in \{1, 2\}$, then it follows that $\mathfrak{I}(\{\rho_k\} \cup \mathfrak{M})$ is no longer small, hence there exist $\mu_k \in \mathfrak{M}$ and a_k such that $\rho_k \vee \mu_k \geq a_k > 0$. Since

$$\begin{aligned}(\rho_1 \wedge \rho_2) \vee \mu_1 \vee \mu_2 &\geq (\rho_1 \vee \mu_1) \wedge (\rho_2 \vee \mu_2) \\ &\geq a_1 \wedge a_2 > 0\end{aligned}$$

it follows that $\rho_1 \wedge \rho_2 \notin \mathfrak{M}$. □

We will write $I_s(\mathfrak{L})$ for the set of all small ideals, $B_s(\mathfrak{L})$ for the set of all sets with the finite-sup property and $I_{sm}(\mathfrak{L}) \subseteq I_s(\mathfrak{L})$ for those elements in $I_s(\mathfrak{L})$which are maximal.

If a set with the finite-sup property is contained in a subbasis $\mathfrak{B}$ for $\mathfrak{L}$ then we will say it has the finite-sup property in $\mathfrak{B}$ and the set of all such sets is denoted by $B_s(\mathfrak{B})$. The following lemma and the third equality in 4.3.9 are of the same flavour as Alexander's subbasis lemmá (see e.g. Kelley 1955).

Not to overload the notations, in the following results we write $\sup \mathfrak{A}$ for $\sup_{\mu \in \mathfrak{A}} \mu$.

4.3.6 Lemma *If $\mathfrak{B}$ is a subbasis for $\mathfrak{L}$, and if $\alpha \in \mathbb{P}$ is such that* $\inf_{x \in X} \sup \mathfrak{I}(x) \leq \alpha$ *for each $\mathfrak{I} \in B_s(\mathfrak{B})$, then also* $\inf_{x \in X} \sup \mathfrak{I}(x) \leq \alpha$ *for each $\mathfrak{I} \in B_s(\mathfrak{L})$.*

Proof Take $\mathfrak{I} \in B_s(\mathfrak{L})$ and let $\mathfrak{M} \subseteq \mathfrak{L}$ be a maximal set with the finite-sup property containing $\mathfrak{I}$. Then $\sup \mathfrak{I} \leq \sup \mathfrak{M}$, which proves that

$$\inf_{x \in X} \sup \mathfrak{I}(x) \leq \inf_{x \in X} \sup \mathfrak{M}(x).$$

We now prove that $\sup \mathfrak{M} = \sup(\mathfrak{M} \cap \mathfrak{B})$. Note that $\sup(\mathfrak{M} \cap \mathfrak{B}) \leq \sup \mathfrak{M}$ is trivial.

Take $x \in X$ and let $\sup \mathfrak{M}(x) = \beta$. Then for $\varepsilon > 0$, there exists $\mu \in \mathfrak{M}$ such that $\mu(x) > \beta - \varepsilon/2$, and since $\mathfrak{B}$ is a subbasis for $\mathfrak{L}$, we can write $\mu = \sup_{j\in J} \inf_{k\in K_j} \mu_{j,k}$ with K_j finite and all $\mu_{j,k} \in \mathfrak{B}$. Hence there is a $j_0 \in J$ such that $\inf_{k\in K_{j_0}} \mu_{j_0,k}(x) > \beta - \varepsilon$, and since $\mathfrak{M}$ is an ideal, we have $\inf_{k\in K_{j_0}} \mu_{j_0,k} \in \mathfrak{M}$. From the foregoing lemma it follows that $\mu_{j_0,k_0} \in \mathfrak{M}$, and hence $\mu_{j_0,k_0} \in \mathfrak{M} \cap \mathfrak{B}$, for some $k_0 \in K_{j_0}$. Since

$$\mu_{j_0,k_0}(x) \geq \inf_{k\in K_{j_0}} \mu_{j_0,k}(x) > \beta - \varepsilon,$$

we obtain $\sup(\mathfrak{M} \cap \mathfrak{B})(x) \geq \beta - \varepsilon$, hence $\sup(\mathfrak{M} \cap \mathfrak{B})(x) \geq \beta$.

Since $\mathfrak{M} \in B_s(\mathfrak{L})$, it follows that $\mathfrak{M} \cap \mathfrak{B} \in B_s(\mathfrak{B})$ and therefore

$$\inf_{x\in X} \sup \mathfrak{I}(x) \leq \inf_{x\in X} \sup \mathfrak{M}(x) = \inf_{x\in X} \sup(\mathfrak{M} \cap \mathfrak{B})(x) \leq \alpha. \qquad \square$$

If $\mathfrak{I} \in I_s(\mathfrak{L})$, it is clear that $\{\{\mu \leq \varepsilon\} \mid \mu \in \mathfrak{I}, \varepsilon > 0\}$ is a filterbasis on X, and we will write $\mathsf{f}^*(\mathfrak{I})$ for the filter generated by it. Furthermore, if $\mathcal{F} \in \mathsf{F}(X)$, then $\mathsf{i}^*(\mathcal{F}) := \{\varphi \in \mathfrak{L} \mid \exists F \in \mathcal{F}, \phi \leq \delta_F\}$ is a small ideal as is readily seen.

4.3.7 Lemma *For any $\mathcal{F} \in \mathsf{F}(X)$ and $\mathfrak{I} \in I_s(\mathfrak{L})$ the following properties hold.*

1. $\alpha\mathcal{F} = \sup \mathsf{i}^*(\mathcal{F})$.
2. $\sup \mathfrak{I} \leq \alpha\mathsf{f}^*(\mathfrak{I})$.

Proof 1. This is evident from the definition of adherence-operator.

2. Take $\mu \in \mathfrak{I}, x \in X, 0 < \varepsilon < \mu(x)$. It is clear that

$$\alpha\mathsf{f}^*(\mathfrak{I})(x) = \sup_{\mu\in\mathfrak{I},\rho>0} \delta(x, \{\mu \leq \rho\}).$$

If $A = \{\mu \leq \varepsilon\}$, then it follows from the contractivity of μ (see 1.3.5) that

$$0 < \mu(x) - \varepsilon \leq \mu(x) - \sup \mu(A) \leq \delta(x, A) \leq \alpha\mathsf{f}^*(\mathfrak{I})(x).$$

The result then follows from the arbitrariness of ε, x and μ. $\square$

4.3.8 Example The inequality in the foregoing result is strict in general as is immediately seen by taking $X := \mathbb{R}$ with the Euclidean topology and $\mathfrak{I}$ the ideal generated by $\mu(x) := |x|$.

4.3.9 Theorem *For an approach space X and a subbasis $\mathfrak{B}$ for the lower regular function frame $\mathfrak{L}$ we have*

$$\chi_c(X) = \sup_{\mathfrak{I}\in I_s(\mathfrak{L})} \inf_{x\in X} \sup \mathfrak{I}(x)$$
$$= \sup_{\mathfrak{I}\in I_{sm}(\mathfrak{L})} \inf_{x\in X} \sup \mathfrak{I}(x)$$
$$= \sup_{\mathfrak{I}\in I_s(\mathfrak{B})} \inf_{x\in X} \sup \mathfrak{I}(x).$$

Proof The first equality follows from the definition of the index of compactness and a straightforward application of 4.3.7. The second equality follows from 4.3.4 and the third equality then follows from 4.3.6. □

Note that the formulas in the foregoing theorem can be reformulated in terms of sets with the finite-sup property and we will freely do so whenever useful.

4.3.10 Proposition *The index of compactness χ_c is $(1, \chi_c)$-layered.*

Proof This follows from 4.2.1, 4.1.5 and 4.3.2. □

4.3.11 Proposition (Top) *A topological approach space is 0-compact if and only if the underlying topology is compact.*

Proof In view of the fact that the adherence operator in topological approach spaces attains only the values 0 and ∞, by definition of χ_c and upon invoking 2.1.1, the fact that $\chi_c(X) = 0$ precisely means that every filter has an adherence point. □

4.3.12 Proposition (Met) *A metric approach space is 0-compact if and only if the metric is totally bounded and it has a finite index of compactness if and only if the metric is bounded.*

Proof Let (X, d) be a metric space. For the first claim, it follows from 2.3.1 that the collection of all sets $\mathscr{B}_d(x) := \{d(x, \cdot)\}$, $x \in X$, is a basis for the approach system of (X, δ_d). Applying 4.3.2 it therefore follows that

$$\chi_c(X) = \inf_{Y\in 2^{(X)}} \sup_{z\in X} \inf_{x\in Y} d(x, z).$$

Consequently the fact that $\chi_c(X) = 0$ means that for each $\varepsilon > 0$ there exists a finite set $Y \in 2^{(X)}$ such that $X = \bigcup_{y\in Y} B(y, \varepsilon)$, i.e. that X is totally bounded.

The second claim is perfectly analogous. □

Note that in a metric space where boundedness and total boundedness coincide, e.g. in a finite-dimensional Euclidean space, just as in the case of a topological space, the index of compactness can only attain the two extreme values 0 and ∞.

4.3.13 Proposition *If X is a uniform approach space with symmetric gauge basis $\mathscr{D}$ and X is 0-compact, then $(X, \mathscr{U}(\mathscr{D}))$ is totally bounded.*

Proof By 4.3.2, the fact that $\chi_c(X) = 0$ means that

$$0 = \sup_{\varphi \in \mathcal{D}^X} \inf_{Y \in 2^{(X)}} \sup_{z \in X} \inf_{x \in Y} \varphi(x)(x, z)$$
$$\geq \sup_{d \in \mathcal{D}} \inf_{Y \in 2^{(X)}} \sup_{z \in X} \inf_{x \in Y} d(x, z),$$

which in turn implies that, for any $d \in \mathcal{D}$, there exists a finite subset $Y \subseteq X$ such that

$$\bigcup_{x \in Y} B_d(x, \varepsilon) = X.$$

Hence, all spaces (X, d) are totally bounded which means that $(X, \mathcal{U}(\mathcal{D}))$ is totally bounded. □

The converse of the foregoing result does not hold.

4.3.14 Example Consider $X := [0, 1] \cap \mathbb{Q}$ equipped with $\{\alpha d_{\mathbb{E}} \mid \alpha > 0\}$ as basis for a gauge. Then the generated approach space is actually given by the usual topology and hence is not 0-compact. However all $\alpha d_{\mathbb{E}}$ are totally bounded metrics and hence $\mathcal{U}(\mathcal{D})$ is totally bounded.

4.3.15 Proposition *An approach space with a compact topological coreflection is* 0-*compact.*

Proof This follows immediately from the fact that if the topological coreflection is compact, every ultrafilter $\mathcal{U}$ converges and hence its limit operator attains its minimum 0. □

Here too, the converse of the foregoing result does not hold.

4.3.16 Example $X := [0, 1] \cap \mathbb{Q}$ equipped with the Euclidean metric does not have an underlying compact topology but does have $\chi_c(X) = 0$.

4.3.17 Theorem *If $f : X \longrightarrow X'$ is a function between approach spaces and $A \subseteq X$, then*

$$\chi_c(f(A)) \leq \chi_c(A) + \chi_c(f)$$

and in particular if f is a contraction and A is 0-*compact, then $f(A)$ is* 0-*compact.*

Proof Since every ultrafilter on $f(A)$ is the image by f of an ultrafilter on A it follows from 4.2.1 that

$$\begin{aligned}\chi_c(f(A)) &= \sup_{\mathcal{U}\in \mathsf{U}(f(A))} \inf_{y\in f(A)} \lambda'\mathcal{U}(y)\\ &\leq \sup_{\mathcal{U}\in \mathsf{U}(A)} \inf_{x\in A} \lambda' f(\mathcal{U})(f(x))\\ &\leq \sup_{\mathcal{U}\in \mathsf{U}(A)} \inf_{x\in A} \lambda\mathcal{U}(x) + \chi_c(f)\\ &= \chi_c(A) + \chi_c(f).\end{aligned}$$

□

4.3.18 Corollary (Top, Met) *The continuous image of a compact topological space is compact, the nonexpansive image of a totally bounded metric space is totally bounded and the nonexpansive image of a bounded metric space is bounded.*

4.3.19 Theorem (Tychonoff) *If $(X_j)_{j\in J}$ is a family of approach spaces, then*

$$\chi_c(\prod_{j\in J} X_j) = \sup_{j\in J} \chi_c(X_j)$$

and in particular a product of approach spaces is 0-compact if and only if all factors are 0-compact.

Proof It follows from 1.3.12 that if $\mathcal{U}$ is an ultrafilter on $\prod_{j\in J} X_j$, then $\alpha\mathcal{U} = \sup_{j\in J} \alpha_j(\mathrm{pr}_j(\mathcal{U})) \circ \mathrm{pr}_j$. Using this fact, the proof of the theorem now follows from the following calculation:

$$\begin{aligned}\chi_c(\prod_{j\in J} X_j) &= \sup_{\mathcal{U}\in \mathsf{U}(\prod_{j\in J} X_j)} \inf_{x\in \prod_{j\in J} X_j} \sup_{j\in J} \alpha_j(\mathrm{pr}_j(\mathcal{U}))(x_j)\\ &= \sup_{\mathcal{U}\in \mathsf{U}(\prod_{j\in J} X_j)} \sup_{j\in J} \inf_{z\in X_j} \alpha_j(\mathrm{pr}_j(\mathcal{U}))(z)\\ &= \sup_{j\in J} \sup_{\mathcal{U}\in \mathsf{U}(X_j)} \inf_{z\in X_j} \alpha_j\mathcal{U}(z)\\ &= \sup_{j\in J} \chi_c(X_j).\end{aligned}$$

□

4.3.20 Corollary (**Tychonoff in** Top) *The product of a family of topological spaces is compact if and only if each factor space is compact.*

4.3.21 Corollary (Met) *The product (in* Met*) of a finite family of metric spaces is totally bounded if and only if each factor space is totally bounded and the product of an arbitrary family of metric spaces (in* App*) is 0-compact if and only if all spaces are totally bounded.*

4.3.22 Proposition *If X is an approach space and if $X = \bigcup_{k=1}^{n} X_k$, then*

$$\chi_c(X) \leq \max_{1 \leq k \leq n} \chi_c(X_k).$$

Proof Put $\mathsf{U}_k := \{\mathcal{U} \in \mathsf{U}(X) \mid X_k \in \mathcal{U}\}$, then $\mathsf{U}(X) = \cup_{k=1}^{n} \mathsf{U}_k$. Further, for any $\mathcal{U} \in \mathsf{U}_k$ let $\mathcal{U}(k) := \{U \dot{\in} \mathcal{U} \mid U \subseteq X_k\}$ then for any $k = 1, \ldots, n$

$$\sup_{\mathcal{U} \in \mathsf{U}_k} \inf_{x \in X} \lambda \mathcal{U}(x) \leq \sup_{\mathcal{U} \in \mathsf{U}_k} \inf_{x \in X_k} \sup_{A \in \mathcal{U}(k)} \delta(x, A) = \sup_{\mathcal{U} \in \mathsf{U}_k} \inf_{x \in X_k} \sup_{A \in \mathcal{U}} \delta(x, A) = \chi_c(X_k)$$

and hence

$$\chi_c(X) = \sup_{k=1}^{n} \sup_{\mathcal{U} \in \mathsf{U}_k} \inf_{x \in X} \lambda \mathcal{U}(x) \leq \sup_{k=1}^{n} \chi_c(X_k).$$ □

Note that the inequality in the foregoing result is strict in general, as is easily seen already from the topological case.

We now consider the following ultrafilter spaces (see 1.2.63) where we have simplified the notation somewhat because here we are dealing with a special case. Fix a set X and an ultrafilter $\mathcal{U}$ on X. Take $\omega \notin X$, let $X_{\mathcal{U}} := X \cup \{\omega\}$ and put $\mathcal{U}_\omega$ the ultrafilter on $X_{\mathcal{U}}$ generated by $\mathcal{U}$. The following defines a topology on $X_{\mathcal{U}}$: the only convergent ultrafilters are the point filters, which converge to their defining points and the filter $\mathcal{U}_\omega$ which converges to ω. The approach space generated by this topology has as limit operator (on ultrafilters)

$$\lambda^{\mathcal{U}} \mathcal{V}(x) := \begin{cases} 0 & \mathcal{V} = \dot{x} \text{ or } (\mathcal{V} = \mathcal{U}_\omega \text{ and } x = \omega) \\ \infty & \text{all other cases,} \end{cases}$$

and as distance

$$\delta^{\mathcal{U}}(x, A) := \begin{cases} 0 & x \in A \text{ or } (A \in \mathcal{U}_\omega \text{ and } x = \omega) \\ \infty & \text{all other cases.} \end{cases}$$

This gives a special (even topological) case of the filter spaces considered in 1.2.63 where the underlying set is $X_{\mathcal{U}}$, the filter $\mathcal{F}$ is $\mathcal{U}_\omega$ and the function f is identically zero. In 1.2.36 the limit operator was denoted $\lambda_{(\mathcal{U}_\omega, 0)}$.

In the following few results we put $\Delta := \{(x, x) \in X \times X_{\mathcal{U}} \mid x \in X\}$.

4.3.23 Theorem *If $f : X \longrightarrow X'$ is a function between approach spaces, then*

$$\chi_p(f) = \sup\{\chi_{ce}(f \times 1_Z) \mid Z \in \mathsf{App}\} = \sup\{\chi_{ce}(f \times 1_Z) \mid Z \in \mathsf{Top}\}.$$

Proof Obviously we only need to show two inequalities. Let Z be an arbitrary approach space and consider the map $f \times 1_Z : X \times Z \longrightarrow X' \times Z$. We denote the structures on the domain of $f \times 1_Z$ by a superscript d and those on the range by a superscript r. First, for any ultrafilter $\mathcal{U} \in \mathsf{U}(X \times Z)$ and $(x', z) \in X' \times Z$ we have

$$\begin{aligned}
f \times 1_Z(\lambda^d \mathcal{U})(x', z) &= \inf_{x \in f^{-1}(x')} \lambda^d \mathcal{U}(x, z) \\
&= \inf_{x \in f^{-1}(x')} \lambda_X \mathrm{pr}_1 \mathcal{U}(x) \vee \lambda_Z \mathrm{pr}_2 \mathcal{U}(z) \\
&= f(\lambda_X \mathrm{pr}_1 \mathcal{U})(x') \vee 1_Z(\lambda_Z \mathrm{pr}_2 \mathcal{U})(z) \\
&\leq (\lambda_{X'}(f \mathrm{pr}_1 \mathcal{U})(x') + \chi_p(f)) \vee (\lambda_Z(1_Z \mathrm{pr}_2 \mathcal{U})(z) + \chi_p(1_Z)) \\
&\leq \lambda_{X'}(\mathrm{pr}_1(f \times 1_Z \mathcal{U}))(x') \vee \lambda_Z(\mathrm{pr}_2(f \times 1_Z \mathcal{U}))(z) + \chi_p(f) \\
&= \lambda^r f \times 1_Z(\mathcal{U})(x', z) + \chi_p(f).
\end{aligned}$$

Consequently, if $A \subseteq X \times Z$ and $(x', z) \in X' \times Z$ it follows that

$$\begin{aligned}
f \times 1_Z(\delta^d_A)(x', z) &= \inf_{x \in f^{-1}(x')} \delta^d((x, z), A) \\
&= \inf_{x \in f^{-1}(x')} \inf_{\mathcal{U} \in \mathsf{U}(A)} \lambda^d \mathcal{U}(x, z) \\
&= \inf_{\mathcal{U} \in \mathsf{U}(A)} f \times 1_Z(\lambda^d \mathcal{U})(x', z) \\
&\leq \inf_{\mathcal{U} \in \mathsf{U}(A)} \lambda^r(f \times 1_Z \mathcal{U})(x', z) + \chi_p(f) \\
&\leq \inf_{\mathcal{W} \in \mathsf{U}(f \times 1_Z(A))} \lambda^r \mathcal{W}(x', z) + \chi_p(f) \\
&= \delta^r_{f \times 1_Z(A)}(x', z) + \chi_p(f)
\end{aligned}$$

which proves that $\chi_{ce}(f \times 1_Z) \leq \chi_p(f)$ and hence by arbitrariness of $Z \in \mathsf{App}$ that $\sup\{\chi_{ce}(f \times 1_Z) \mid Z \in \mathsf{App}\} \leq \chi_p(f)$.

Now consider an ultrafilter $\mathcal{U} \in \mathsf{U}(X)$, the associated ultrafilter space $X_{\mathcal{U}}$ and the function

$$f \times 1_{X_{\mathcal{U}}} : X \times X_{\mathcal{U}} \longrightarrow X' \times X_{\mathcal{U}}.$$

For any set $U \subseteq X$ we put $U_\Delta := (U \times U) \cap \Delta$. Let $\mathcal{U}_\Delta$ be the ultrafilter on $X \times X_{\mathcal{U}}$ generated by the sets U_Δ for $U \in \mathcal{U}$. By 4.2.9 we have

$$\sup_{U \in \mathcal{U}} f \times 1_{X_{\mathcal{U}}}(\delta^d_{U_\Delta}) \leq \lambda^r(f \times 1_{X_{\mathcal{U}}}(\mathcal{U}_\Delta)) + \chi_{ce}(f \times 1_{X_{\mathcal{U}}}).$$

Claim 1: $\forall x' \in X',\ \ \lambda^r(f \times 1_{X_{\mathcal{U}}}(\mathcal{U}_\Delta))(x', \omega) = \lambda_{X'} f(\mathcal{U})(x')$. From the characterization of product limit in 1.3.12 we obtain

$$\lambda^r(f \times 1_{X_{\mathcal{U}}})(\mathcal{U}_\Delta)(x', \omega) = \lambda_{X'} f(\mathcal{U})(x') \vee \lambda^{\mathcal{U}} \mathcal{U}(\omega)$$
$$= \lambda_{X'} f(\mathcal{U})(x').$$

Claim 2: $\forall x' \in X', \ \ f(\lambda_X \mathcal{U})(x') = \sup_{U \in \mathcal{U}} f \times 1_{X_{\mathcal{U}}}(\delta^d_{U_\Delta})(x', \omega)$. Indeed

$$\sup_{U \in \mathcal{U}} f \times 1_{X_{\mathcal{U}}}(\delta^d_{U_\Delta})(x', \omega) = \sup_{U \in \mathcal{U}} \inf_{x \in f^{-1}(x')} \delta^d_{U_\Delta}(x, \omega).$$

Now by 1.3.17

$$\delta^d_{U_\Delta}(x, \omega) = \sup_{\mathcal{P} \in \mathsf{R}(U_\Delta)} \min_{P \in \mathcal{P}} \max\{\delta_X(x, \mathrm{pr}_X P), \delta^{\mathcal{U}}(\omega, \mathrm{pr}_{X_{\mathcal{U}}} P)\}$$

where we recall that $\mathsf{R}(U_\Delta)$ is the set of all finite partitions of U_Δ. Now from any such partition exactly one element belongs to the ultrafilter $\mathcal{U}_\Delta$ and for all other elements $\delta^{\mathcal{U}}(\omega, \mathrm{pr}_{X_{\mathcal{U}}} P) = \infty$. Moreover, if $V \in \mathcal{U}$ then $\delta^{\mathcal{U}}(\omega, \mathrm{pr}_{X_{\mathcal{U}}} V_\Delta) = 0$. Hence the expression reduces to

$$\delta^d_{U_\Delta}(x, \omega) = \sup_{W \in \mathcal{U}, W \subseteq U} \delta_X(x, W)$$
$$= \lambda_X \mathcal{U}(x).$$

Consequently, substituting gives

$$\sup_{U \in \mathcal{U}} f \times 1_{X_{\mathcal{U}}}(\delta^d_{U_\Delta})(x', \omega) = \sup_{U \in \mathcal{U}} \inf_{x \in f^{-1}(x')} \lambda_X \mathcal{U}(x)$$
$$= \sup_{U \in \mathcal{U}} f(\lambda_X \mathcal{U})(x')$$
$$= f(\lambda_X \mathcal{U})(x').$$

Finally, substituting the result of both claims in the inequality preceding the first claim gives

$$f(\lambda_X \mathcal{U})(x') \leq \lambda_{X'} f(\mathcal{U})(x') + \chi_{ce}(f \times 1_{X_{\mathcal{U}}})$$

which by 4.2.23 proves that $\chi_p(f) \leq \sup\{\chi_{ce}(f \times 1_Z) \mid Z \in \mathsf{Top}\}$. □

4.3.24 Corollary (Top) *A function $f : X \longrightarrow X'$ between topological spaces is proper if and only if for any topological space Z the map $f \times 1_Z : X \times Z \longrightarrow X' \times Z$ is closed.*

The inequality in the following lemma is actually an equality as we will show in 4.3.34.

4.3.25 Lemma *Consider a one-point space P and the unique morphism* $\pi : X \longrightarrow P$. *Then*

$$\chi_c(X) \le \chi_p(\pi)$$

and in particular, if π *is proper then X is 0-compact.*

Proof Let $\mathscr{U}$ be an ultrafilter on X and consider the diagram below, where i is the canonical isomorphism.

$$
\begin{array}{ccc}
X \times X_{\mathscr{U}} & \xrightarrow{\;\mathrm{pr}_2\;} & X_{\mathscr{U}} \\
{\scriptstyle \pi \times 1_{X_{\mathscr{U}}}} \downarrow & \nearrow {\scriptstyle i} & \\
P \times X_{\mathscr{U}} & &
\end{array}
$$

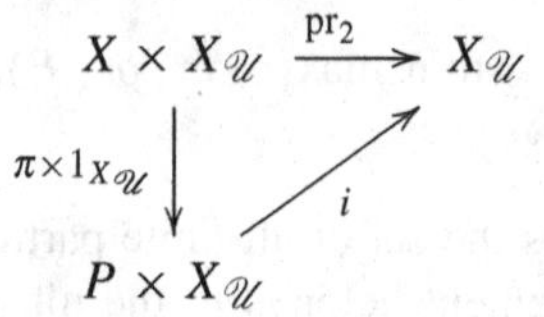

Then by the foregoing theorem

$$\chi_{ce}(\mathrm{pr}_2) = \chi_{ce}(\pi \times 1_{X_{\mathscr{U}}}) \le \chi_p(\pi).$$

This implies that

$$\mathrm{pr}_2(\delta_\Delta)(\omega) \le \delta^{\mathscr{U}}_{\mathrm{pr}_2(\Delta)}(\omega) + \chi_{ce}(\mathrm{pr}_2) \le \delta^{\mathscr{U}}_{\mathrm{pr}_2(\Delta)}(\omega) + \chi_p(\pi).$$

For the righthand side, since $X \in \mathscr{U}_\omega$, we clearly have

$$\delta^{\mathscr{U}}_{\mathrm{pr}_2(\Delta)}(\omega) = \delta^{\mathscr{U}}(\omega, X) = 0.$$

We now calculate the lefthand side.

$$
\begin{aligned}
\mathrm{pr}_2(\delta_\Delta)(\omega) &= \inf_{(x,y)\in(\mathrm{pr}_2)^{-1}(\omega)} \delta_\Delta(x, y) \\
&= \inf_{x\in X} \delta_\Delta(x, \omega) \\
&= \inf_{x\in X} \inf_{\mathscr{V}\in U(\Delta)} \lambda\mathscr{V}(x, \omega) \\
&= \inf_{x\in X} \inf_{\mathscr{V}\in U(\Delta)} \lambda_X \mathrm{pr}_1 \mathscr{V}(x) \vee \lambda^{\mathscr{U}} \mathrm{pr}_2 \mathscr{V}(\omega).
\end{aligned}
$$

Note that the only way $\lambda^{\mathscr{U}} \mathrm{pr}_2 \mathscr{V}(\omega)$ can be different from ∞ (and then necessarily equal to 0) is if $\mathrm{pr}_2 \mathscr{V} = \mathscr{U}_\omega$. This happens precisely when $\mathscr{V} = \mathscr{U}_\Delta$ where $\mathscr{U}_\Delta$ is the filter generated by $\{(U \times U) \cap \Delta \mid U \in \mathscr{U}\}$. Hence we find that

$$\mathrm{pr}_2(\delta_\Delta)(\omega) = \inf_{x\in X} \lambda_X \mathrm{pr}_1 \mathscr{U}_\Delta(x) = \inf_{x\in X} \lambda_X \mathscr{U}(x).$$

Thus $\inf_{x\in X}\lambda_X\mathcal{U}(x)\leq\chi_p(\pi)$, and the result now follows from the arbitrariness of $\mathcal{U}$. □

4.3.26 Lemma *If $\mathcal{U}$ is a filter on X and $A\subseteq X$, then*

$$\inf_{x\in A}\sup_{U\in\mathcal{U}}\delta(x,U)\leq\sup_{U\in\mathcal{U}}\inf_{x\in A}\delta(x,U)+\chi_c(A).$$

Proof Suppose that $\sup_{U\in\mathcal{U}}\inf_{x\in A}\delta(x,U)<\gamma$ where $\gamma<\infty$. Then this implies that $\{U^{(\gamma)}\cap A\mid U\in\mathcal{U}\}$ generates a filter $\mathcal{F}$ on A and since by 4.3.1, $\inf_{x\in A}\alpha\mathcal{F}(x)\leq\chi_c(A)$ it follows that, for any $\varepsilon>0$, there exists $x\in A$ such that

$$\begin{aligned}\sup_{U\in\mathcal{U}}\delta(x,U)&\leq\sup_{U\in\mathcal{U}}\delta(x,U^{(\gamma)})+\gamma\\&\leq\sup_{U\in\mathcal{U}}\delta(x,U^{(\gamma)}\cap A)+\gamma\\&\leq\alpha\mathcal{F}(x)+\gamma\\&\leq\chi_c(A)+\varepsilon+\gamma\end{aligned}$$

which by the arbitrariness of ε and γ proves the lemma. □

4.3.27 Theorem *If $f:X\longrightarrow X'$ is a function between approach spaces, then*

$$\chi_{ce}(f)\vee\sup_{x'\in X'}\chi_c(f^{-1}(x'))\leq\chi_p(f)\leq\chi_{ce}(f)+\sup_{x'\in X'}\chi_c(f^{-1}(x')).$$

Proof Let $\mathcal{U}\in\mathsf{U}(X)$ and $x'\in X'$, then by the foregoing lemma

$$\begin{aligned}f(\lambda\mathcal{U})(x')&=\inf_{x\in f^{-1}(x')}\sup_{U\in\mathcal{U}}\delta(x,U)\\&\leq\sup_{U\in\mathcal{U}}\inf_{x\in f^{-1}(x')}\delta(x,U)+\chi_c(f^{-1}(x'))\\&=\sup_{U\in\mathcal{U}}f(\delta_U)(x')+\chi_c(f^{-1}(x'))\\&\leq\sup_{U\in\mathcal{U}}\delta'_{f(U)}(x')+\chi_{ce}(f)+\chi_c(f^{-1}(x'))\\&\leq\lambda'f(\mathcal{U})(x')+\chi_{ce}(f)+\sup_{x'\in X'}\chi_c(f^{-1}(x'))\end{aligned}$$

which by 4.2.23 proves the inequality on the right.

That $\chi_{ce}(f)\leq\chi_p(f)$ was shown in 4.2.25 and by 4.2.27 and 4.3.25 we have

$$\sup_{x'\in X'}\chi_c(f^{-1}(x'))\leq\sup_{x'\in X'}\chi_p(f_{|f^{-1}(x')})\leq\chi_p(f)$$

which proves the inequality on the left. □

In order to see that the two expressions on the extreme sides of the inequalities in the foregoing result can indeed be different it suffices to give an example where the two quantities appearing in the expressions are neither zero nor infinite. The following provides such an example.

4.3.28 Example Let B_3 and B_1 be the closed balls in l^2 with radii respectively 3 and 1. We consider l^2 to be equipped with its natural metric approach structure and consider the following function:

$$f : B_3 \longrightarrow B_1 : x \mapsto \begin{cases} 0 & x \in B_1, \\ \frac{1}{2}(x - \frac{x}{\|x\|}) & x \in B_3 \setminus B_1. \end{cases}$$

Then $\sup_{y \in B_1} \chi_c(f^{-1}(y)) = \chi_c(B_1) = 1$ and a tedious calculation considering all possible cases shows that $\chi_{ce}(f) = 4$.

4.3.29 Corollary *For a function $f : X \longrightarrow X'$ between approach spaces the following are equivalent.*

1. *f is a proper contraction.*
2. *f is a closed expansion and a contraction and for each $x' \in X'$, $f^{-1}(x')$ is 0-compact.*
3. *for each $\mathcal{U} \in \mathsf{U}(X)$: $f(\lambda\mathcal{U}) = \lambda' f(\mathcal{U})$.*

The following result answers the question which was left open in 2.2.9.

4.3.30 Corollary (Top) *If X and X' are topological spaces and $f : X \longrightarrow X'$ is a function, then f is proper if and only if it is a proper contraction between the associated approach spaces.*

4.3.31 Corollary (Top) *A function $f : X \longrightarrow X'$ between topological spaces is proper if and only if it is closed and has compact fibres.*

4.3.32 Theorem *If $f : X \longrightarrow Y$ is a surjective function between approach spaces, then*

$$\chi_c(X) \leq \chi_c(Y) + \chi_p(f).$$

Proof This follows from

$$\begin{aligned} \chi_c(X) &= \sup_{\mathcal{U} \in \mathsf{U}(X)} \inf_{x \in X} \lambda\mathcal{U}(x) \\ &= \sup_{\mathcal{U} \in \mathsf{U}(X)} \inf_{y \in Y} \inf_{x \in f^{-1}(y)} \lambda\mathcal{U}(x) \\ &= \sup_{\mathcal{U} \in \mathsf{U}(X)} \inf_{y \in Y} f(\lambda\mathcal{U})(y) \end{aligned}$$

$$
\begin{aligned}
&= \sup_{\mathcal{U} \in \mathsf{U}(X)} \inf_{y \in Y} \lambda' f(\mathcal{U})(y) + \chi_p(f) \\
&\leq \sup_{\mathcal{W} \in \mathsf{U}(Y)} \inf_{y \in Y} \lambda' \mathcal{W}(y) + \chi_p(f) = \chi_c(Y) + \chi_p(f).
\end{aligned}
$$
□

4.3.33 Corollary (Top) *If $f : X \longrightarrow Y$ is a surjective and proper function between topological spaces and Y is compact, then X is compact.*

4.3.34 Theorem *For an approach space X, any one-point space P and the unique morphism $\pi : X \longrightarrow P$ we have*

$$\chi_c(X) = \chi_p(\pi).$$

Proof Since $\pi : X \longrightarrow P$ is a closed expansion this follows from 4.3.27. □

4.3.35 Proposition *If $f : X \longrightarrow X'$ is a function between approach spaces and $B \subseteq X'$, then*

$$\chi_c(f^{-1}(B)) \leq \chi_c(B) + \chi_p(f).$$

Proof Consider the diagram

$$f^{-1}(B) \xrightarrow{f|_{f^{-1}(B)}} B \xrightarrow{\pi} P, \qquad \rho = \pi \circ f|_{f^{-1}(B)} : f^{-1}(B) \longrightarrow P.$$

Then it follows from 4.3.34 that

$$
\begin{aligned}
\chi_c(f^{-1}(B)) &= \chi_p(\rho) \\
&\leq \chi_p(f|_{f^{-1}(B)}) + \chi_p(\pi) \\
&\leq \chi_p(f) + \chi_c(B).
\end{aligned}
$$
□

4.3.36 Corollary *If $f : X \longrightarrow Y$ is a proper contraction between approach spaces, and $B \subseteq Y$ is 0-compact, then also $f^{-1}(B)$ is 0-compact.*

4.3.37 Corollary *An approach space X is 0-compact if and only if for any one-point space P the unique morphism $\pi : X \longrightarrow P$ is a proper contraction.*

This result proves that the concept of 0-compactness coincides with $\mathcal{F}$-compactness in the sense of Clementino et al. (2003).

4.3.38 Theorem (**Kuratowski-Mrówka**) *For any approach space X*

$$
\begin{aligned}
\chi_c(X) &= \sup\{\chi_{ce}(\mathrm{pr}_Z) \mid Z \in \mathsf{App}, \mathrm{pr}_Z : X \times Z \longrightarrow Z\} \\
&= \sup\{\chi_{ce}(\mathrm{pr}_Z) \mid Z \in \mathsf{Top}, \mathrm{pr}_Z : X \times Z \longrightarrow Z\}.
\end{aligned}
$$

In particular, an approach space X is 0-compact if and only if for any approach space (respectively any topological space) Z the projection $\mathrm{pr}_Z : X \times Z \longrightarrow Z$ *is closed expansive.*

Proof Let P be a one-point space and for any space Z consider the commutative diagram

$$
\begin{array}{ccc}
X \times Z & \xrightarrow{\ \mathrm{pr}_Z\ } & Z \\[2mm]
{\scriptstyle \pi \times 1_Z}\Big\downarrow & \swarrow{\scriptstyle i} & \\[2mm]
P \times Z & &
\end{array}
$$

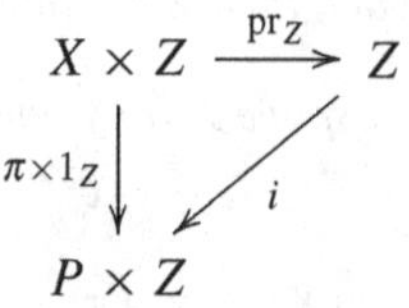

where i is the evident isomorphism. Then the result now follows immediately from 4.3.23 and 4.3.34.   □

4.3.39 Corollary (**Kuratowski-Mrówka in** Top) *A topological space X is compact if and only if for any topological space Z the projection* $\mathrm{pr}_Z : X \times Z \longrightarrow Z$ *is closed.*

In 3.5.9 and 4.3.13 we have seen that as far as completeness and total boundedness are concerned, when these properties are considered independently only one-sided implications are possible for the properties with respect to the various structures involved. When combined, however, we obtain the following result.

4.3.40 Proposition *If X is a uniform approach space, then the following properties are equivalent.*

1. $(X, \mathcal{T}_\delta)$ *is compact.*
2. *X is complete and* 0-*compact.*

Proof $1 \Rightarrow 2$. Let $\mathscr{D}$ be a symmetric gauge basis for X. If $(X, \mathcal{T}_\delta)$ is compact, then $(X, \mathscr{U}(\mathscr{D}))$ is complete and it follows from 3.5.9 that (X, δ) is complete. Furthermore it then follows from 4.3.15 that $\chi_c(X) = 0$.

$2 \Rightarrow 1$. Conversely, if $\mathscr{U}$ is an ultrafilter on X it follows from the fact that $\chi_c(X) = 0$ that $\mathscr{U}$ is a Cauchy filter. From the completeness of (X, δ) it then follows that $\mathscr{U}$ converges. Hence, $(X, \mathcal{T}_\delta)$ is compact.   □

In the sequel we will also require indices of other forms of compactness. We will not make an equally extensive study of these various alternatives as we did for the index of compactness, but since we will require some of them later on we introduce them here and give some basic properties.

4.3.41 Definition (**Index of relative compactness**) Given an approach space X and a subset $A \subseteq X$, we define the *index of relative compactness of A* (*with respect to X*) as

$$
\boxed{\ \chi_{rc}(A) := \sup_{\mathscr{F} \in \mathsf{F}(A)} \ \inf_{x \in X} \ \alpha.\mathscr{F}(x)\ }
$$

In precisely the same way as for the index of compactness we can then prove the following theorem .

4.3.42 Theorem *For any approach space X and subset $A \subseteq X$, we have*

$$\begin{aligned}\chi_{rc}(A) &= \sup_{\mathcal{U}\in \mathsf{U}(A)} \inf_{x\in X} \alpha\mathcal{U}(x)\\ &= \sup_{\mathcal{U}\in \mathsf{U}(A)} \inf_{x\in X} \lambda\mathcal{U}(x)\\ &= \sup_{\varphi\in \prod_{x\in X}\mathcal{B}(x)} \inf_{Y\in 2^{(X)}} \sup_{z\in A} \inf_{x\in Y} \varphi(x)(z)\\ &= \sup_{\varphi\in \mathcal{H}^X} \inf_{Y\in 2^{(X)}} \sup_{z\in A} \inf_{x\in Y} \varphi(x)(x,z) .\end{aligned}$$

where $(\mathcal{B}(x))_{x\in X}$ is a basis for the approach system and $\mathcal{H}$ is a basis for the gauge.

Proof This is analogous to 4.3.2 and we leave this to the reader. □

4.3.43 Proposition *The index of relative compactness χ_{rc} is $(1, \chi_c)$-layered.*

Proof This follows from 4.2.1, 4.1.5 and the definition of χ_{rc}. □

4.3.44 Proposition (Top) *If $(X, \delta_{\mathcal{T}})$ is a topological approach space and $A \subseteq X$, then the following properties hold.*

1. *If A is relatively compact in $(X, \mathcal{T})$ we have $\chi_{rc}(A) = 0$.*
2. *If $(X, \mathcal{T})$ is moreover regular and $\chi_{rc}(A) = 0$ then A is relatively compact.*

Proof If A is relatively compact then since an ultrafilter on A is also an ultrafilter on $\overline{A}$ it follows at once from the definitions that $\chi_{rc}(A) = 0$.

Conversely, if $\mathcal{U}$ is an ultrafilter on $\overline{A}$ it follows that there exists an ultrafilter $\mathcal{W}$ on A such that the filter $\overline{\mathcal{W}}$ generated by the closures of elements of $\mathcal{W}$ is coarser than $\mathcal{U}$. Indeed, suppose not, then it follows that for any ultrafilter $\mathcal{W}$ on A there exists $W_{\mathcal{W}} \in \mathcal{W}$ such that $\overline{W_{\mathcal{W}}} \notin \mathcal{U}$. From 1.1.4 it then follows that we can find a finite number of ultrafilters $\mathcal{W}_k$, $k = 1, \ldots, n$ on A such that $A \subset \cup_{k=1}^n W_{\mathcal{W}_k}$ and hence $\overline{A} \subseteq \cup_{k=1}^n \overline{W_{\mathcal{W}_k}}$. Since $\overline{A} \in \mathcal{U}$ this would imply that there is a k such that $\overline{W_{\mathcal{W}_k}} \in \mathcal{U}$ which is a contradiction. Hence take an ultrafilter $\mathcal{W}$ on A such that $\overline{\mathcal{W}} \subseteq \mathcal{U}$. Then it follows from $\chi_{rc}(A) = 0$ that $\mathcal{W}$ converges. By regularity also $\overline{\mathcal{W}}$ converges and hence $\mathcal{U}$ converges. □

The regularity in the condition cannot be omitted as shown by the following example.

4.3.45 Example Consider an infinite set X and a fixed point $a \in X$ equipped with the topology for which a set is open if it either contains the point a or is empty. Take any other point $b \in X$, then the set $A := \{a, b\}$ is not relatively compact but since it is finite we do have that $\chi_{rc}(A) = 0$.

4.3.46 Corollary (Met) *If X is a metric approach space, then a subset $A \subseteq X$ is totally bounded if and only if $\chi_{rc}(A) = 0$ and it is bounded if and only if $\chi_{rc}(A) < \infty$.*

Proof This follows from 4.3.42. □

4.3.47 Theorem *If $f : X \longrightarrow X'$ is a function between approach spaces and $A \subseteq X$, then*

$$\chi_{rc}(f(A)) \leq \chi_{rc}(A) + \chi_c(f)$$

and in particular if f is a contraction and A is 0-relatively compact, then $f(A)$ is 0-relatively compact.

Proof This is analogous to 4.3.17 and we leave this to the reader. □

4.3.48 Corollary (Top) *The continuous image of a relatively compact set is relatively compact.*

The index of relative compactness has an interesting advantage over the index of compactness.

4.3.49 Proposition *If X is an approach space and $A \subseteq B \subseteq X$, then $\chi_{rc}(A) \leq \chi_{rc}(B)$.*

Proof This follows from the definition. □

Furthermore, analogously to 4.3.22 we have the following stronger result.

4.3.50 Proposition *If X is an approach space and $X = \bigcup_{k=1}^{n} X_k$, then*

$$\chi_{rc}(X) = \max_{1 \leq k \leq n} \chi_{rc}(X_k).$$

Proof This is analogous to 4.3.22 and we leave this to the reader. □

We put $\mathsf{S}(X)$ for the set of all sequences in X.

4.3.51 Definition (Index of sequential compactness) Given an approach space X we define the *index of sequential compactness* as

$$\boxed{\chi_{sc}(X) := \sup_{(x_n)_n \in \mathsf{S}(X)} \inf_{k\uparrow} \inf_{x \in X} \lambda\langle (x_{k_n})_n \rangle(x)}$$

4.3.52 Proposition *The index of sequential compactness χ_{sc} is $(1, \chi_c)$-layered.*

Proof This follows from 4.2.1, 4.1.5 and the definition of χ_{sc}. □

4.3.53 Proposition (Top, Met) *A topological approach space is 0-sequentially compact if and only if the underlying topology is sequentially compact and a metric approach space is 0-sequentially compact if and only if the underlying metric is totally bounded.*

Proof The first claim follows at once from the definition. For the second claim, let (X, d) be a metric space. To show the only-if part, let $(x_n)_n$ be an arbitrary sequence, then it follows from $\chi_{sc}(X) = 0$ that there exists a subsequence $(x_{k_n})_n$, an index m and a point $x \in X$ such that for all $l \geq m$, $d(x_{k_l}, x) \leq \frac{1}{2}$. From this it follows immediately that there is a further subsequence $(x_{k_1(n)})_n$ such that $d(x_{k_1(n)}, x_{k_1(m)}) \leq 1$ for all $n, m \in \mathbb{N}$. Now consider this sequence and reason exactly in the same way to obtain a further subsequence $(x_{k_1 \circ k_2(n)})_n$ such that $d(x_{k_1 \circ k_2(n)}, x_{k_1 \circ k_2(m)}) \leq \frac{1}{2}$ for all $n, m \in \mathbb{N}$. Continuing this procedure we obtain an infinite sequence of subsequences of which the diagonal subsequence clearly is a Cauchy sequence. Hence X is totally bounded.

Conversely, to show the if part, if X is totally bounded then every sequence has a Cauchy subsequence. Now we only need to note that by 3.5.6, if $(y_n)_n$ is a Cauchy sequence in (X, d) then $\inf_{x \in X} \lambda\langle(y_n)_n\rangle(x) = 0$. Hence $\chi_{sc}(X) = 0$. □

4.3.54 Theorem *If $f : X \longrightarrow X'$ is a function between approach spaces and $A \subseteq X$, then*

$$\chi_{sc}(f(A)) \leq \chi_{sc}(A) + \chi_c(f)$$

and in particular if f is a contraction and A is 0-sequentially compact, then $f(A)$ is 0-sequentially compact.

Proof This is analogous to 4.3.17 and we leave this to the reader. □

4.3.55 Corollary (Top) *The continuous image of a sequentially compact set is sequentially compact.*

Since we will require this in the chapter on applications in probability theory, we also introduce a relative index of sequential compactness.

4.3.56 Definition **(Index of relative sequential compactness)** Given an approach space X and a subset $A \subseteq X$ we define the *index of relative sequential compactness* of A (with respect to X) as

$$\boxed{\chi_{rsc}(A) := \sup_{(x_n)_n \in \mathsf{S}(A)} \inf_{k\uparrow} \inf_{x \in X} \lambda\langle(x_{k_n})_n\rangle(x)}$$

4.3.57 Proposition *The index of relative sequential compactness χ_{rsc} is $(1, \chi_c)$-layered.*

Proof This follows from 4.2.1, 4.1.5 and the definition of χ_{rsc}. □

4.3.58 Proposition (Top, Met) *If X is a topological approach space then $A \subseteq X$ is 0-relatively sequentially compact if and only if A is relatively sequentially compact in the underlying topology and if X is a metric approach space then $A \subseteq X$ is 0-relatively sequentially compact if and only if A is totally bounded in the underlying metric.*

Proof The first claim follows immediately from the definition (note that the alleged convergent subsequence necessarily converges in $\overline{A}$).

The proof of the second claim is completely analogous to the proof of 4.3.53. □

4.3.59 Example The same example as 4.3.16 does not have an underlying (relatively) sequentially compact topology but does have $\chi_{sc}(X) = \chi_{rsc}(X) = 0$.

4.3.60 Theorem *If $f : X \longrightarrow X'$ is a function between approach spaces and $A \subseteq X$, then*

$$\chi_{rsc}(f(A)) \leq \chi_{rsc}(A) + \chi_c(f)$$

and in particular if f is a contraction and A is 0-relatively sequentially compact, then $f(A)$ is 0-relatively sequentially compact.

Proof This is analogous to 4.3.17 and we leave this to the reader. □

4.3.61 Corollary (Top) *The continuous image of a relatively sequentially compact set is relatively sequentially compact.*

We put $\mathsf{F}_c(X)$ for the set of all filters on X which have a countable basis.

4.3.62 Definition **(Index of countable compactness)** Given an approach space X we define the *index of countable compactness of X* as

$$\boxed{\chi_{cc}(X) := \sup_{\mathcal{F} \in \mathsf{F}_c(X)} \inf_{x \in X} \alpha\mathcal{F}(x)}$$

4.3.63 Theorem *For any approach space X, we have*

$$\begin{aligned}
\chi_{cc}(X) &= \sup_{(x_n)_n \in \mathsf{S}(X)} \inf_{x \in X} \alpha\langle (x_n)_n \rangle (x) \\
&= \sup_{\varphi \in \prod_{x \in X} \mathcal{B}(x)} \sup_{(x_n)_n \in \mathsf{S}(X)} \inf_{x \in X} \liminf_{n \to \infty} \varphi(x)(x_n) \\
&= \sup_{\psi \in \mathcal{H}^X} \sup_{(x_n)_n \in \mathsf{S}(X)} \inf_{x \in X} \liminf_{n \to \infty} \psi(x)(x, x_n).
\end{aligned}$$

Proof In order to prove the first equality it is sufficient to note that for any filter $\mathcal{F}$ with a countable basis we can find a sequence $(x_n)_n$ such that the elementary filter generated by it is finer than $\mathcal{F}$. The second equality follows from

$$\begin{aligned}
\sup_{(x_n)_n \in \mathsf{S}(X)} \inf_{x \in X} \alpha\langle (x_n)_n \rangle (x) &= \sup_{(x_n)_n \in \mathsf{S}(X)} \inf_{x \in X} \sup_{\varphi \in \mathcal{B}(x)} \liminf_{n \to \infty} \varphi(x_n) \\
&= \sup_{(x_n)_n \in \mathsf{S}(X)} \sup_{\varphi \in \prod_{x \in X} \mathcal{B}(x)} \inf_{x \in X} \liminf_{n \to \infty} \varphi(x)(x_n)
\end{aligned}$$

and the third one is an immediate consequence. □

4.3.64 Proposition *The index of countable compactness χ_{cc} is $(1, \chi_c)$-layered.*

Proof This follows from 4.2.1, 4.1.5 and the definition of χ_{cc}. □

4.3.65 Proposition (Top, Met) *A topological approach space is 0-countably compact if and only if the underlying topology is countably compact and a metric approach space is 0-countably compact if and only if the underlying metric is totally bounded.*

Proof This follows from the definitions. □

4.3.66 Theorem *If $f : X \longrightarrow X'$ is a function between approach spaces and $A \subseteq X$, then*

$$\chi_{cc}(f(A)) \leq \chi_{cc}(A) + \chi_c(f)$$

and in particular if f is a contraction and A is 0-countably compact, then $f(A)$ is 0-countably compact.

Proof This is analogous to 4.3.17 and we leave this to the reader. □

4.3.67 Corollary (Top) *The continuous image of a countably compact set is countably compact.*

4.3.68 Definition **(Lindelöf index)** Given an approach space X we define the *Lindelöf index of X* as

$$\boxed{\chi_l(X) := \sup_{\varphi \in \prod_{x \in X} \mathcal{A}(x)} \inf_{Y \in 2^{[X]}} \sup_{z \in X} \inf_{x \in Y} \varphi(x)(z)}$$

where $2^{[X]}$ stands for the set of countable subsets of X.

We denote the set of all filters on X which have the countable intersection property by $\mathsf{F}_\omega(X)$.

4.3.69 Theorem *For an approach space X we have*

$$\chi_l(X) = \sup_{\mathcal{F} \in \mathsf{F}_\omega(X)} \inf_{x \in X} \alpha\mathcal{F}(x).$$

Proof To prove that $\sup_{\mathcal{F} \in \mathsf{F}_\omega(X)} \inf_{x \in X} \alpha\mathcal{F}(x) \geq \chi_l(X)$ suppose that $\chi_l(X) \neq 0$ and fix $c < \chi_l(X)$ and ϕ such that $\inf_{Y \in 2^{[X]}} \sup_{z \in X} \inf_{x \in Y} \varphi(x)(z) > c$. For any $Y \in 2^{[X]}$ put

$$B_Y = \{z \in X | \inf_{x \in Y} \phi(x)(z) > c\}$$

then $B_Y \neq \emptyset$ and $\bigcap_{n \in \mathbb{N}} B_{Y_n} = B_{\bigcup_n Y_n}$. Hence $\mathcal{B}_\varepsilon = \{B_Y | Y \in 2^{[X]}\}$ is a basis for a filter with the countable intersection property. The rest of the proof now goes along the same lines as the proof of 4.3.2 and we leave this to the reader. □

4.3.70 Proposition *The Lindelöf index χ_l is $(1, \chi_c)$-layered.*

Proof This follows from 4.2.1, 4.1.5 and the definition of χ_l. □

4.3.71 Corollary (Top, Met) *A topological approach space is 0-Lindelöf if and only if the underlying topology is Lindelöf and a metric approach space (X, δ_d) is 0-Lindelöf if and only if for any $\varepsilon > 0$ there exists a countable subset $Y \subseteq X$ such that $\cup_{y \in Y} B_d(y, \varepsilon) = X$.*

Proof This follows from the definitions. □

Now that we have seen all indices of types of compactnesses we can prove some relations between them.

4.3.72 Theorem *If X is an approach space, then the following properties hold.*

1. $\chi_{cc}(X) \leq \chi_{sc}(X)$.
2. $\chi_{cc}(X) \leq \chi_c(X)$.
3. $\chi_l(X) \leq \chi_c(X)$.
4. *If X is locally countable, then the indices of countable compactness and of sequential compactness coincide.*

Proof We leave 1, 2 and 3 to the reader. As for 4 suppose that for each $x \in X$, $\mathscr{B}(x) := \{\varphi_n \mid n \in \mathbb{N}\}$ is a countable basis for the approach system and let $(x_n)_n$ be an arbitrary sequence. It follows from 1.2.69 that

$$\alpha\langle(x_n)_n\rangle(x) = \sup_m \sup_l \inf_{n \geq l} \varphi_m(x_n).$$

From this it follows that, given $\varepsilon > 0$, we can then find $k \uparrow$ such that for all $l \in \mathbb{N}$, $\varphi_l(x_{k_l}) \leq \alpha\langle(x_n)_n\rangle(x) + \varepsilon$. Now it is easily deduced that

$$\lambda\langle(x_{k_n})_n\rangle(x) \leq \alpha\langle(x_n)_n\rangle(x) + 2\varepsilon.$$

Since this holds for any sequence, any $x \in X$ and any $\varepsilon > 0$ it follows

$$\begin{aligned}\chi_{sc}(X) &= \sup_{(x_n)_n \in \mathsf{S}(X)} \inf_{k\uparrow} \inf_{x \in X} \lambda\langle(x_{k_n})_n\rangle(x) \\ &\leq \alpha\langle(x_n)_n\rangle(x) \\ &\leq \sup_{(x_n)_n \in \mathsf{S}(X)} \inf_{x \in X} \alpha\langle(x_n)_n\rangle(x) = \chi_{cc}(X).\end{aligned}$$ □

4.3.73 Corollary (Top) *In a topological space a sequentially compact or compact set is countably compact, a compact set is Lindelöf and if the space is first countable then countable compactness and sequential compactness coincide.*

4.3.74 Theorem *For an approach space X we have*

$$\chi_{cc}(X) \vee \chi_l(X) \leq \chi_c(X) \leq \chi_{cc}(X) + \chi_l(X).$$

Proof The first inequality follows at once from the foregoing result. As to the second one, suppose that $0 < \varepsilon < \chi_c(X)$. Then it follows from the definition of χ_c that there exists a filter $\mathcal{F}$ such that

$$\forall y \in X, \exists d_y \in \mathcal{G}, \exists F_y \in \mathcal{F}, \forall x \in F_y : d_y(y, x) > \chi_c(X) - \varepsilon$$

and consequently it follows from the definition of χ_l that there exists a countable set $Y \subseteq X$ such that

$$\forall x \in X, \exists y \in Y : d_y(y, x) < \chi_l(X) + \varepsilon.$$

Now consider the filter $\mathcal{H} \subseteq \mathcal{F}$ generated by the set $\{F_y \mid y \in Y\}$ then it follows from the definition of χ_{cc} that there exists an $x_0 \in X$ such that

$$\sup_{d \in \mathcal{G}} \sup_{y \in Y} \inf_{t \in F_y} d(x_0, t) < \chi_{cc}(X) + \varepsilon$$

and thus there exists $y_0 \in Y$ such that $d_{y_0}(y_0, x_0) < \chi_l(X) + \varepsilon$. Then it follow that there exist $z_0 \in F_{y_0}$ such that $d_{y_0}(x_0, z_0) < \chi_{cc}(X) + 2\varepsilon$. All together this implies that

$$\begin{aligned}\chi_c(X) - \varepsilon &< d_{y_0}(y_0, z_0)\\ &\leq d_{y_0}(y_0, x_0) + d_{y_0}(x_0, z_0)\\ &\leq \chi_l(X) + \varepsilon + \chi_{cc}(X) + 2\varepsilon\end{aligned}$$

which by the arbitrariness of ε proves the result. □

4.3.75 Corollary (Top) *A topological space is compact if and only if it is countably compact and Lindelöf.*

4.3.76 Theorem *If $f : X \longrightarrow X'$ is a function between approach spaces and $A \subseteq X$, then*

$$\chi_l(f(A)) \leq \chi_l(A) + \chi_c(f)$$

and in particular, if f is a contraction and A is 0-Lindelöf, then $f(A)$ is 0-Lindelöf.

Proof This is analogous to 4.3.17 and we leave this to the reader. □

4.3.77 Corollary (Top) *The continuous image of a Lindelöf space is Lindelöf.*

4.4 Local Compactness Index

4.4.1 Definition **(Index of local compactness)** In topological spaces, local compactness can be characterized by stating that any convergent filter contains a compact set. Hence, if X is an approach space and $\mathcal{F}$ is a filter on X, then we put $\chi_c(\mathcal{F}) := \inf_{F\in\mathcal{F}} \chi_c(F)$ and we define the *index of local compactness* of X as

$$\boxed{\chi_{lc}(X) := \sup_{\mathcal{F}\in\mathsf{F}(X)} (\chi_c(\mathcal{F}) \ominus \inf_{x\in X} \lambda\mathcal{F}(x))}$$

If $\chi_{lc}(X) = 0$, then we will say that X is 0-*locally compact*. Note that this index can also be written as

$$\chi_{lc}(X) = \min\{c \mid \forall \mathcal{F} \in \mathsf{F}(X) : \chi_c(\mathcal{F}) \leq \inf_{x\in X} \lambda\mathcal{F}(x) + c\}.$$

4.4.2 Theorem *For any approach space X, let* $(\mathcal{V}_\theta(x))_\theta$ *be the neighbourhood tower, then we have*

$$\begin{aligned}\chi_{lc}(X) &= \sup_{\mathcal{U}\in\mathsf{U}(X)} (\chi_c(\mathcal{U}) \ominus \inf_{x\in X} \lambda\mathcal{U}(x))\\ &= \sup_{x\in X,\theta\geq 0} (\chi_c(\mathcal{V}_\theta(x)) \ominus \lambda\mathcal{V}_\theta(x)(x))\\ &= \sup_{x\in X,\theta\geq 0} (\chi_c(\mathcal{V}_\theta(x)) \ominus \theta).\end{aligned}$$

Proof To prove the first equality we only need to prove one inequality as the other one is of course trivial. Let $\mathcal{F} \in \mathsf{F}(X)$ and suppose that $\sup_{\mathcal{U}\in\mathsf{U}(X)}(\chi_c(\mathcal{U}) \ominus \inf_{x\in X} \lambda\mathcal{U}(x)) < \alpha$. Then we have

$$\forall\mathcal{U} \in \mathsf{U}(\mathcal{F}), \exists\sigma(\mathcal{U}) \in \mathcal{U} : \chi_c(\sigma(\mathcal{U})) \leq \inf_{x\in X} \lambda\mathcal{U}(x) + \alpha.$$

By 1.1.4 there exists a finite subset $\mathsf{U}_\sigma \subseteq \mathsf{U}(\mathcal{F})$ such that $\bigcup_{\mathcal{U}\in\mathsf{U}_\sigma} \sigma(\mathcal{U}) \in \mathcal{F}$, and then we obtain

$$\begin{aligned}\chi_c(\bigcup_{\mathcal{U}\in\mathsf{U}_\sigma} \sigma(\mathcal{U})) &\leq \max_{\mathcal{U}\in\mathsf{U}_\sigma} \chi_c(\sigma(\mathcal{U}))\\ &\leq \max_{\mathcal{U}\in\mathsf{U}_\sigma} \inf_{x\in X} \lambda\mathcal{U}(x) + \alpha\\ &\leq \inf_{x\in X} \max_{\mathcal{U}\in\mathsf{U}_\sigma} \lambda\mathcal{U}(x) + \alpha\\ &\leq \inf_{x\in X} \lambda\mathcal{F}(x) + \alpha\end{aligned}$$

which proves the remaining inequality.

For the second equality, from the definition of χ_{lc} it follows that we again only need to prove one inequality. Consider a filter $\mathcal{F}$ on X and $\varepsilon > 0$ such that

$$\theta := \inf_{z \in X} \lambda\mathcal{F}(z) < \theta + \varepsilon < \infty,$$

then it follows that there exists $x \in X$ such that $\mathcal{V}_{\theta+\varepsilon}(x) \subseteq \mathcal{F}$. For this neighbourhood filter it is evident that $\chi_c(\mathcal{F}) \leq \chi_c(\mathcal{V}_{\theta+\varepsilon}(x))$ and further it follows from 1.2.54 that

$$\lambda\mathcal{V}_{\theta+\varepsilon}(x)(x) \leq \theta + \varepsilon = \inf_{z \in X} \lambda\mathcal{F}(z) + \varepsilon.$$

The missing inequality follows.

The third equality follows from the fact that if

$$\varepsilon_x := \min\{\theta \mid \mathcal{V}_\theta(x) = \mathcal{V}_{\varepsilon_x}(x)\},$$

then, making use of 1.2.54

$$\chi_c(\mathcal{V}_\theta(x)) \ominus \lambda\mathcal{V}_\theta(x)(x) \leq \chi_c(\mathcal{V}_\theta(x)) \ominus \varepsilon_x = \chi_c(\mathcal{V}_{\varepsilon_x}(x)) \ominus \varepsilon_x.$$

□

4.4.3 Example For $\mathbb{Q}$ with the Euclidean topology $\chi_{lc}(\mathbb{Q}) = \infty$ and for $\mathbb{Q}$ with the Euclidean metric $\chi_{lc}(\mathbb{Q}) = 0$ (see also 4.4.8).

4.4.4 Proposition *The index of local compactness χ_{lc} is $(2, \chi_c)$-layered.*

Proof Consider spaces $U := (X, \delta)$ and $V := (X, \delta')$ and fix a filter $\mathcal{F} \in \mathsf{F}(X)$. It follows from 4.2.1, 4.3.2 and 4.3.17 that

$$\chi'_c(\mathcal{F}) \leq \chi_c(\mathcal{F}) + \chi_c(1_{UV}) \text{ and } \inf_{x \in X} \lambda'\mathcal{F}(x) \leq \inf_{x \in X} \lambda\mathcal{F}(x) + \chi_c(1_{UV}).$$

Interchanging the roles of U and V we similarly have that

$$\chi_c(\mathcal{F}) \leq \chi'_c(\mathcal{F}) + \chi_c(1_{VU}) \text{ and } \inf_{x \in X} \lambda\mathcal{F}(x) \leq \inf_{x \in X} \lambda'\mathcal{F}(x) + \chi_c(1_{VU}),$$

from which we obtain

$$|\chi_c(\mathcal{F}) - \chi'_c(\mathcal{F})| \leq \chi_c(1_{UV}) \vee \chi_c(1_{VU})$$

and

$$|\inf_{x \in X} \lambda\mathcal{F}(x) - \inf_{x \in X} \lambda'\mathcal{F}(x)| \leq \chi_c(1_{UV}) \vee \chi_c(1_{VU}).$$

From these inequalities it now follows that

$$|\chi_c(\mathscr{F}) \ominus \inf_{x\in X} \lambda\mathscr{F}(x) - \chi'_c(\mathscr{F}) \ominus \inf_{x\in X} \lambda'\mathscr{F}(x)| \leq 2\left(\chi_c(1_{UV}) \vee \chi_c(1_{VU})\right)$$

which upon taking the suprema over all $\mathscr{F} \in \mathsf{F}(X)$ and invoking 4.1.5, gives us the required conclusion. □

4.4.5 Theorem *If X is an approach space then $\chi_{lc}(X) \leq \chi_c(X)$.*

Proof This follows from the definitions. □

4.4.6 Corollary (Top) *A compact topological space is locally compact.*

4.4.7 Proposition (Top) *A topological approach space is 0-locally compact if and only if the underlying topology is locally compact.*

Proof This follows from the definitions. □

4.4.8 Proposition (qMet) *A quasi-metric approach space is always 0-locally compact.*

Proof This follows from the fact that in a quasi-metric space (X, d) for all $x \in X$ and $\theta < \infty$ we have $\chi_c(B_d(x, \theta)) \leq \theta$. □

4.4.9 Theorem *If $f : X \longrightarrow X'$ is a surjective function between approach spaces, then*

$$\chi_{lc}(X') \leq \chi_{lc}(X) + \chi_c(f) + \chi_{oe}(f).$$

Proof For any $\mathscr{U} \in \mathsf{U}(X)$ by 4.2.1 and 4.2.15 we have

$$\begin{aligned}\chi_c(f(\mathscr{U})) &= \inf_{U\in\mathscr{U}} \sup_{\mathscr{G}'\in\mathsf{U}(f(U))} \inf_{y\in f(U)} \lambda'\mathscr{G}'(y)\\ &= \inf_{U\in\mathscr{U}} \sup_{\mathscr{G}\in\mathsf{U}(U)} \inf_{x\in U} \lambda' f(\mathscr{G})(f(x))\\ &\leq \inf_{U\in\mathscr{U}} \sup_{\mathscr{G}\in\mathsf{U}(U)} \inf_{x\in U} \lambda\mathscr{G}(x) + \chi_c(f)\\ &= \chi_c(\mathscr{U}) + \chi_c(f).\end{aligned}$$

We now use the variant given after 4.4.2, so let $\mathscr{U}'$ be an ultrafilter on X', then it follows from the inequality we just proved and from 4.2.15 that

$$\begin{aligned}\chi_c(\mathscr{U}') &= \inf_{\mathscr{U}\in\mathsf{U}(f^{-1}(\mathscr{U}'))} \chi_c(f(\mathscr{U}))\\ &\leq \inf_{\mathscr{U}\in\mathsf{U}(f^{-1}(\mathscr{U}'))} \chi_c(\mathscr{U}) + \chi_c(f)\\ &\leq \chi_{lc}(X) + \inf_{\mathscr{U}\in\mathsf{U}(f^{-1}(\mathscr{U}'))} \inf_{x\in X} \lambda\mathscr{U}(x) + \chi_c(f)\end{aligned}$$

$$
\begin{aligned}
&= \chi_{lc}(X) + \inf_{x\in X} \inf_{\mathcal{U}\in \mathsf{U}(f^{-1}(\mathcal{U}'))} \lambda\mathcal{U}(x) + \chi_c(f) \\
&\leq \chi_{lc}(X) + \inf_{x\in X} \lambda'\mathcal{U}'(f(x)) + \chi_{oe}(f) + \chi_c(f) \\
&= \chi_{lc}(X) + \inf_{x'\in X'} \lambda'\mathcal{U}'(x') + \chi_{oe}(f) + \chi_c(f),
\end{aligned}
$$

from which the desired inequality follows. □

4.4.10 Corollary (Top) *The open continuous image of a locally compact topological space is locally compact.*

4.4.11 Theorem *If $(X_i)_{i\in I}$ is a family of approach spaces, then*

$$\chi_{lc}(\prod_{i\in I} X_i) = \sup_{i\in I} \chi_{lc}(X_i) \vee \inf_{J\in 2^{(I)}} \sup_{i\in I\setminus J} \chi_c(X_i).$$

Proof We put $X := \prod_{i\in I} X_i$. We first prove that

$$\chi_{lc}(X) \leq \sup_{i\in I} \chi_{lc}(X_i) \vee \inf_{J\in 2^{(I)}} \sup_{i\in I\setminus J} \chi_c(X_i)$$

and hereto suppose that all terms on the right are finite. Fix an arbitrary finite subset J of I such that $\sup_{i\in I\setminus J} \chi_c(X_i)$ is finite. Take $\mathcal{U} \in \mathsf{U}(X)$, $x = (x_i)_{i\in I} \in X$ and $\varepsilon > 0$. Then, for any $j \in J$

$$\inf_{U\in\mathcal{U}} \chi_c(\mathrm{pr}_j\, U) - \lambda_j(\mathrm{pr}_j\,\mathcal{U})(x_j) = \chi_c(\mathrm{pr}_j\,\mathcal{U}) - \lambda_j(\mathrm{pr}_j\,\mathcal{U})(x_j) \leq \sup_{i\in I} \chi_{lc}(X_i),$$

so there exists $U^j \in \mathcal{U}$ such that

$$\chi_c\left(\mathrm{pr}_j(U^j)\right) - \lambda_j(\mathrm{pr}_j\,\mathcal{U})(x_j) \leq \sup_{i\in I} \chi_{lc}(X_i) + \varepsilon.$$

If we now define

$$U := \prod_{j\in J} \mathrm{pr}_j(U^j) \times \prod_{i\in I\setminus J} X_i$$

then $U \in \mathcal{U}$ and, making use of 4.3.19, it follows that

$$
\begin{aligned}
\chi_c(U) - \lambda\mathcal{U}(x) &= \left(\sup_{j\in J} \chi_c\left(\mathrm{pr}_j(U^j)\right) \vee \sup_{i\in I\setminus J} \chi_c(X_i)\right) - \lambda\mathcal{U}(x) \\
&= \left(\sup_{j\in J} \chi_c(\mathrm{pr}_j(U^j)) - \lambda\mathcal{U}(x)\right) \vee \left(\sup_{i\in I\setminus J} \left(\chi_c(X_i) - \lambda\mathcal{U}(x)\right)\right) \\
&\leq \sup_{j\in J} \left(\chi_c\left(\mathrm{pr}_j(U^j)\right) - \lambda_j(\mathrm{pr}_j\,\mathcal{U})(x_j)\right) \vee \sup_{i\in I\setminus J} \chi_c(X_i)
\end{aligned}
$$

$$\leq \left(\sup_{i \in I} \chi_{lc}(X_i) + \varepsilon \right) \vee \sup_{i \in I \setminus J} \chi_c(X_i)$$

$$\leq \left(\sup_{i \in I} \chi_{lc}(X_i) \vee \sup_{i \in I \setminus J} \chi_c(X_i) \right) + \varepsilon$$

and the result follows from the arbitrariness of $\mathcal{U}$, x and ε.

Second, that $\sup_{i \in I} \chi_{lc}(X_i) \leq \chi_{lc}(X)$ is an immediate consequence of 4.4.9 and of the fact that projections are contractive open expansions.

Finally we prove that

$$\inf_{J \in 2^{(I)}} \sup_{i \in I \setminus J} \chi_c(X_i) \leq \chi_{lc}(X).$$

For $x \in X$ and $\varepsilon > 0$, consider the neighbourhood filters $\mathcal{V}_\varepsilon(x)$ of the tower and $\mathcal{V}(x) = \vee_{\varepsilon > 0} \mathcal{V}_\varepsilon(x)$ of the topological coreflection. Then, by 4.4.2 we have

$$\chi_{lc}(X) = \sup_{x \in X} \sup_{\varepsilon > 0} \inf_{V \in \mathcal{V}_\varepsilon(x)} (\chi_c(V) \ominus \varepsilon)$$

$$\geq \sup_{x \in X} \inf_{V \in \mathcal{V}(x)} \chi_c(V).$$

Now take $x \in X$ arbitrary and fix $\varepsilon > 0$. Then there exists a finite subset $J \subseteq I$, $d_j \in \mathcal{G}_j$ for $j \in J$ and $\gamma > 0$ such that with

$$V := \{y \mid \forall j \in J : d_j(x_j, y_j) < \gamma\} = \prod_{j \in J} \mathrm{pr}_j^{-1}(B_{d_j}(x_j, \gamma)) \times \prod_{i \in I \setminus J} X_i$$

we have $\chi_c(V) \leq \chi_{lc}(X) + \varepsilon$ and hence by 4.3.19 also $\sup_{i \in I \setminus J} \chi_c(X_i) \leq \chi_{lc}(X) + \varepsilon$ which by the arbitrariness of ε proves the result. □

Note that one cannot deduce that if a product is 0-locally compact then all but a finite number of the spaces involved are 0-compact. The best one has is the following corollary.

4.4.12 Corollary *Let $(X_i)_{i \in I}$ be a family of approach spaces. If all the X_i are 0-locally compact and all but a finite number of the X_i's are 0-compact, then $\prod_{i \in I} X_i$ is 0-locally compact. Conversely, if $\prod_{i \in I} X_i$ is 0-locally compact then all the X_i's are 0-locally compact and all but an at most countable number of X_i's are 0-compact.*

For topological spaces however we can deduce the usual result.

4.4.13 Corollary (Top) *The product of a family of topological spaces is locally compact if and only if all the spaces are locally compact and all but a finite number are compact.*

4.5 Connectedness Index

Our approach to connectedness follows the general theory of connectedness and disconnectedness in topological categories, as developed by Preuss (1970). For more information on connectedness concepts we also refer to the work of Arhangel'skii and Wiegandt in (1975) and of Mrówka and Pervin in (1964).

In order to formulate our definition of an index of connectedness we need some preliminary concepts.

For each $\varepsilon > 0$, we define the space D_ε to be the two-point set $\{0, \infty\}$ equipped with the metric d_ε, where $d_\varepsilon(0, \infty) := \varepsilon$. The space D_ε serves as a prototype for a disconnected space and the disconnectedness is quantified by the distance between the points 0 and ∞.

4.5.1 Definition (**Index of connectedness**) An approach space X is called ε-*connected* if the only contractions from X to D_ε are constant functions. Given an approach space X, we then define the *index of connectedness* of X as

$$\boxed{\chi_{cn}(X) := \inf\{\varepsilon > 0 \mid X \text{ is } \varepsilon\text{-connected}\}}$$

If X is an approach space which is ε-connected and $\varepsilon' > \varepsilon$, then clearly X is ε'-connected. The space D_ε for example is ε'-connected, for all $\varepsilon' > \varepsilon$, but it is not ε-connected. Consequently $\chi_{cn}(D_\varepsilon) = \varepsilon$.

4.5.2 Proposition (Top) *A topological approach space is* 0*-connected if and only if the underlying topology is connected.*

Proof Let $(X, \mathscr{T})$ be a topological space. Suppose that X is not ε-connected for some $\varepsilon > 0$. Then there exists a contraction $f : X \longrightarrow D_\varepsilon$ which is not constant. Let $X_0 := f^{-1}(\{0\})$ and $X_\infty := f^{-1}(\{\infty\})$. Then, for any $x \in X_0$, it follows that $\varepsilon \leq \delta_{\mathscr{T}}(x, X_\infty)$ which implies that $x \notin \mathrm{cl}_{\mathscr{T}}(X_\infty)$. Thus $X_0 \cap \mathrm{cl}_{\mathscr{T}}(X_\infty) = \emptyset$. Analogously $X_\infty \cap \mathrm{cl}_{\mathscr{T}}(X_0) = \emptyset$, which proves that both X_0 and X_∞ are open and thus $(X, \mathscr{T})$ is not connected.

Conversely, let X be partitioned into two open sets X_0 and X_∞. Then it is easily verified that, for any $\varepsilon > 0$, the map $f : X \longrightarrow D_\varepsilon$ defined by $f(X_0) := \{0\}$ and $f(X_\infty) := \{\infty\}$ is a contraction. Thus $\chi_{cn}(X) = \infty$. □

In order to prove our next result we recall that a metric space (X, d) is called *Cantor-connected*, Cantor (1883), if, for any $\varepsilon > 0$, any two points $x, y \in X$ can be "connected" by a so-called ε-chain , i.e. a finite number of points $x_0 = x, x_1, \ldots, x_n = y$ such that for all $i \in \{1, \ldots, n\}$: $d(x_{i-1}, x_i) \leq \varepsilon$. It is well known that a space is Cantor-connected if and only if it cannot be partitioned into sets A and B such that $d(A, B) > 0$ (see e.g. Herrlich 1986).

4.5.3 Proposition (Met) *A metric approach space is* 0*-connected if and only if the underlying metric is Cantor-connected.*

Proof Let (X, d) be a metric space. Suppose that X is not ε-connected for some $\varepsilon > 0$. Then there exists a contraction $f : X \longrightarrow D_\varepsilon$ which is not constant. Let $X_0 := f^{-1}(\{0\})$ and $X_\infty := f^{-1}(\{\infty\})$. Then it follows that

$$\begin{aligned} d(X_0, X_\infty) &= \inf_{x\in X_0} \inf_{y\in X_\infty} d(x, y) \\ &\geq \inf_{x\in X_0} \inf_{y\in X_\infty} d_\varepsilon(f(x), f(y)) = \varepsilon. \end{aligned}$$

Conversely, let X be partitioned into two sets X_0 and X_∞ such that $d(X_0, X_\infty) \geq \varepsilon$. Then the map $f : X \longrightarrow D_\varepsilon$, defined by $f(X_0) := \{0\}$ and $f(X_\infty) := \{\infty\}$, is a contraction and thus $\chi_{cn}(X) \geq \varepsilon > 0$. □

4.5.4 Example If we consider the set of rational numbers $\mathbb{Q}$ to be equipped with the Euclidean topology, then this space is not connected; hence $\chi_{cn}(\mathbb{Q}) = \infty$. If on the other hand we consider it to be equipped with the Euclidean metric, then it is Cantor-connected and $\chi_{cn}(\mathbb{Q}) = 0$.

4.5.5 Example If we consider the natural numbers $\mathbb{N}$ to be equipped with the Euclidean topology, then this space is not connected; hence $\chi_{cn}(\mathbb{N}) = \infty$. If we consider it to be equipped with the Euclidean metric, then it is also not Cantor-connected but the index of connectedness is $\chi_{cn}(\mathbb{N}) = 1$.

An alternative characterization for the index of connectedness is given by the following result.

4.5.6 Theorem *For an approach space X we have*

$$\chi_{cn}(X) = \sup_{A\in 2^X\setminus\{\emptyset,X\}} \min\left\{\inf_{x\in A} \delta(x, X\setminus A),\ \inf_{x\in X\setminus A} \delta(x, A)\right\}.$$

Proof That X is not ε-connected is equivalent with the existence of a non-constant contraction $f : X \longrightarrow D_\varepsilon$. This in turn, however, is equivalent with the existence of a set A such that $A \neq \emptyset$, $A \neq X$, and such that, for all $x \in A$, $\delta(x, X \setminus A) \geq \varepsilon$ and, for all $x \in X \setminus A$, $\delta(x, A) \geq \varepsilon$. □

4.5.7 Proposition *The index of connectedness χ_{cn} is $(1, \chi_c)$-layered.*

Proof This follows from 4.2.1, 4.1.5 and 4.5.6. □

We recall that a uniform space, with uniformity generated by a collection $\mathscr{D}$ of metrics, is said to be *uniformly connected* (see Mrowka and Pervin 1964, Arhangel'skii and Wiegandt 1975) if for all $A \subseteq X$, $d \in \mathscr{D}$, and $\varepsilon > 0$ such that $A \neq \emptyset$ and $A \neq X$, we have

$$\{x \in X \mid \delta_d(x, X \setminus A) \leq \varepsilon\} \bigcap \{x \in X \mid \delta_d(x, A) \leq \varepsilon\} \neq \emptyset.$$

4.5.8 Proposition *If X is a uniform approach space with symmetric gauge basis $\mathcal{G}$ which is 0-connected, then $(X, \mathcal{U}(\mathcal{G}))$ is uniformly connected.*

Proof Suppose that $(X, \mathcal{U}(\mathcal{G}))$ is not uniformly connected and let $A \subseteq X, d \in \mathcal{G}$, and $\varepsilon > 0$ be such that $A \neq \emptyset$, $A \neq X$, and

$$\{x \in X \mid \delta_d(x, X \setminus A) \leq \varepsilon\} \bigcap \{x \in X \mid \delta_d(x, A) \leq \varepsilon\} = \emptyset.$$

Then it follows from 4.5.6 that

$$\begin{aligned} \chi_{cn}(X) &\geq \min\left\{\inf_{x\in A}\sup_{e\in\mathcal{G}} \delta_e(x, X \setminus A),\ \inf_{x\in X\setminus A}\sup_{e\in\mathcal{G}} \delta_e(x, A)\right\} \\ &\geq \min\left\{\inf_{x\in A} \delta_d(x, X \setminus A),\ \inf_{x\in X\setminus A} \delta_d(x, A)\right\} \geq \varepsilon. \end{aligned}$$

□

The following example shows that the converse of 4.5.8 does not hold.

4.5.9 Example Let $X := \mathbb{Q}$ and let $d_{\mathbb{E}}$ stand for the Euclidean metric on X. If we let $\mathcal{G} := \{\alpha d_{\mathbb{E}} \mid \alpha > 0\}$, then δ is the topological distance generated by the Euclidean topology and hence $\chi_{cn}(X) = \infty$. However, $\mathcal{U}(\mathcal{G})$ is the usual uniformity and, hence, $(X, \mathcal{U}(\mathcal{G}))$ is uniformly connected.

4.5.10 Theorem *If $f : X \longrightarrow X'$ is a function between approach spaces and $A \subseteq X$, then*

$$\chi_{cn}(f(A)) \leq \chi_{cn}(A) + \chi_c(f)$$

and in particular if f is a contraction and A is 0-connected, then $f(A)$ is 0-connected.

Proof This follows from 4.2.1 and 4.5.6. □

4.5.11 Corollary (Top, Met) *The continuous image of a connected topological space is connected and the nonexpansive image of a Cantor-connected metric space is Cantor-connected.*

In order to give a product theorem for the index of connectedness we need some more preliminary results, which are interesting in their own right.

4.5.12 Proposition *If X is an approach space and $Y \subseteq Z \subseteq Y^{(c)} \subseteq X$ then*

$$\chi_{cn}(Z) \leq \chi_{cn}(Y) + c.$$

Proof Let $f : Z \longrightarrow D_{\varepsilon+c}$ be a contraction. Then consider the composition

$$Y \xrightarrow{f|_Y} D_{\varepsilon+c} \xrightarrow{i} D_\varepsilon, \qquad \rho := i \circ f|_Y : Y \longrightarrow D_\varepsilon$$

where i is the identity. Then since ρ is a contraction it is constant and thus there exists $p \in \{0, \infty\}$ such that $f(Y) = \rho(Y) = \{p\}$. Now if there exists $z \in Z$ such that $f(z) \neq p$ i.e. $\delta_{d_{\varepsilon+c}}(f(z), \{p\}) = \varepsilon + c$ then, on the other hand

$$\delta_{d_{\varepsilon+c}}(f(z), \{p\}) \leq \delta(z, Y) \leq c$$

which is a contradiction. Hence f is constant. □

If $\mathscr{A}$ is a collection of subsets of a set X, then $\mathscr{A}$ is called *chained* if, for any pair of sets $A, B \in \mathscr{A}$, there exists a finite collection $\{A_1, \ldots, A_n\} \subseteq \mathscr{A}$ such that $A_1 = A$, $A_n = B$, and for all $i \in \{1, \ldots, n-1\}$: $A_i \cap A_{i+1} \neq \emptyset$.

4.5.13 Proposition *If X is an approach space and $\mathscr{A}$ is a chained collection of subsets of X, then*

$$\chi_{cn}(\bigcup \mathscr{A}) \leq \sup_{A \in \mathscr{A}} \chi_{cn}(A).$$

Proof Suppose that $A \in \mathscr{A}$ is ε_A-connected and that $Y := \bigcup \mathscr{A}$ is not $(\sup_{A \in \mathscr{A}} \varepsilon_A)$-connected. Then there exists a subset $B \subseteq Y$ such that $\emptyset \neq B \neq Y$ and such that if we let $\varepsilon := \sup_{A \in \mathscr{A}} \varepsilon_A$ then

$$\inf_{x \in B} \delta(x, Y \setminus B) \geq \varepsilon \text{ and } \inf_{x \in Y \setminus B} \delta(x, B) \geq \varepsilon.$$

Now suppose that there exists a set $A \in \mathscr{A}$ such that $A \not\subseteq B$ and $A \not\subseteq Y \setminus B$. Then it follows that

$$\inf_{x \in A \cap B} \delta(x, A \setminus B) \geq \inf_{x \in B} \delta(x, Y \setminus B) \geq \varepsilon \geq \varepsilon_A,$$
$$\text{and}$$
$$\inf_{x \in A \setminus B} \delta(x, A \cap B) \geq \inf_{x \in Y \setminus B} \delta(x, B) \geq \varepsilon \geq \varepsilon_A,$$

which is impossible. Consequently if we let $\mathscr{C} := \{A \in \mathscr{A} \mid A \subseteq B\}$ and $\mathscr{D} := \{A \in \mathscr{A} \mid A \subseteq Y \setminus B\}$, then $\mathscr{A} = \mathscr{C} \cup \mathscr{D}$. Since $\mathscr{A}$ is chained, this implies that either $\mathscr{C} = \emptyset$ or $\mathscr{D} = \emptyset$. Suppose, for example, that $\mathscr{D} = \emptyset$. Then $\mathscr{A} = \mathscr{C}$ and it follows that $Y = B$ which is a contradiction. □

4.5.14 Theorem *If $(X_j)_{j \in J}$ is a family of approach spaces then*

$$\chi_{cn}(\prod_{j \in J} X_j) = \sup_{j \in J} \chi_{cn}(X_j).$$

Proof That $\chi_{cn}(\prod_{j \in J} X_j) \geq \sup_{j \in J} \chi_{cn}(X_j)$ follows from 4.5.10.

Conversely, let $(z_j)_{j\in J}$ be a fixed point in the product space $\prod_{j\in J} X_j$. For any subset $K \subseteq J$, let

$$X_K := \left\{(x_j)_{j\in J} \mid \forall i \in J \setminus K : x_i = z_i\right\}.$$

By induction on the cardinality of K, making use of 4.5.13, one sees that, for any finite set K, X_K is $(\sup_{k\in K} \varepsilon_k)$-connected. Now the collection

$$\mathscr{B} := \{X_K \mid K \subseteq J, K \text{finite}\}$$

is chained and, hence, it once again follows from 4.5.13 that $\bigcup \mathscr{B}$ is $(\sup_{j\in J} \varepsilon_j)$-connected. Since $\bigcup \mathscr{B}$ is dense in $\prod_{j\in J} X_j$ it follows from 4.5.12 that $\prod_{j\in J} X_j$ is $(\sup_{j\in J} \varepsilon_j)$-connected. □

4.5.15 Corollary (Top) *The product of a family of topological spaces is connected if and only if each factor space is connected.*

4.5.16 Corollary (Met) *The product of a finite family of metric spaces is Cantor-connected if and only if each space is Cantor-connected and an arbitrary product of metric spaces (in* App*) is 0-connected if and only if each space is Cantor-connected.*

4.5.17 Example If we consider the Cantor set C to be equipped with the Euclidean topology, then it is not connected and hence $\chi_c(\mathrm{C}) = \infty$. If we consider it to be equipped with the Euclidean metric, then it is also not Cantor-connected and it is easily seen that $\chi_{cn}(\mathrm{C}) = \frac{1}{3}$. If, however, we identify the Cantor set with the product $\{0, 1\}^{\mathbb{N}}$ and equip $\{0, 1\}$ with the Euclidean metric, then we can also equip C with the product distance. In that case, by 4.5.14, we have $\chi_{cn}(\mathrm{C}) = \sup_{n\in\mathbb{N}} \chi_{cn}(\{0, 1\}) = 1$. There is no contradiction in this result. The Cantor set can topologically be described in two homeomorphic ways. Both are also naturally derived from approach structures, however these approach structures, although having homeomorphic underlying topologies are not isomorphic and hence there is no reason why the indices of connectedness should coincide.

4.6 Comments

1. **Limit and adherence operator in the literature**

There are many traces of numerical indices to be found in the literature. Let us give two examples concerning the basic concepts of limit operator (index of convergence) and adherence operator (index of adherence).

Opial's condition plays an important role in the study of convergence of iterates of nonexpansive mappings and of the asymptotic behaviour of nonlinear semigroups (see e.g. Lami Dozo 1973; Karlovitz 1976; Kuczumow 1985; Opial 1967; Gornicki 1991). This condition says that if a sequence $(x_n)_n$ in a normed space converges weakly to x then $\liminf_n \|x-x_n\| < \liminf_n \|y-x_n\|$ for every $y \neq x$ or equivalently $\limsup_n \|x - x_n\| < \limsup_n \|y - x_n\|$ for every $y \neq x$ but the first condition is nothing else than saying that x is the unique point where $\alpha_{\|\ \|}\langle(x_n)_n\rangle(\cdot)$ attains its minimal value and the second one is nothing else than saying that x is the unique point where $\lambda_{\|\ \|}\langle(x_n)_n\rangle(\cdot)$ attains its minimal value, i.e. a unique asymptotic center. All Hilbert spaces satisfy Opial's condition as do all spaces l^p for $1 < p < \infty$ and any infinite dimensional Banach space admits a renorming such that the new norm satisfies Opial's condition (see e.g. Kirk and Sims 2001).

In Brezis and Lieb (1983) it is proved that if $(f_n)_n$ is a sequence of L^p-uniformly bounded functions on a measure space, and if $f_n \to f$ almost everywhere, then $\liminf_n \|f_n\|_p = \liminf_n \|f_n - f\|_p + \|f\|_p$ which says that

$$\alpha_p\langle(f_n)_n\rangle(0) = \alpha_p\langle(f_n)_n\rangle(f) + \|f\|_p$$

and this is a strengthening in a special case of the general result on limit and adherence operators in 1.2.70.

2. **Near-isometries and the index of contractivity in the literature**

The idea of measuring the deviation which maps have from being either contractive or expansive, or both, in metric spaces is not new and goes back to work of Hyers and Ulam (1945, 1947), Bourgin (1946). A survey is given in the paper by Väisälä (2002). There are many links with concepts introduced in the literature in various areas. In Edwards (1975) a map $f : X \longrightarrow Y$ between metric spaces is called an ε-isometry if for all $x, y \in X$ we have $|d(f(x), f(y)) - d(x, y)| \leq \varepsilon$. If the map f is a bijection then it is easily seen that this is equivalent to $\chi_c(f) \leq \varepsilon$ and $\chi_{oe}(f) = \chi_{ce}(f) \leq \varepsilon$. The same concept was later re-defined in Alestalo et al. (2001) and in Mémoli (2008). There the above concept is referred to as being an ε-*nearisometry* and the quantity $\text{dist}(f) := \sup_{x,y\in X} |d(f(x), f(y) - d(x, y)|$ is referred to as the *distortion* of f. For an arbitrary function $\text{dist}(f) \leq \varepsilon$ implies $\chi_c(f) \leq \varepsilon$ and for a bijection $\text{dist}(f) = \chi_c(f) \vee \chi_{oe}(f) = \chi_c(f) \vee \chi_{ce}(f)$.

Near- or ε-isometries encompass as special cases also the so-called *quasi-isometries* which are used in geometric group theory, see e.g. de la Harpe (2000). In particular we mention the Gromov program to study quasi-isometric invariants of groups via their actions on metric spaces.

In a paper by Colebunders et al. (2014) the notion of index of contractivity, there called default, was also used.

3. **Measures of noncompactness in the literature**

There is a vast literature on "measures of noncompactness" and their use in various areas of mathematics such as operator theory, functional equations, approximation theory and fixpoint theory. We refer to Akhmerov et al. (1992), Banaś (1997), Banaś and Goebel (1980), Benavides (1986), de Malafosse and Rakočević (2006), de Pagter and Schep (1988), Horvath (1985), In-Sook and Martin (2007), Malkowsky and Rakočević (1997, 1998, 2001, Preprint), Wiśnicki and Wośko (1996).

4. **Small ideals and functional ideal convergence**

The attentive reader will have noticed that our treatment of small ideals in order to arrive at a characterization of the index of compactness via lower regular function frames shows some resemblance with functional ideals. The treatment could indeed have been done via the concept of functional ideals but since we only required this characterization in one section of the applications to hyperspaces we made it entirely self-contained.

5. **Index of relative countable compactness**

Although we have not done so, it is perfectly feasible to define a relative version also of the index of countable compactness. Given an approach space X and a subset $A \subseteq X$, the *index of relative countable compactness of* A (with respect to X) would be $\chi_{rcc}(A) := \sup_{\mathscr{F} \in \mathsf{F}_c(A)} \inf_{x \in X} \alpha\mathscr{F}(x)$ and we then have the following alternative formulas

$$\begin{aligned}\chi_{rcc}(X) &= \sup_{(x_n)_n \in \mathsf{S}(A)} \inf_{x \in X} \alpha(x_n)(x) \\ &= \sup_{\varphi \in \prod_{x \in X} \mathscr{B}(x)} \sup_{(x_n)_n \in \mathsf{S}(A)} \inf_{x \in X} \liminf_{n \to \infty} \varphi(x)(x_n) \\ &= \sup_{\varphi \in \mathscr{H}^X} \sup_{(x_n)_n \in \mathsf{S}(A)} \inf_{x \in X} \liminf_{n \to \infty} \varphi(x)(x, x_n).\end{aligned}$$

If $(X, \delta_{\mathscr{T}})$ is a topological approach space and $A \subseteq X$ then $\chi_{rcc}(A) = 0$ if and only if A is relatively countably compact and if (X, δ_d) is a metric approach space and $A \subseteq X$ then $\chi_{rcc}(A) = 0$ if and only if A is totally bounded. Most alternative indices of types of compactness were originally introduced in the PhD thesis of Baekeland (1992), and we refer to that for more information. See also Baekeland and Lowen (1995).

6. **Index of basis-local compactness**

Again, although we have not done so, it is possible to define a "basis"-version of the index of local compactness. A plausible definition, based on the formula in 4.4.2, would be

$$\chi_{blc}(X) = \sup_{x \in X, \theta \geq 0} \left(\liminf_{V \in \mathscr{V}_\theta(x)} \chi_c(V) \ominus \theta \right)$$

where

$$\liminf_{V \in \mathscr{V}_\theta(x)} \chi_c(V) := \sup_{V \in \mathscr{V}_\theta(x)} \inf_{W \in \mathscr{V}_\theta(x), W \subseteq V} \chi_c(W)$$

and which can also be written as

$$\chi_{blc}(X) = \min\{k \mid \forall x \in X, \forall \theta \geq 0 : \liminf_{V \in \mathscr{V}_\theta(x)} \chi_c(V) \leq \theta + k\}.$$

7. **Measures of connectedness in the literature**

Various other types of measures of connectedness and disconnectedness were introduced in the paper by Holgate and Sioen (2007) as fixed points of a Galois connection. In that paper measures of connectedness found in the literature are put in a general categorical framework. The index of connectedness χ_{cn} fits their general framework.

Another concept related to Cantor connectedness is that of Lipschitz- and uniform Lipschitz-connectedness, see Baboolal and Pillay (2009).

8. **Topological versus metric properties**

In this chapter we have seen that topological properties have remarkable metric counterparts. Compactness for instance is a topological property which in the case of a metrizable topological space is closely linked to the metric concept of total boundedness. But more strikingly, both properties have similar characterizations. For compactness one starts with arbitrary open covers and for total boundedness one starts with collections of all balls with an arbitrary fixed radius, and in both cases, for a space to be respectively compact or totally bounded, one asserts the existence of a finite sub-collection which still covers the whole space.

A similar situation presents itself for connectedness. Topological connectedness of a space X can be characterized by the fact that X cannot be split into two nontrivial closed parts. Similarly, Cantor-connectedness in metric spaces (introduced by Cantor in 1883 and extended to the realm of uniform spaces by Mrowka and Pervin in 1964) is characterized by the fact that the space cannot be split into two nontrivial parts which lie at a strictly positive distance from each other.

All these similarities become canonical in the setting of index analysis.

Chapter 5
Uniform Gauge Spaces

Science is the attempt to make the chaotic diversity of our sense experience correspond to a logically uniform system of thought.
(Albert Einstein)

As is clear from the foregoing chapters, approach spaces form a local theory. A local theory which, quite contrary to topology or other local theories in the literature, nevertheless allows for a concept of completeness, which one usually encounters only in the realm of uniform theories such as uniform or metric spaces, uniform convergence spaces in the sense of Preuss (2002) and nearness spaces in the sense of Herrlich (1974b). As we will see in Chap. 6 this "local" version of completeness also allows for the construction of a completion, which moreover is categorically well-behaved.

Nevertheless there is also a natural uniform notion of completeness and completion and in this chapter we will define the appropriate setting hereto. Not surprisingly this setting is linked to uniform spaces. The category UG which we will construct is the approach counterpart of Unif in the same way that approach spaces are the counterpart to topological spaces. Just as was the case for approach spaces here UG will be a supercategory of both the categories Unif of uniform spaces and uniformly continuous maps and of Met (as before with non-expansive maps). Since, as before, Met will only be concretely coreflectively embedded and not reflectively, this enlargement will again make it possible to consider arbitrary initial structures of metric spaces or of metrizable uniform spaces preserving numerical information.

We will also study the relationship between UG and UAp and we will introduce various natural indices in the setting of UG.

Finally we also consider the non-symmetric variant $q\mathsf{UG}$ of UG.

R. Lowen, *Index Analysis*, Springer Monographs in Mathematics,
DOI 10.1007/978-1-4471-6485-2_5

5.1 The Structures, Objects and Morphisms

Given a set X, a collection $\mathcal{G} \subseteq \mathsf{Met}(X)$ and a metric $d \in \mathsf{Met}(X)$, we will say that d is *dominated by* $\mathcal{G}$, or that $\mathcal{G}$ *dominates* d, if for all $\varepsilon > 0$ and $\omega < \infty$ there exists a $d^{\varepsilon,\omega} \in \mathcal{G}$ such that

$$d \wedge \omega \leq d^{\varepsilon,\omega} + \varepsilon.$$

We will then also say that the family $(d^{\varepsilon,\omega})_{\varepsilon>0,\omega<\infty}$ dominates d.

Further we will say that a collection of metrics $\mathcal{G}$ is *saturated*, if any metric d which is dominated by $\mathcal{G}$ already belongs to $\mathcal{G}$.

In spite of the fact that these concepts are very similar to the analogous concepts for approach spaces there is a fundamental difference, namely the fact that here we have inequalities with metrics as functions of two variables, whereas in the case of approach spaces the analogous inequalities were formulated for localized versions where the first variable was held constant.

5.1.1 Definition (**Uniform gauge**) A subset $\mathcal{G}$ of $\mathsf{Met}(X)$ is called a *uniform gauge* if it is an ideal in $\mathsf{Met}(X)$ which fulfils the following property.

(UG1) $\mathcal{G}$ is saturated.

As was the case for approach spaces, here too, it regularly happens that one has a collection of metrics which would be a natural candidate to form a uniform gauge but not all conditions are fulfilled. The following type of collection will often be encountered. We recall that a subset $\mathcal{H}$ of $\mathsf{Met}(X)$ is an ideal basis in $\mathsf{Met}(X)$ if for any $d, e \in \mathcal{H}$ there exists $c \in \mathcal{H}$ such that $d \vee e \leq c$.

5.1.2 Definition A subset $\mathcal{H}$ of $\mathsf{Met}(X)$ is called a *uniform gauge basis* if it is an ideal basis in $\mathsf{Met}(X)$.

By definition, a uniform gauge is also a uniform gauge basis, and similarly to the situation for approach spaces, here, too, any result shown to hold for uniform gauge bases will also hold for uniform gauges.

In order to derive the uniform gauge from a uniform gauge basis we will require a *saturation operation* which is similar to the local one for approach systems and gauges. Given a subset $\mathcal{D} \subseteq \mathsf{Met}(X)$ we define

$$\widetilde{\mathcal{D}} := \{d \in \mathsf{Met}(X) \mid \mathcal{D} \text{ dominates } d\}.$$

We call $\widetilde{\mathcal{D}}$ the *(uniform) saturation* of $\mathcal{D}$.

5.1.3 Definition An ideal basis $\mathcal{H}$ in $\mathsf{Met}(X)$ is called a *basis for a uniform gauge* $\mathcal{G}$ if $\widetilde{\mathcal{H}} = \mathcal{G}$. In this case we also say that $\mathcal{H}$ *generates* $\mathcal{G}$ or that $\mathcal{G}$ is *generated by* $\mathcal{H}$.

5.1.4 Proposition *If $\mathcal{H}$ is a uniform gauge basis, then $\widetilde{\mathcal{H}}$ is a uniform gauge with $\mathcal{H}$ as basis and if $\mathcal{H}$ is a basis for a uniform gauge $\mathcal{G}$, then it is a uniform gauge basis.*

Proof This is analogous to 1.1.16 and we leave this to the reader. □

5.1.5 Definition A pair $(X, \mathcal{G})$ where $\mathcal{G}$ is a uniform gauge on X is called a *uniform gauge space* or shortly a UG-space.

Note that this terminology differs from the one used in Lowen and Windels (1998). The associated morphisms are defined in the same way as the characterization of morphisms with gauges in the approach case.

5.1.6 Definition Let $(X, \mathcal{G})$ and $(X', \mathcal{G}')$ be uniform gauge spaces and let $f : X \longrightarrow X'$ be a function. Then f is called a *uniform contraction* if

$$\forall d \in \mathcal{G}' : d \circ (f \times f) \in \mathcal{G}.$$

From the saturation condition it is easily seen that this is equivalent to the statement that

$$\forall d \in \mathcal{G}', \forall \varepsilon > 0, \forall \omega < \infty, \exists e \in \mathcal{G} : d \circ (f \times f) \wedge \omega \leq e + \varepsilon.$$

There is no contradiction in the fact that contractions and uniform contractions are defined in precisely the same way. The difference lies in the different saturation conditions for gauges and uniform gauges.

Uniform gauge spaces and uniform contractions form a category which we denote UG.

5.1.7 Theorem UG *is a topological category.*

Proof This goes along the same lines as 1.3.9. In particular given uniform gauge spaces $(X_j)_{j\in J}$, consider the source

$$(f_j : X \longrightarrow X_j)_{j\in J}$$

in UG. If, for each $j \in J$, $\mathcal{H}_j$ is a basis for the uniform gauge of X_j, then a basis for the initial uniform gauge on X is given by

$$\mathcal{H} := \left\{ \sup_{j\in K} d_j \circ (f_j \times f_j) | K \in 2^{(J)}, \forall j \in K : d_j \in \mathcal{H}_j \right\}.$$ □

Just as uniformities defined by entourages are equivalent to uniformities defined by gauges (called uniform structures in Gillman and Jerison 1976), so are uniform gauges equivalent to towers of semi-uniformities in a way similar to the equivalence between towers and gauges in the first chapter.

5.1.8 Definition (**Uniform tower**) Let X be a set. A family of filters $(\mathcal{U}_\varepsilon)_{\varepsilon\in\mathbb{R}^+}$ on $X \times X$ is called a *uniform tower (on X)* if it fulfils the following properties.

(UT1) $\forall \varepsilon \in \mathbb{R}^+, \forall U \in \mathscr{U}_\varepsilon : \Delta_X \subseteq U$.
(UT2) $\forall \varepsilon \in \mathbb{R}^+, \forall U \in \mathscr{U}_\varepsilon : U^{-1} \in \mathscr{U}_\varepsilon$.
(UT3) $\forall \varepsilon, \varepsilon' \in \mathbb{R}^+ : \mathscr{U}_\varepsilon \circ \mathscr{U}_{\varepsilon'} \supseteq \mathscr{U}_{\varepsilon+\varepsilon'}$.
(UT4) $\forall \varepsilon \in \mathbb{R}^+ : \mathscr{U}_\varepsilon = \bigcup_{\alpha>\varepsilon} \mathscr{U}_\alpha$.

Thus, a uniform tower is a family of semi-uniformities satisfying (UT3) and (UT4). We recall that a semi-uniformity on X is a filter on $X \times X$ such that all members of the filter contain the diagonal and such that with every U in the filter also U^{-1} belongs to the filter. Also note that by (UT3), $\mathscr{U}_0$ is actually a uniformity.

5.1.9 Theorem **(UT $\rightrightarrows$ UG)** *If $(\mathscr{U}_\varepsilon)_{\varepsilon\in\mathbb{R}^+}$ is a uniform tower on X, then*

$$\mathscr{G} := \{d \in \mathsf{Met}(X) \mid \forall \varepsilon \in \mathbb{R}^+, \forall \alpha > \varepsilon : \{d < \alpha\} \in \mathscr{U}_\varepsilon\}$$

is a uniform gauge and for any ε, $\mathscr{U}_\varepsilon$ is generated by $\{\{d < \alpha\} \mid d \in \mathscr{G}, \alpha > \varepsilon\}$.

Proof We only verify that $\mathscr{G}$ is saturated, i.e. that it fulfils (UG1), leaving the remaining points to the reader. Let e be a metric such that for all $\omega < \infty$ and $\varepsilon > 0$ there exists $d \in \mathscr{G}$ such that $e \wedge \omega \leq d + \varepsilon$. Take θ and $\alpha > \theta$ fixed and choose $\omega > \alpha$ and $\varepsilon := \alpha - \theta$. Now considering $d \in \mathscr{G}$ as above we find that

$$\{d < \theta + \frac{\varepsilon}{2}\} \subseteq \{e < \alpha\}$$

and hence $\{e < \alpha\} \in \mathscr{U}_\theta$ which, by the arbitrariness of α and θ proves that $e \in \mathscr{G}$. □

5.1.10 Theorem **(UG $\rightrightarrows$ UT)** *If $\mathscr{G}$ is a uniform gauge on X, then the family $(\mathscr{U}_\varepsilon)_{\varepsilon\in\mathbb{R}^+}$, where for every $\varepsilon \in \mathbb{R}^+$, $\mathscr{U}_\varepsilon$ is the semi-uniformity generated by*

$$\{\{d < \alpha\} \mid d \in \mathscr{G}, \alpha > \varepsilon\}$$

is a uniform tower and $\mathscr{G} = \{d \in \mathsf{Met}(X) \mid \forall \varepsilon \in \mathbb{R}^+, \forall \alpha > \varepsilon : \{d < \alpha\} \in \mathscr{U}_\varepsilon\}$.

Proof We only prove that (UT3) holds, again leaving the remaining points to the reader. Fix $d \in \mathscr{G}$, $\varepsilon_1, \varepsilon_2 \in \mathbb{R}^+$ and take $\alpha > \varepsilon_1 + \varepsilon_2$. Now if $\alpha_1 > \varepsilon_1$ and $\alpha_2 > \varepsilon_2$ are such that $\alpha_1 + \alpha_2 = \alpha$, we have

$$\{d < \alpha_1\} \circ \{d < \alpha_2\} \subseteq \{d < \alpha\}.$$

□

5.1.11 Theorem *If $f : X \longrightarrow X'$ is a function between uniform gauge spaces then, in terms of the associated uniform towers, it is a uniform contraction if and only if $f : (X, \mathscr{U}_\varepsilon) \longrightarrow (X', \mathscr{U}'_\varepsilon)$ is uniformly continuous for each $\varepsilon \in \mathbb{R}^+$.*

Proof This is analogous to 1.3.3 and we leave this to the reader. □

5.2 Embedding Unif and Met in UG

Given a uniform space $(X, \mathscr{U})$, we associate with it a natural uniform gauge space by simply taking as uniform gauge the collection of all uniformly continuous metrics.

5.2.1 Proposition (Unif) *If* $(X, \mathscr{U})$ *is a uniform space, then the collection* $\mathscr{G}_{\mathscr{U}}$ *of all uniformly continuous metrics is a uniform gauge and the associated uniform tower is* $(\mathscr{U}_\varepsilon := \mathscr{U})_{\varepsilon \in \mathbb{R}^+}$.

Proof The collection $\mathscr{G}_{\mathscr{U}}$ of all uniformly continuous metrics for $\mathscr{U}$ is an ideal which satisfies the condition that if e is a metric and

$$\forall \varepsilon > 0, \exists \delta > 0, \exists d \in \mathscr{G}_{\mathscr{U}} : \{d < \delta\} \subseteq \{e < \varepsilon\}$$

then $e \in \mathscr{G}_{\mathscr{U}}$ (Gillman and Jerison 1976). Clearly this implies that $\mathscr{G}_{\mathscr{U}}$ satisfies (UG1) i.e. that it is saturated according to Definition 5.1.1. The second claim follows immediately from the fact that if d is a uniformly continuous metric then so is αd for any $0 < \alpha < \infty$. □

5.2.2 Definition A uniform gauge space of type $(X, \mathscr{G}_{\mathscr{U}})$, for some uniformity $\mathscr{U}$ on X, is called a *uniform gauge space* or somewhat less unfortunate, a uniform UG-space.

It is evident that if $(X, \mathscr{U})$ and $(X', \mathscr{U}')$ are uniform spaces and $f : X \longrightarrow X'$ is uniformly continuous then for any uniformly continuous metric d on X' we have that $d \circ (f \times f)$ is a uniformly continuous metric on X. Hence we immediately have that there is a concrete functor from Unif to UG which takes $(X, \mathscr{U})$ to $(X, \mathscr{G}_{\mathscr{U}})$ and which is a full embedding of Unif into UG.

As was the case for Top in App, here too we can show that the above embedding is both concretely reflective and concretely coreflective.

5.2.3 Theorem Unif *is embedded as a concretely reflective subcategory of* UG.

Proof It follows at once from the characterization of initial structures in UG in 5.1.7 that uniform UG-spaces are closed under the formation of initial structures. Hence Unif is concretely reflectively embedded in UG. □

5.2.4 Corollary Unif *is closed under the formation of limits and initial structures in* UG. *In particular, a product in* UG *of a family of uniform spaces is uniform and, likewise, a subspace in* UG *of a uniform space is a uniform space.*

As was the case for App, the Unif-reflection of a UG-space is not particularly interesting, hence we do not describe it here.

5.2.5 Proposition *If* $(X, \mathscr{G})$ *is a uniform gauge space, then the collection*

$$\mathscr{U}(\mathscr{G}) := \{U \subseteq X \times X \mid \exists \varepsilon > 0, \exists d \in \mathscr{G} : \{d < \varepsilon\} \subseteq U\}$$

is a uniformity on X. This uniformity also is simply the uniformity generated by the collection of metrics $\mathscr{G}$ and it is also $\mathscr{U}_0$ in the associated tower.

Proof See e.g. Gillman and Jerison (1976). □

5.2.6 Theorem Unif *is embedded as a concretely coreflective subcategory of* UG. *For any uniform gauge space $(X, \mathscr{G})$, its* Unif*-coreflection is determined by $\mathscr{G}_{\mathscr{U}(\mathscr{G})}$.*

Proof That $1_X : (X, \mathscr{G}_{\mathscr{U}(\mathscr{G})}) \longrightarrow (X, \mathscr{G})$ is a uniform contraction follows at once from the observation that $\mathscr{G} \subseteq \mathscr{G}_{\mathscr{U}(\mathscr{G})}$. Suppose that $(Y, \mathscr{H})$ is a uniform UG-space and that

$$f : (Y, \mathscr{H}) \longrightarrow (X, \mathscr{G})$$

is a uniform contraction, then it immediately follows that

$$f : (Y, \mathscr{U}(\mathscr{H})) \longrightarrow (X, \mathscr{U}(\mathscr{G}))$$

is uniformly continuous and hence that

$$f : (Y, \mathscr{H}) \longrightarrow (X, \mathscr{G}_{\mathscr{U}(\mathscr{G})})$$

is a uniform contraction. □

5.2.7 Corollary Unif *is closed under the formation of colimits and final structures in* UG. *In particular, a coproduct in* UG *of a family of uniform spaces is a uniform space and, likewise, a quotient in* UG *of a uniform space is a uniform space.*

5.2.8 Proposition (Unif) *A uniform gauge space $(X, \mathscr{G})$ is a uniform* UG*-space if and only if $\mathscr{G}$ satisfies the stronger uniform saturation condition which says that if e is a metric and*

$$\forall \varepsilon > 0, \exists \delta > 0, \exists d \in \mathscr{G} : \{d < \delta\} \subseteq \{e < \varepsilon\}$$

then $e \in \mathscr{G}$.

Proof This follows from the definitions (see also Gillman and Jerison 1976). □

Given a metric space (X, d), we associate with it a natural uniform gauge space in exactly the same way as for App.

5.2.9 Proposition *If (X, d) is a metric space, then the collection*

$$\mathscr{G}_d := \{e \in \mathsf{Met}(X) \mid e \leq d\}$$

is a uniform gauge.

Proof This follows from the definitions. □

Metric spaces thus have three different forms, as a metric space, as a uniform approach space or as a uniform gauge space, and in the last two cases the gauge and uniform gauge are exactly the same.

5.2.10 Definition A uniform gauge space of type $(X, \mathcal{G}_d)$, for some metric d on X, is called a *metric uniform gauge space* or shortly a metric UG-space.

If (X, d) and (X', d') are metric spaces and $f : X \longrightarrow X'$ is non-expansive then for any metric $e' \leq d'$ on X' we obviously have that $e' \circ (f \times f) \leq d$. Hence there is a concrete functor from Met to UG which takes (X, d) to $(X, \mathcal{G}_d)$ and which is a full embedding of Met in UG.

Given a uniform gauge space $(X, \mathcal{G})$ we define the metric

$$d_{\mathcal{G}} := \sup \mathcal{G}.$$

5.2.11 Theorem Met *is embedded as a concretely coreflective subcategory of* UG. *For any uniform gauge space $(X, \mathcal{G})$, its* Met*-coreflection is determined by $\mathcal{G}_{d_{\mathcal{G}}}$.*

Proof That $1_X : (X, \mathcal{G}_{d_{\mathcal{G}}}) \longrightarrow (X, \mathcal{G})$ is a uniform contraction follows from the fact that for any $d \in \mathcal{G}$ we have $d \leq d_{\mathcal{G}}$. Now suppose that (Y, d) is a metric space and that

$$f : (Y, \mathcal{G}_d) \longrightarrow (X, \mathcal{G})$$

is a uniform contraction. Then for all $e \in \mathcal{G}$ we have $e \circ (f \times f) \leq d$ and hence also $d_{\mathcal{G}} \circ (f \times f) \leq d$ which proves that

$$f : (Y, \mathcal{G}_d) \longrightarrow (X, \mathcal{G}_{d_{\mathcal{G}}})$$

is also a uniform contraction. □

5.2.12 Proposition (Met) *A uniform gauge space $(X, \mathcal{G})$ is a metric uniform gauge space if and only if $\mathcal{G}$ satisfies any of the following equivalent properties.*

1. $\sup \mathcal{G} \in \mathcal{G}$.
2. *$\mathcal{G}$ is closed under the formation of arbitrary suprema.*

Proof In one direction this follows immediately from the definition of a metric UG-space and in the other direction it suffices to remark that given a uniform gauge $\mathcal{G}$, $\sup \mathcal{G}$ is a metric. □

Just as in the case of App, Met is not embedded reflectively in UG. In particular it is not stable under the formation of infinite products. Taking an infinite product of metric uniform gauge spaces in UG, provides the underlying product set with a uniform gauge structure which, in general, is neither metric nor uniform but which has as uniform coreflection precisely the product uniformity of the uniformities underlying the metrics. More precisely we have the following theorem.

5.2.13 Theorem *Any uniform gauge space $(X, \mathcal{G})$ is a subspace of a product of metric spaces in* UG*, i.e.* UG *is the epireflective hull of the subcategory of metric uniform gauge spaces.*

Proof Let $(X, \mathcal{G})$ be a uniform gauge space then the injective map

$$f : (X, \mathcal{G}) \longrightarrow (X^{\mathcal{G}}, \prod_{d \in \mathcal{G}} \mathcal{G}_d) : x \mapsto (x_d := x)_d$$

is an embedding as can easily be seen making use of 5.1.7. □

5.3 The Relation Between UAp and UG

5.3.1 Definition If $(X, \mathcal{G})$ is a uniform gauge space then the local saturation $\widehat{\mathcal{G}}$ is a gauge on X and $(X, \widehat{\mathcal{G}})$ is called the *underlying approach space* of $(X, \mathcal{G})$.

By definition, obviously the underlying approach space of a uniform gauge space is a uniform approach space.

If $\mathcal{H} \subseteq \mathsf{Met}(X)$ is an ideal, then it is at the same time a basis for a uniform gauge and for a gauge. Since the various other associated structures can be derived from a basis for the gauge we immediately obtain the same formulas for the various associated approach structures underlying a uniform gauge space. Thus, if $(X, \mathcal{G})$ is a uniform gauge space with basis $\mathcal{H}$ for $\mathcal{G}$ then e.g.

$$\delta_{\widehat{\mathcal{G}}}(x, A) = \sup_{d \in \mathcal{H}} \inf_{y \in A} d(x, y)$$

and

$$\lambda_{\widehat{\mathcal{G}}} \mathcal{F}(x) = \sup_{d \in \mathcal{H}} \inf_{F \in \mathcal{F}} \sup_{y \in F} d(x, y)$$

and analogously for the other structures.

Just as in topology, if an approach property is attributed to a uniform gauge space then it will always be meant for the underlying approach space, and analogously for indices.

5.3.2 Proposition *If X and X' are uniform gauge spaces and $f : X \longrightarrow X'$ is a uniform contraction, then it is a contraction.*

Proof This follows from the definitions, 1.3.3 and the fact that any uniform gauge is a basis for the gauge of the associated approach space. □

5.3.3 Theorem UAp *is a full subcategory of* UG.

Proof This follows from the fact that any gauge is also a uniform gauge and that the definition of uniform contraction is precisely the same as the definition of a contraction (see 1.3.3). Hence if $(X, \mathcal{G})$ and $(X', \mathcal{G}')$ are uniform approach spaces and $f : (X, \mathcal{G}) \longrightarrow (X', \mathcal{G}')$ is a contraction it is obviously also a uniform contraction when the spaces are considered to be uniform gauge spaces. □

Of course this situation also holds for CReg and Unif, but there this is much less apparent because a completely regular space is seldom considered via its collections of pointwise continuous metrics, whereas in our case the description with gauges, both for approach spaces and for uniform gauge spaces, is one of the most important ones. In the case of topology it requires considering what is called the fine uniformity associated with a completely regular space.

5.3.4 Theorem *The concrete functor from* UG *to* UAp *which takes* $(X, \mathcal{G})$ *to* $(X, \widehat{\mathcal{G}})$ *is a forgetful functor and is left adjoint to the embedding of* UAp *in* UG. *In other words,* UAp *is coreflectively embedded in* UG.

Proof If $(X, \mathcal{G})$ is a uniform gauge space, $(X', \mathcal{G}')$ is a uniform approach space and if $f : (X', \mathcal{G}') \longrightarrow (X, \mathcal{G})$ is a uniform contraction then obviously $f : (X', \mathcal{G}') \longrightarrow (X, \widehat{\mathcal{G}})$ is also a uniform contraction. □

The relation among the categories which we have seen is depicted in the following diagram.

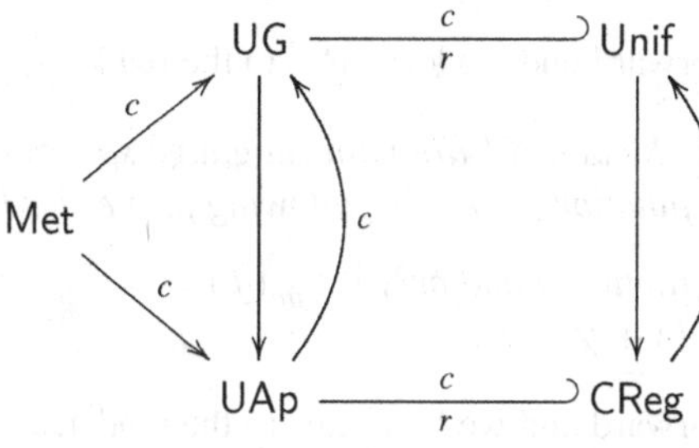

The well-known functorial relation between CReg and Unif, to a large extent, carries over to UAp and UG.

5.4 Indices in Uniform Gauge Spaces

Analogously as for approach spaces, in order to treat uniform indices in the correct way, we need to consider the following setting. We define UG_{Set} to be the category with objects uniform gauge spaces and morphisms functions between the underlying sets.

5.4.1 Definition (Index of uniform contractivity) Given a function $f : X \longrightarrow X'$ between uniform gauge spaces we define the *index of uniform contractivity* of f by

$$\boxed{\chi_{uc} : \mathsf{MorUGSet} \longrightarrow \mathbb{P} : f \mapsto \min\{c \mid \forall d' \in \mathcal{G}', \exists d \in \mathcal{G} : d' \circ (f \times f) \leq d + c\}}$$

Note that the minimum is well-defined and that for any function $f : X \longrightarrow X'$ between uniform gauge spaces

$$\forall d' \in \mathcal{G}', \exists d \in \mathcal{G} : d' \circ (f \times f) \leq d + \chi_{uc}(f).$$

Note further that, by the saturation property we also have

$$\chi_{uc}(f) := \min\{c \mid \forall d' \in \mathcal{G}', \forall \varepsilon > 0, \forall \omega < \infty : \exists d \in \mathcal{G} : d' \circ (f \times f) \wedge \omega \leq d + \varepsilon + c\}$$

and hence $\chi_{uc}(f)$ is also characterized by the claim that for any $d' \in \mathcal{G}'$, $\varepsilon > 0$ and $\omega < \infty$ there exists $d \in \mathcal{G}$ such that

$$d' \circ (f \times f) \wedge \omega \leq d + \varepsilon + \chi_{uc}(f).$$

That this definition indeed provides us with a morphism-index follows from the following two results.

5.4.2 Proposition *If X, Y, U and V are uniform gauge spaces and $f : X \longrightarrow Y$ and $g : U \longrightarrow V$ are uniform gauge isomorphic then $\chi_{uc}(f) = \chi_{uc}(g)$.*

Proof This is straightforward and we leave this to the reader. □

5.4.3 Proposition *If X, X' and X'' are uniform gauge spaces and $f : X \longrightarrow X'$ and $g : X' \longrightarrow X''$ are functions, then the following properties hold.*

1. *f is a uniform contraction if and only if $\chi_{uc}(f) = 0$.*
2. *$\chi_{uc}(g \circ f) \leq \chi_{uc}(f) + \chi_{uc}(g)$.*

Proof This is straightforward and we leave this to the reader. □

5.4.4 Theorem *If $f : X \longrightarrow X'$ is a function between uniform gauge spaces, then*

$$\chi_c(f) \leq \chi_{uc}(f) \leq \chi_c(f) + 2\chi_c(X)$$

and in particular, if f is uniformly contractive, then it is also contractive and if f is a contraction and X is 0-compact, then f is a uniform contraction.

Proof The first inequality follows at once from the definitions and for the second one, take $d' \in \mathcal{G}'$ and fix $\varepsilon > 0$ and $\omega < \infty$. By definition of index of contractivity there exists $e \in \widehat{\mathcal{G}}$ such that $d' \circ (f \times f) \leq e + \chi_c(f)$. For any $x \in X$ we can then find $e_x \in \mathcal{G}$ such that

$$e(x, .) \wedge \omega \leq e_x(x, .) + \varepsilon.$$

If we define $\varphi \in \mathcal{G}^X$ by $\varphi(x) := e_x$ then it follows from 4.3.2 that there exists $Y \in 2^{(X)}$ such that for all $z \in X$ there exists $y_z \in Y$ such that

$$e_{y_z}(y_z, z) \leq \varepsilon + \chi_c(X).$$

Since Y is finite we have $d := \sup_{y \in Y} e_y \in \mathcal{G}$. We then have, for any $x, z \in X$

$$\begin{aligned}
d' \circ (f \times f)(x, z) \wedge \omega &\leq (e(x, z) + \chi_c(f)) \wedge \omega \\
&\leq e(x, z) \wedge \omega + \chi_c(f) \\
&\leq e(y_z, x) \wedge \omega + e(y_z, z) \wedge \omega + \chi_c(f) \\
&\leq e_{y_z}(y_z, x) + \varepsilon + e_{y_z}(y_z, z) + \varepsilon + \chi_c(f) \\
&\leq e_{y_z}(y_z, x) + \varepsilon + \varepsilon + \chi_c(X) + \varepsilon + \chi_c(f) \\
&\leq e_{y_z}(y_z, z) + e_{y_z}(z, x) + 3\varepsilon + \chi_c(X) + \chi_c(f) \\
&\leq \varepsilon + \chi_c(X) + e_{y_z}(z, x) + 3\varepsilon + \chi_c(X) + \chi_c(f) \\
&\leq d(x, z) + 2\chi_c(X) + \chi_c(f) + 4\varepsilon.
\end{aligned}$$

This, by 5.4.1 and the arbitrariness of ε and ω proves our claim. □

Note that the first inequality in the foregoing theorem implies that the forgetful functor from UG to UAp is index-true.

5.4.5 Corollary (Unif) *A uniformly continuous map is continuous and if X and X' are uniform spaces and X is compact then a continuous map $f : X \longrightarrow X'$ is uniformly continuous.*

One of the forms of the index of compactness $\chi_c(X)$ making use of a basis $\mathcal{H}$ for the gauge, is

$$\chi_c(X) := \sup_{\varphi \in \mathcal{H}^X} \inf_{Y \in 2^{(X)}} \sup_{z \in X} \inf_{x \in Y} \varphi(x)(x, z).$$

In the underlying approach space $(X, \widehat{\mathcal{G}})$ of a uniform gauge space $(X, \mathcal{G})$ with basis $\mathcal{H}$ the foregoing formula for $\chi_c(X)$ need not be changed since $\mathcal{H}$ is then a basis both for $\mathcal{G}$ and for $\widehat{\mathcal{G}}$. Thus for instance $\mathcal{G}$ itself (which is a basis for $\widehat{\mathcal{G}}$) can be taken.

5.4.6 Definition (Index of precompactness) For a uniform gauge space $(X, \mathcal{G})$ its *index of precompactness* is defined as

$$\boxed{\chi_{pc}(X) = \sup_{d \in \mathcal{G}} \inf_{Y \in 2^{(X)}} \sup_{z \in X} \inf_{x \in Y} d(x, z)}$$

If $\chi_{pc}(X) = 0$ then we will say that X is *0-precompact.*

It is interesting to compare the formulas for χ_c (for the underlying approach space) and χ_{pc} (for the uniform gauge space itself). What we see is that they are entirely the same except for the first supremum, which in the case of compactness ranges over the set $\mathcal{G}^X$ and in the case of precompactness ranges over the set $\mathcal{G}$.

In the approach case, which is a local theory, the metrics must be allowed to vary from point to point, and in the uniform gauge case, which is a global or uniform theory, the same metric has to be chosen in every point.

5.4.7 Proposition *The index of precompactness χ_{pc} is $(1, \chi_{uc})$-layered.*

Proof This follows from 5.4.1, 4.1.5 and 5.4.6. □

5.4.8 Lemma *If $\mathcal{G}$ is a uniform gauge on X, and if $\mathcal{H}$ and $\mathcal{H}'$ are bases for $\mathcal{G}$, then for a filter $\mathcal{F}$*

$$\sup_{d\in\mathcal{H}} \inf\{r \mid \exists x : B_d(x,r) \in \mathcal{F}\} = \sup_{d\in\mathcal{H}'} \inf\{r \mid \exists x : B_d(x,r) \in \mathcal{F}\}.$$

Proof It is sufficient to prove one inequality. Suppose that

$$\sup_{d\in\mathcal{H}'} \inf\{r \mid \exists x : B_d(x,r) \in \mathcal{F}\} < \beta$$

and let $d \in \mathcal{H}$. Take $\omega > \beta$ and let $\omega - \beta > \varepsilon > 0$. Choose $d' \in \mathcal{H}'$ such that $d \wedge \omega \leq d' + \varepsilon$. Then there exist $r < \beta$ and $x \in X$ such that $B_{d'}(x,r) \in \mathcal{F}$. Now if $y \in B_{d'}(x,r)$ then

$$d(x,y) \wedge \omega \leq d'(x,y) + \varepsilon < r + \varepsilon$$

and thus $d(x,y) < r + \varepsilon$ and hence $B_d(x, r+\varepsilon) \in \mathcal{F}$. By the arbitrariness of ε this proves that also $\sup_{d\in\mathcal{H}} \inf\{r \mid \exists x : B_d(x,r) \in \mathcal{F}\} \leq \beta$. □

5.4.9 Definition **(Cauchy index)** If X is a uniform gauge space and if $\mathcal{H}$ is a gauge basis, then for $\mathcal{F} \in \mathsf{F}(X)$ we define the *Cauchy index* of $\mathcal{F}$ as

$$\boxed{\chi_{cy}(\mathcal{F}) := \sup_{d\in\mathcal{H}} \inf\{r \mid \exists x \in X : B_d(x,r) \in \mathcal{F}\}}$$

If $\chi_{cy}(\mathcal{F}) = 0$ we will say that $\mathcal{F}$ is *Cauchy (for the* UG*-structure)*. By the foregoing lemma this number is well-defined as it is independent of the chosen basis. Note that another way of writing the Cauchy index is

$$\chi_{cy}(\mathcal{F}) := \sup_{d\in\mathcal{H}} \inf_{x\in X} \lambda_d \mathcal{F}(x).$$

5.4.10 Proposition *In a uniform gauge space a filter is Cauchy if and only if it is Cauchy for the uniform coreflection.*

Proof This follows from the definition. □

In the foregoing we have already mentioned the asymptotic radius of a sequence. We recall that for a sequence $(x_n)_n$ in a metric space (X,d) this is $\inf_{x\in X} \lambda_d \langle (x_n)_n \rangle (x)$

and as we have seen in the section on completeness in App, it turns out that this value, in the general setting of uniform approach spaces, is actually also a kind of Cauchy index of the sequence, which, in hindsight and keeping in mind the actual meaning of an asymptotic radius, makes perfect sense.

5.4.11 Definition **(Local Cauchy index)** Let X be a uniform approach space and $\mathscr{F} \in \mathsf{F}(X)$. Then we define the *local Cauchy index of* $\mathscr{F}$ as

$$\boxed{\chi_{lcy}(\mathscr{F}) := \inf_{x \in X} \lambda \mathscr{F}(x)}$$

Note that in a metric space the Cauchy index and the local Cauchy index coincide and that in a uniform gauge space

$$\chi_{cy}(\mathscr{F}) = \sup_{d \in \mathscr{G}} \chi_{cy}^{d}(\mathscr{F}) = \sup_{d \in \mathscr{G}} \chi_{lcy}^{d}(\mathscr{F}).$$

Given a specific symmetric gauge basis $\mathscr{H}$, we define the $\mathscr{H}$*-width* of $\mathscr{F}$ as

$$\omega_{\mathscr{H}}(\mathscr{F}) := \sup_{d \in \mathscr{H}} \inf_{F \in \mathscr{F}} \operatorname{diam}_d(F).$$

5.4.12 Example In contrast with the result of 5.4.8, the value $\omega_{\mathscr{H}} \mathscr{F}$ is dependent on the particular choice of gauge basis. It is shown in 3.1.20 that the families $\mathscr{D} := \{d_a \mid a > 0\}$ and $\mathscr{D}' := \{d_{\mathbb{E}}\}$ where $d_{\mathbb{E}}$ stands for the Euclidean metric on $\mathbb{R}$ and where, for each $a > 0$, d_a is defined by

$$d_a(x, y) := |((-a) \vee (x \wedge a)) - ((-a) \vee (y \wedge a))|$$

generate the same gauge. However, if we consider the filter $\mathscr{F}$ generated by the collection $\{[n, \infty[\mid n \in \mathbb{N}\}$, then it is immediately verified that $\omega_{\mathscr{D}}(\mathscr{F}) = 0$, whereas $\omega_{\mathscr{D}'}(\mathscr{F}) = \infty$.

5.4.13 Proposition *If X is a uniform approach space with symmetric gauge basis $\mathscr{H}$ and $\mathscr{F} \in \mathsf{F}(X)$, then*

$$\lambda \mathscr{F} \leq \alpha \mathscr{F} + \omega_{\mathscr{H}}(\mathscr{F}).$$

Proof We prove this for a metric space, the general result then is an immediate consequence since in the general case all three terms are obtained taking suprema over the metrics in the gauge. If $\omega_d(\mathscr{F}) = \infty$ there is nothing to show. Suppose therefore that $\omega_d(\mathscr{F}) < \infty$, let $x \in X$, and let $\varepsilon > 0$. Choose $F_\varepsilon \in \mathscr{F}$ such that

$$\operatorname{diam}_d(F_\varepsilon) \leq \omega_d(\mathscr{F}) + \varepsilon.$$

For all $y, z \in F_\varepsilon$ we then have that

$$d(x, y) \leq d(x, z) + \omega_d(\mathscr{F}) + \varepsilon$$

and thus

$$\sup_{y \in F_\varepsilon} d(x, y) \leq \delta_d(x, F_\varepsilon) + \omega_d(\mathscr{F}) + \varepsilon,$$

which implies that

$$\begin{aligned} \lambda_d \mathscr{F}(x) &= \inf_{F \in \mathscr{F}} \sup_{y \in F} d(x, y) \\ &\leq \sup_{y \in F_\varepsilon} d(x, y) \\ &\leq \delta_d(x, F_\varepsilon) + \omega_d(\mathscr{F}) + \varepsilon \\ &\leq \alpha_d \mathscr{F} + \omega_d(\mathscr{F}) + \varepsilon, \end{aligned}$$

which, by the arbitrariness of ε, proves the result. □

5.4.14 Corollary *If X is a uniform approach space with symmetric gauge basis $\mathscr{H}$ and $\mathscr{F} \in \mathsf{F}(X)$ is $\mathscr{U}(\mathscr{H})$-Cauchy, then $\lambda\mathscr{F} = \alpha\mathscr{F}$.*

5.4.15 Theorem *If X is a uniform approach space and $\mathscr{F} \in \mathsf{F}(X)$, then*

$$\lambda\mathscr{F} \leq \alpha\mathscr{F} + 2\chi_{lcy}(\mathscr{F}).$$

Proof By 5.4.13 we only need to remark that $\omega_\mathscr{G}(\mathscr{F}) \leq 2\chi_{lcy}(\mathscr{F})$. □

The foregoing result generalizes what was already shown in 3.5.5. Note also that the index of local compactness of X can now be written as

$$\chi_{lc}(X) = \sup_{\mathscr{F} \in \mathsf{F}(X)} (\chi_c(\mathscr{F}) \ominus \chi_{lcy}(\mathscr{F})).$$

5.4.16 Proposition *If X is a uniform gauge space and $\mathscr{F} \in \mathsf{F}(X)$, then*

$$\chi_{cy}(\mathscr{F}) \leq \chi_{lcy}(\mathscr{F}).$$

Proof If $\mathscr{G}$ is the uniform gauge of the space then the result follows from the obvious fact that $\sup_{d \in \mathscr{G}} \inf_{x \in X} \lambda_d \mathscr{F}(x) \leq \inf_{x \in X} \sup_{d \in \mathscr{G}} \lambda_d \mathscr{F}(x)$. □

5.4.17 Example Let $X := \mathbb{R}\backslash\{0\}$ be the uniform UG-space with the usual uniformity. The uniform gauge is then generated by the basis $\{nd_\mathbb{E} \mid n \geq 1\}$. Consider the filter $\mathscr{F} := \{V \setminus \{0\} \mid V \in \mathscr{V}_\mathbb{E}(0)\}$. Then $\chi_{cy}\mathscr{F} = 0$ and $\chi_{lcy}\mathscr{F} = \infty$.

5.4.18 Corollary *In a uniform gauge space a filter which is Cauchy for the underlying approach structure is Cauchy in the uniform gauge space.*

5.4.19 Theorem *If X is a uniform gauge space and $\mathscr{F} \in \mathsf{F}(X)$, then*

$$\lambda\mathscr{F} \leq \alpha\mathscr{F} + 2\chi_{cy}\mathscr{F}.$$

Proof Again, it is sufficient to prove this for a metric space since all terms in the above inequality are obtained as a supremum over the metrics in the gauge. For λ and α this is by definition and for χ_{cy} we use the remark following Definition 5.4.11. The result then follows from 5.4.15. □

5.4.20 Example The inequality in the foregoing result can, in general, be strict. Consider the real line $\mathbb{R}$ equipped with the usual Euclidean metric and consider the sequence $(x_n)_n$ where

$$x_n := \begin{cases} 1 & n \text{ even,} \\ -1 & n \text{ odd.} \end{cases}$$

Then $\lambda\langle(x_n)_n\rangle(0) = \alpha\langle(x_n)_n\rangle(0) = 1$ and $\chi_{cy}\langle(x_n)_n\rangle = 2$. On the other hand the inequality is also best possible since $\lambda\langle(x_n)_n\rangle(1) = 2$ and $\alpha\langle(x_n)_n\rangle(1) = 0$.

5.4.21 Corollary (Unif) *In a uniform space a Cauchy filter adheres to a point if and only if it converges to that point and a convergent filter is Cauchy.*

In particular cases we can improve the general bounds given above. In the following few results we are always considering the approach structure and uniform gauge structure generated by the metric and the set of adherence points of a filter is always considered in the topology determined by the metric.

5.4.22 Proposition *If (X, d) is a metric space and $\mathcal{F} \in \mathsf{F}(X)$, then*

$$\frac{1}{2}\omega_d\mathcal{F} \leq \lambda\mathcal{F} \leq \alpha\mathcal{F} + \omega_d\mathcal{F}.$$

Proof Let $x \in X$ and $\varepsilon > 0$. To prove the first inequality let $F \in \mathcal{F}$ and take $z, y \in F$ such that $d(z, y) + 2\varepsilon \geq \mathrm{diam}_d(F)$. Then it follows that

$$\mathrm{diam}_d(F) \leq 2(\sup_{t\in F} d(x, t) + \varepsilon)$$

and the result follows from 2.3.1.

To prove the second inequality let $F \in \mathcal{F}$ be such that $\mathrm{diam}_d(F) \leq \omega_d\mathcal{F} + \varepsilon$. Then we have that

$$\sup_{y\in F} d(x, y) \leq \delta_d(x, \mathrm{F}) + \omega_d\mathcal{F} + \varepsilon$$

and again the result follows from 2.3.1. □

5.4.23 Proposition *If (X, d) is a metric space and $\mathcal{F} \in \mathsf{F}(X)$ is total, then*

$$\frac{1}{2}\mathrm{diam}_d(\mathrm{adh}\mathcal{F}) \leq \lambda\mathcal{F} \leq \alpha\mathcal{F} + \mathrm{diam}_d(\mathrm{adh}\mathcal{F}).$$

Proof It follows from Vaughan (1976a) that for any $\varepsilon > 0$ we have $(\mathrm{adh}\mathcal{F})^{(\varepsilon)} \in \mathcal{F}$, hence for a total filter it follows that $\omega_{\{d\}}\mathcal{F} = \mathrm{diam}_d(\mathrm{adh}\mathcal{F})$ and the result follows from 5.4.22. □

In the following results conv(A) stands for the convex hull of A.

5.4.24 Theorem *If X is a real Hilbert space and $\mathscr{F} \in \mathsf{F}(X)$ is total, then for any $x \in X$ there exists $x^* \in \mathrm{conv}(\mathrm{adh}\mathscr{F})$ such that $\lambda\mathscr{F}(x^*) \leq \lambda\mathscr{F}(x)$.*

Proof Since $\mathscr{F}$ is total it follows from Vaughan (1976b) that adh$\mathscr{F}$ and hence also conv(adh$\mathscr{F}$) is compact. Now it suffices to take x^* the projection of x on conv(adh$\mathscr{F}$) and the result then follows from 3.2.6. □

The previous result shows that in a Hilbert space best convergence is achieved on the convex hull of the set of adherence points. An old result of Jung (1901) allows us to improve this further.

5.4.25 Theorem *If X is a real Hilbert space of dimension n and $\mathscr{F} \in \mathsf{F}(X)$ is total, then there exists $x \in \mathrm{conv}(\mathrm{adh}\mathscr{F})$ such that*

$$\lambda\mathscr{F}(x) \leq \left(\frac{n}{2(n+1)}\right)^{\frac{1}{2}} \mathrm{diam}_d(\mathrm{adh}\mathscr{F})$$

and if X is infinite dimensional, then there exists $x \in \mathrm{conv}(\mathrm{adh}\mathscr{F})$ such that

$$\lambda\mathscr{F}(x) \leq \frac{1}{\sqrt{2}}\mathrm{diam}_d(\mathrm{adh}\mathscr{F}).$$

Proof If X has dimension n then Jung's theorem, (Jung 1901), says we can find a ball B with radius less than

$$\left(\frac{n}{2(n+1)}\right)^{\frac{1}{2}} \mathrm{diam}_d(\mathrm{adh}\mathscr{F})$$

such that adh$\mathscr{F} \subseteq B$. Since we can take the center x of this ball in conv(adh$\mathscr{F}$) the result follows from 3.2.6.

If X is infinite dimensional then, since adh$\mathscr{F}$ is compact, for each $n \in \mathbb{N}_0$ we can find a finite subset $A_n \subseteq \mathrm{adh}\mathscr{F}$ such that

$$\mathrm{adh}\mathscr{F} \subseteq \bigcup_{a \in A_n} B(a, \frac{1}{n}).$$

Let X_n be the finite dimensional subspace of X spanned by A_n with dimension $m(n)$. Then again by Jung's theorem, we can find $x_n \in \mathrm{conv}(A_n) \subseteq X_n$ such that

$$A_n \subseteq B\left(x_n, \left(\frac{m(n)}{2(m(n)+1)}\right)^{\frac{1}{2}} \mathrm{diam}_d(A_n)\right) \subseteq B\left(x_n, \frac{1}{\sqrt{2}}\mathrm{diam}_d(A_n)\right).$$

Then it follows that

$$\text{adh}\mathscr{F} \subseteq B\left(x_n, \frac{1}{n} + \frac{1}{\sqrt{2}}\text{diam}_d(\text{adh}\mathscr{F})\right).$$

Since $\text{conv}(A_n) \subseteq \text{conv}(\text{adh}\mathscr{F})$ and since the latter set is compact there exists a subsequence $(x_{k_n})_n$ which converges to some $x \in \text{conv}(\text{adh}\mathscr{F})$. Then for any $y \in \text{adh}\mathscr{F}$ and $n \in \mathbb{N}_0$ it follows that

$$\|x - y\| \leq \|x - x_{k_n}\| + \frac{1}{k_n} + \frac{1}{\sqrt{2}}\text{diam}_d(\text{adh}\mathscr{F})$$

and the result follows letting $n \longrightarrow \infty$ and applying 3.2.6. □

5.4.26 Example All foregoing results are in general best possible. For 5.4.22 and 5.4.23 this is seen taking $X = \mathbb{R}$ equipped with the usual Euclidean metric and taking the filter generated by a sequence for which $x_{2n} = a$, $x_{2n+1} = b$ and $a \neq b$ (see also 5.4.20). For 5.4.25 this follows from the fact that the bounds in Jung's theorem are best possible.

5.4.27 Theorem *If X is a uniform gauge space, then*

$$\chi_{pc}(X) = \sup_{\mathscr{U} \in \mathsf{U}(X)} \chi_{cy}(\mathscr{U}).$$

Proof Suppose that $\chi_{pc}(X) > \alpha$ then there exists $d \in \mathscr{G}$ such that

$$\{X \setminus B_d(x, \alpha) \mid x \in X\}$$

has the finite intersection property and hence there exists an ultrafilter $\mathscr{U}$ containing this collection. Now since $B_d(x, \alpha) \in \mathscr{U}$ would yield a contradiction, it follows that $\chi_{cy}(\mathscr{U}) > \alpha$. Conversely, if $\chi_{pc}(X) < \alpha$ then for each $d \in \mathscr{G}$ there exists a finite subset $Y \subseteq X$ such that $\bigcup_{y \in Y} B_d(y, \alpha) = X$. If $\mathscr{U}$ is an ultrafilter then it follows that for some $y \in Y$, $B_d(y, \alpha) \in \mathscr{U}$ and hence it follows that $\chi_{cy}(\mathscr{U}) < \alpha$. □

5.4.28 Corollary (Unif) *A uniform space is precompact if and only if each ultrafilter is Cauchy.*

5.4.29 Proposition (Unif, Met) *A uniform* UG*-space* $(X, \mathscr{G}_{\mathscr{U}})$ *is* 0*-precompact if and only if* $(X, \mathscr{U})$ *is precompact and analogously a metric uniform gauge space* $(X, \mathscr{G}_d)$ *is* 0*-precompact if and only if* (X, d) *is precompact and* $\chi_{pc}(X) < \infty$ *if and only if X is bounded.*

Proof This follows from the definitions and is analogous to 4.3.11 and 4.3.12. □

5.4.30 Theorem *If X is a uniform gauge space, then*

$$\chi_{pc}(X) \leq \chi_c(X).$$

Proof This follows from 5.4.16 and 5.4.27 since

$$\begin{aligned}\chi_{pc}(X) &= \sup_{\mathscr{U}\in\mathsf{U}(X)} \chi_{cy}(\mathscr{U}) \\ &\le \sup_{\mathscr{U}\in\mathsf{U}(X)} \chi_{lcy}(\mathscr{U}) \\ &= \sup_{\mathscr{U}\in\mathsf{U}(X)} \inf_{x\in X} \lambda\mathscr{U}(x) \\ &\le \inf_{x\in X} \sup_{\mathscr{U}\in\mathsf{U}(X)} \lambda\mathscr{U}(x) = \chi_c(X).\end{aligned}$$

□

5.4.31 Corollary (Unif) *If a uniform space has a compact underlying topology, then it is precompact.*

5.4.32 Proposition *If X is a uniform gauge space and $A_1, \ldots, A_n$ are subsets of X, then*

$$\chi_{pc}(\cup_{i=1}^n A_i) \le \sup_{i=1}^n \chi_{pc}(A_i).$$

Proof This is analogous to 4.3.22 and is left to the reader. □

5.4.33 Proposition *If $f : X \longrightarrow X'$ is a function between uniform gauge spaces and $\mathscr{F} \in \mathsf{F}(X)$, then*

$$\chi_{cy}(f(\mathscr{F})) \le \chi_{cy}(\mathscr{F}) + \chi_{uc}(f).$$

Proof Let $d' \in \mathscr{G}'$ and $\varepsilon > 0$ then we can find $d \in \mathscr{G}$ such that

$$d' \circ (f \times f) \le d + \chi_{uc}(f)$$

and $x \in X$ and $r < \chi_{cy}(\mathscr{F}) + \varepsilon$ such that $B_d(x, r) \in \mathscr{F}$. It now suffices to note that

$$f(B_d(x, r)) \subseteq B_{d'}(f(x), r + \chi_{uc}(f)).$$

□

5.4.34 Corollary (Unif) *The uniformly continuous image of a Cauchy filter is Cauchy.*

5.4.35 Theorem *If $f : X \longrightarrow X'$ is a function between uniform gauge spaces and $A \subseteq X$, then*

$$\chi_{pc}(f(A)) \le \chi_{pc}(A) + \chi_{uc}(f)$$

and in particular if f is a uniform contraction and A is 0-precompact then $f(A)$ is 0-precompact.

Proof Since every ultrafilter containing $f(A)$ is the image of an ultrafilter containing A this follows at once from 5.4.27 and 5.4.33. □

All inequalities in 5.4.30–5.4.35 are in general strict as can already be seen from the classical case.

5.4.36 Corollary (Unif) *The uniformly continuous image of a precompact set is precompact.*

5.5 Quasi-UG Spaces, the Non-symmetric Variant

Just as for uniform spaces there is a non-symmetric variant to UG. The basic concepts concerning saturation remain unaltered with the proviso that we are now not dealing with metrics but with quasi-metrics.

Given a set X, a collection $\mathscr{G} \subseteq q\mathsf{Met}(X)$ and a metric $d \in q\mathsf{Met}(X)$, we will say that d is *dominated by* $\mathscr{G}$, or that $\mathscr{G}$ *dominates* d, if for all $\varepsilon > 0$ and $\omega < \infty$ there exists a $d^{\varepsilon,\omega} \in \mathscr{G}$ such that

$$d \wedge \omega \leq d^{\varepsilon,\omega} + \varepsilon.$$

We will then also say that the family $(d^{\varepsilon,\omega})_{\varepsilon>0,\omega<\infty}$ dominates d.

Further we will say that a collection of quasi-metrics $\mathscr{G}$ is *saturated*, if any quasi-metric d which is dominated by $\mathscr{G}$ already belongs to $\mathscr{G}$.

5.5.1 Definition **(Quasi-uniform gauge)** An ideal in $q\mathsf{Met}(X)$ is called a *quasi-uniform gauge* if it fulfils the following property.

(qUG1) $\mathscr{G}$ is saturated.

A subset $\mathscr{H}$ of $q\mathsf{Met}(X)$ is called a *quasi-uniform gauge basis* if it is an ideal basis in $q\mathsf{Met}(X)$.

In order to derive the quasi-uniform gauge from a quasi-uniform gauge basis, as before, we require the same saturation operation as for uniform gauge spaces.

Given a subset $\mathscr{D} \subseteq q\mathsf{Met}(X)$ we define

$$\widetilde{\mathscr{D}} := \{d \in \mathsf{Met}(X) \mid \mathscr{D} \text{ dominates } d\}.$$

We call $\widetilde{\mathscr{D}}$ the *(quasi-uniform) saturation* of $\mathscr{D}$.

An ideal basis $\mathscr{H}$ in $\mathsf{Met}(X)$ is said to be a *basis for a quasi-uniform gauge* $\mathscr{G}$ if $\widetilde{\mathscr{H}} = \mathscr{G}$. In this case we also say that $\mathscr{H}$ *generates* $\mathscr{G}$ or that $\mathscr{G}$ is *generated by* $\mathscr{H}$.

5.5.2 Proposition *If $\mathscr{H}$ is a quasi-uniform gauge basis, then $\widetilde{\mathscr{H}}$ is a quasi-uniform gauge with $\mathscr{H}$ as basis and if $\mathscr{H}$ is a basis for a quasi-uniform gauge $\mathscr{G}$, then it is a quasi-uniform gauge basis.*

Proof This is analogous to 1.1.16 and we leave this to the reader. □

5.5.3 Definition A pair $(X, \mathscr{G})$ where $\mathscr{G}$ is a quasi-uniform gauge on X is called a *quasi-uniform gauge space* or shortly a qUG-space.

5.5.4 Definition Let $(X, \mathcal{G})$ and $(X', \mathcal{G}')$ be quasi-uniform gauge spaces and let $f : X \longrightarrow X'$ be a function. Then f is a called a *quasi-uniform contraction* if

$$\forall d \in \mathcal{G}' : d \circ (f \times f) \in \mathcal{G}.$$

From the saturation condition it is easily seen that this is equivalent to the statement that

$$\forall d \in \mathcal{G}', \ \forall \varepsilon > 0, \ \forall \omega < \infty, \ \exists e \in \mathcal{G} : d \circ (f \times f) \wedge \omega \leq e + \varepsilon.$$

Quasi-uniform gauge spaces and quasi-uniform contractions form a category which we denote $q\mathsf{UG}$.

5.5.5 Theorem *$q\mathsf{UG}$ is a topological category.*

Proof Given quasi-uniform gauge spaces $(X_j)_{j \in J}$, consider the source

$$(f_j : X \longrightarrow X_j)_{j \in J}$$

in $q\mathsf{UG}$. If, for each $j \in J$, $\mathcal{H}_j$ is a basis for the quasi-uniform gauge of X_j, then a basis for the initial quasi-uniform gauge on X is given by

$$\mathcal{H} := \left\{ \sup_{j \in K} d_j \circ (f_j \times f_j) | K \in 2^{(J)}, \forall j \in K : d_j \in \mathcal{H}_j \right\}.$$

□

5.5.6 Definition **(Quasi-uniform tower)** Let X be a set. A family of filters $(\mathcal{U}_\varepsilon)_{\varepsilon \in \mathbb{R}^+}$ on $X \times X$ is called a *quasi-uniform tower (on X)* if it fulfils the following properties.

(qUT1) $\forall \varepsilon \in \mathbb{R}^+, \forall U \in \mathcal{U}_\varepsilon : \Delta_X \subseteq U$.
(qUT2) $\forall \varepsilon, \varepsilon' \in \mathbb{R}^+ : \mathcal{U}_\varepsilon \circ \mathcal{U}_{\varepsilon'} \supseteq \mathcal{U}_{\varepsilon+\varepsilon'}$.
(qUT3) $\forall \varepsilon \in \mathbb{R}^+ : \mathcal{U}_\varepsilon : \bigcup_{\alpha > \varepsilon} \mathcal{U}_\alpha$.

Thus, a quasi-uniform tower is a family of quasi-semi-uniformities satisfying (qUT2) and (qUT3). We recall that a quasi-semi-uniformity on X is a filter on $X \times X$ such that all members of the filter contain the diagonal. Also note that by (qUT2), $\mathcal{U}_0$ is actually a quasi-uniformity.

5.5.7 Theorem **(qUT $\rightrightarrows$ qUG)** *If $(\mathcal{U}_\varepsilon)_{\varepsilon \in \mathbb{R}^+}$ is a quasi-uniform tower on X, then*

$$\mathcal{G} := \{d \in q\mathsf{Met}(X) \mid \forall \varepsilon \in \mathbb{R}^+, \forall \alpha > \varepsilon : \{d < \alpha\} \in \mathcal{U}_\varepsilon\}$$

is a quasi-uniform gauge and for any ε, the quasi-uniformity $\mathcal{U}_\varepsilon$ is generated by $\{\{d < \alpha\} \mid d \in \mathcal{G}, \alpha > \varepsilon\}$.

Proof This is analogous to 5.1.9 and we leave this to the reader. □

5.5.8 Theorem **(qUG $\rightrightarrows$ qUT)** *If $\mathscr{G}$ is a quasi-uniform gauge on X, then the family $(\mathscr{U}_\varepsilon)_{\varepsilon\in\mathbb{R}^+}$, where for every $\varepsilon \in \mathbb{R}^+$, $\mathscr{U}_\varepsilon$ is the quasi-semi-uniformity generated by*

$$\{\{d < \alpha\} \mid d \in \mathscr{G}, \alpha > \varepsilon\}$$

is a quasi-uniform tower and $\mathscr{G} = \{d \in q\mathsf{Met}(X) \mid \forall\varepsilon \in \mathbb{R}^+, \forall\alpha > \varepsilon : \{d < \alpha\} \in \mathscr{U}_\varepsilon\}$.

Proof This is analogous to 5.1.10 and we leave this to the reader. □

5.5.9 Theorem *If $f : X \longrightarrow X'$ is a function between quasi-uniform gauge spaces, then, in terms of the associated towers, it is a quasi-uniform contraction if and only if $f : (X, \mathscr{U}_\varepsilon) \longrightarrow (X', \mathscr{U}'_\varepsilon)$ is quasi-uniformly continuous for each $\varepsilon \in \mathbb{R}^+$.*

Proof This is analogous to 5.1.11 and we leave this to the reader. □

Given a quasi-uniform space $(X, \mathscr{U})$, we associate with it a natural quasi-uniform gauge space by simply taking as quasi-uniform gauge the collection of all quasi-uniformly continuous metrics.

Suppose $(X, \mathscr{U})$ is a quasi-uniform space. Then the collection $\mathscr{G}_\mathscr{U}$ of all quasi-uniformly continuous metrics is a quasi-uniform gauge and the associated quasi-uniform tower is $(\mathscr{U}_\varepsilon := \mathscr{U})_\varepsilon$.

A quasi-uniform gauge space of type $(X, \mathscr{G}_\mathscr{U})$, for some quasi-uniformity $\mathscr{U}$ on X, will be called a *quasi-uniform quasi-uniform gauge space* or somewhat less unfortunate, a quasi-uniform qUG-space.

It is evident that if $(X, \mathscr{U})$ and $(X', \mathscr{U}')$ are quasi-uniform spaces and $f : X \longrightarrow X'$ is quasi-uniformly continuous then for any quasi-uniformly continuous metric d on X' we have that $d \circ (f \times f)$ is a quasi-uniformly continuous metric on X. Hence we immediately have that the concrete functor from qUnif to qUG which takes $(X, \mathscr{U})$ to $(X, \mathscr{G}_\mathscr{U})$ is a full embedding of qUnif into qUG.

As was the case for Unif in UG, here too we will be able to show that the above embedding is both concretely reflective and concretely coreflective.

5.5.10 Theorem *qUnif is embedded as a concretely reflective subcategory of qUG.*

Proof This is analogous to 5.2.3 and we leave this to the reader. □

As was the case for UG, again the qUnif-reflection of a qUG-space is not particularly interesting, hence we also do not describe it here.

Given a quasi-uniform gauge space $(X, \mathscr{G})$ the collection

$$\mathscr{U}(\mathscr{G}) := \{U \subseteq X \times X \mid \exists\varepsilon > 0, \exists d \in \mathscr{G} : \{d < \varepsilon\} \subseteq U\}$$

is a quasi-uniformity on X. This quasi-uniformity also is simply the quasi-uniformity generated by the collection of metrics $\mathscr{G}$ and it is also $\mathscr{U}_0$ in the associated tower.

5.5.11 Theorem *$q\mathsf{Unif}$ is embedded as a concretely coreflective subcategory of $q\mathsf{UG}$. For any quasi-uniform gauge space $(X, \mathcal{G})$, its $q\mathsf{Unif}$-coreflection is given by*

$$1_X : (X, \mathcal{G}_{\mathcal{U}(\mathcal{G})}) \longrightarrow (X, \mathcal{G}).$$

Proof This is analogous to 5.2.6 and we leave this to the reader. □

Given a quasi-metric space (X, d), we associate with it a natural quasi-uniform gauge space in exactly the same way as for UG.

Suppose (X, d) is a quasi-metric space. Then the collection

$$\mathcal{G}_d := \{e \in q\mathsf{Met}(X) \mid e \leq d\}$$

is a quasi-uniform gauge.

A quasi-uniform gauge space of type $(X, \mathcal{G}_d)$, for some quasi-metric d on X, will be called a *quasi-metric quasi-uniform gauge space* or shortly a quasi-metric $q\mathsf{UG}$-space. If (X, d) and (X', d') are quasi-metric spaces and $f : X \longrightarrow X'$ is non-expansive then for any quasi-metric $e' \leq d'$ on X' we obviously have that $e' \circ (f \times f) \leq d$. Hence the concrete functor from $q\mathsf{Met}$ to $q\mathsf{UG}$ which takes (X, d) to $(X, \mathcal{G}_d)$ is a full embedding of $q\mathsf{Met}$ in $q\mathsf{UG}$.

Given a quasi-uniform gauge space $(X, \mathcal{G})$ we define the metric $d_{\mathcal{G}} := \sup \mathcal{G}$.

5.5.12 Theorem *$q\mathsf{Met}$ is embedded as a concretely coreflective subcategory of $q\mathsf{UG}$. For any quasi-uniform gauge space $(X, \mathcal{G})$, its $q\mathsf{Met}$-coreflection is determined by $d_{\mathcal{G}}$.*

Proof This is analogous to 5.2.11 and we leave this to the reader. □

As before, a quasi-uniform gauge space $(X, \mathcal{G})$ is a quasi-metric quasi-uniform gauge space if and only if $\mathcal{G}$ satisfies any of the following equivalent conditions

1. $\sup \mathcal{G} \in \mathcal{G}$.
2. $\mathcal{G}$ is closed under the formation of arbitrary suprema.

Just as in the case of UG, $q\mathsf{Met}$ is not embedded reflectively in $q\mathsf{UG}$. In particular it is not stable under the formation of infinite products. Taking an infinite product of metric uniform gauge spaces in $q\mathsf{UG}$, provides the underlying product set with a quasi-uniform gauge structure which, in general, is neither quasi-metric nor quasi-uniform but which has as quasi-uniform coreflection precisely the product quasi-uniformity of the quasi-uniformities underlying the quasi-metrics. In particular we again have the following theorem.

5.5.13 Theorem *Any quasi-uniform gauge space $(X, \mathcal{G})$ is a subspace of a product of quasi-metric spaces in $q\mathsf{UG}$, i.e. $q\mathsf{UG}$ is the epireflective hull of the subcategory of quasi-metric quasi-uniform gauge spaces.*

Proof This is analogous to 5.2.13 and we leave this to the reader. □

5.5.14 Definition If $(X, \mathcal{G})$ is a quasi-uniform gauge space then the local saturation $\widehat{\mathcal{G}}$ is a gauge on X. We call $(X, \widehat{\mathcal{G}})$ the *underlying approach space* of $(X, \mathcal{G})$.

In contrast with the UG-case, now this underlying approach space obviously need not be uniform.

If $\mathcal{H} \subseteq q\mathsf{Met}(X)$ is an ideal, then it is at the same time a basis for a quasi-uniform gauge and for a gauge. Again, since the various other associated structures can be derived from a basis for the gauge we immediately obtain the same formulas for the various associated approach structures underlying a quasi-uniform gauge space. Thus, if $(X, \mathcal{G})$ is a quasi-uniform gauge space with basis $\mathcal{H}$ for $\mathcal{G}$ then e.g.

$$\delta_{\widehat{\mathcal{G}}}(x, A) = \sup_{d \in \mathcal{H}} \inf_{y \in A} d(x, y)$$

and

$$\lambda_{\widehat{\mathcal{G}}}\mathcal{F}(x) = \sup_{d \in \mathcal{H}} \inf_{F \in \mathcal{F}} \sup_{y \in F} d(x, y)$$

and analogously for the other structures.

Suppose that X and X' are quasi-uniform gauge spaces and that $f : X \longrightarrow X'$ is a function. If f is a quasi-uniform contraction, then it is a contraction.

5.5.15 Theorem App *is a full subcategory of* $q\mathsf{UG}$.

Proof This is analogous to 5.3.3 and we leave this to the reader. □

5.5.16 Theorem *The concrete functor from* $q\mathsf{UG}$ *to* App *which takes* $(X, \mathcal{G})$ *to* $(X, \widehat{\mathcal{G}})$ *is a forgetful functor and is left adjoint to the embedding of* App *in* $q\mathsf{UG}$. *In other words,* App *is coreflectively embedded in* $q\mathsf{UG}$.

Proof This is analogous to 5.3.4 and we leave this to the reader. □

The relation among the categories which we have seen is depicted in the following diagram.

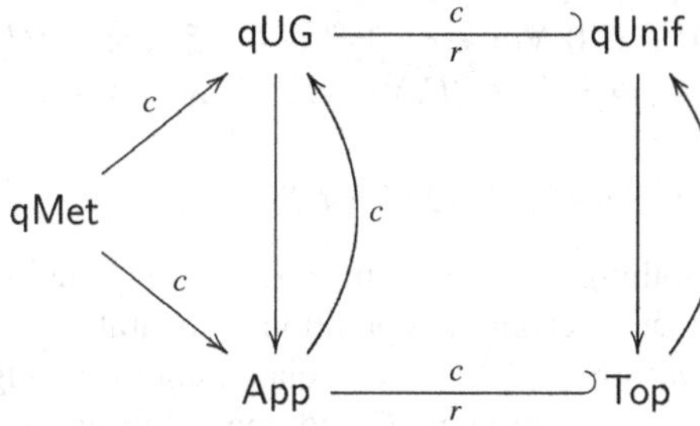

Finally we see that UG and qUG fill in the place of respectively the third and fourth question marks in the diagram of the introduction. At the end of this section we can now moreover see the complete picture of the situation of both the local and

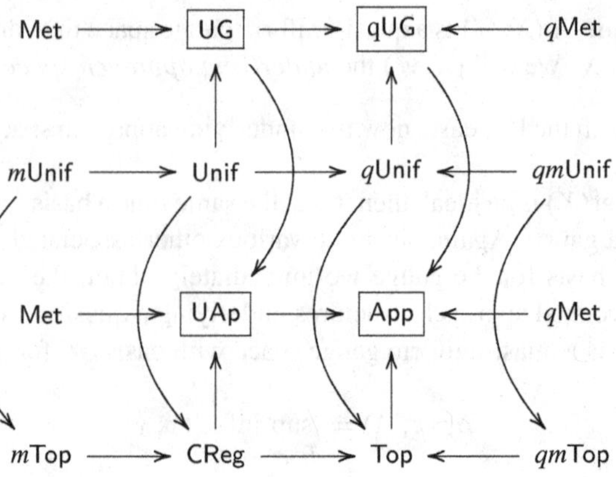

Fig. 5.1 The categorical position of UG and qUG with regard to UAp and App

the uniform approach theories and how they relate to topological, uniform and metric theories.

The curved arrows are forgetful functors. All other arrows are embeddings, which depending case by case (see text) can be either concretely reflective, concretely coreflective or both (Fig. 5.1).

5.6 Comments

1. **Approach uniformity**

It is possible to give a description of the structure of uniform gauge spaces in a similar way as approach system do for approach spaces. This leads to the definition of what is called (in Lowen and Windels 1998; Windels 1997) an *approach uniformity*. An approach uniformity is an ideal Γ in $\mathbb{P}^{X\times X}$ satisfying the following axioms:

(AU1) $\forall \gamma \in \Gamma, \forall x \in X : \gamma(x,x) = 0$.
(AU2) $\forall \xi \in \mathbb{P}^{X\times X} : (\forall \varepsilon > 0, \forall \omega < \infty, \exists \gamma_\varepsilon^\omega \in \Gamma : \xi \wedge \omega \leq \gamma_\varepsilon^\omega + \varepsilon) \Rightarrow \xi \in \Gamma$.
(AU3) $\forall \gamma \in \Gamma, \forall \omega < \infty, \exists \gamma^\omega \in \Gamma, \forall x, y, z \in X : \gamma(x,z) \wedge \omega \leq \gamma^\omega(x,y) + \gamma^\omega(y,z)$.
(AU4) $\forall \gamma \in \Gamma : \gamma^s \in \Gamma$ where $\gamma^s(x,y) := \gamma(y,x)$.

Note that (AU2) is nothing else but a typical saturation condition, comparable to (A2) and that (AU3) is an interlinking condition comparable to (A3). The way to go from an approach uniformity Γ to its associated uniform gauge $\mathcal{G}$ is by simply taking $\mathcal{G}$ to be the set of all metrics in Γ and conversely to go from the uniform gauge to the approach uniformity by taking Γ to be the saturation of $\mathcal{G}$ in $\mathbb{P}^{X\times X}$.

2. **Approach quasi-uniformity**

Note that the same comment as 1 holds for the "quasi"-case. There too the structure can be defined by what we could call an *approach quasi-uniformity*. Obviously this then is a system satisfying all the conditions above except (AU4)

(AqU1) $\forall \gamma \in \Gamma, \forall x \in X : \gamma(x,x) = 0.$
(AqU2) $\forall \xi \in \mathbb{P}^{X\times X} : (\forall \varepsilon > 0, \forall \omega < \infty, \exists \gamma_\varepsilon^\omega \in \Gamma : \xi \wedge \omega \leq \gamma_\varepsilon^\omega + \varepsilon) \Rightarrow \xi \in \Gamma.$
(AqU3) $\forall \gamma \in \Gamma, \forall \omega < \infty, \exists \gamma^\omega \in \Gamma, \forall x, y, z \in X : \gamma(x,z) \wedge \omega \leq \gamma^\omega(x,y) + \gamma^\omega(y,z).$

See also the PhD thesis of Windels (1997).

3. **Approach covering structures**

In his PhD thesis Bart Windels also considered another characterization of uniform gauge space by what he called "approach covering structures", a description similar to the way uniform structures can be characterized by certain covering structures (Windels 1997). Given a set X an *approach covering* on X is a collection of functions $\mathscr{A} \subseteq \mathbb{P}^X$ satisfying the following properties.

(C1) $\forall x \in X, \exists f \in \mathscr{A} : f(x) = 0$
(C2) $\forall f \in \mathscr{A}, \exists x \in X : f(x) = 0$

If $\mathscr{A}$ and $\mathscr{B}$ are approach coverings on X then $\mathscr{A} \prec \mathscr{B}$ if for all $f \in \mathscr{A}$ there exists a $g \in \mathscr{B}$ such that $g \leq f$. For $\varepsilon > 0$ and $\omega < \infty$, $\mathscr{A}$ is called an (ε, ω)-refinement of $\mathscr{B}$, denoted by $\mathscr{A} \prec_\omega^\varepsilon \mathscr{B}$ if for all $f \in \mathscr{A}$ there exists $g \in \mathscr{B}$ such that $g \wedge \omega \leq f + \varepsilon$. Further, $\mathscr{A}$ is called an ω-*star-refinement* of $\mathscr{B}$, denoted by $\mathscr{A} \prec_\omega^* \mathscr{B}$ if for all $f \in \mathscr{A}$ there exists $g \in \mathscr{B}$ such that for all $h \in \mathscr{A}$ we have $g \wedge \omega \leq \inf(f+h) + h$. An *approach covering structure* on X is a collection Φ of approach coverings satisfying the following conditions.

(CS1) $\mathscr{A} \in \Phi$ and $\mathscr{A} \prec \mathscr{B} \Rightarrow \mathscr{B} \in \Phi.$
(CS2) $\mathscr{A}, \mathscr{B} \in \Phi \Rightarrow \mathscr{A} \wedge \mathscr{B} \in \Phi.$
(CS3) $\forall \mathscr{A} \in \Phi$ and $\omega < \infty, \exists \mathscr{B} \in \Phi : \mathscr{B} \prec_\omega^* \mathscr{A}.$
(CS4) $(\forall \varepsilon > 0, \forall \omega < \infty, \exists \mathscr{A} \in \Phi : \mathscr{A} \prec_\omega^\varepsilon \mathscr{B}) \Rightarrow \mathscr{B} \in \Phi.$

Approach quasi-uniformity

Note that the [illegible] holds for the [illegible] case. [illegible] the structure [illegible] what we call [illegible] approach [illegible]. Obviously, [illegible] then a system satisfying all the [illegible] except (AU1).

(AQU1) [illegible]

(AQU2) [illegible]

(AQU3) [illegible]

See also the PhD thesis of Windels (1997).

Approach covering structures

In his PhD thesis [illegible] also considered another characterization of uniform [illegible] spaces [illegible] what he called [illegible] approach covering structures, a description similar to the way uniform structures can be characterized [illegible] uniform covering structures [illegible] (1997). [illegible] is a collection of [illegible] satisfying the following conditions:

(C1) [illegible]

(C2) [illegible]

[illegible] and [illegible] an approach covering structure [illegible] if for all [illegible] there exists [illegible] such that [illegible] and [illegible], [illegible] is called an [illegible] refinement of [illegible] such that [illegible]. Further, [illegible] is called an approach covering structure on X [illegible] such that for all [illegible] we have [illegible]. An approach covering structure on X is a collection of approach coverings satisfying the following conditions:

(CS1) [illegible]

(CS2) [illegible]

(CS3) [illegible]

(CS4) [illegible]

Chapter 6
Extensions of Spaces and Morphisms

Creativity is a natural extension of our enthusiasm.
(Earl Nightingale)

Every extension of knowledge arises from making the conscious the unconscious.
(Friedrich Nietzsche)

As we have seen, the fact that App contains both Top and Met as full and isomorphism-closed subcategories has as a consequence that approach spaces have both a topological and a metric side to them. Thus, it for instance turns out that we can consider both a notion of completion and a notion of compactification, at least for uniform approach spaces. In the first section of the present chapter we construct a completion in App which coincides with the usual completion in the case of metric spaces and which is a firm epireflection from the subcategory of Hausdorff uniform approach spaces to the subcategory of complete Hausdorff uniform approach spaces.

In the second section we consider the case of UG. Here the situation is far more simple. Completeness, almost by definition, is equivalent to the completeness of the underlying uniform space and hence the construction of a completion follows the same lines as the construction of completion for uniform spaces. In particular the underlying set of the completion is just the same as the underlying set of the completion of the uniform coreflection. Here too we obtain a firm epireflection from the subcategory of Hausdorff uniform gauge spaces to the subcategory of complete Hausdorff uniform gauge spaces.

In the third section, the compactification which we will construct in App is a Čech-Stone-type compactification. By this we mean that it is an epireflection from the subcategory of Hausdorff uniform approach spaces to the subcategory of compact Hausdorff uniform approach spaces, in particular it is the unique compactification of a uniform approach space X to which contractions from X to Hausdorff compact uniform approach spaces have unique extensions. It is well known that compactifications do not often preserve metrizability. For instance a nontrivial Čech-Stone compactification is never metrizable and even if a compactification is metrizable, then it need not be metrizable by an extension of the given metric. This is the case for instance with the Alexandroff compactification of $\mathbb{R}$. In UAp, as we will see the situation is better. We also give an alternative characterization of this compactification. It turns

R. Lowen, *Index Analysis*, Springer Monographs in Mathematics,
DOI 10.1007/978-1-4471-6485-2_6

out that the underlying set of the compactification coincides with the underlying set of the Smirnov compactification of an associated proximity space (see Naimpally and Warrack 1970). This then provides us with a tangible description of the points of the compactification comparable to the description of the points of the classical Čech-Stone compactification as being maximal zero-filters (see Gillman and Jerison 1976). Finally we also see that our Čech-Stone compactification can be described as the underlying approach space of the completion of a particular uniform gauge space.

6.1 Completion in UAp

In this section, Cauchy filters are always meant in the sense of 3.5.1.

In order to construct a completion for uniform approach spaces we proceed as in the case of the metric or the uniform completion, by first enlarging the space and then extending the structure (see Chap. 3 for the definition of completeness).

Let X be a uniform approach space. For any Cauchy filter $\mathscr{F}$, put

$$C(\mathscr{F}) := \{\mathscr{G} \in \mathsf{F}(X) | \mathscr{G} \text{ Cauchy}, \mathscr{G} \subseteq \mathscr{F}\},$$

and let

$$\mathscr{M}_{\mathscr{F}} := \bigcap_{\mathscr{G} \in C(\mathscr{F})} \mathscr{G}.$$

Given Cauchy filters $\mathscr{F}$ and $\mathscr{G}$ we will say that they are *equivalent* if $\mathscr{F} \cap \mathscr{G}$ is a Cauchy filter, and we denote this by $\mathscr{F} \sim \mathscr{G}$. This is indeed an equivalence relation. Symmetry and reflexivity are evident. If $\mathscr{F} \sim \mathscr{G}$ and $\mathscr{G} \sim \mathscr{H}$, then it follows from 3.5.5 that $\lambda\mathscr{F} = \lambda(\mathscr{F} \cap \mathscr{G}) = \lambda\mathscr{G} = \lambda(\mathscr{G} \cap \mathscr{H}) = \lambda\mathscr{H}$. Hence, $\lambda(\mathscr{F} \cap \mathscr{H}) = \lambda\mathscr{F} \vee \lambda\mathscr{H} = \lambda\mathscr{F}$ which implies that $\mathscr{F} \sim \mathscr{H}$.

6.1.1 Proposition *Let X be a uniform approach space and let $\mathscr{F}$ and $\mathscr{G}$ be Cauchy filters. Then the following properties hold.*

1. *$\mathscr{M}_{\mathscr{F}}$ is the smallest Cauchy filter coarser than $\mathscr{F}$ and $\lambda\mathscr{M}_{\mathscr{F}} = \lambda\mathscr{F}$.*
2. *$\mathscr{F}$ and $\mathscr{G}$ are equivalent if and only if $\lambda\mathscr{F} = \lambda\mathscr{G}$.*
3. *$\mathscr{M}_{\mathscr{F}}$ is the minimal element of the set of all Cauchy filters which are equivalent with $\mathscr{F}$.*

Proof 1. It follows from (L2) and 3.5.5 that

$$\lambda\mathscr{M}_{\mathscr{F}} = \sup_{\mathscr{G} \in C(\mathscr{F})} \lambda\mathscr{G} = \sup_{\mathscr{G} \in C(\mathscr{F})} \lambda\mathscr{F} = \lambda\mathscr{F}.$$

By definition this then also implies that $\mathscr{M}_{\mathscr{F}}$ is a Cauchy filter, which by construction is of course the smallest one coarser than $\mathscr{F}$.

2. If $\mathcal{F} \sim \mathcal{G}$, then it follows from 3.5.5 that $\lambda\mathcal{F} = \lambda(\mathcal{F} \cap \mathcal{G}) = \lambda\mathcal{G}$. Conversely if $\lambda\mathcal{F} = \lambda\mathcal{G}$, then $\lambda(\mathcal{F} \cap \mathcal{G}) = \lambda\mathcal{F} \vee \lambda\mathcal{G} = \lambda\mathcal{F}$ which implies that $\mathcal{F} \cap \mathcal{G}$ is a Cauchy filter and hence that $\mathcal{F} \sim \mathcal{G}$.

3. That $\mathcal{M}_{\mathcal{F}}$ is equivalent with $\mathcal{F}$ follows from 1 and 2. If $\mathcal{G}$ is equivalent with $\mathcal{F}$ then by 1, $\mathcal{M}_{\mathcal{F}} \subseteq \mathcal{F} \cap \mathcal{G} \subseteq \mathcal{G}$ and, hence, $\mathcal{M}_{\mathcal{F}}$ is minimal. □

A filter of type $\mathcal{M}_{\mathcal{F}}$ will be called a *minimal Cauchy filter*.

6.1.2 Proposition *If X is a uniform approach space, then, for any $x \in X$, the neighbourhood filter $\mathcal{V}(x)$ in the* Top*-coreflection is a minimal Cauchy filter.*

Proof Since $\lambda\mathcal{V}(x)(x) = 0$ it follows that $\mathcal{V}(x)$ is a Cauchy filter. If $\mathcal{G} \subseteq \mathcal{V}(x)$ is a Cauchy filter, then it follows from 3.5.5 that $\lambda\mathcal{G}(x) = \lambda\mathcal{V}(x)(x) = 0$ and hence $\mathcal{G}$ converges to x in the Top-coreflection of (X, δ). Consequently, $\mathcal{G} \supseteq \mathcal{V}(x)$ which proves that $\mathcal{V}(x)$ is minimal. □

Now let $\mathcal{D}$ be a symmetric gauge basis for X and let $\mathcal{F}$ be a Cauchy filter, then it follows from 3.5.3 that it is also a $\mathcal{U}(\mathcal{D})$-Cauchy filter. Consequently, there is a minimal $\mathcal{U}(\mathcal{D})$-Cauchy filter contained in $\mathcal{F}$. This filter, which we denote by $\mathcal{F}^{\mathcal{D}}$, is generated by the basis

$$\left\{F_d^{(\varepsilon)} \mid d \in \mathcal{D}, F \in \mathcal{F}, \varepsilon > 0\right\},$$

where, for any $d \in \mathcal{D}$, $F \in \mathcal{F}$, and $\varepsilon > 0$,

$$F_d^{(\varepsilon)} := \left\{x \in X \mid \delta_d(x, F) \leq \varepsilon\right\}.$$

6.1.3 Proposition *If X is a uniform approach space, $\mathcal{D}$ is a symmetric gauge basis, and $\mathcal{F}$ is a Cauchy filter, then $\mathcal{F}^{\mathcal{D}} = \mathcal{M}_{\mathcal{F}}$.*

Proof Let $\mathcal{F}$ be a Cauchy filter. Then, for any $x \in X$, we have

$$\begin{aligned}
\lambda\mathcal{F}^{\mathcal{D}}(x) &= \sup_{d\in\mathcal{D}} \inf_{d'\in\mathcal{D}} \inf_{\varepsilon>0} \inf_{F\in\mathcal{F}} \sup_{z\in F_{d'}^{(\varepsilon)}} d(x,z) \\
&\leq \sup_{d\in\mathcal{D}} \inf_{\varepsilon>0} \inf_{F\in\mathcal{F}} \sup_{z\in F_d^{(\varepsilon)}} d(x,z) \\
&\leq \sup_{d\in\mathcal{D}} \inf_{F\in\mathcal{F}} \inf_{\varepsilon>0} (\sup_{z\in F} d(x,z) + \varepsilon) \\
&= \lambda\mathcal{F}(x).
\end{aligned}$$

Consequently, it follows that $\mathcal{F}^{\mathcal{D}}$ too is a Cauchy filter. Since $\mathcal{F}^{\mathcal{D}} \subseteq \mathcal{F}$, the minimality of $\mathcal{M}_{\mathcal{F}}$ implies that $\mathcal{M}_{\mathcal{F}} \subseteq \mathcal{F}^{\mathcal{D}}$. Conversely, $\mathcal{M}_{\mathcal{F}}$ is also a $\mathcal{U}(\mathcal{D})$-Cauchy filter, and thus it follows from the minimality of $\mathcal{F}^{\mathcal{D}}$ that $\mathcal{F}^{\mathcal{D}} \subseteq \mathcal{M}_{\mathcal{F}}$. □

Given a uniform approach space X we denote by $\widehat{X}$ the set of all minimal Cauchy filters. Note that it follows from 6.1.3 that a minimal Cauchy filter has an open basis in the Top-coreflection.

If $\mathscr{D}$ is a symmetric gauge basis for X then for any $d \in \mathscr{D}$ we define $\widehat{d}$ by

$$\widehat{d}(\mathscr{F}, \mathscr{G}) = \sup_{F \in \mathscr{F}} \sup_{G \in \mathscr{G}} \inf_{x \in F} \inf_{y \in G} d(x, y).$$

It follows from 6.1.3 that minimal Cauchy filters are also minimal $\mathscr{U}(\mathscr{D})$-Cauchy filters and hence the function $\widehat{d}$ is well defined.

6.1.4 Proposition *If X is a uniform approach space and $\mathscr{D}$ is a symmetric gauge basis then the set*

$$\widehat{\mathscr{D}} := \{\widehat{d} \mid d \in \mathscr{D}\}$$

is a symmetric gauge basis on $\widehat{X}$.

Proof It suffices to prove that $\widehat{\mathscr{D}}$ is upwards directed. This follows from the fact that, for any $d, e \in \mathscr{D}$, we have $\widehat{d} \vee \widehat{e} \leq \widehat{d \vee e}$, as is easily verified. □

The distance generated by $\widehat{\mathscr{D}}$ will be denoted by $\widehat{\delta}$ and analogously for the associated structures.

6.1.5 Theorem *If X is a Hausdorff uniform approach space, then the map*

$$e_X : X \longrightarrow \widehat{X} : x \mapsto \mathscr{V}(x)$$

is an embedding and thus X can be identified with the subspace $e_X(X)$ of $\widehat{X}$. In particular, if X is moreover complete then it is isomorphic to $\widehat{X}$.

Proof Clearly e_X is an injective map via which we can identify X with $e_X(X)$. It now suffices to note that if $\mathscr{D}$ is a symmetric gauge basis which generates δ, then, for any $d \in \mathscr{D}$, we have

$$\widehat{d}(\mathscr{V}(x), \mathscr{V}(y)) = \sup_{F \in \mathscr{V}(x)} \sup_{G \in \mathscr{V}(y)} \inf_{u \in F} \inf_{v \in G} d(u, v) = d(x, y). \qquad \square$$

6.1.6 Proposition *If (X, δ) is a uniform approach space, $\mathscr{C} \in \widehat{X}$, and $x \in X$, then*

$$\lambda\mathscr{C}(x) = \widehat{\delta}(\mathscr{C}, \{\mathscr{V}(x)\}).$$

Proof Let $\mathscr{D}$ be a symmetric gauge basis which generates δ. Applying 3.5.5 we have

$$\begin{aligned}\widehat{\delta}(\mathscr{C}, \{\mathscr{V}(x)\}) &= \sup_{d\in\mathscr{D}} \sup_{d'\in\mathscr{D}} \sup_{\varepsilon>0} \sup_{A\in\mathscr{C}} \inf_{z\in B_{d'}(x,\varepsilon)} \inf_{y\in A} d(z, y)\\ &= \sup_{d\in\mathscr{D}} \sup_{\varepsilon>0} \sup_{A\in\mathscr{C}} \inf_{z\in B_d(x,\varepsilon)} \inf_{y\in A} d(z, y)\\ &= \sup_{d\in\mathscr{D}} \sup_{A\in\mathscr{C}} \inf_{y\in A} d(x, y)\\ &= \alpha\mathscr{C}(x)\\ &= \lambda\mathscr{C}(x).\end{aligned}$$

□

6.1.7 Proposition *The subset $e_X(X)$ is dense in $(\widehat{X}, \mathscr{T}_{d_{\widehat{\delta}}})$, i.e. with respect to the topology generated by the* Met*-coreflection of $\widehat{\delta}$.*

Proof By 6.1.6 we have, for any $x \in X$ and $\mathscr{C} \in \widehat{X}$,

$$\begin{aligned}d_{\widehat{\delta}}(\mathscr{C}, \mathscr{V}(x)) &= \widehat{\delta}(\mathscr{C}, \{\mathscr{V}(x)\})\\ &= \lambda\mathscr{C}(x),\end{aligned}$$

and thus, since $\mathscr{C}$ is δ-Cauchy,

$$\begin{aligned}d_{\widehat{\delta}}(\mathscr{C}, e_X(X)) &= \inf_{x\in X} d_{\widehat{\delta}}(\mathscr{C}, \mathscr{V}(x))\\ &= \inf_{x\in X} \lambda\mathscr{C}(x) = 0.\end{aligned}$$

□

For any filter $\mathscr{F}$ on X, we will denote the filter generated on $\widehat{X}$ by the basis $e_X(\mathscr{F})$ by $\widehat{\mathscr{F}}$. Note that if $\mathscr{F}$ is a δ-Cauchy filter, then $\widehat{\mathscr{F}}$ is a $\widehat{\delta}$-Cauchy filter.

6.1.8 Proposition *If (X, δ) is a uniform approach space and $\mathscr{C} \in \widehat{X}$, then*

$$\widehat{\lambda}\widehat{\mathscr{C}}(\mathscr{C}) = 0.$$

Proof Let $\mathscr{D}$ be a symmetric gauge basis which generates δ. Fix $\varepsilon > 0$. Then, since $\mathscr{C}$ is Cauchy, it follows from 3.5.2 that we can find $x \in X$ such that, for all $d' \in \mathscr{D}$, we have $B_{d'}(x, \varepsilon) \in \mathscr{C}$. Consequently, we obtain

$$\begin{aligned}\widehat{\lambda}\widehat{\mathscr{C}}(\mathscr{C}) &= \sup_{d\in\mathscr{D}} \inf_{A\in\mathscr{C}} \sup_{a\in A} \widehat{d}(\mathscr{V}(a), \mathscr{C})\\ &= \sup_{d\in\mathscr{D}} \inf_{A\in\mathscr{C}} \sup_{a\in A} \sup_{A'\in\mathscr{C}} \inf_{b\in A'\cap A} d(a, b)\\ &\le \sup_{d\in\mathscr{D}} \inf_{d'\in\mathscr{D}} \sup_{a\in B_{d'}(x,\varepsilon)} \sup_{A'\in\mathscr{C}} \inf_{b\in A'\cap B_{d'}(x,\varepsilon)} d(a, b)\\ &\le \sup_{d\in\mathscr{D}} \sup_{a\in B_d(x,\varepsilon)} \sup_{A'\in\mathscr{C}} \inf_{b\in A'\cap B_d(x,\varepsilon)} d(a, b)\\ &\le 2\varepsilon\end{aligned}$$

which by the arbitrariness of ε proves our claim. □

In what follows we will denote the full subcategory of UAp consisting of all Hausdorff spaces by UAp_2 and the full subcategory of UAp_2 consisting of all complete spaces by $c\mathsf{UAp}_2$. In order to prove our main theorem of this section we first have to characterize the epimorphisms in UAp_2.

6.1.9 Theorem *A function is an epimorphism in* UAp_2 *if and only if it is a contraction which has a dense image (with regard to the* Top*-coreflection).*

Proof The if part follows from the analogous topological result. Let X, Y and Z be UAp_2-spaces and consider the diagram $X \xrightarrow{f} Y \underset{s}{\overset{r}{\rightrightarrows}} Z$ where f has a dense image and $r \circ f = s \circ f$. Applying the concrete coreflector to CReg_2 we can conclude that $r = s$.

To show the only-if part suppose that (X, δ_X) and (Y, δ_Y) are uniform approach spaces, where δ_X is generated by $\mathcal{D}_X$ and δ_Y is generated by the collection of bounded metrics $\mathcal{D}_Y$. Further let

$$f : (X, \delta_X) \longrightarrow (Y, \delta_Y)$$

be a non-dense contraction. Then there exists $y_0 \in Y$ such that

$$\delta_Y(y_0, f(X)) > 0$$

and hence there further exists $d_0 \in \mathcal{D}_Y$ such that

$$\inf_{x \in X} d_0(y_0, f(x)) > 0.$$

Now consider $[0, \infty[$ equipped with the Euclidean metric $d_{\mathbb{E}}$ and define the map

$$\theta : (Y, \delta_Y) \longrightarrow ([0, \infty[, \delta_{d_{\mathbb{E}}}) : y \longrightarrow \inf_{x \in X} d_0(y, f(x)).$$

Since, for all $z \in Y$, $Z \in 2^Y$, we have

$$\begin{aligned} \delta_{d_{\mathbb{E}}}(\theta(z), \theta(Z)) &= \inf_{z' \in Z} \left| \inf_{x \in X} d_0(z, f(x)) - \inf_{x \in X} d_0(z', f(x)) \right| \\ &\leq \inf_{z' \in Z} \sup_{x \in X} \left| d_0(z, f(x)) - d_0(z', f(x)) \right| \\ &\leq \inf_{z' \in Z} d_0(z, z') \\ &\leq \delta_Y(z, Z), \end{aligned}$$

it follows that θ is a contraction. Clearly $\theta \circ f = 0$ whereas $\theta(y_0) \neq 0$. The constant map $\eta = 0$ is also a contraction and, since $\theta \neq \eta$ whereas $\theta \circ f = \eta \circ f$, it follows that f is not an epimorphism. □

6.1.10 Theorem *A function is an extremal monomorphism in* UAp_2 *if and only if it is an injective contraction which has a closed image.*

Proof This goes in exactly the same way as the proof for Top_2 (the full subcategory of Top with objects all Hausdorff spaces) and we therefore leave this to the reader. □

We define $\mathscr{E}_m$ to be the class of all embeddings in UAp_2 which are dense for the metric coreflection. Note that in 6.1.7 we showed that e_X belongs to this class.

6.1.11 Theorem *$c\mathsf{UAp}_2$ is an $\mathscr{E}_m$-reflective subcategory of UAp_2. For any Hausdorff uniform approach space (X, δ),*

$$e_X : (X, \delta) \longrightarrow (\widehat{X}, \widehat{\delta}) : x \longrightarrow \mathscr{V}(x)$$

is an epireflection of (X, δ) in $c\mathsf{UAp}_2$.

Proof 1. *$(\widehat{X}, \widehat{\delta})$ is Hausdorff.* This is clear since, by construction, its Top-coreflection is a subspace of the topological space underlying the uniform completion of $(X, \mathscr{U}(\mathscr{D}))$.

2. *$(\widehat{X}, \widehat{\delta})$ is complete.* To prove this, first of all note that we can restrict our attention to showing that minimal $\widehat{\delta}$-Cauchy filters are convergent. Let Φ be such a minimal $\widehat{\delta}$-Cauchy filter. Then, since it has an open basis, it has a trace on X, say $\mathscr{F}$. Let $\varepsilon > 0$. Then it follows from 6.1.8 that we can find $\mathscr{C} \in \widehat{X}$ such that $\widehat{\lambda}\Phi(\mathscr{C}) \leq \varepsilon$ and next we can find $x \in X$ such that $\lambda\mathscr{C}(x) \leq \varepsilon$. Then from the fact that $\Phi \subseteq \widehat{\mathscr{F}}$, that $\lambda\mathscr{F}(x) = \widehat{\lambda}\widehat{\mathscr{F}}(\mathscr{V}(x))$, and upon applying 1.2.70 and 6.1.6 it follows that

$$\begin{aligned}\lambda\mathscr{F}(x) &\leq \widehat{\lambda}\Phi(\mathscr{V}(x))\\ &\leq \widehat{\lambda}\Phi(\mathscr{C}) + \widehat{\delta}(\mathscr{C}, \{\mathscr{V}(x)\})\\ &\leq 2\varepsilon,\end{aligned}$$

which by the arbitrariness of ε implies that $\mathscr{F}$ is a δ-Cauchy filter.

Now, since Φ and $\widehat{\mathscr{F}}$ are $\widehat{\delta}$ -Cauchy filters and since $\widehat{\mathscr{M}_{\mathscr{F}}} \subseteq \widehat{\mathscr{F}}$ and $\Phi \subseteq \widehat{\mathscr{F}}$, it follows from 3.5.5 and 6.1.8 that

$$\begin{aligned}\widehat{\lambda}\Phi(\mathscr{M}_{\mathscr{F}}) &= \widehat{\lambda}\widehat{\mathscr{F}}(\mathscr{M}_{\mathscr{F}})\\ &= \widehat{\lambda}(\widehat{\mathscr{M}_{\mathscr{F}}})(\mathscr{M}_{\mathscr{F}})\\ &= 0.\end{aligned}$$

This proves that $(\widehat{X}, \widehat{\delta})$ is complete.

3. *$e_X : (X, \delta) \longrightarrow (\widehat{X}, \widehat{\delta}) : x \longrightarrow \mathscr{V}(x)$ determines an $\mathscr{E}_m$-reflection.* That e_X belongs to $\mathscr{E}_m$ follows from 6.1.7.

Now suppose that (Y, δ_Y) is a complete Hausdorff uniform approach space and that

$$f : (X, \delta) \longrightarrow (Y, \delta_Y)$$

is a contraction. If $\mathscr{C} \in \widehat{X}$, then $f(\mathscr{C})$ is a δ_Y-Cauchy filter and thus there exists a unique point $y_c \in Y$ such that $\lambda_Y f(\mathscr{C})(y_c) = 0$. We define

$$\widehat{f} : \widehat{X} \longrightarrow Y : \mathscr{C} \longrightarrow y_c.$$

Since, for any $x \in X$, we have that $\lambda_Y f(\mathscr{V}(x))(f(x)) = 0$ it follows that the diagram

$$\begin{array}{ccc} (X, \delta) & \xrightarrow{\ f\ } & (Y, \delta_Y) \\ {\scriptstyle e_X}\big\downarrow & \nearrow{\scriptstyle \widehat{f}} & \\ (\widehat{X}, \widehat{\delta}) & & \end{array}$$

commutes. Moreover, since e_X is an epimorphism, $\widehat{f}$ is the only function which makes the diagram commute. So all that remains to be shown is that it is a contraction. Let $\varepsilon > 0$ and let us denote by

$$\overline{\varepsilon} : \widehat{X} \longrightarrow X$$

an arbitrary fixed map which for each $x \in X$ and $\mathscr{C} \in \widehat{X}$ fulfils

$$\overline{\varepsilon}(\mathscr{V}(x)) = x \text{ and } \lambda\mathscr{C}(\overline{\varepsilon}(\mathscr{C})) \leq \varepsilon.$$

Now, if $\mathscr{C} \in \widehat{X}$ and $\Gamma \in 2^{\widehat{X}}$, then

$$\begin{aligned} \delta_Y(f(\overline{\varepsilon}(\mathscr{C})), f(\overline{\varepsilon}(\Gamma))) &\leq \delta(\overline{\varepsilon}(\mathscr{C}), \overline{\varepsilon}(\Gamma)) \\ &= \widehat{\delta}(\mathscr{V}(\overline{\varepsilon}(\mathscr{C})), e_X(\overline{\varepsilon}(\Gamma))) \\ &\leq \widehat{\delta}(\mathscr{V}(\overline{\varepsilon}(\mathscr{C})), \{\mathscr{C}\}) + \widehat{\delta}(\mathscr{C}, \Gamma) + \sup_{\mathscr{C}' \in \Gamma} \widehat{\delta}(\mathscr{C}', e_X(\overline{\varepsilon}(\Gamma))). \end{aligned}$$

From the definition of $\overline{\varepsilon}$ it follows that

$$\widehat{\delta}(\mathscr{V}(\overline{\varepsilon}(\mathscr{C})), \{\mathscr{C}\}) \leq \varepsilon$$
$$\text{and}$$
$$\sup_{\mathscr{C}' \in \Gamma} \widehat{\delta}(\mathscr{C}', e_X(\overline{\varepsilon}(\Gamma))) \leq \varepsilon,$$

and consequently

$$\delta_Y(f(\overline{\varepsilon}(\mathscr{C})), f(\overline{\varepsilon}(\Gamma))) \leq \widehat{\delta}(\mathscr{C}, \Gamma) + 2\varepsilon.$$

On the other hand we have that

$$\begin{aligned} \delta_Y(\widehat{f}(\mathscr{C}), \widehat{f}(\Gamma)) \leq \delta_Y(\widehat{f}(\mathscr{C}), \{f(\overline{\varepsilon}(\mathscr{C}))\}) + \delta_Y(f(\overline{\varepsilon}(\mathscr{C})), f(\overline{\varepsilon}(\Gamma))) \\ + \sup_{\mathscr{C}' \in \Gamma} \delta_Y(f(\overline{\varepsilon}(\mathscr{C}')), \widehat{f}(\Gamma)). \end{aligned}$$

It follows from 3.5.5 and 6.1.6 that, for any $\mathscr{C}' \in \widehat{X}$,

$$\begin{aligned}\delta_Y(\widehat{f}(\mathscr{C}'), \{f(\overline{\varepsilon}(\mathscr{C}'))\}) &= \lambda_Y f(\mathscr{C}')(f(\overline{\varepsilon}(\mathscr{C}'))) \\ &\leq \lambda\mathscr{C}'(\overline{\varepsilon}(\mathscr{C}')) \\ &\leq \varepsilon.\end{aligned}$$

Combining the foregoing inequalities we thus obtain that

$$\begin{aligned}\delta_Y(\widehat{f}(\mathscr{C}), \widehat{f}(\Gamma)) &\leq \delta_Y(f(\overline{\varepsilon}(\mathscr{C})), f(\overline{\varepsilon}(\Gamma))) + 2\varepsilon \\ &\leq \widehat{\delta}(\mathscr{C}, \Gamma) + 4\varepsilon,\end{aligned}$$

which by the arbitrariness of ε shows that $\widehat{f}$ is indeed a contraction. □

6.1.12 Corollary *The subcategory* cUAp$_2$ *is epireflective in* UAp$_2$ *and therefore is closed under the formation of products and closed subspaces.*

Proof This is a consequence of 6.1.10 and 6.1.11. □

6.1.13 Corollary *If (Y, δ_Y) is a complete Hausdorff uniform approach space, (X, δ_X) is a Hausdorff uniform approach space, $Z \subseteq X$ is dense with respect to the underlying* Met*-coreflection of (X, δ_X), and $f : (Z, \delta_{X|Z}) \longrightarrow (Y, \delta_Y)$ is a contraction, then there exists a unique extension*

$$\overline{f} : (X, \delta_X) \longrightarrow (Y, \delta_Y)$$

which is also a contraction.

Proof The proof goes along exactly the same lines as the proof of the epireflectivity in 6.1.11 and we leave this verification to the reader. □

6.1.14 Theorem *If (X, δ) is a Hausdorff uniform approach space, (X_i, δ_i) is a complete Hausdorff uniform approach space for $i \in \{1, 2\}$, and the maps*

$$e_i : (X, \delta) \longrightarrow (X_i, \delta_i)$$

belong to $\mathscr{E}_m$, then (X_1, δ_1) and (X_2, δ_2) are isomorphic.

Proof It follows from 6.1.13 that there exist unique contractions

$$\varphi_1 : (X_2, \delta_2) \longrightarrow (X_1, \delta_1) \text{ and } \varphi_2 : (X_1, \delta_1) \longrightarrow (X_2, \delta_2),$$

such that $e_1 = \varphi_1 \circ e_2$ and $e_2 = \varphi_2 \circ e_1$. Since e_1 and e_2 are dense embeddings also with respect to the Top-coreflection of (X, δ) they are epimorphisms in UAp$_2$ and it thus follows that $\varphi_1 \circ \varphi_2 = 1_{X_1}$ and $\varphi_2 \circ \varphi_1 = 1_{X_2}$. □

6.1.15 Definition Given a Hausdorff uniform approach space (X, δ) the space $(\widehat{X}, \widehat{\delta})$ is called the *completion of* (X, δ).

Let us now verify that in case (X, d) is a metric space, this completion indeed coincides with the usual metric completion which we denote by $(\widetilde{X}, \widetilde{d})$.

6.1.16 Proposition (Met) *If (X, d) is a metric space, then $(\widehat{X}, \widehat{\delta_d}) = (\widetilde{X}, \delta_{\widetilde{d}})$.*

Proof It follows from 3.5.6 that Cauchy filters coincide, and from 6.1.3 that also minimal Cauchy filters coincide since $\mathscr{D} := \{d\}$ is a basis for the gauge. Hence $\widetilde{X} = \widehat{X}$. Then it is immediate from the definition of $\widehat{d}$ that this function coincides with $\widetilde{d}$ and hence it follows from 6.1.4 that $\widehat{\delta_d} = \delta_{\widetilde{d}}$. □

6.2 Completeness and Completion in UG

In this section Cauchy filters are always meant in the sense of 5.4.9 and 5.4.10.

6.2.1 Definition A uniform gauge space is called *complete* if its uniform coreflection is complete.

In contrast with the completion in UAp, for underlying set of the completion in UG we simply take the underlying set of the uniform completion of the uniform coreflection. We will therefore freely use all that is known about this underlying set and its structure from the known results in uniform spaces. The main step consists in extending the structure and the morphisms.

Two Cauchy filters $\mathscr{F}$ and $\mathscr{H}$ in a uniform gauge space are called *equivalent* iff $\mathscr{F} \cap \mathscr{H}$ is Cauchy. Note that by 5.4.10, this is the usual equivalence of Cauchy filters in uniform spaces.

6.2.2 Proposition *Let $(X, \mathscr{G})$ be a uniform gauge space, and let $\mathscr{F}$ be a Cauchy filter on X, then*

$$\mathscr{B} := \left\{ \{y \mid \inf_{x \in F} d(x, y) < \alpha\} \mid d \in \mathscr{G}, \alpha > 0, F \in \mathscr{F} \right\}$$

is a filterbasis for a Cauchy filter $\mathscr{M}^{u}_{\mathscr{F}}$, coarser than $\mathscr{F}$. Moreover, $\mathscr{M}^{u}_{\mathscr{F}}$ is the coarsest Cauchy filter with that property and it is the coarsest filter in the equivalence class of $\mathscr{F}$.

Proof This follows from the fact that $\mathscr{B} = \{U(F) \mid U \in \mathscr{U}_0, F \in \mathscr{F}\}$. □

From Theorem 6.2.2, it follows that we can define $\widehat{X}$ to be the set of all minimal Cauchy filters on X. Again, note that we are simply using the notion of minimal Cauchy filter in uniform spaces and consequently, for any $x \in X$, the neighbourhood filter $\mathscr{V}(x)$ is a minimal Cauchy filter.

6.2.3 Proposition *Let X be a uniform gauge space. If $d \in \mathscr{G}$, we define*

$$\widehat{d} : \widehat{X} \times \widehat{X} \longrightarrow \mathbb{P}$$

by

$$\widehat{d}(\mathcal{M},\mathcal{N}) := \sup_{M\in\mathcal{M}} \sup_{N\in\mathcal{N}} \inf_{x\in M} \inf_{y\in N} d(x,y) = \inf_{M\in\mathcal{M}} \inf_{N\in\mathcal{N}} \sup_{x\in M} \sup_{y\in N} d(x,y).$$

Then $\widehat{\mathcal{G}} := \{\widehat{d} \mid d \in \mathcal{G}\}$ *is a uniform gauge basis on* $\widehat{X}$.

Proof This is straightforward and we leave this to the reader. □

Note that if $\mathcal{H}$ is a basis for $\mathcal{G}$ then $\widehat{\mathcal{H}} := \{\widehat{d} \mid d \in \mathcal{H}\}$ is a basis for $\widehat{\mathcal{G}}$. Also note that the definition of $\widehat{d}$ is precisely the same as in the case of UAp in the foregoing section.

6.2.4 Proposition *If* $(\mathcal{U}_\varepsilon)_\varepsilon$ *and* $(\widehat{\mathcal{U}_\varepsilon})_\varepsilon$ *are the towers of* $(X,\mathcal{G})$ *and* $(\widehat{X},\widehat{\mathcal{G}})$ *respectively, then* $(\widehat{X},\widehat{\mathcal{U}_0})$ *coincides with the uniform completion of* $(X,\mathcal{U}_0)$.

Proof This is straightforward and we leave this to the reader. □

6.2.5 Theorem *Let* $(X,\mathcal{G})$ *be a uniform gauge space. Then*

$$i_X : (X,\mathcal{G}) \longrightarrow (\widehat{X},\widehat{\mathcal{G}}) : x \mapsto \mathcal{V}(x)$$

is initial, and $i_X(X)$ *is dense in* $\widehat{X}$. *Moreover, if* $(X,\mathcal{G})$ *is Hausdorff, then* i_X *is an embedding.*

Proof In order to show initiality, first note that if $d \in \mathcal{G}$, then for all $x, y \in X$:

$$\widehat{d} \circ (i_X \times i_X)(x,y) = \widehat{d}(\mathcal{V}(x),\mathcal{V}(y)) = \inf_{e\in\mathcal{G}} \inf_{\varepsilon>0} \sup_{a,b\in B_e(x,\varepsilon)\cup B_e(y,\varepsilon)} d(a,b).$$

Then on the one hand

$$\begin{aligned}\widehat{d} \circ (i_X \times i_X)(x,y) &= \widehat{d}(\mathcal{V}(x),\mathcal{V}(y))\\ &= \inf_{e\in\mathcal{G}} \inf_{\varepsilon>0} \sup_{a,b\in B_e(x,\varepsilon)\cup B_e(y,\varepsilon)} d(a,b)\\ &\le \inf_{\varepsilon>0} \sup_{a,b\in B_d(x,\varepsilon)\cup B_d(y,\varepsilon)} d(a,b)\\ &= d(x,y)\end{aligned}$$

and on the other hand

$$\widehat{d} \circ (i_X \times i_X)(x,y) \ge \inf_{e\in\mathcal{G}} \inf_{\varepsilon>0} \sup_{a,b\in\{x\}\cup\{y\}} d(a,b) = d(x,y).$$

□

In what follows we will denote the full subcategory of UG consisting of all Hausdorff spaces by UG_2 and the full subcategory of UG_2 consisting of all complete spaces by $c\mathsf{UG}_2$. In order to prove our main theorem of this section we first have to characterize the epimorphisms in UG_2.

6.2.6 Theorem *A function is an epimorphism in* UG_2 *if and only if it is a uniform contraction which has a dense image (with regard to the* Top*-coreflection).*

Proof The if part is the same as in 6.1.9. To show the only-if part suppose that $(X, \mathcal{G}_X)$ and $(Y, \mathcal{G}_Y)$ are uniform gauge spaces. Further let

$$f : (X, \mathcal{G}_X) \longrightarrow (Y, \mathcal{G}_Y)$$

be a non-dense uniform contraction. Then there exists $y_0 \in Y$ such that

$$\delta_Y(y_0, f(X)) > 0$$

and hence there further exists $d_0 \in \mathcal{G}_Y$ such that

$$\inf_{x\in X} d_0(y_0, f(x)) > 0.$$

Now consider $[0, \infty[$ equipped with the Euclidean metric $d_{\mathbb{E}}$ and define the map

$$\theta : (Y, \mathcal{G}_Y) \longrightarrow ([0, \infty[, d_{\mathbb{E}}) : y \longrightarrow \inf_{x\in X} d_0(y, f(x)).$$

Since, for all $z, y \in Y$, we have

$$\begin{aligned} d_{\mathbb{E}}(\theta(z), \theta(y)) &= |\inf_{x\in X} d_0(z, f(x)) - \inf_{x\in X} d_0(y, f(x))| \\ &\leq \sup_{x\in X} |d_0(z, f(x)) - d_0(y, f(x))| \\ &\leq d_0(z, y) \end{aligned}$$

it follows that θ is a uniform contraction. Clearly $\theta \circ f = 0$ whereas $\theta(y_0) \neq 0$. The constant map $\eta = 0$ is also a uniform contraction and, since $\theta \neq \eta$ whereas $\theta \circ f = \eta \circ f$, it follows that f is not an epimorphism. □

6.2.7 Theorem *A function is an extremal monomorphism in* UG_2 *if and only if it is an injective uniform contraction which has a closed image.*

Proof This goes in exactly the same way as the proof for Top_2 and we therefore leave this to the reader. □

6.2.8 Theorem *Let* $(X, \mathcal{G})$ *be a uniform gauge space. Then* $(\widehat{X}, \widehat{\mathcal{G}})$ *is a complete Hausdorff uniform gauge space.*

Proof This follows from 6.2.4.

We define $\mathscr{E}_t$ to be the class of all embeddings in UG_2 which are dense for the topological coreflection. Note that in 6.2.5 we showed that i_X belongs to this class.

6.2.9 Theorem *$c\mathsf{UG}_2$ is an $\mathscr{E}_t$-reflective subcategory of UG_2. For any Hausdorff uniform gauge space $(X, \mathscr{G})$,*

$$i_X : (X, \mathscr{G}) \longrightarrow (\widehat{X}, \widehat{\mathscr{G}}) : x \longrightarrow \mathscr{V}(x)$$

is an epireflection of $(X, \mathscr{G})$ in $c\mathsf{UG}_2$.

Proof Let $(Y, \mathscr{H})$ be a Hausdorff complete uniform gauge space. If $f : (X, \mathscr{G}) \longrightarrow (Y, \mathscr{H})$ is a uniform contraction, then there is a unique uniform contraction

$$\widehat{f} : (\widehat{X}, \widehat{\mathscr{G}}) \longrightarrow (Y, \mathscr{H})$$

such that $\widehat{f} \circ i_X = f$.

From the classical theory of completions of uniform spaces it follows that given a uniform contraction $(X, \mathscr{G}) \longrightarrow (Y, \mathscr{H})$ there exists a unique uniformly continuous extension $\widehat{f}$ making the diagram

$$\begin{array}{ccc} (X, \mathscr{U}_0^{\mathscr{G}}) & \xrightarrow{f} & (Y, \mathscr{U}_0^{\mathscr{H}}) \\ {\scriptstyle i_X}\big\downarrow & \nearrow {\scriptstyle \widehat{f}} & \\ (\widehat{X}, \mathscr{U}_0^{\widehat{\mathscr{G}}}) & & \end{array}$$

commute. Here $(\mathscr{U}_\varepsilon^{\mathscr{G}})_\varepsilon$ (respectively $(\mathscr{U}_\varepsilon^{\mathscr{H}})_\varepsilon$ and $(\mathscr{U}_\varepsilon^{\widehat{\mathscr{G}}})_\varepsilon$) stand for the uniform towers associated with $\mathscr{G}$ (respectively $\mathscr{H}$ and $\widehat{\mathscr{G}}$).

Now let $\mathscr{M}, \mathscr{N} \in \widehat{X}$ and put

$$m := \widehat{f}(\mathscr{M}) = \lim f(\mathscr{M}) \text{ and } n := \widehat{f}(\mathscr{N}) = \lim f(\mathscr{N})$$

then it follows that for any $\varepsilon > 0$ and $d \in \mathscr{H}$ there exist $M_\varepsilon \in \mathscr{M}$ and $N_\varepsilon \in \mathscr{N}$ such that

$$f(M_\varepsilon) \subseteq B_d(m, \varepsilon) \text{ and } f(N_\varepsilon) \subseteq B_d(n, \varepsilon).$$

Consequently

$$\begin{aligned} d \circ (\widehat{f} \times \widehat{f})(\mathscr{M}, \mathscr{N}) &= d(m, n) \\ &= \sup_{\varepsilon > 0} d(B_d(m, \varepsilon), B_d(n, \varepsilon)) \\ &\leq \sup_{\varepsilon > 0} d(f(M_\varepsilon), f(N_\varepsilon)) \end{aligned}$$

$$\leq \sup_{M \in \mathcal{M}} \sup_{N \in \mathcal{N}} d(f(M), f(N))$$
$$= \widehat{d \circ (f \times f)}(\mathcal{M}, \mathcal{N})$$

which proves that $\widehat{f} : (\widehat{X}, \widehat{\mathcal{G}}) \longrightarrow (Y, \mathcal{H})$ is a uniform contraction. □

6.2.10 Corollary *The subcategory* cUG_2 *is closed under the formation of products and closed subspaces.*

6.2.11 Corollary *If* $(Y, \mathcal{G}_Y)$ *is a complete Hausdorff uniform gauge space,* $(X, \mathcal{G}_X)$ *is a Hausdorff uniform gauge space,* $Z \subseteq X$ *is dense with respect to the underlying* Top*-coreflection, and* $f : (Z, \mathcal{G}_{X|_Z}) \longrightarrow (Y, \mathcal{G}_Y)$ *is a uniform contraction, then there exists a unique extension*

$$\overline{f} : (X, \mathcal{G}_X) \longrightarrow (Y, \mathcal{G}_Y)$$

which is also a uniform contraction.

Proof This is analogous to 6.1.13 and we leave this to the reader. □

6.2.12 Theorem *If* $(X, \mathcal{G})$ *is a Hausdorff uniform gauge space,* $(X_i, \mathcal{G}_i)$ *is a complete Hausdorff uniform gauge space for* $i \in \{1, 2\}$*, and the maps*

$$e_i : (X, \mathcal{G}) \longrightarrow (X_i, \mathcal{G}_i)$$

belong to $\mathcal{E}_t$*, then* $(X_1, \mathcal{G}_1)$ *and* $(X_2, \mathcal{G}_2)$ *are isomorphic.*

Proof This is analogous to 6.1.14 and we leave this to the reader. □

6.2.13 Definition The uniform gauge space $(\widehat{X}, \widehat{\mathcal{G}})$ is called the *uniform completion* of $(X, \mathcal{G})$.

Let us now verify that the completion constructed in this section coincides with the usual uniform completion in the case of a uniform UG-space. Given a uniform space $(X, \mathcal{U})$ we denote by $(\widetilde{X}, \widetilde{\mathcal{U}})$ its usual completion.

6.2.14 Proposition (Unif) *If* $(X, \mathcal{U})$ *is a uniform space then* $(\widehat{X}, \widehat{\mathcal{G}_{\mathcal{U}}}) = (\widetilde{X}, \mathcal{G}_{\widetilde{\mathcal{U}}})$.

Proof That $\widehat{X} = \widetilde{X}$ follows at once from the definitions. Further, that $\widehat{\mathcal{G}_{\mathcal{U}}} = \mathcal{G}_{\widetilde{\mathcal{U}}}$ follows from the fact that a metric on $\widetilde{X}$ is uniformly continuous for $\widetilde{\mathcal{U}}$ if and only if it is the extension (as defined in 6.2.3) of a metric on X which is uniformly continuous for $\mathcal{U}$ and from the fact that any constant multiple of a uniformly continuous metric is uniformly continuous. □

6.3 Compactification

We recall that, given an approach space (X, δ), we denote by $\mathscr{K}^*(X)$ the set of all bounded contractions from (X, δ) to $(\mathbb{R}, \delta_{d_{\mathbb{E}}})$. Further, given any set of contractions $\mathscr{F} \subseteq \mathscr{K}^*(X)$, $\mathscr{D}_{\mathscr{F}}$ stands for the set of all metrics

$$d_{\mathscr{F}_0} : X \times X \longrightarrow \mathbb{P} : (x, y) \longrightarrow \sup_{f \in \mathscr{F}_0} |f(x) - f(y)|,$$

where $\mathscr{F}_0$ ranges over the finite subsets of $\mathscr{F}$.

Similarly to 3.1.5 we then obtain the following characterizations of the objects in UAp_2.

6.3.1 Proposition *For a uniform approach space X the following properties are equivalent.*

1. *X is Hausdorff.*
2. *$\mathscr{G}$ is point-separating.*
3. *Every generating set $\mathscr{F} \subseteq \mathscr{K}^*(X)$ is point-separating.*
4. *$\mathscr{K}^*(X)$ is point-separating.*
5. *X is isomorphic to a subspace of a power of $\mathbb{R}$ (equipped with $\delta_{d_{\mathbb{E}}}$).*

Proof 1 $\Rightarrow$ 2. Since $(X, \mathscr{T}_\delta)$ is Hausdorff, for any $x, y \in X$, $x \neq y$, we can find $d, d' \in \mathscr{G}$, $\varepsilon > 0$, and $\varepsilon' > 0$ such that

$$B_d(x, \varepsilon) \cap B_d(y, \varepsilon') = \emptyset.$$

Hence, if we put $d'' := d \vee d'$, then $d''(x, y) > 0$.

2 $\Rightarrow$ 3. If $\mathscr{F} \subseteq \mathscr{K}^*(X)$ generates δ, then it follows from 1.2.9 that $\mathscr{D}_{\mathscr{F}}$ is a basis for $\mathscr{G}$. Hence, given $x, y \in X, x \neq y$, there exists $\mathscr{F}_0 \in 2^{(\mathscr{F})}$ such that $d_{\mathscr{F}_0}(x, y) > 0$ from which it follows that there exists $f \in \mathscr{F}_0$ such that $f(x) \neq f(y)$.

3 $\Rightarrow$ 4. This is trivial.

4 $\Rightarrow$ 5. Consider the product space $\mathbb{R}^{\mathscr{K}^*(X)}$ where each copy of $\mathbb{R}$ is endowed with $\delta_{d_{\mathbb{E}}}$. Then it follows from the fact that $\mathscr{K}^*(X)$ is point-separating that the map

$$e : X \longrightarrow \mathbb{R}^{\mathscr{K}^*(X)} : x \longrightarrow (f(x))_{f \in \mathscr{K}^*(X)},$$

is injective and hence an embedding.

5 $\Rightarrow$ 1. This follows from the fact that $\mathbb{R}$, endowed with $\delta_{d_{\mathbb{E}}}$, is a Hausdorff uniform approach space and from 3.1.10. □

6.3.2 Proposition *A Hausdorff uniform approach space is compact if and only if it is isomorphic to a closed subspace of a product of compact subsets of $\mathbb{R}$ (each endowed with the restriction of $\delta_{d_{\mathbb{E}}}$).*

Proof That the second property implies the first one is a consequence of the fact that an isomorphism between approach spaces induces a homeomorphism between their topological coreflections. To see that the first property implies the second one, let X be a Hausdorff compact uniform approach space. Then, from 6.3.1 and the fact that, for each $f \in \mathscr{K}^*(X)$, $f(X) \subseteq \mathbb{R}$ is compact, it follows that

$$e : X \longrightarrow \prod_{f \in \mathscr{K}^*(X)} f(X) \subseteq \mathbb{R}^{\mathscr{K}^*(X)} : x \longrightarrow (f(x))_{f \in \mathscr{K}^*(X)}$$

is an embedding into a product of compact subsets of $\mathbb{R}$. Since X itself is compact, it follows that the embedding is also closed. □

In what follows we denote by $k\mathsf{UAp}_2$ the full and isomorphism-closed subcategory of UAp_2 consisting of all Hausdorff compact uniform approach spaces.

6.3.3 Definition If X is a Hausdorff uniform approach space then a pair (X', e), where X' is an object in $k\mathsf{UAp}_2$, i.e. a Hausdorff compact uniform approach space, and where $e : X \longrightarrow X'$ is a dense embedding is called a (UAp_2-)*compactification*. Usually we will refer to X' simply as being a compactification of X.

Each Hausdorff uniform approach space has a compactification. Consider a subset $\mathscr{F}$ of $\mathscr{K}^*(X)$ which generates δ. For each $f \in \mathscr{F}$, let I_f be a compact subset of $\mathbb{R}$ containing the image of f. Let

$$(X', \delta_{\mathscr{F}}) := (\prod_{f \in \mathscr{F}} I_f, \prod_{f \in \mathscr{F}} \delta_{d_{\mathbb{E}}}).$$

Then

$$e_{\mathscr{F}} : (X, \delta) \longrightarrow (X', \delta_{\mathscr{F}}) : x \longrightarrow (f(x))_{f \in \mathscr{F}}$$

is an embedding and, obviously, if we let $\beta_{\mathscr{F}}$ be the same map as $e_{\mathscr{F}}$ but with range $\mathrm{cl}_{\mathscr{T}_{\delta_{\mathscr{F}}}}(e_{\mathscr{F}}(X))$, then $((\mathrm{cl}_{\mathscr{T}_{\delta_{\mathscr{F}}}}(e_{\mathscr{F}}(X)), \delta_{\mathscr{F}}), \beta_{\mathscr{F}})$ is a compactification of (X, δ). We will denote this compactification by $(\beta_{\mathscr{F}} X, \delta_{\mathscr{F}})$ or by $\beta_{\mathscr{F}} X$ for short. The appropriateness of this notation will become clear in the sequel.

Notice that $\delta_{\mathscr{F}}$ is generated by the symmetric gauge basis

$$\mathscr{D}_{\mathscr{F}} := \left\{ \sup_{f \in \mathscr{H}} d_f \mid \mathscr{H} \in 2^{(\mathscr{F})} \right\},$$

where for each $f \in \mathscr{F}$

$$d_f : \beta_{\mathscr{F}} X \times \beta_{\mathscr{F}} X \longrightarrow \mathbb{P} : (p, q) \longrightarrow \left|\mathrm{pr}_f(p) - \mathrm{pr}_f(q)\right|.$$

Notice also that if (X, δ) is compact, then $\beta_{\mathscr{F}}$ is an isomorphism.

In order not to overload the notation, when we construct a compactification as above, starting from the entire set of contractions $\mathcal{K}^*(X)$, we will drop all subscripts and simply put $\beta^* X := \beta_{\mathcal{K}^*(X)} X$, $\beta^*\delta := \delta_{\mathcal{K}^*(X)}$, and $\beta^* := \beta_{\mathcal{K}^*(X)}$.

6.3.4 Theorem *For any Hausdorff uniform approach space (X, δ) the compactification $\beta^* X$ has the following equivalent properties.*

1. *Every $f \in \mathcal{K}^*(X)$ has a unique extension $f^* \in \mathcal{K}^*(\beta^* X)$, i.e. there exists a unique bounded contraction f^* such that the diagram*

$$\begin{array}{ccc} (X, \delta) & \xrightarrow{\ f\ } & (\mathbb{R}, \delta_{\mathbb{E}}) \\ {\scriptstyle \beta^*}\downarrow & \nearrow {\scriptstyle f^*} & \\ (\beta^* X, \beta^* \delta) & & \end{array}$$

commutes.

2. *For any Hausdorff compact uniform approach space (X', δ') and any contraction $f : (X, \delta) \longrightarrow (X', \delta')$, there exists a unique extension to $\beta^* X$, i.e. there exists a unique contraction $f^* : (\beta^* X, \beta^* \delta) \longrightarrow (X', \delta')$ such that the diagram*

$$\begin{array}{ccc} (X, \delta) & \xrightarrow{\ f\ } & (X', \delta') \\ {\scriptstyle \beta^*}\downarrow & \nearrow {\scriptstyle f^*} & \\ (\beta^* X, \beta^* \delta) & & \end{array}$$

commutes.

This compactification is essentially unique in the sense that any other compactification which is also a reflection is isomorphic to $\beta^* X$.

Proof The proof goes in the same way as the analogous proof for the Čech-Stone compactification in Top. For example, if $f \in \mathcal{K}^*(X)$, then $f^* \in \mathcal{K}^*(\beta^*(X))$ is defined as the restriction of the projection pr_f to $\beta^* X$. If h is another such extension, then it suffices to note that

$$\{x \in \beta^* X \mid h(x) = f^*(x)\}$$

is a closed subset of $\beta^* X$ containing $e_{\mathcal{K}^*(X)}(X)$. □

The following result is an immediate consequence of 6.3.4.

6.3.5 Corollary *For any Hausdorff uniform approach space (X, δ), the map*

$$\beta^* : (X, \delta) \longrightarrow (\beta^* X, \beta^* \delta)$$

is an epireflection of (X, δ) in $k\mathsf{UAp}_2$.

Notice that, in the same way as the Čech-Stone compactification, $\beta^* X$ is the largest compactification of (X, δ). For any other compactification $((X', \delta'), e)$ there exists a contraction

$$c : (\beta^* X, \beta^* \delta) \longrightarrow (X', \delta')$$

such that $e = c \circ \beta^*$.

6.3.6 Proposition (Top) *If (X, δ) is a Tychonoff topological approach space, then $\beta^* X$ is the Čech-Stone compactification.*

Proof By our previous results, it is sufficient to show that $\beta^* X$ is topological. Let $x \in \beta^* X$ and $A \subseteq \beta^* X$. Then

$$\beta^* \delta(x, A) = \sup_{\mathscr{H} \in 2^{(\mathscr{K}^*(X))}} \inf_{a \in A} \sup_{f \in \mathscr{H}} \left| \mathrm{pr}_f(x) - \mathrm{pr}_f(a) \right| .$$

Since X is topological we know that $\mathscr{K}^*(X) = \mathscr{C}^*(X)$, where $\mathscr{C}^*(X)$ is the set of all bounded continuous real-valued functions on X. Consequently, if $f \in \mathscr{K}^*(X)$ and $\alpha \in \mathbb{R}$, then $\alpha f \in \mathscr{K}^*(X)$. Moreover, since both $\mathrm{pr}_{\alpha f}$ and $\alpha\, \mathrm{pr}_f$ are contractions which extend αf to $\beta^* X$, we also have that $\alpha\, \mathrm{pr}_f = \mathrm{pr}_{\alpha f}$. Now, if $\beta^* \delta(x, A) > 0$, then there exists $\mathscr{H} \in 2^{(\mathscr{K}^*(X))}$ such that

$$\inf_{a \in A} \sup_{f \in \mathscr{H}} \left| \mathrm{pr}_f(x) - \mathrm{pr}_f(a) \right| > 0,$$

and then it follows that

$$\begin{aligned} \beta^* \delta(x, A) &= \sup_{\mathscr{H} \in 2^{(\mathscr{K}^*(X))}} \inf_{a \in A} \sup_{f \in \mathscr{H}} \left| \mathrm{pr}_f(x) - \mathrm{pr}_f(a) \right| \\ &\geq \sup_{\alpha \geq 1} \inf_{a \in A} \sup_{f \in \mathscr{H}} \left| \mathrm{pr}_{\alpha f}(x) - \mathrm{pr}_{\alpha f}(a) \right| \\ &= \sup_{\alpha \geq 1} \inf_{a \in A} \sup_{f \in \mathscr{H}} \alpha \left| \mathrm{pr}_f(x) - \mathrm{pr}_f(a) \right| \\ &= \infty. \end{aligned}$$

Thus $\beta^* \delta$ attains only the values 0 and ∞ which proves that $\beta^* X$ is topological. □

It is clear that the topological coreflection of $\beta^* X$ is a compactification (in the topological sense) of the topological coreflection of X. In general it is smaller than the Čech-Stone compactification. We will now proceed to give a more concrete characterization of $\beta^* X$ which will clarify the situation. In this description we will make use of proximity spaces and related concepts. We refer the reader to Naimpally and Warrack (1970) for an account of the basic concepts of proximity spaces which we require.

The following result gives a basic relationship which we will need.

6.3.7 Proposition *If (X, δ) is a uniform approach space, generated by the symmetric gauge basis $\mathscr{D}$, then the relation $\Delta_{\mathscr{D}}$ on $2^X \times 2^X$, defined by*

$$A \Delta_{\mathscr{D}} B \Leftrightarrow \sup_{d \in \mathscr{D}} \inf_{a \in A} \inf_{b \in B} d(a, b) = 0,$$

is a proximity relation which is compatible with the topological coreflection of (X, δ), i.e. for any $x \in X$ and $A \subseteq X$, we have

$$\{x\} \Delta_{\mathscr{D}} A \Leftrightarrow \delta(x, A) = 0.$$

Proof This is straightforward and we leave this to the reader. □

6.3.8 Example The same example as the third one in 3.1.20 proves that different symmetric gauge bases can generate the same distance but different proximities. Again consider the real line $\mathbb{R}$ and the two collections of metrics $\mathscr{D} := \{d_a \mid a > 0\}$ and $\mathscr{D}' := \{d_{\mathbb{E}}\}$, where for each real number $a > 0$, $f_a := (1_{\mathbb{R}} \wedge a) \vee (-a)$, and

$$d_a(x, y) := |f_a(x) - f_a(y)|$$

and where $d_{\mathbb{E}}$ stands for the Euclidean metric. Then $\mathscr{D}$ and $\mathscr{D}'$ are bases for the same symmetric gauge, but $\Delta_{\mathscr{D}} \neq \Delta_{\mathscr{D}'}$. For instance for the sets $2\mathbb{N} + 1$ and $2\mathbb{N}$ of odd and even integers we have

$$\sup_{\alpha > 0} \inf_{a \in 2\mathbb{N}+1} \inf_{b \in 2\mathbb{N}} d_\alpha(a, b) = 0$$

whereas obviously A is not $\Delta_{\mathscr{D}'}$-proximal to B.

We recall that, given a proximity space (X, Δ), a set p of subsets of X is called a *cluster* if the following properties are fulfilled:

1. $\forall A, B \in p : A \Delta B$.
2. if $A \Delta B$ for every $B \in p$ then $A \in p$.
3. if $A \cup B \in p$ then $A \in p$ or $B \in p$.

Given a Hausdorff uniform approach space (X, δ) and a subset $\mathscr{F} \subseteq \mathscr{K}^*(X)$ which generates (X, δ), we will denote the set of all clusters in the proximity space $(X, \Delta_{\mathscr{D}_{\mathscr{F}}})$ by $\kappa_{\mathscr{F}} X$. Further we let

$$c_{\mathscr{F}} : X \longrightarrow \kappa_{\mathscr{F}} X : x \mapsto \big\{A \subseteq X \mid \delta(x, A) = 0\big\}$$

be the canonical injection (see e.g. Naimpally and Warrack 1970; Naimpally 2009). We identify the points of X with their images under the map $c_{\mathscr{F}}$. The *Smirnov compactification* of $(X, \Delta_{\mathscr{D}_{\mathscr{F}}})$ is the set $\kappa_{\mathscr{F}} X$ equipped with the proximity relation for which two subsets $\mathscr{A}$ and $\mathscr{B}$ of $\kappa_{\mathscr{F}} X$ are proximal if any subsets A and B of X

which absorb $\mathscr{A}$ and $\mathscr{B}$ respectively are proximal in X. We recall that A is said to *absorb* $\mathscr{A}$ if, for all $q \in \mathscr{A}$, we have that $A \in q$, i.e. if $A \in \bigcap_{q \in \mathscr{A}} q$. Again we refer to Naimpally and Warrack (1970) for more details.

For any nonempty subset $A \subseteq X$, we now define

$$\widehat{\delta}_A : \kappa_{\mathscr{F}} X \longrightarrow \mathbb{P} : p \mapsto \inf \left\{ \varepsilon \geq 0 \mid \forall d \in \mathscr{D}_{\mathscr{F}} : A_d^{(\varepsilon)} \in p \right\},$$

where, for any $d \in \mathscr{D}_{\mathscr{F}}$ and $\varepsilon \geq 0$,

$$A_d^{(\varepsilon)} := \left\{ x \in X \mid \delta_d(x, A) \leq \varepsilon \right\}.$$

Then we define

$$\delta^*_{\mathscr{F}} : \kappa_{\mathscr{F}} X \times 2^{\kappa_{\mathscr{F}} X} \longrightarrow \mathbb{P}$$

as follows. For any $p \in \kappa_{\mathscr{F}} X$ and $\mathscr{A} \subseteq \kappa_{\mathscr{F}} X$,

$$\delta^*_{\mathscr{F}}(p, \mathscr{A}) := \begin{cases} \infty & \mathscr{A} = \emptyset, \\ \sup \left\{ \widehat{\delta}_A(p) \mid A \text{ absorbs } \mathscr{A} \right\} & \mathscr{A} \neq \emptyset. \end{cases}$$

Further, if $A \subseteq X$ and $\varepsilon \geq 0$, and if we consider A as a subset of $\kappa_{\mathscr{F}} X$ then we will put

$$A^{(\varepsilon)^*} := \left\{ p \in \kappa_{\mathscr{F}} X \mid \delta^*_{\mathscr{F}}(p, A) \leq \varepsilon \right\}.$$

6.3.9 Lemma *Let X be a uniform approach space, generated by the symmetric gauge basis $\mathscr{D}$. Then, for any $A \subseteq X$ and any cluster $p \in \kappa_{\mathscr{F}} X$, the following properties are equivalent.*

1. *$A \in p$.*
2. *$\forall d \in \mathscr{D}, \forall \varepsilon > 0 : A_d^{(\varepsilon)} \in p$.*

Proof We only need to prove that the second condition implies the first one. Suppose that $A \notin p$. Then there exists $C \in p$ such that A is not $\Delta_{\mathcal{D}}$-proximal to C. Hence, there also exists $d \in \mathscr{D}$ and $\varepsilon > 0$ such that $A_d^{(\varepsilon)} \cap C_d^{(\varepsilon)} = \emptyset$, which implies that $A_d^{(\varepsilon)} \notin p$. □

6.3.10 Theorem *If (X, δ) is a Hausdorff uniform approach space, and $\mathscr{F} \subseteq \mathscr{K}^*(X)$ generates δ, then the following properties hold.*

1. *$(\kappa_{\mathscr{F}} X, \delta^*_{\mathscr{F}})$ is an approach space.*
2. *$c_{\mathscr{F}} : (X, \delta) \longrightarrow (\kappa_{\mathscr{F}} X, \delta^*_{\mathscr{F}})$ is a dense embedding.*
3. *The topological coreflection of $(\kappa_{\mathscr{F}} X, \delta^*_{\mathscr{F}})$ is (homeomorphic to) the topological space underlying the Smirnov compactification of $(X, \Delta_{\mathscr{D}_{\mathscr{F}}})$.*

Proof 1. First of all note that, for all $A, B \subseteq X$, we have that $A \in \bigcap \{q \mid q \in B\}$ if and only if $B \subseteq \overline{A}$. From this we conclude that

$$\begin{aligned}\delta^*_{\mathcal{F}}(p, B) &= \sup \left\{ \widehat{\delta}_A(p) | A \in \bigcap_{q \in B} q \right\} \\ &= \widehat{\delta}_B(p).\end{aligned}$$

It is clear that $\delta^*_{\mathcal{F}}$ satisfies (D1) and (D2). Moreover, if $p \in \kappa_{\mathcal{F}} X$ and $\mathcal{A} \subseteq \mathcal{A}' \subseteq \kappa_{\mathcal{F}} X$, then it is clear from the definition that $\delta^*_{\mathcal{F}}(p, \mathcal{A}') \leq \delta^*_{\mathcal{F}}(p, \mathcal{A})$. Consequently, in order to prove (D3), only one inequality has to be shown. Now, if $A, B \subseteq X$, then (omitting the explicit notation of $c_{\mathcal{F}}$)

$$\delta^*_{\mathcal{F}}(p, A \cup B) = \delta^*_{\mathcal{F}}(p, A) \wedge \delta^*_{\mathcal{F}}(p, B),$$

and the general case follows from this since, if A absorbs $\mathcal{A}$ and B absorbs $\mathcal{B}$, then $A \cup B$ absorbs $\mathcal{A} \cup \mathcal{B}$. To prove that (D4) is satisfied, we first make some preliminary observations.

Claim 1: For any $\alpha, \beta \in \mathbb{R}^+$, $d \in \mathcal{D}_{\mathcal{F}}$, and $A \subseteq X$, we have

$$(A_d^{(\alpha)})_d^{(\beta)} \subseteq A_d^{(\alpha+\beta)}.$$

Indeed, this is nothing more than (T4) for δ_d.

Claim 2: For any $A \subseteq X$ and $\varepsilon > 0$, we have

$$A^{(\varepsilon)^*} = \bigcap_{\varepsilon' > \varepsilon} \left\{ q \mid \forall d \in \mathcal{D}_{\mathcal{F}} : A_d^{(\varepsilon')} \in q \right\}.$$

Indeed, this follows from the definitions of $\delta^*_{\mathcal{F}}$ and $\widehat{\delta}_A$.

Claim 3: If A absorbs $\mathcal{A}$, then for any $\varepsilon > 0$ we have $\mathcal{A}^{(\varepsilon)} \subseteq A^{(\varepsilon)^*}$.

Indeed, if $p \in \mathcal{A}^{(\varepsilon)}$ then it follows from the second claim that, for all $\varepsilon' > \varepsilon$, for all $d \in \mathcal{D}_{\mathcal{F}}$, and for any B which absorbs $\mathcal{A}$, we have $B_d^{(\varepsilon')} \in p$. Since A absorbs $\mathcal{A}$ it follows again from the second claim that $p \in A^{(\varepsilon)^*}$.

Now, consider first the case where $p \in \kappa_{\mathcal{F}} X$ and where $A \subseteq X$. From the first and second claims we can then deduce that, for any $\varepsilon \geq 0$,

$$\widehat{\delta}_A(p) \leq \delta^*_{\mathcal{F}}(p, A^{(\varepsilon)^*}) + \varepsilon.$$

For the general case, let $\mathcal{A} \subseteq \kappa_{\mathcal{F}} X$. Then it follows from the foregoing inequality that

$$\delta^*_{\mathcal{F}}(p, \mathcal{A}) \leq \sup \left\{ \delta^*_{\mathcal{F}}(p, A^{(\varepsilon)^*}) \mid A \text{ absorbs } \mathcal{A} \right\} + \varepsilon$$

and the result follows from the third claim.

This proves that $(\kappa_{\mathcal{F}} X, \delta^*_{\mathcal{F}})$ is indeed an approach space.

2. That $c_{\mathcal{F}}$ is an embedding follows from the fact that, if $x \in X$ and $A \subseteq X$, then

$$\begin{aligned}\delta^*_{\mathcal{F}}(x, A) &= \inf\left\{\varepsilon \geq 0 \mid \forall d \in \mathcal{D}_{\mathcal{F}} : x \in A_d^{(\varepsilon)}\right\}\\ &= \delta(x, A).\end{aligned}$$

That the embedding is dense follows from the fact that X belongs to any cluster $p \in \kappa_{\mathcal{F}} X$.

3. Given $p \in \kappa_{\mathcal{F}} X$ and $\mathcal{A} \subseteq \kappa_{\mathcal{F}} X$, p is in the closure of $\mathcal{A}$ for the topology of the Smirnov compactification if $A \in p$, for any A which absorbs $\mathcal{A}$. By the definition of $\delta^*_{\mathcal{F}}$, p is in the closure of $\mathcal{A}$ for the topological coreflection of $(\kappa_{\mathcal{F}} X, \delta^*_{\mathcal{F}})$ if $A_d^{(\varepsilon)} \in p$, for any $\varepsilon > 0$, any $d \in \mathcal{D}_{\mathcal{F}}$, and any A which absorbs $\mathcal{A}$. It follows from 6.3.9 that these two conditions are equivalent. □

Before proceeding with the alternative description of our compactification we need another lemma.

6.3.11 Lemma *If (X, δ) is a uniform approach space and Y is a dense subspace of X, then, for any $x \in X$ and $A \subseteq X$*

$$\delta(x, A) = \left\{\sup \delta(x, B) | B \subseteq Y, A \subseteq \overline{B}\right\}.$$

Proof Let $\mathcal{D}$ be a symmetric gauge for (X, δ). If $\delta(x, A) > \varepsilon > 0$, then there exists $d_0 \in \mathcal{D}$ and $\theta > 0$ such that $\delta_{d_0}(x, A) > \varepsilon + \theta$. If we let

$$B' := \left\{y \in X \mid \delta_{d_0}(y, A) < \theta\right\},$$

and $B := B' \cap Y$, then it follows from the fact that B' is open and that Y is dense that $A \subseteq \overline{B}$. Since further

$$\begin{aligned}\delta(x, B) + \theta &\geq \delta_{d_0}(x, B) + \theta\\ &\geq \delta_{d_0}(x, B) + \sup_{b \in B} \delta_{d_0}(b, A)\\ &\geq \delta_{d_0}(x, A)\\ &\geq \varepsilon + \theta,\end{aligned}$$

this proves one inequality. As the other inequality is trivial, we are finished. □

We now define the map

$$\Psi : \kappa_{\mathcal{F}} X \longrightarrow \beta_{\mathcal{F}} X,$$

which assigns to each cluster $p \in \kappa_{\mathcal{F}} X$ the unique adherence point of the unique cluster in $\beta_{\mathcal{F}} X$ which contains p.

6.3.12 Theorem *If (X, δ) is a Hausdorff uniform approach space which is generated by the set of contractions $\mathscr{F} \subseteq \mathscr{K}^*(X)$, then*

$$\Psi : (\kappa_{\mathscr{F}} X, \delta^*_{\mathscr{F}}) \longrightarrow (\beta_{\mathscr{F}} X, \delta_{\mathscr{F}})$$

is an isomorphism.

Proof Clearly the map Ψ is well defined and bijective and the diagram

$$\begin{array}{ccc} (X, \delta) & & \\ {\scriptstyle c_{\mathscr{F}}}\big\downarrow & \overset{e_{\mathscr{F}}}{\searrow} & \\ (\kappa_{\mathscr{F}} X, \delta^*_{\mathscr{F}}) & \xrightarrow[\Psi]{} & (\beta_{\mathscr{F}} X, \delta_{\mathscr{F}}) \end{array}$$

is commutative. Now first suppose that $p \in \kappa_{\mathscr{F}} X$ and that $A \subseteq X$. Then it follows that

$$\begin{aligned} \delta_{\mathscr{F}}(\Psi(p), A) &= \sup_{\mathscr{H} \in 2^{(\mathscr{F})}} \inf_{a \in A} \sup_{f \in \mathscr{H}} \left|\mathrm{pr}_f(\Psi(p)) - \mathrm{pr}_f(a)\right| \\ &= \sup_{d^* \in \mathscr{D}^*_{\mathscr{F}}} \inf_{a \in A} d^*(\Psi(p), a) \\ &= \sup_{d^* \in \mathscr{D}^*_{\mathscr{F}}} \delta_{d^*}(\Psi(p), A) \\ &= \inf \left\{ \varepsilon \geq 0 \mid \forall d^* \in \mathscr{D}^*_{\mathscr{F}} : \Psi(p) \in A^{(\varepsilon)}_{d^*} \right\} \\ &= \inf \left\{ \varepsilon \geq 0 \mid \forall d \in \mathscr{D}_{\mathscr{F}} : A^{(\varepsilon)}_{d} \in p \right\} \\ &= \delta^*_{\mathscr{F}}(p, A). \end{aligned}$$

Notice that the second-last equality follows from the fact that $\Psi(p) \in A^{(\varepsilon)}_{d^*}$ implies that, for all $\varepsilon' > \varepsilon$, we have $A^{(\varepsilon')}_{d} \in p$ and similarly that $A^{(\varepsilon)}_{d} \in p$ implies that, for all $\varepsilon' > \varepsilon$, we have $\Psi(p) \in A^{(\varepsilon')}_{d^*}$. It follows, in particular, that $p \in \overline{A}$ if and only if $\Psi(p) \in \overline{A}$. By definition of the Smirnov compactification it further follows that if $A \subseteq X$ and $\mathscr{A} \subseteq \kappa_{\mathscr{F}} X$, then A absorbs $\mathscr{A}$ if and only if $\Psi(A) \subseteq \overline{A}$. So if $p \in \kappa_{\mathscr{F}} X$ and $\mathscr{A} \subseteq \kappa_{\mathscr{F}} X$ then, by definition of $\delta^*_{\mathscr{F}}$ and 6.3.11, we have

$$\begin{aligned} \delta^*_{\mathscr{F}}(p, \mathscr{A}) &= \sup \left\{ \delta^*_{\mathscr{F}}(p, A) \mid A \in \bigcap_{q \in \mathscr{A}} q \right\} \\ &= \sup \left\{ \delta_{\mathscr{F}}(\Psi(p), A) \mid \Psi(A) \subseteq \overline{A} \right\} \\ &= \delta_{\mathscr{F}}(\Psi(p), \Psi(A)). \end{aligned}$$

This proves that Ψ is an isomorphism, and we are finished. □

The Čech-Stone compactification of a topological space can be described in at least two different ways: as a closed subspace of a power of the unit interval [0, 1] and as the underlying topological space of the completion of the initial uniform space for all continuous functions into [0, 1] (see e.g. Gillman and Jerison 1976). The alternative description mentioned above can be generalized in the context of UG. Suppose [0, 1] is equipped with the usual, i.e. Euclidean, uniform gauge.

6.3.13 Theorem *Let (X, δ) be a Hausdorff uniform approach space. If $\mathcal{G}^*(X)$ is the initial uniform gauge structure for the source*

$$(X \xrightarrow{f} [0, 1])_{f \in \mathcal{K}^*(X)}$$

and $\widehat{\delta}$ is the underlying distance of $\widehat{\mathcal{G}^*(X)}$, then $(\widehat{X}, \widehat{\delta}) = (\beta^* X, \beta^* \delta)$.

Proof By definition, the source

$$((X, \mathcal{G}^*(X)) \xrightarrow{f} [0, 1])_{f \in \mathcal{K}^*(X)}$$

is initial. Hence we obtain that $(\widehat{X}, \widehat{\mathcal{G}^*(X)}) = (\overline{e(X)}, \mathcal{H})$ where $\mathcal{H}$ is the restriction of the product approach uniformity on $[0, 1]^{\mathcal{K}^*(X)}$ and the result follows. □

6.3.14 Corollary *Let (X, δ) be a uniform approach space. Then (X, δ) is compact if and only if $(X, \mathcal{G}^*(X))$ is complete.*

6.4 Comments

1. **Completion in metrically generated theories**

The constructions of completion in UAp and in UG as explained in this chapter are special cases of a more general approach in the setting of symmetric metrically generated theories (see Colebunders and Lowen 2005; Colebunders et al. 2007).

2. **Completion in non-symmetric metrically generated theories**

UAp and UG are symmetric theories in the sense that the gauges involved have bases consisting of metrics. For non-symmetric theories a completion theory, which generalizes the one applicable to UAp and UG, was developed in Colebunders and Vandersmissen (2010).

3. **Completion for larger subcategories of** App

In Lowen et al. (1999, 2003) the theory of completions as developed here for Hausdorff uniform approach spaces is extended to the realm of T_0 approach spaces. This shows some surprising relationship to completion in the setting of nearness spaces (see Herrlich 1974b) and also shows that the completion of a quasi-metric space need no longer be quasi-metric.

In a different, and more canonical way a theory of bicompletions was developed by Brümmer and Sioen (2006). This completion is proved to be firm with respect to a certain class of dense embeddings.

4. **General theory of completion of objects**
In the terminology introduced in Brümmer and Giuli (1992) and Brümmer et al. (1992) the completion in UAp_2 is firm with respect to the class $\mathscr{E}_m$ and the completion in UG_2 is firm with respect to the class $\mathscr{E}_t$.

5. **Wallman-Shanin compactification**
We have restricted ourselves to the categorically well-behaved Čech-Stone compactification for uniform approach spaces but there is also a Wallman-Shanin-type compactification developed by Lowen and Sioen (2000b), the categorical behaviour of which was studied by Sioen (2000), for so-called *weakly symmetric* T_1-approach spaces (an approach space is called weakly symmetric if $\delta(x, \{y\}) = \delta(y, \{x\})$ for all $x, y \in X$).

Chapter 7
Approach Theory Meets Topology

> *Always topologize.*
>
> (Marshal Stone)
>
> *Topology is the property of something that doesn't change when you bend it or stretch it as long as you don't break anything.*
>
> (Edward Witten)

This chapter and the following ones up to and including Chap. 11 will be devoted to showing in which other areas of mathematics approach spaces arise naturally and how in these areas the use of index analysis leads to general numerical results of which many classical ones are simple consequences.

In this chapter we highlight two areas in topology, namely function spaces and the Čech-Stone compactification, especially $\beta\mathbb{N}$.

Clearly any non-trivial application of approach theory has to involve numerical data. Hence in the first section where we treat function spaces, we do not start merely with topological and uniform spaces, but rather with approach spaces and uniform gauge spaces. The results therefore are applicable to any situation where for instance a choice has been made of a particular metric for a topology or a uniformity. Numerous indexed theorems are proved, e.g. 7.1.18, 7.1.11, 7.1.20, 7.1.22, in particular an indexed version of an Ascoli and a Dini theorem 7.1.25 and 7.1.27. We end this section with a result concerning completion of a function space where we start with a metric space Y and consider the approach structure of pointwise convergence on Y^X.

In the second section we study the Čech-Stone compactification of an Atsuji space, and in particular $\mathbb{N}$, in more detail. We precisely describe the distance $\delta_{\beta\mathbb{N}}$ and show some basic properties (see 7.2.3 and following).

7.1 Function Spaces

In this section $\mathcal{G}_X$ and $\mathcal{G}_Y$ will always stand for either a gauge or a uniform gauge on X and on Y.

R. Lowen, *Index Analysis*, Springer Monographs in Mathematics,
DOI 10.1007/978-1-4471-6485-2_7

7.1.1 Definition (1) If X is an approach space and Y a uniform gauge space, then $\mathcal{H} \subseteq Y^X$ is called *equicontractive at* $a \in X$ if for all $d \in \mathcal{G}_Y$ the function

$$X \longrightarrow \mathbb{P} : x \mapsto \sup_{f \in H} d(f(a), f(x))$$

is an element of $\widehat{\mathcal{G}_X(a)}$ where we recall that $\mathcal{G}_X(a) = \{d(x, \cdot) \mid d \in \mathcal{G}_X\}$. It is clear that this condition is equivalent with

$$\forall d \in \mathcal{G}_Y, \exists \varphi \in \widehat{\mathcal{G}_X(a)}, \forall f \in \mathcal{H} : d \circ (f \times f)(a, .) \leq \varphi,$$

and it is also clear that it is sufficient that the condition is satisfied for all d in a basis for $\mathcal{G}_Y$. $\mathcal{H}$ is called *equicontractive* if it is equicontractive at all $a \in X$. In line with the definition of the index of contractivity this immediately allows us to associate a natural *index of equicontractivity of* $\mathcal{H}$ as

$$\boxed{\chi_{ec}(\mathcal{H}) := \min\{c \mid \forall a \in X, \forall d \in \mathcal{G}_Y, \exists \varphi \in \widehat{\mathcal{G}_X(a)}, \forall f \in \mathcal{H} : d \circ (f \times f)(a, .) \leq \varphi + c\}}$$

For a given collection $\mathcal{H}$ this index is hence the smallest value such that

$$\forall a \in X, \forall d \in \mathcal{G}_Y, \exists \varphi \in \widehat{\mathcal{G}_X(a)}, \forall f \in \mathcal{H} : d \circ (f \times f)(a, \cdot) \leq \varphi + \chi_{ec}(\mathcal{H}).$$

(2) If X and Y are uniform gauge spaces, then $\mathcal{H} \subseteq Y^X$ is called *uniformly equicontractive* if for all $d \in \mathcal{G}_Y$ the function

$$X \times X \longrightarrow \mathbb{P} : (x, y) \mapsto \sup_{f \in \mathcal{H}} d(f(x), f(y))$$

is in $\mathcal{G}_X$. It is clear that this condition is equivalent with

$$\forall d \in \mathcal{G}_Y, \exists e \in \mathcal{G}_X, \forall f \in \mathcal{H} : d \circ (f \times f) \leq e,$$

and that it is sufficient that the condition is satisfied for all d in a basis for $\mathcal{G}_Y$.

This definition too allows for a natural *index of uniform equicontractivity of* $\mathcal{H}$ defined by

$$\boxed{\chi_{uec}(\mathcal{H}) := \min\{c \mid \forall d \in \mathcal{G}_Y, \exists e \in \mathcal{G}_X, \forall f \in \mathcal{H} : d \circ (f \times f) \leq e + c\}}$$

Again, note that for any collection $\mathcal{H} \subseteq Y^X$ this index is the smallest value such that

$$\forall d \in \mathcal{G}_Y, \exists e \in \mathcal{G}_X, \forall f \in \mathcal{H} : d \circ (f \times f) \leq e + \chi_{uec}(\mathcal{H}).$$

7.1.2 Proposition (Top, Unif) *In topological spaces the notion of equicontractivity coincides with equicontinuity and in uniform spaces the notion of uniform equicontractivity coincides with uniform equicontinuity.*

Proof This follows from the definitions. □

7.1.3 Proposition (Met) *In metric spaces any collection of contractions (= uniform contractions) is equicontractive (= uniformly equicontractive).*

Proof This follows from the definitions. □

7.1.4 Proposition *If X is an approach space (respectively a uniform gauge space), Y is a uniform gauge space and $\mathcal{H} \subseteq Y^X$, then*

$$\sup_{f\in\mathcal{H}} \chi_c(f) \leq \chi_{ec}(\mathcal{H}) \textit{ (respectively } \sup_{f\in\mathcal{H}} \chi_{uc}(f) \leq \chi_{uec}(\mathcal{H}).$$

Proof This follows from the definitions. □

It already follows from the classical case that the inequalities in the foregoing result, in general, can be strict.

In this section we study some basic approach structures and uniform gauge structures on function spaces. The setting will always be that of a set or approach space X and a uniform gauge space Y and we consider the function space Y^X or subspaces thereof. In particular we denote the set of all contractions between approach spaces X and Y by $\mathcal{K}(X, Y)$ and if X and Y are moreover uniform gauge spaces then we denote the set of all uniform contractions by $\mathcal{K}^u(X, Y)$.

First let X be a set. For any $A \subseteq X$ and $d \in \mathcal{G}_Y$ we define the metric

$$\gamma_{d,A} : Y^X \times Y^X \longrightarrow \mathbb{P} : (f, g) \mapsto \sup_{x\in A} d(f(x), g(x)).$$

Further we suppose given a collection Σ of subsets of X which satisfies two conditions: (1) Σ is closed under the formation of finite unions and (2) Σ is a cover of X. We will refer to such a collection Σ as a *tiling (of X)*. If $\mathcal{H}$ is a basis for $\mathcal{G}_Y$ then

$$\{\gamma_{d,A} \mid d \in \mathcal{H}, A \in \Sigma\}$$

in turn is a basis for a uniform gauge $\mathcal{D}_\Sigma$ on Y^X. It is easily seen that this uniform gauge is independent of the choice of basis for $\mathcal{G}_Y$.

The subscript Σ will be replaced by u if Σ is generated by $\{X\}$ i.e. $\Sigma = 2^X$, by p if Σ is generated by $\{\{x\} \mid x \in X\}$ i.e. $\Sigma = 2^{(X)}$ and by c if Σ consists of all 0-compact subsets of X (in which case X is supposed to be an approach space). The same subscripts will of course be used to denote any of the associated structures. The u of course refers to the structure of uniform convergence, the p to the structure of pointwise convergence and the c to uniform convergence on compact subsets.

Sometimes a collection Σ also fulfils the condition that for any $A \in \Sigma$ and $B \subseteq A$ also $B \in \Sigma$. This happens for instance with two of the examples above. However, this need not be the case, for instance, if Σ consists of all sets which are 0-compact, then this collection does fulfil (1) and (2), but not the subset condition. If necessary it is no problem to add smaller sets to Σ since $B \subseteq A$ implies $\gamma_{d,B} \leq \gamma_{d,A}$ and so these metrics will appear in the uniform gauge $\mathscr{D}_\Sigma$ anyway.

Being closed or taking a closure will always refer to the topology underlying the uniform coreflection of a uniform gauge structure or the topological coreflection of an approach structure.

7.1.5 Proposition *If X is a set, Y a uniform gauge space and Σ a tiling of X, then the uniform coreflection of $(Y^X, \mathscr{D}_\Sigma)$ is Y^X equipped with the uniformity of uniform convergence on Σ-sets and the metric coreflection is Y^X equipped with the uniform metric*

$$d_{\mathscr{D}_\Sigma}(f, g) := \sup_{x \in X} \sup_{d \in \mathscr{G}_Y} d(f(x), g(x))$$

i.e. with the metric of uniform convergence with respect to the metric coreflection of Y.

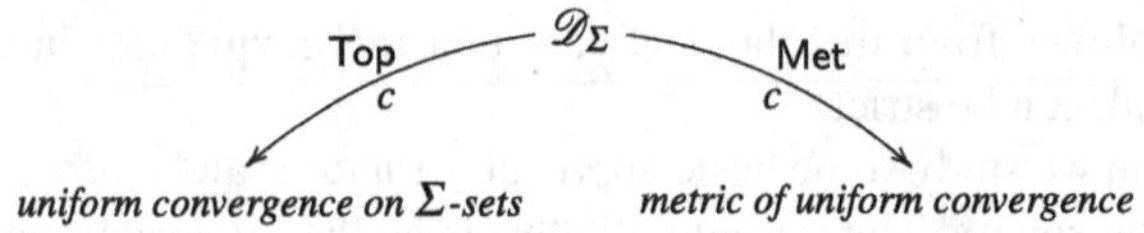

Proof This follows from the definitions. □

7.1.6 Proposition *If X is a set, Y a uniform gauge space and Σ a tiling of X, then the following properties hold.*

1. *The map $c : Y \longrightarrow Y^X : y \mapsto (c_y : X \longrightarrow Y : x \mapsto y)$ is a uniform embedding of Y into $(Y^X, \mathscr{D}_\Sigma)$.*
2. *For any $x \in X$ the evaluation map $\mathrm{ev}_x : (Y^X, \mathscr{D}_\Sigma) \longrightarrow Y$ is a uniform contraction.*

Proof 1. This follows from the fact that $\gamma_{d,A}(c_y, c_z) = d(y, z)$ for any d and where c_z stands for the constant map with value z.

2. This follows from the fact that the maps ev_x are nothing else but the projections and these are uniform contractions for the coarsest structure, namely $\mathscr{D}_p$, and hence also for all the other structures. □

In the following theorem the Cauchy index is taken with respect to the uniform gauge structure $\mathscr{D}_\Sigma$.

7.1.7 Theorem *If X is a set, Y is a uniform gauge space, Σ is a tiling of X, $\mathfrak{F} \in \mathsf{F}(Y^X)$ and $f \in Y^X$, then the following properties hold.*

1. $\sup_{x \in X} \lambda_{\mathscr{G}_Y} \mathrm{ev}_x(\mathfrak{F})(\mathrm{ev}_x(f)) \vee \chi_{cy}(\mathfrak{F}) \leq \lambda_{\mathscr{D}_\Sigma} \mathfrak{F}(f)$.
2. $\lambda_{\mathscr{D}_\Sigma} \mathfrak{F}(f) \leq \sup_{x \in X} \lambda_{\mathscr{G}_Y} \mathrm{ev}_x(\mathfrak{F})(\mathrm{ev}_x(f)) + 2\chi_{cy}(\mathfrak{F})$.

Proof 1. That $\sup_{x\in X} \lambda_{\mathscr{G}_Y} \mathrm{ev}_x(\mathfrak{F})(\mathrm{ev}_x(f)) \leq \lambda_{\mathscr{D}_\Sigma}\mathfrak{F}(f)$ follows from 7.1.6 and that $\chi_{cy}(\mathfrak{F}) \leq \lambda_{\mathscr{D}_\Sigma}\mathfrak{F}(f)$ follows from 5.4.16.

2. We consider the basis $\mathscr{H}_\Sigma := \{\gamma_{d,A} \mid d \in \mathscr{H}, A \in \Sigma\}$ for $\mathscr{D}_\Sigma$ and note that $\omega_{\mathscr{H}_\Sigma}(\mathfrak{F}) \leq 2\chi_{cy}(\mathfrak{F})$ as can be easily verified (see also the proof of 5.4.15). Now suppose that $\sup_{x\in X} \lambda_{\mathscr{G}_Y} \mathrm{ev}_x(\mathfrak{F})(\mathrm{ev}_x(f)) < \alpha$ and that $\omega_{\mathscr{H}_\Sigma}(\mathfrak{F}) < \beta$, both finite. For any $d \in \mathscr{G}_Y$ this implies that

$$\forall z \in X, \exists G_z \in \mathfrak{F}, \forall g \in G_z : d(f(z), g(z)) < \alpha,$$

and

$$\forall A \in \Sigma, \exists F \in \mathfrak{F}, \forall u, v \in F, \forall y \in A : d(u(y), v(y)) < \beta.$$

Now take $A \in \Sigma$ and choose $F \in \mathfrak{F}$ as in the second claim above. If $g \in F$ and $x \in A$ then choose $G_x \in \mathfrak{F}$ as in the first claim above and let $h \in F \cap G_x$ be arbitrary. It then follows that $d(h(x), g(x)) < \beta$ and $d(f(x), h(x)) < \alpha$ and consequently $d(f(x), g(x)) < \alpha + \beta$. Consequently

$$\lambda_{\mathscr{D}_\Sigma}\mathfrak{F}(f) = \sup_{d\in\mathscr{G}_Y} \sup_{A\in\Sigma} \inf_{F\in\mathfrak{F}} \sup_{g\in F} \sup_{x\in A} d(f(x), g(x)) < \alpha + \beta$$

from which the required result follows. □

7.1.8 Corollary (Top, Unif) *If X is a set, Y is a uniform space and Σ is a tiling on X, then a filter $\mathfrak{F}$ on Y^X will converge uniformly on Σ-sets to f if and only if it is a Cauchy filter in the uniformity of uniform convergence on Σ-sets and it is pointwise convergent to f.*

7.1.9 Theorem *If X is an approach space, Y is a uniform gauge space, $f : X \longrightarrow Y$ is a function and $\mathfrak{F}$ is a filter on Y^X, then*

$$\chi_c(f) \leq 2\lambda_u\mathfrak{F}(f) + \sup_{H\in\mathfrak{F}} \inf_{g\in H} \chi_c(g).$$

Proof We suppose that all terms on the righthand side are finite. Fix $d \in \mathscr{G}_Y$ and $\varepsilon > 0$. By definition of λ_u there exists $H_d \in \mathfrak{F}$ such that for all $h \in H_d$ and $z \in X$

$$d(f(z), h(z)) \leq \lambda_u\mathfrak{F}(f) + \varepsilon.$$

Suppose that $\sup_{H\in\mathfrak{F}} \inf_{h\in H} \chi_c(h) < \beta$, then by definition of χ_c, for any $H \in \mathfrak{F}$ there exists $h \in H$ such that for all $d' \in \mathscr{G}_Y$

$$d' \circ (h \times h) \ominus \beta \in \mathscr{G}_X$$

Consequently we can choose $h_0 \in H_d$ such that $d \circ (h_0 \times h_0) \ominus \beta \in \mathscr{G}_X$ and then for any $x, y \in X$ we have

$$\begin{aligned} d(f(x), f(y)) &\leq d(f(x), h_0(x)) + d(h_0(x), h_0(y)) + d(h_0(y), f(y)) \\ &\leq 2(\lambda_u\mathfrak{F}(f) + \varepsilon) + d \circ (h_0 \times h_0) \ominus \beta(x, y) + \beta \end{aligned}$$

which proves that

$$d \circ (f \times f) \ominus (2\lambda_u \mathfrak{F}(f) + \varepsilon + \beta) \in \mathcal{G}_X$$

and the result now follows from the arbitrariness of ε and β. □

7.1.10 Corollary (Top, Unif) *The uniform limit of a filter of continuous functions is continuous.*

7.1.11 Theorem *If X and Y are uniform gauge spaces, $f : X \longrightarrow Y$ is a function and $\mathfrak{F}$ is a filter on Y^X, then*

$$\chi_{uc}(f) \leq 2\lambda_u \mathfrak{F}(f) + \sup_{H \in \mathfrak{F}} \inf_{g \in H} \chi_{uc}(g).$$

Proof This is analogous to 7.1.9. □

7.1.12 Corollary (Top, Unif) *The uniform limit of a filter of uniformly continuous functions is uniformly continuous.*

In the following proposition, if X and Y are approach spaces, we denote by $\mathcal{K}_\gamma(X, Y)$ the set of functions $f \in Y^X$ for which $\chi_c(f) \leq \gamma$ and if X and Y are uniform gauge spaces we denote by $\mathcal{K}^u_\gamma(X, Y)$ the set of functions $f \in Y^X$ for which $\chi_{uc}(f) \leq \gamma$.

7.1.13 Proposition *If X is an approach space (respectively a uniform gauge space) and Y is a uniform gauge space then for any $\gamma \in \mathbb{P}$, $\mathcal{K}_\gamma(X, Y)$ (respectively $\mathcal{K}^u_\gamma(X, Y)$) is closed for the topology of uniform convergence.*

Proof This follows from 7.1.9 (respectively 7.1.11). □

7.1.14 Corollary (Top, Unif) *If X is a topological space (respectively a uniform space) and Y is a uniform space then $\mathcal{C}(X, Y)$ (respectively $\mathcal{C}^u(X, Y)$) is closed for the topology of uniform convergence.*

If X, Y and Z are sets and $f : X \times Y \longrightarrow Z$ then we put $\widetilde{f} : X \longrightarrow Z^Y$ the function defined as $\widetilde{f}(x)(y) := f(x, y)$. In particular for the evaluation $\mathrm{ev} : X \times \mathcal{H} \longrightarrow Y$ where $\mathcal{H} \subseteq Y^X$ we have $\widetilde{\mathrm{ev}} : X \longrightarrow Y^{\mathcal{H}} : x \mapsto \mathrm{ev}_x$ (see 7.1.6).

7.1.15 Proposition *If X is an approach space (respectively a uniform gauge space), Y is a uniform gauge space, $\mathcal{H} \subseteq Y^X$ and $Y^{\mathcal{H}}$ is equipped with the uniform gauge structure of uniform convergence then*

$$\chi_{ec}(\mathcal{H}) = \chi_c(\widetilde{\mathrm{ev}}) \text{ (respectively } \chi_{uec}(\mathcal{H}) = \chi_{uc}(\widetilde{\mathrm{ev}})).$$

Proof This follows from the definitions, since it suffices to note that for any $x, y \in X$ and $d \in \mathcal{G}_Y$ we have

$$\gamma_d \circ (\widetilde{\mathrm{ev}} \times \widetilde{\mathrm{ev}})(x, y) = \sup_{f \in \mathcal{H}} d(f(x), f(y)).$$ □

7.1.16 Theorem *If X and Y are uniform gauge spaces and $\mathcal{H} \subseteq Y^X$ then*

$$\chi_{uec}(\mathcal{H}) \leq \chi_{ec}(\mathcal{H}) + 2\chi_c(X)$$

Proof Making use of 7.1.15 the claim reduces to

$$\chi_{uc}(\widetilde{\text{ev}}) \leq \chi_c(\widetilde{\text{ev}}) + 2\chi_c(X)$$

which we know to be true from 5.4.4. □

7.1.17 Corollary (Top, Unif) *If X and Y are uniform spaces and X is compact then any equicontinuous set of functions is uniformly equicontinuous.*

7.1.18 Proposition *If X is an approach space (respectively a uniform gauge space), Y is a uniform gauge space and $\mathcal{H} \subseteq Y^X$, then for any $\varepsilon \in \mathbb{R}^+$*

$$\chi_{ec}(\mathcal{H}^{(\varepsilon)_p}) \leq \chi_{ec}(\mathcal{H}) + 2\varepsilon \ (\textit{respectively}\ \chi_{uec}(\mathcal{H}^{(\varepsilon)_p}) \leq \chi_{uec}(\mathcal{H}) + 2\varepsilon).$$

Proof We only prove the first claim. Let $a \in X$ and put

$$c_a := \min\{c \mid \forall d \in \mathcal{G}_Y, \exists \varphi \in \widehat{\mathcal{G}_X(a)}, \forall f \in \mathcal{H} : d \circ (f \times f)(a, .) \leq \varphi + c\}.$$

Let $d \in \mathcal{G}_Y$ and take $\varphi \in \widehat{\mathcal{G}_X(a)}$ such that for all $f \in \mathcal{H}$: $d \circ (f \times f)(a, \cdot) \leq \varphi + c_a$. Since for any $g \in \mathcal{H}^{(\varepsilon)_p}$ and $\theta > 0$ we can find $h_x \in \mathcal{H}$ such that

$$d(g(x), h_x(x)) < \varepsilon + \theta \text{ and } d(g(a), h_x(a)) < \varepsilon + \theta$$

it follows that

$$\begin{aligned} d(g(a), g(x)) &\leq d(g(a), h_x(a)) + d(h_x(a), h_x(x)) + d(h_x(x), g(x)) \\ &\leq 2(\varepsilon + \theta) + \varphi(x) + c_a \end{aligned}$$

and the result now follows from the definition of χ_{ec} and the arbitrariness of $a \in X$, $d \in \mathcal{G}_Y$ and $\theta > 0$. □

7.1.19 Corollary (Top, Unif) *If X is a topological space (respectively a uniform space), Y a uniform space and $\mathcal{H} \subseteq Y^X$ is equicontinuous (respectively uniformly equicontinuous) then also $\overline{\mathcal{H}}^p$ is equicontinuous (respectively uniformly equicontinuous).*

Conditions under which, in the uniform case, the topologies of uniform convergence on Σ-sets and pointwise convergence coincide are well known. This can of course be expressed by stating that an identity map is an isomorphism. In the following two theorems we see that in the case of uniform gauge spaces, making use of indices, we can quantify this. Obviously, given a subset $\mathcal{H} \subseteq Y^X$, the map $1_{\mathcal{H}} : (\mathcal{H}, \mathcal{D}_\Sigma|_{\mathcal{H}}) \longrightarrow (\mathcal{H}, \mathcal{D}_p|_{\mathcal{H}})$ is a contraction. In order to quantify to what extent the two structures deviate from being isomorphic it therefore suffices to gauge to what extent the inverse map deviates from being a contraction.

7.1.20 Theorem *If X is an approach space, Y is a uniform gauge space, $\mathcal{H} \subseteq Y^X$, Σ is a tiling of X and we consider the identity function*

$$1_{\mathcal{H}} : (\mathcal{H}, \mathcal{D}_p|_{\mathcal{H}}) \longrightarrow (\mathcal{H}, \mathcal{D}_\Sigma|_{\mathcal{H}}),$$

then

$$\chi_{uc}(1_{\mathcal{H}}) \leq 2 \sup_{A \in \Sigma} \chi_c(A) + 2\chi_{ec}(\mathcal{H}).$$

Proof Fix $A \in \Sigma$ and $d \in \mathcal{G}_Y$. Then, by definition of the index of equicontractivity we have

$$\forall a \in A, \exists \varphi_a \in \widehat{\mathcal{G}_X(a)}, \forall f \in \mathcal{H}, \forall x \in A : d(f(a), f(x)) \leq \varphi_a(x) + \chi_{ec}(\mathcal{H}).$$

By definition of the index of compactness, given any $\theta > 0$ there exists a finite set $B \subseteq A$ such that

$$\forall a \in A, \exists b_a \in B : \varphi_{b_a}(a) < \chi_c(A) + \theta.$$

Then for any $f, g \in \mathcal{H}$ we have

$$\begin{aligned}
\gamma_{d,A}(f, g) &\leq \sup_{a \in A} d(f(a), f(b_a)) + \sup_{a \in A} d(f(b_a), g(b_a)) + \sup_{a \in A} d(g(b_a), g(a)) \\
&\leq \gamma_{d,B}(f, g) + 2 \sup_{a \in A} \varphi_{b_a}(a) + 2\chi_{ec}(\mathcal{H}) \\
&\leq \gamma_{d,B}(f, g) + 2(\chi_c(A) + \theta) + 2\chi_{ec}(\mathcal{H})
\end{aligned}$$

which by arbitrariness of θ concludes the proof. □

7.1.21 Corollary (Top, Unif) *If X is a topological space, Y is a uniform space, $\mathcal{H} \subseteq Y^X$ and Σ is a tiling of X, then, if all sets in Σ are compact and $\mathcal{H}$ is equicontinuous, the uniformities of uniform convergence on Σ-sets and of pointwise convergence coincide on $\mathcal{H}$.*

The same reasoning as that given before 7.1.20 holds in case the space X is a uniform gauge space rather than an approach space, with the difference that here we are dealing with uniform contractions rather than with contractions.

7.1.22 Theorem *If X and Y are uniform gauge spaces, $\mathcal{H} \subseteq Y^X$, Σ is a tiling of X and we we consider the identity function*

$$1_{\mathcal{H}} : (\mathcal{H}, \mathcal{D}_p|_{\mathcal{H}}) \longrightarrow (\mathcal{H}, \mathcal{D}_\Sigma|_{\mathcal{H}}),$$

then

$$\chi_{uc}(1_{\mathcal{H}}) \leq 2 \sup_{A \in \Sigma} \chi_{pc}(A) + 2\chi_{uec}(\mathcal{H}).$$

Proof This is analogous to 7.1.20 and we leave this to the reader. □

7.1.23 Corollary (Top, Unif) *If X and Y are uniform spaces, $\mathscr{H} \subseteq Y^X$ and Σ is a tiling of X, then, if all sets in Σ are precompact and $\mathscr{H}$ is uniformly equicontinuous, the uniformities of uniform convergence on Σ-sets and of pointwise convergence coincide on $\mathscr{H}$.*

The following theorem is the uniform gauge generalization of the converse of Ascoli's theorem.

7.1.24 Theorem *If X and Y are uniform gauge spaces, $\mathscr{H} \subseteq Y^X$, Σ a tiling of X and Y^X is equipped with $\mathscr{D}_\Sigma$, then the following properties hold.*

1. $\sup_{x\in X} \chi_{pc}(\mathrm{ev}_x(\mathscr{H})) \leq \chi_{pc}(\mathscr{H})$.
2. $\chi_{uec}(\mathscr{H}|_A) \leq 2\chi_{pc}(\mathscr{H}) + \sup_{f\in\mathscr{H}} \chi_{uc}(f|_A)$.

Proof 1. For any $x \in X$ choose $A \in \Sigma$ such that $x \in A$. Then it follows that $d \circ (\mathrm{ev}_x \times \mathrm{ev}_x) \leq \gamma_{d,A}$. Hence, for any $x \in X$, the map

$$\mathrm{ev}_x : \mathscr{H} \longrightarrow Y$$

is a uniform contraction. Consequently the result follows from 5.4.35.

2. We suppose that all values on the right-hand side are finite and we choose α such that

$$\chi_{pc}(\mathscr{H}) = \sup_{d\in\mathscr{G}_Y, A\in\Sigma} \inf_{\mathscr{K}\in 2^{(\mathscr{H})}} \sup_{f\in\mathscr{H}} \inf_{g\in\mathscr{K}} \sup_{a\in A} d(f(a), g(a)) < \alpha.$$

Next we also choose β such that for all $h \in \mathscr{H}$

$$\chi_{uc}(h|_A) = \inf\{\delta \mid \forall d \in \mathscr{G}_Y, \exists e \in \mathscr{G}_X : d \circ (h|_A \times h|_A) \leq e + \delta\} < \beta.$$

Now fix $d \in \mathscr{G}_Y$ and $A \in \Sigma$. Then it follows that there exists a finite subset $\mathscr{K} \subseteq \mathscr{H}$ such that

$$\sup_{f\in\mathscr{H}} \inf_{g\in\mathscr{K}} \sup_{a\in A} d(f(a), g(a)) < \alpha.$$

Further, for any $g \in \mathscr{K}$ there exists $e_g \in \mathscr{G}_X$ such that $d \circ (g|_A \times g|_A) \leq e_g + \beta$. Put $e := \sup_{g\in\mathscr{K}} e_g \in \mathscr{G}_X$. Then, if $f \in \mathscr{H}$ we can find $g \in \mathscr{K}$ such that for all $a \in A$, $d(f(a), g(a)) < \alpha$. Hence it follows that for all $x, y \in A$:

$$\begin{aligned} d(f(x), f(y)) &\leq d(f(x), g(x)) + d(g(x), g(y)) + d(g(y), f(y)) \\ &\leq \alpha + e_g(x, y) + \beta + \alpha \\ &\leq e(x, y) + (2\alpha + \beta). \end{aligned}$$

□

7.1.25 Theorem (Ascoli) *If X and Y are uniform gauge spaces, $\mathscr{H} \subseteq Y^X$, Σ is a tiling of X and Y^X is equipped with $\mathscr{D}_\Sigma$, then*

$$\frac{1}{2}\chi_{pc}(\mathscr{H}) \leq \sup_{x\in X} \chi_{pc}(\mathrm{ev}_x(\mathscr{H})) + \sup_{A\in\Sigma} \chi_{pc}(A) + \sup_{A\in\Sigma} \chi_{uec}(\mathscr{H}|_A).$$

Proof We suppose again that all values on the right-hand side are finite, and we let α and β be such that, for all $x \in X$ and $A \in \Sigma$:

$$\chi_{pc}(\mathrm{ev}_x(\mathscr{H})) = \sup_{d\in\mathscr{G}_Y} \inf_{\mathscr{F}\in 2^{(\mathscr{H})}} \sup_{f\in\mathscr{H}} \inf_{g\in\mathscr{F}} d(f(x), g(x)) < \alpha,$$

$$\chi_{pc}(A) = \sup_{e\in\mathscr{G}_X} \inf_{B\in 2^{(A)}} \sup_{a\in A} \inf_{b\in B} e(a, b) < \beta.$$

Next we also choose γ such that for all $A \in \Sigma$, $\chi_{uec}(\mathscr{H}|_A) < \gamma$, which implies that

$$\forall d \in \mathscr{G}_Y, \exists e \in \mathscr{G}_X, \forall k \in \mathscr{H}|_A : d \circ (k \times k) \leq e + \gamma.$$

Now fix $d \in \mathscr{G}_Y$ and $A \in \Sigma$. Then it follows that there exists $e \in \mathscr{G}_X$ such that for all $f \in \mathscr{H}$:

$$d \circ (f|_A \times f|_A) \leq e + \gamma.$$

For this e, it then follows that there exists a finite subset $B \subseteq A$ and a function $A \longrightarrow B : a \mapsto b_a$ such that $e(a, b_a) < \beta$.

Let $Z := \bigcup_{b\in B} \mathrm{ev}_b(\mathscr{H}) \subseteq Y$. Then it follows that

$$\chi_{pc}(Z) = \chi_{pc}(\bigcup_{b\in B} \mathrm{ev}_b(\mathscr{H})) \leq \sup_{b\in B} \chi_{pc}(\mathrm{ev}_b(\mathscr{H})) < \alpha.$$

Hence, there exists a finite subset $C \subseteq Z$ and a function $Z \longrightarrow C : z \mapsto c_z$ such that $d(z, c_z) < \alpha$ for all $z \in Z$. For any $h \in C^B$ let

$$\mathscr{B}(h) := \{f \in \mathscr{H} \mid \forall b \in B : d(f(b), h(b)) < \alpha\}.$$

Now, fix $f \in \mathscr{H}$, and consider the function

$$h_f : B \longrightarrow C : b \mapsto c_{f(b)}.$$

It then follows that $f \in \mathscr{B}(h_f)$. Let

$$\mathscr{K} := \{h \in C^B \mid \mathscr{B}(h) \neq \emptyset\}.$$

Then the foregoing shows that the collection $\{\mathscr{B}(h) \mid h \in \mathscr{K}\}$ is a finite cover of $\mathscr{H}$. Now for each $h \in \mathscr{K}$ we choose an arbitrary function $g_h \in \mathscr{B}(h)$ and we let

$\mathcal{F} := \{g_h \mid h \in \mathcal{K}\}$. Then $\mathcal{F}$ is a finite subset of $\mathcal{H}$ and, by the foregoing, we obtain that for any $a \in A$

$$\begin{aligned}
&d(f(a), g_{h_f}(a)) \\
&\quad \leq d(f(a), f(b_a)) + d(f(b_a), h_f(b_a)) + d(h_f(b_a), g_{h_f}(b_a)) + d(g_{h_f}(b_a), g_{h_f}(a)) \\
&\quad = d(f(a), f(b_a)) + d(f(b_a), c_{f(b_a)}) + d(h_f(b_a), g_{h_f}(b_a)) + d(g_{h_f}(b_a), g_{h_f}(a)) \\
&\quad \leq (e(a, b_a) + \gamma) + \alpha + \alpha + (e(a, b_a) + \gamma) \\
&\quad \leq 2(\alpha + \beta + \gamma),
\end{aligned}$$

which by the arbitrariness of respectively $a \in A$, $d \in \mathcal{G}_Y$ and $A \in \Sigma$ shows that $\chi_{pc}(\mathcal{H}|_A) \leq 2(\alpha + \beta + \gamma)$. □

The foregoing two theorems generalize the well-known Ascoli theorem.

7.1.26 Corollary (Ascoli in Unif**)** *If X and Y are uniform spaces, Σ is a cover of X and $\mathcal{C}(X, Y)$ is endowed with the uniformity of uniform convergence on Σ-sets then for $\mathcal{H} \subseteq \mathcal{C}(X, Y)$ the following properties hold.*

1. *If $\mathcal{H}$ is precompact then $\mathrm{ev}_x(\mathcal{H})$ is precompact for every $x \in X$.*
2. *If $\mathrm{ev}_x(\mathcal{H})$ is precompact for every $x \in X$, all sets in Σ are precompact and for any set $A \in \Sigma$, $\mathcal{H}|_A$ is precompact then $\mathcal{H}$ is precompact.*
3. *For any set $A \in \Sigma$, if $\mathcal{H}$ is precompact and the restriction of every function in $\mathcal{H}$ to A is uniformly continuous then $\mathcal{H}|_A$ is uniformly equicontinuous.*

7.1.27 Theorem (Dini) *If X is an approach space, X' is a metric space, $f_n : X \longrightarrow X'$, $n \in \mathbb{N}$ and $f : X \longrightarrow X'$ are functions such that for all $x \in X$, $d'(f_n(x), f(x))$ is decreasing, then*

$$\lambda_u\langle(f_n)_n\rangle(f) \leq \lambda_p\langle(f_n)_n\rangle(f) + \chi_c(f) + \sup_n \chi_c(f_n) + 2\chi_c(X).$$

Proof We suppose that all terms on the righthand side are finite. Let $\varepsilon > 0$. From the definition of λ_p it follows that for any $x \in X$ there exists n_x such that for all $k \geq n_x$

$$d'(f(x), f_k(x)) \leq \lambda_p(f_n)(f) + \varepsilon.$$

Also, by 4.2.2, for all $x \in X$ we have that

$$d' \circ (f \times f) \ominus \chi_c(f) \in \mathcal{G}_X \text{ and } d' \circ (f_{n_x} \times f_{n_x}) \ominus \chi_c(f_{n_x}) \in \mathcal{G}_X.$$

Put

$$e_x := (d' \circ (f \times f) \ominus \chi_c(f)) \vee (d' \circ (f_{n_x} \times f_{n_x}) \ominus \chi_c(f_{n_x}))$$

then by 4.3.2 there exists a finite set $A \subseteq X$ such that

$$X = \bigcup_{a \in A} B_{e_a}(a, \chi_c(X) + \varepsilon).$$

Put $n_0 := \max\{n_a \mid a \in A\}$ and fix $x \in X$ and $n \geq n_0$ then there exists $a \in A$ such that

$$(d'(f(x), f(a)) \ominus \chi_c(f)) \vee (d'(f_{n_a}(x), f_{n_a}(a)) \ominus \chi_c(f_{n_a})) \leq \chi_c(X) + \varepsilon$$

and then it follows that

$$\begin{aligned} d'(f(x), f_n(x)) &\leq d'(f(x), f_{n_a}(x)) \\ &\leq d'(f(a), f_{n_a}(a)) + d'(f(x), f(a)) + d'(f_{n_a}(x), f_{n_a}(a)) \\ &\leq \lambda_p(f_n)(f) + \chi_c(f) + \sup_n \chi_c(f_n) + 2\chi_c(X) + 3\varepsilon \end{aligned}$$

which by the arbitrariness of x, $n \geq n_0$ and ε and the definition of λ_u proves the result. □

7.1.28 Corollary (Dini in Top) *Let X be a compact topological space and X' be a metric space. If $f_n : X \longrightarrow X'$, $n \in \mathbb{N}$, and $f : X \longrightarrow X'$ are continuous functions such that for all $x \in X$, $d'(f_n(x), f(x))$ converges decreasingly to 0 then the sequence $(f_n)_n$ converges uniformly to f.*

7.1.29 Theorem *If X is a set, (Y, δ_d) is a complete metric approach space, and $\mathscr{X} \subseteq Y^X$ is endowed with the approach structure δ_p of pointwise convergence on Y^X, then the* Met*-coreflection of the completion of $(\mathscr{X}, \delta)$ is isomorphic to the usual metric completion of $(\mathscr{X}, d_{\delta_p})$.*

Proof Let $\mathscr{D}$ be a symmetric gauge basis which generates δ_p. Further let $(\mathscr{C}(\mathscr{X}), \widetilde{d}_{\delta_p})$ stand for the usual metric completion of $(\mathscr{X}, d_{\delta_p})$, where $\mathscr{C}(\mathscr{X})$ stands for the set of minimal Cauchy filters in $\mathscr{X}$. Clearly, if $\mathsf{H} \in \mathscr{C}(\mathscr{X})$, then $\mathsf{H}^{\mathscr{D}} \in \widehat{\mathscr{X}}$. Thus we obtain a map

$$\psi : \mathscr{C}(\mathscr{X}) \longrightarrow \widehat{\mathscr{X}} : \mathsf{H} \longrightarrow \mathsf{H}^{\mathscr{D}}.$$

This map is onto. Indeed, if $\mathsf{F} \in \widehat{\mathscr{X}}$, then F is a $\mathscr{U}(\mathscr{D})$-Cauchy filter, i.e. for all $x \in X$, we have that

$$\mathsf{F}(x) := \{\mathscr{F}(x) \mid \mathscr{F} \in \mathsf{F}\},$$

where $\mathscr{F}(x) := \{f(x) \mid f \in \mathscr{F}\}$ is Cauchy in (Y, d). Since Y is complete, there exists $f \in Y^X$ such that, for all $x \in X$, $\mathsf{F}(x) \longrightarrow f(x)$. This implies that

$$\forall n \in \mathbb{N}_0, \forall x \in X, \exists \mathscr{F}_0 \in \mathsf{F} : \mathscr{F}_0(x) \subseteq B_d(f(x), \frac{1}{2n}).$$

Also, since F is δ_p-Cauchy, we have that

$$\forall n \in \mathbb{N}_0, \exists g_n \in \mathscr{X}, \forall x \in X : \left\{h \in \mathscr{X} \mid d(g_n(x), h(x)) < \frac{1}{2n}\right\} \in \mathsf{F}.$$

Consequently, for all $n \in \mathbb{N}_0$ and for all $x \in X$, we have

$$\left\{h \in \mathscr{Z} \mid d(g_n(x), h(x)) < \frac{1}{2n}\right\} \cap \mathscr{F}_0 \neq \emptyset,$$

and thus, for all $x \in X$, $d(g_n(x), f(x)) \leq \frac{1}{n}$, i.e. $(g_n)_n \longrightarrow f$ uniformly. Let H' stand for the filter generated by the sequence $(g_n)_n$ and let H be the minimal d_{δ_p}-Cauchy filter contained in H'. Since both

$$\lambda_{\delta_p}\mathsf{F}(f) = 0 \text{ and } \lambda_{\delta_p}\mathsf{H}(f) = 0,$$

F and H are $\mathscr{U}(\mathscr{D})$-equivalent. Since F is $\mathscr{U}(\mathscr{D})$-minimal this implies that $\psi(\mathsf{H}) = \mathsf{H}^{\mathscr{D}} = \mathsf{F}$.

Now let $\mathsf{F}, \mathsf{H} \in \widehat{\mathscr{Z}}$. Then

$$\begin{aligned}
\widetilde{d}_{\delta_p}(\mathsf{F}, \mathsf{H}) &= \sup_{\mathscr{F} \in \mathsf{F}} \sup_{\mathscr{H} \in \mathsf{H}} \inf_{f \in \mathscr{F}} \inf_{h \in \mathscr{H}} \sup_{x \in X} d(f(x), h(x)) \\
&\geq \sup_{x \in X} \widehat{d}_{\{x\}}(\mathsf{F}^{\mathscr{D}}, \mathsf{H}^{\mathscr{D}}) \\
&= d_{\widehat{\delta}_p}(\mathsf{F}^{\mathscr{D}}, \mathsf{H}^{\mathscr{D}}).
\end{aligned}$$

Conversely, if

$$d_{\widehat{\delta}_p}(\mathsf{F}^{\mathscr{D}}, \mathsf{H}^{\mathscr{D}}) < \varepsilon,$$

then, given $\theta > 0$, we can find $f, h \in \mathscr{Z}$ such that

$$\lambda_{d_{\delta_p}}\mathsf{F}(f) < \frac{\theta}{2} \text{ and } \lambda_{d_{\delta_p}}\mathsf{H}(h) < \frac{\theta}{2},$$

which implies that, for any $x \in X$,

$$B_{d_{\{x\}}}(f, \theta) \in \mathsf{F}^{\mathscr{D}} \text{ and } B_{d_{\{x\}}}(h, \theta) \in \mathsf{H}^{\mathscr{D}}.$$

It then follows that, for any $x \in X$, $d_{\{x\}}(f, h) \leq \varepsilon + 2\theta$ and thus $d_{\delta_p}(f, h) \leq \varepsilon + 2\theta$. Consequently,

$$\begin{aligned}
\lambda_{d_{\delta_p}}\mathsf{H}(f) &\leq \lambda_{d_{\delta_p}}\mathsf{H}(h) + d_{\delta_p}(h, f) \\
&\leq \varepsilon + 3\theta,
\end{aligned}$$

which by the arbitrariness of θ finally implies that

$$\begin{aligned}
\widetilde{d}_{\delta_p}(\mathsf{F}, \mathsf{H}) &= \inf_{f \in \mathscr{Z}} (\lambda_{d_{\delta_p}}\mathsf{F}(f) + \lambda_{d_{\delta_p}}\mathsf{H}(f)) \\
&\leq \varepsilon.
\end{aligned}$$

This proves at the same time that ψ is injective and an isomorphism and we are finished. □

7.2 The Čech-Stone Compactification

Our description via clusters (see 6.3.12), allows us to answer the question as to when the topological coreflection of our compactification β^*X coincides with the Čech-Stone compactification of the topological coreflection of X.

If X is a Tychonoff space, then the finest compatible proximity relation Δ_{fine} is given by

$$A \Delta_{fine} B \Leftrightarrow A \text{ and } B \text{ cannot be completely separated.}$$

This proximity relation is called the *fine proximity*. We recall that two sets A and B are said to be *completely separated* if there exists a continuous function $f : X \longrightarrow [0, 1]$ such that $f(A) = 0$ and $f(B) = 1$. This proximity is also completely determined by the property that a pair of zero-sets is proximal if and only if they intersect. If X is normal then $A \Delta_{fine} B$ is equivalent with $\overline{A} \cap \overline{B} \neq \emptyset$.

Let $\mathsf{Z}(X)$ stand for the set of all zero sets in X.

We recall that a metric space X is called a *UC space* or an *Atsuji space* (see e.g. Atsuji 1958; Chavez 1985) if any continuous real-valued function on X is uniformly continuous. Many equivalent characterizations of Atsuji spaces have been given in the literature.

7.2.1 Theorem *If X is a Hausdorff uniform approach space, then the topological coreflection of β^*X is isomorphic to the Čech–Stone compactification of the topological coreflection of X if and only if $\Delta_{\mathscr{D}_{\mathscr{K}^*(X)}}$ is the fine proximity.*

Proof The Čech-Stone compactification of a space is characterized as the compactification with the property that disjoint zero sets have disjoint closures. We have the following equivalences for any pair of zero sets $A, B \in \mathsf{Z}(X)$:

$$\begin{aligned} A \cap B = \emptyset &\Leftrightarrow A \text{ and } B \text{ have disjoint closures in } \beta^*X \\ &\Leftrightarrow \text{no } \Delta_{\mathscr{D}_{\mathscr{K}^*(X)}} \text{ cluster contains both } A \text{ and } B \\ &\Leftrightarrow A \text{ is not } \Delta_{\mathscr{D}_{\mathscr{K}^*(X)}} \text{ proximal to } B. \end{aligned}$$
□

We can now immediately derive a result which determines which metric approach spaces have a compactification such that the Čech–Stone compactification of the underlying topological space is quantified by a distance which extends the metric. In other words, for which the diagram below commutes.

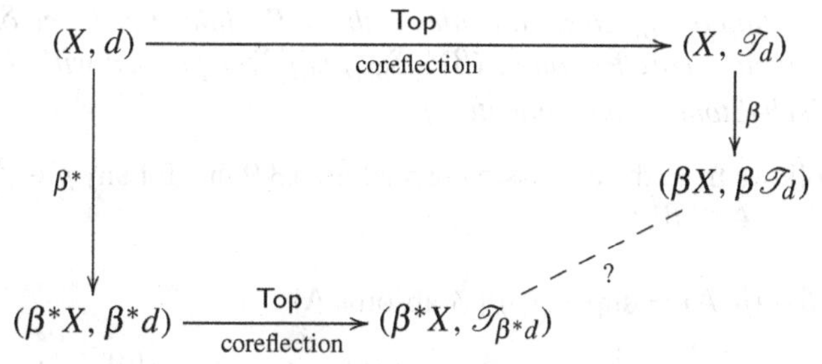

7.2.2 Corollary *For a metric space (X, d) the following properties are equivalent.*

1. *The topological coreflection of β^*X coincides with the Čech–Stone compactification of $(X, \mathcal{T}_d)$.*
2. *(X, d) is an Atsuji space.*

Proof This follows from 7.2.1 since a metric space (X, d) is an Atsuji space if and only if for any pair of closed subsets $A, B \subseteq X$, we have that

$$A \cap B = \emptyset \Leftrightarrow \inf_{a \in A} \inf_{b \in B} d(a, b) > 0. \qquad \square$$

The foregoing result is applicable to $\mathbb{N}$ since, equipped with the usual Euclidean metric, $\mathbb{N}$ is clearly an Atsuji space. This implies that, although $\beta\mathbb{N}$ is not metrizable, it is in a natural way quantifiable with a canonical approach structure. We note that the fine proximity relation in this case is

$$A \Delta_{fine} B \Leftrightarrow A \cap B \neq \emptyset$$

and that clusters hence are indeed just the ultrafilters on $\mathbb{N}$ as we already know from 7.2.2. We will therefore maintain the notation $\beta\mathbb{N}$. The distance on $\mathbb{N}$, generated by the Euclidean metric, is also simply denoted by $\delta_{d_{\mathbb{E}}}$. The distance $\beta^* \delta_{d_{\mathbb{E}}}$ will be denoted by $\delta_{\beta\mathbb{N}}$ for short. In keeping with the usual practice in the case of $\beta\mathbb{N}$ and with our way of denoting clusters, we will denote the points of $\beta\mathbb{N}$ by lower case letters $p, q, r, \ldots$.

We give some main properties of this distance in the following result.

7.2.3 Theorem *The following properties hold.*

1. *The distance $\delta_{\beta\mathbb{N}}$ on $\beta\mathbb{N}$ is given by*

$$\begin{aligned} \delta_{\beta\mathbb{N}}(p, \mathsf{A}) &= \sup_{\{A \in \cap q \mid q \in \mathsf{A}\}} \inf\{\alpha \mid A^{(\alpha)}_{d_{\mathbb{E}}} \in p\} \\ &= \sup_{\{A \in \cap q \mid q \in \mathsf{A}\}} \sup_{F \in p} \inf_{a \in A} \inf_{y \in F} |a - y| \end{aligned}$$

for all $p \in \beta\mathbb{N}$ and $\mathsf{A} \subseteq \beta\mathbb{N}$.

2. *The restriction of* $\delta_{\beta\mathbb{N}}$ *to* $\mathbb{N}$ *coincides with the Euclidean distance* $\delta_{d_{\mathbb{E}}}$.
3. *The topological coreflection of* $(\beta\mathbb{N}, \delta_{\beta\mathbb{N}})$ *is* $\beta\mathbb{N}$ *(equipped with the topology of the Čech–Stone compactification).*

Proof 1. It follows from the discussion preceding 6.3.9 that for any $p \in \beta\mathbb{N}$ and any nonempty subset $\mathsf{A} \subseteq \beta\mathbb{N}$

$$\delta_{\beta\mathbb{N}}(p, \mathsf{A}) = \sup\{\widehat{\delta_A}(p) \mid A \text{ absorbs } \mathsf{A}\}$$

$$= \sup_{\{A \in \cap q \mid q \in \mathsf{A}\}} \inf\{\alpha \mid \forall d \in \mathscr{D}_{\mathscr{K}^*(\mathbb{N})} : A_d^{(\alpha)} \in p\}.$$

Now put

$$\mathscr{K} := \{\delta_d(\cdot, A) \mid d \in \mathscr{G}_b, A \subseteq \mathbb{N}\}$$

then clearly

$$\mathscr{D}_{\mathscr{K}} \subseteq \mathscr{D}_{\mathscr{K}^*(\mathbb{N})} \subseteq \mathscr{G}_b.$$

Since for any $d \in \mathscr{G}_b$ and $A \subseteq \mathbb{N}$

$$d^A : \mathbb{N} \times \mathbb{N} \longrightarrow \mathbb{R} : (x, y) \mapsto |\delta_d(x, A) - \delta_d(y, A)|$$

is in $\mathscr{D}_{\mathscr{K}}$ and since for any $\varepsilon \in \mathbb{R}^+$ we have $A_d^{(\varepsilon)} = A_{d^A}^{(\varepsilon)}$ it follows that in the above formula we can replace $\mathscr{D}_{\mathscr{K}^*(X)}$ by $\mathscr{G}_b$. Furthermore, we can obviously replace $\mathscr{G}_b$ by the canonical basis $\{d_{\mathbb{E}} \wedge \omega \mid \omega \in \mathbb{P}\}$. But since for any $\omega \in \mathbb{P}$ and $A \subseteq \mathbb{N}$ we have $A_{d_{\mathbb{E}}}^{(\omega)} = A_{d_{\mathbb{E}} \wedge (\omega+1)}^{(\omega)}$ it finally follows that we can replace $\mathscr{D}_{\mathscr{K}^*(X)}$ by the unique metric $d_{\mathbb{E}}$ and hence

$$\delta_{\beta\mathbb{N}}(p, \mathsf{A}) = \sup_{\{A \in \cap q \mid q \in \mathsf{A}\}} \inf\{\alpha \mid A_{d_{\mathbb{E}}}^{(\alpha)} \in p\}$$

which proves the first formula.

Now, for the second formula, if there exists $B \in p$ such that $B \subseteq A_{d_{\mathbb{E}}}^{(\alpha)}$ then since any $B' \in p$ meets B it follows that $\inf_{x \in A} \inf_{y \in B'} |x - y| \leq \alpha$. Conversely if for any $B \in p$ we have that $\inf_{x \in A} \inf_{y \in B} |x - y| < \alpha$ then for each such B we can find $y_B \in B$ such that $\inf_{a \in A} |x - y_B| < \alpha$ and then $B_0 := \{y_B \mid B \in p\} \in p$ and $B_0 \subseteq A_{d_{\mathbb{E}}}^{(\alpha)}$. Hence we finally obtain that

$$\delta_{\beta\mathbb{N}}(p, \mathsf{A}) := \sup_{\{A \in \cap q \mid q \in \mathsf{A}\}} \sup_{F \in p} \inf_{a \in A} \inf_{y \in F} |a - y| .$$

2. This is immediate.
3. This follows from 7.2.2. □

In the sequel, for simplicity, we denote $A_{d_{\mathbb{E}}}^{(\alpha)}$ simply by $A^{(\alpha)}$.

That $\delta_{\beta\mathbb{N}}$ is not trivial on the remainder $\mathbb{N}^* := \beta\mathbb{N} \setminus \mathbb{N}$ of $\beta\mathbb{N}$ is shown by the following result. Given an ultrafilter p on $\mathbb{N}$, we define

$$p \oplus n := \{U \oplus n \mid U \in p\} \text{ and } p \ominus n := \{U \ominus n \mid U \in p\}$$

where, for any $n \in \mathbb{N}$ and $A \subseteq \mathbb{N}$,

$$A \oplus n := \{a + n \mid a \in A\} \text{ and } A \ominus n := \{(a - n) \vee 0 \mid a \in A\}.$$

7.2.4 Proposition *If p and q are ultrafilters on $\mathbb{N}$, then for any $n \in \mathbb{N}$*

$$\delta_{\beta\mathbb{N}}(q, \{p\}) = n \Leftrightarrow p = q \oplus n \text{ or } p = q \ominus n.$$

Proof Fix q and n. That $\delta_{\beta\mathbb{N}}(q, \{q \oplus n\}) \leq n$ follows from the fact that $F \oplus n \subseteq F^{(n)}$, for all $F \in q$ and the first formula in 7.2.3. To show the other inequality, for any $i \in \{0, 1, \dots, 2n - 1\}$, let

$$Z_i := \{2kn + i \mid k \in \mathbb{N}\}.$$

Then $(Z_i)_{i=0}^{2n-1}$ is a finite partition of $\mathbb{N}$ and thus there exists a unique $j \in \{0, 1, \dots, 2n - 1\}$ such that $Z_j \in q$. If we let $j' := j + n (\text{mod } 2n)$ then $Z_{j'} \in q \oplus n$ and

$$\delta_{\beta\mathbb{N}}(q, \{q \oplus n\}) \geq \inf_{x \in Z_j} \inf_{y \in Z_{j'}} |x - y| = n.$$

This shows that $\delta_{\beta\mathbb{N}}(q, \{q \oplus n\}) = n$. The result for $\delta_{\beta\mathbb{N}}(q, \{q \ominus n\})$ is of course an immediate consequence.

To prove the necessity, assume that for some $p, q \in \beta\mathbb{N}$ we have $\delta_{\beta\mathbb{N}}(q, \{p\}) \leq n < \infty$. Choose $i, j \in \{0, \dots, 2n\}$ such that

$$D_i := (2n + 1)\mathbb{N} \oplus i \in q$$

and

$$D_j := (2n + 1)\mathbb{N} \oplus j \in p.$$

Then

$$\mathscr{B}_q := \{F \cap D_i \mid F \in q\}$$

and

$$\mathscr{B}_p := \left\{G \cap D_j \mid G \in p\right\}$$

are bases for respectively q and p. For any $A \in \mathscr{B}_q$, we can find $B \in \mathscr{B}_p$ such that $B \subseteq A^{(n)}$. Then there exists a unique $l \in \{-n, \dots, n\}$ such that $i + l = j (\text{mod } 2n + 1)$. It then follows that $B \subseteq A \oplus l$. Consequently $q \oplus l \subseteq p$ and thus $q \oplus l = p$. □

It follows from the foregoing proposition that unless q and p are translates of each other, $\delta_{\beta\mathbb{N}}(q, \{p\}) = \infty$, i.e. ultrafilters which are not translates of each other contain sets which lie arbitrarily far apart. Consequently we immediately obtain the following corollary.

7.2.5 Corollary *The metric coreflection of* $(\beta\mathbb{N}, \delta_{\beta\mathbb{N}})$ *is a metric which generates the discrete topology on* $\beta\mathbb{N}$.

The foregoing allows us to deduce some interesting results. Since for each $n \in \mathbb{N}$ the set $A_n := \{k \in \mathbb{N} \mid k \geq n\}$ belongs to every non-principal ultrafilter it follows that for any $m \in \mathbb{N}$

$$\delta_{\beta\mathbb{N}}(m, \mathbb{N}^*) \geq \sup_{n\in\mathbb{N}} \inf_{k\in A_n} |m-k| = \infty.$$

In Van Douwen (1991) an addition was defined on $\beta\mathbb{N}$ in the following way. For any $p \in \mathbb{N}^*$ and $n \in \mathbb{N}$ we already know what $p \oplus n$ and $p \ominus n$ mean. It then is easily seen that always

$$(p \oplus n) \ominus n = p \text{ and } (p \ominus n) \oplus n = p.$$

Now for any fixed $p \in \mathbb{N}^*$, making use of the second formula in 7.2.3, the map

$$\tau_p : \mathbb{N} \longrightarrow \beta\mathbb{N} : n \mapsto p \oplus n$$

is easily seen to preserve distances, and since it is clearly an injection it is an embedding. Hence it follows from 6.3.4 that it has a unique extension

$$\tau_p^* : \beta\mathbb{N} \longrightarrow \beta\mathbb{N} : q \mapsto p \oplus q$$

which is a contraction. That way we obtain a binary operation on $\beta\mathbb{N}$

$$\oplus : \beta\mathbb{N} \times \beta\mathbb{N} \longrightarrow \beta\mathbb{N} : (p, q) \mapsto \tau_p^*(q).$$

This is an associative extension of ordinary addition of natural numbers (see e.g. Hindman 1979).

Van Douwen proved that $\mathbb{N}^* \oplus \mathbb{N}^*$ is nowhere dense in $\mathbb{N}^*$. However the following actually holds.

7.2.6 Proposition $\bigcup_{n\in\mathbb{N}}(\mathbb{N}^* \oplus \mathbb{N}^*)^{(n)}$ *is nowhere dense in* $\mathbb{N}^*$.

Proof A basis for the topology of $\mathbb{N}^*$ is given by

$$\{\mathrm{cl}_{\beta\mathbb{N}}(A) \cap \mathbb{N}^* \mid A \subseteq \mathbb{N} \text{ infinite}\}.$$

If $A \subseteq \mathbb{N}$ is an infinite subset we take an increasing sequence $(s_n)_n$ in A such that $s_{n+1} > 2s_n + 3n$ for each $n \in \mathbb{N}$ and we let

$$S := \{s_n \mid n \in \mathbb{N}\} \text{ and } S_n := \{s_m \mid m \geq n\}.$$

Suppose that for some $p, q \in \mathbb{N}^*$ and $n \in \mathbb{N}$ we have $S^{(n)} \in p \oplus q$ then, since $S^{(n)} \setminus S_n^{(n)}$ is finite, it follows that also $S_n^{(n)} \in p \oplus q$. Then we can find $B \in q$ and for all $b \in B$ a set $F_b \in p$ such that $F_b + b \subseteq S_n^{(n)}$.

Now, fix k and $l \in B$ such that $k + 2n < l$ and let $C, D \in p$ be such that

$$C + k \subseteq S_n^{(n)} \text{ and } D + l \subseteq S_n^{(n)}.$$

As $C \cap D \in p$ is an infinite set there is an $i \in C \cap D$ such that $i \geq l$. Then $i + k \in S_n^{(n)}$ and thus, for some $m \geq n$ we have

$$s_m - n \leq i + k \leq s_m + n.$$

From the fact that $k + 2n \leq l$ it now follows that

$$s_m + n < i + l.$$

On the other hand we have

$$l \leq i \leq i + k \leq s_m + n \quad \text{and thus} \quad i + l \leq 2i \leq 2s_m + 2n < s_{m+1} - n.$$

Consequently we obtain

$$s_m + n < i + l < s_{m+1} - n$$

and hence $i + l \notin S_n^{(n)}$ which contradicts the fact that $i + l \in D + l$.

This contradiction shows that $\mathbb{N} \setminus S^{(n)}$ belongs to each element of $\mathbb{N}^* \oplus \mathbb{N}^*$. Consequently, if $p \in \cup_{n \in \mathbb{N}} (\mathbb{N}^* \oplus \mathbb{N}^*)^{(n)}$ then $S \notin p$. In other words, S is an infinite subset of A such that

$$\mathrm{cl}_{\beta\mathbb{N}}(S) \cap (\cup_{n \in \mathbb{N}} (\mathbb{N}^* \oplus \mathbb{N}^*)^{(n)}) = \emptyset$$

which proves the result. □

7.3 Comments

1. **The non-numerical aspect of topological spaces**

Of course, a topological space as such has no numerical information and it would not be possible to produce any meaningful quantitative results just from the qualitative non-numerical data of a topological space. In the first section we hence developed the theory of function spaces for general approach spaces and uniform gauge spaces which as special case of course has the situation for topological and uniform spaces included. In that special case only qualitative classical results ensue. However, if a topological space or a uniform space is endowed with a canonical metric, such as e.g. the real line, then the results are also applicable to that situation and then quantitative results will follow which are related to the underlying topologies and uniformities.

More generally if a topological space, or a uniform space, is defined from an initial construction involving canonical metrics then, as we have seen in the second and fifth chapters, a canonical approach structure or uniform gauge structure overlying the topological or uniform space ensues and the results of this chapter give meaningful quantitative results for those structures.

Analogously in the second section concerning the Čech-Stone compactification of Atsuji spaces, in particular $\mathbb{N}$, here too we take the stance that $\mathbb{N}$ is endowed with a canonical metric, namely the restriction of the Euclidean metric on $\mathbb{R}$. Again this then allows for quantitative results.

2. **Function spaces in metrically generated theories**

Function spaces have been studied also in the more general setting of metrically generated theories in Vandersmissen and Van Geenhoven (2009). Both approach spaces and uniform gauge spaces are special cases of metrically generated spaces, and in particular if the codomain is a uniform gauge space then they obtain the function space structure introduced in this chapter and prove a version of Ascoli's theorem which concords with the results of 7.1.25 and which generalizes the non-indexed version given in Lowen (2004) to the realm of certain metrically generated theories.

Chapter 8
Approach Theory Meets Functional Analysis

In most sciences one generation tears down what another has built ... In mathematics alone each generation builds a new story to the old structure.

(Hermann Hankel)

In this chapter in the first section we start in the classical setting of normed spaces and see how the well-known weak and weak* topologies can be quantified by canonical approach structures and how using index analysis, this allows to obtain quantified results of which several classical results are simple corollaries (see e.g. 8.1.4, 8.1.6, 8.1.8). We also show that the weak and norm distances coincide on weakly complete subsets (8.1.16) and we prove a zero-one law for the index of compactness of the unit ball for the weak approach structure (8.1.34).

In the second section we will see that the construction of the weak and weak* approach structures on normed spaces fits into the wider picture of what we call approach vector spaces and locally convex approach spaces. We define the categories ApVec and lcApVec and investigate their basic properties and see how they relate to topological vector spaces, locally convex vector spaces and seminormed spaces.

8.1 Normed Spaces and Their Duals

We will work with a normed vector space E over the field of real numbers and we write $\| \ \|$ and d for norm and metric on E, with sub- or superscripts if necessary.

To simplify notations, we will write B_E and S_E for the closed unit ball and unit sphere with center 0. The topological dual of E, which is denoted by E', is defined as the space consisting of all continuous real-valued linear maps on E and the dual norm on E' is given by

$$\|f\| := \sup_{x \in B_E} |f(x)|.$$

R. Lowen, *Index Analysis*, Springer Monographs in Mathematics,
DOI 10.1007/978-1-4471-6485-2_8

This norm makes E' into a Banach space. The topological dual of E' is called the bidual of E, and will be denoted by E''. For every $x \in E$ we have a functional

$$\hat{x} : E' \longrightarrow \mathbb{R} : f \mapsto f(x)$$

in E'' and it is a consequence of the Hahn-Banach theorem that

$$J : E \longrightarrow E'' : x \mapsto \hat{x}$$

defines a linear isometrical embedding, meaning that J is linear and that for every $x \in E$

$$\|\hat{x}\| = \sup_{f \in B_{E'}} |f(x)| = \|x\|.$$

For every finite subset $F \subseteq E'$ the function

$$p_F : E \longrightarrow \mathbb{R} : x \mapsto \sup_{f \in F} |f(x)|$$

is a seminorm on E. In the sequel we write $\mathcal{N}_{(E,E')}$ for the set of all such seminorms and $\mathcal{D}_{(E,E')}$ for the set of all metrics derived from these seminorms. Then the weak topology $\sigma(E, E')$ on E is defined as the Hausdorff locally convex topology on E generated by either of these collections, and it owes its name to the fact that it is the initial topology for the source

$$(f : E \longrightarrow (\mathbb{R}, \mathcal{T}_{\mathbb{E}}))_{f \in B_{E'}}.$$

In this section we will introduce a uniform approach structure on E which quantifies the weak topology and we will generalize some well-known basic properties of the weak topology, which can be found in e.g. in Bourbaki (1961) and Brezis (2011), to the quantitative level, yielding the classical results as corollaries. $\mathcal{D}_{(E,E')}$ is saturated for the formation of finite suprema and it generates a distance on E, which we will denote by $\delta_{(E,E')}$ and which we will call the *weak distance*. It is given by the following formula:

$$\delta_{(E,E')} : E \times 2^E \longrightarrow \mathbb{P} : (x, A) \mapsto \sup_{F \in 2^{(B_{E'})}} \inf_{a \in A} p_F(x - a).$$

That the name weak distance for $\delta_{(E,E')}$ is justified follows from the fact that it quantifies the weak topology.

8.1.1 Proposition *The topological coreflection of* $(E, \delta_{(E,E')})$ *is* $(E, \sigma(E, E'))$ *and the metric coreflection of* $(E, \delta_{(E,E')})$ *is* $(E, d_{\| \|})$.

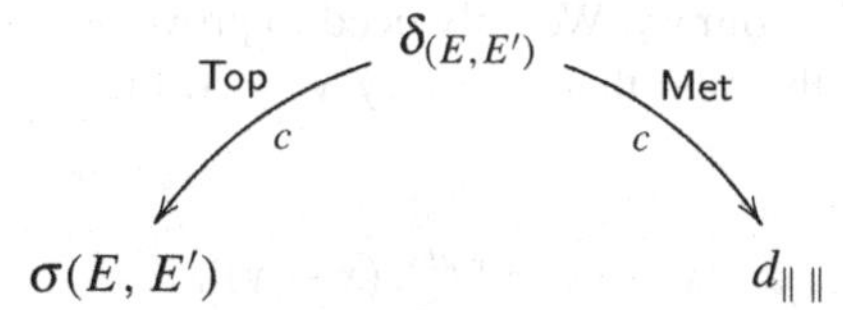

Proof The first claim follows at once from the definitions and 2.2.6 and the second claim follows from the Hahn-Banach theorem since for any $x, y \in E$ we have

$$\begin{aligned} d_{\delta_{(E,E')}}(x, y) &= \delta_{(E,E')}(x, \{y\}) \vee \delta_{(E,E')}(y, \{x\}) \\ &= \sup_{F \in 2^{(B_{E'})}} \sup_{f \in F} |f(x - y)| \\ &= \sup_{f \in B_{E'}} |f(x - y)| \\ &= \|x - y\|. \end{aligned}$$

□

8.1.2 Theorem *The source* $(f : (E, \delta_{(E,E')}) \longrightarrow (\mathbb{R}, \delta_{\mathbb{E}}))_{f \in B_{E'}}$ *is initial.*

Proof If we denote the initial approach distance for the source above by δ then, for every $x \in E$ and every $A \subseteq E$, we have

$$\begin{aligned} \delta(x, A) &= \sup_{F \in 2^{(B_{E'})}} \inf_{a \in A} \sup_{f \in F} (d_{\mathbb{E}}(f(x), \cdot) \circ f)(a) \\ &= \sup_{F \in 2^{(B_{E'})}} \inf_{a \in A} \sup_{f \in F} |f(x - a)| \\ &= \delta_{(E,E')}(x, A). \end{aligned}$$

□

Since coreflectors preserve initiality an immediate consequence is the well-known fact that the source $(f : (E, \sigma_{(E,E')}) \longrightarrow (\mathbb{R}, \mathscr{T}_E))_{f \in E'}$ is initial (in Top). That $\sigma_{(E,E')} \subseteq \mathscr{T}_{\| \|}$ is also an immediate consequence of the fact that $\delta_{(E,E')} \leq \delta_{\| \|}$.

We now show that equality of the strong and weak distances is equivalent to equality of the strong and weak topologies.

8.1.3 Theorem *The following properties are equivalent.*

1. *E is finite-dimensional.*
2. *The weak distance coincides with the norm distance.*
3. *The weak topology coincides with the norm topology.*

Proof 1 ⇒ 2. Assume that E is finite-dimensional and let $\{ e_1, \ldots, e_n \}$ be a basis for E. Then a basis for E' is given by $\{ e'_1, \ldots, e'_n \}$ where for every $i \in \{ 1, \ldots, n \}$

$$e'_i : E \longrightarrow \mathbb{R} : \sum_{k=1}^{n} \alpha_k e_k \mapsto \alpha_i.$$

Take $x \in E$ and $A \subseteq E$ arbitrary. We only need to prove one inequality. It follows from the Hahn-Banach theorem that for every $y \in A$, there exists an $f^{(y)} \in B_{E'}$ such that

$$\|x - y\| = |f^{(y)}(x - y)|.$$

Now define

$$\varphi : E' \longrightarrow \mathbb{R}^n : \sum_{i=1}^{n} \alpha_i e_i' \mapsto (\alpha_1, \ldots, \alpha_n).$$

Then φ is one-to-one and linear and both φ and φ^{-1} are Lipschitz. For every $(\alpha_1, \ldots, \alpha_n) \in \mathbb{R}^n$ we have

$$\|\varphi^{-1}((\alpha_1, \ldots, \alpha_n))\| \leq n \cdot \max_{i=1}^{n} \|e_i'\| \cdot \|(\alpha_1, \ldots, \alpha_n)\|_{\max},$$

and

$$\|\varphi(\sum_{i=1}^{n} \alpha_i e_i')\|_{\max} \leq \max_{i=1}^{n} \|e_i\| \cdot \|\sum_{i=1}^{n} \alpha_i e_i'\|.$$

Analogously of course

$$\psi : E \longrightarrow \mathbb{R}^n : \sum_{i=1}^{n} x_i e_i \mapsto (x_1, \ldots, x_n)$$

is linear and one-to-one and both ψ and ψ^{-1} are Lipschitz. Suppose that $M > 0$ is a Lipschitz constant for ψ. Fix $0 < \varepsilon < 1$. Since $\varphi(B_{E'})$ is compact there exists a finite subset B of $\varphi(B_{E'})$ such that

$$\varphi(B_{E'}) \subseteq \bigcup_{\alpha \in B} B_{d_{\|\cdot\|_{\max}}}(\alpha, \varepsilon')$$

where $\varepsilon' := \varepsilon/(n \cdot M)$. Then $F := \varphi^{-1}(B)$ is a finite subset of $B_{E'}$ and for all $y \in A$

$$p_F(x - y) \geq (1 - \varepsilon) \cdot \|x - y\|.$$

To prove this claim, pick $y \in A$. If we write $x = \sum_{i=1}^{n} x_i e_i$, $y = \sum_{i=1}^{n} y_i e_i$ and $f^{(y)} = \sum_{i=1}^{n} \alpha_i^{(y)} e_i'$, then $\alpha^{(y)} := (\alpha_1^{(y)}, \ldots, \alpha_n^{(y)}) \in \varphi(B_{E'})$, so there exists $\alpha := (\alpha_1, \ldots, \alpha_n) \in B$ with

$$\|\alpha - \alpha^{(y)}\|_{\max} < \varepsilon'.$$

If $g :\in \varphi^{-1}(\alpha)$ we obtain that

$$\begin{aligned}\|x-y\| &= |f^{(y)}(x-y)| \\ &\leq |g(x-y)| + |(f^{(y)}-g)(x-y)| \\ &\leq p_F(x-y) + \sum_{i=1}^{n} |\alpha_i^{(y)} - \alpha_i| \cdot |x_i - y_i| \\ &\leq p_F(x-y) + \varepsilon \cdot \|x-y\|.\end{aligned}$$

This now implies that

$$\delta_{(E,E')}(x,A) \geq \inf_{y\in A} p_F(x-y) \geq (1-\varepsilon)\cdot \delta_{\|\,\|}(x,A)$$

and thus by arbitrariness of ε we are finished.

$2 \Rightarrow 3$. This follows from 8.1.1.

$3 \Rightarrow 1$. This can be found e.g. in Brezis (2011). □

The fact that a weakly convergent sequence is bounded is a consequence of a more powerful quantified result.

8.1.4 Theorem *Let $(x_n)_n$ be a sequence in E. Then the following properties are equivalent.*

1. *For all $x \in E : \lambda_{(E,E')}\langle (x_n)_n \rangle (x) < \infty$.*
2. *There exists $x \in E : \lambda_{(E,E')}\langle (x_n)_n \rangle (x) < \infty$.*
3. *The sequence $(x_n)_n$ is norm bounded.*

Proof $1 \Rightarrow 2$. This is evident.

$2 \Rightarrow 3$. Suppose that

$$M := \lambda_{(E,E')}\langle (x_n)_n\rangle(x) = \sup_{f\in B_{E'}} \limsup_n |f(x-x_n)| < \infty$$

for some $x \in E$. Then we have that

$$\forall f \in B_{E'}, \exists n_f, \forall n > n_f : |f(x) - f(x_n)| \leq M+1,$$

yielding that for every $f \in E'$ and every $n \in \mathbb{N}$

$$|f(x_n)| \leq \left((\|f\|+1)\cdot \left(\left| \frac{f}{\|f\|+1}(x) \right| + M + 1 \right) \right) \vee \sup_{i=1}^{n_{f/(\|f\|+1)}} |f(x_i)|,$$

which shows that $(f(x_n))_n$ is a bounded sequence for every $f \in E'$. Applying a well-known consequence of the Banach-Steinhaus theorem (see e.g. Brezis 2011) now yields that $(x_n)_n$ is norm bounded.

3 ⇒ 1. Note that for each $x \in E$

$$\lambda_{(E,E')}\langle (x_n)_n \rangle (x) \le \lambda_{\|\ \|}\langle (x_n)_n \rangle (x) = \limsup_n \|x - x_n\| \le \|x\| + \sup_n \|x_n\|. \qquad \square$$

8.1.5 Corollary (Top) *Every weakly convergent sequence is norm bounded.*

8.1.6 Proposition *If $(x_\kappa)_\kappa$ is a net in E and $x \in E$, then*

$$\|x\| \le \liminf_\kappa \|x_\kappa\| + \lambda_{(E,E')}\langle (x_\kappa)_\kappa \rangle (x).$$

Proof Take $x \in E$ and let $(x_\kappa)_\kappa$ be a net in E on the directed set $(D, \succeq)$. Assume that all terms on the right-hand side are finite, and let M be such that $\lambda_{(E,E')}\langle (x_\kappa)_\kappa \rangle (x) < M < \infty$. Then it follows that

$$\forall f \in B_{E'}, \exists \eta \in D, \forall \kappa \in D, \kappa \succeq \eta : |f(x)| - |f(x_\kappa)| \le |f(x - x_\kappa)| < M,$$

from which it follows that

$$\forall f \in B_{E'} : |f(x)| \le \liminf_\kappa \|x_\kappa\| + M$$

and taking the supremum over all $f \in B_{E'}$, the result follows once again from the Hahn-Banach theorem. □

8.1.7 Corollary (Top) *If $(x_\kappa)_\kappa$ is a net in E which converges weakly to $x \in E$, then*

$$\|x\| \le \liminf_\kappa \|x_\kappa\|.$$

8.1.8 Proposition *If $(x_\kappa)_\kappa$ and $(f_\kappa)_\kappa$ are nets in E and E', both based on the same directed set, $x \in E$ and $f \in E'$, then*

$$\lambda_{\mathbb{E}}\langle (f_\kappa(x_\kappa))_\kappa \rangle (f(x)) \le (\sup_\kappa \|x_\kappa\|) \cdot \lambda_{\|\ \|}\langle (f_\kappa)_\kappa \rangle (f) + \|f\| \cdot \lambda_{(E,E')}\langle (x_\kappa)_\kappa \rangle (x).$$

Proof We will denote the directed set upon which both nets are based by $(D, \succeq)$. If $f = 0$ and $\sup_\kappa \|x_\kappa\| = 0$ there is nothing to show. If $\sup_\kappa \|x_\kappa\| = 0$ and $f \neq 0$ we have that

$$\begin{aligned}\lambda_{\mathbb{E}}\langle (f_\kappa(x_\kappa))_\kappa \rangle (f(x)) &= |f(x)| \\ &\le \|f\| \cdot \limsup_\kappa \left| \frac{f}{\|f\|}(x - x_\kappa) \right| \\ &\le \|f\| \cdot \lambda_{(E,E')}\langle (x_\kappa)_\kappa \rangle (x).\end{aligned}$$

If $\sup_\kappa \|x_\kappa\| \neq 0$ and $f = 0$, we have that

$$\begin{aligned}\lambda_{\mathbb{E}}\langle (f_\kappa(x_\kappa))_\kappa\rangle(f(x)) &= \limsup_\kappa |f_\kappa(x_\kappa)| \\ &\leq (\sup_\kappa \|x_\kappa\|) \cdot \limsup_\kappa \|f_\kappa - f\| \\ &= (\sup_\kappa \|x_\kappa\|) \cdot \lambda_{\|\ \|}\langle (f_\kappa)_\kappa\rangle(f).\end{aligned}$$

Now suppose that $f \neq 0$ and that $\sup_\kappa \|x_\kappa\| > 0$. If either $\lambda_{\|\ \|}\langle (f_\kappa)_\kappa\rangle(f) = \infty$ or $\lambda_{(E,E')}\langle (x_\kappa)_\kappa\rangle(x) = \infty$ there is nothing to show hence we can suppose that these terms are finite. Fix $\varepsilon > 0$ and take $\kappa_0 \in D$ such that

$$\sup_{\kappa \succeq \kappa_0} \|f - f_\kappa\| \leq \lambda_{\|\ \|}\langle (f_\kappa)_\kappa\rangle(f) + \frac{\varepsilon}{2\sup_\kappa \|x_\kappa\|}$$

and

$$\sup_{\kappa \succeq \kappa_0} \left| \frac{f}{\|f\|}(x_\kappa - x) \right| \leq \limsup_\kappa \left| \frac{f}{\|f\|}(x_\kappa - x) \right| + \frac{\varepsilon}{2\|f\|}.$$

It then follows that

$$\begin{aligned}\sup_{\kappa \succeq \kappa_0} |f_\kappa(x_\kappa) - f(x)| &\leq \sup_{\kappa \succeq \kappa_0} |f_\kappa(x_\kappa) - f(x_\kappa)| + \sup_{\kappa \succeq \kappa_0} |f(x_\kappa) - f(x)| \\ &\leq (\sup_\kappa \|x_\kappa\|) \cdot \sup_{\kappa \succeq \kappa_0} \|f_\kappa - f\| + \|f\| \cdot \sup_{\kappa \succeq \kappa_0} \left| \frac{f}{\|f\|}(x_\kappa - x) \right| \\ &\leq (\sup_\kappa \|x_\kappa\|) \cdot \lambda_{\|\ \|}\langle (f_\kappa)_\kappa\rangle(f) + \|f\| \cdot \limsup_\kappa \left| \frac{f}{\|f\|}(x_\kappa - x) \right| + \varepsilon \\ &\leq (\sup_\kappa \|x_\kappa\|) \cdot \lambda_{\|\ \|}\langle (f_\kappa)_\kappa\rangle(f) + \|f\| \cdot \lambda_{(E,E')}\langle (x_\kappa)_\kappa\rangle(x) + \varepsilon,\end{aligned}$$

which concludes the proof by arbitrariness of ε. □

8.1.9 Corollary (Top) *If $(x_\kappa)_\kappa$ and $(f_\kappa)_\kappa$ are nets in E and E', both based on the same directed set, $x \in E$ and $f \in E'$, $(f_\kappa)_\kappa$ converges to f in the norm topology and $(x_\kappa)_\kappa$ converges weakly to x, then $(f_\kappa(x_\kappa))_\kappa$ converges to x in $\mathbb{R}$.*

8.1.10 Theorem **(Mazur)** *If $C \subseteq E$ is convex then*

$$\delta_{\|\ \|}(\cdot, C) = \delta_{(E,E')}(\cdot, C).$$

Proof If $C = \emptyset$, the equality is trivially fulfilled, so we may assume $C \neq \emptyset$. Fix $x \in E$. Only one inequality needs to be shown. Suppose that $\delta_{\|\ \|}(x, C) > \varepsilon > 0$ and put $B_E(x, \varepsilon)$ the open ball in x with radius ε. Since C and $B_E(x, \varepsilon)$ are nonempty and disjoint the Hahn-Banach theorem yields the existence of $\alpha \in \mathbb{R}$ and $f \in E'$, with $f \neq 0$ such that

$$\forall y \in C, \forall z \in B_E(x, \varepsilon) : f(y) \leq \alpha \leq f(z).$$

Since f is linear this implies that

$$\begin{aligned} f(x) &= \sup_{z \in B_E(x,\varepsilon)} (f(z) + f(x - z)) \\ &\geq \alpha + \sup_{z \in B_E(x,\varepsilon)} f(x - z) \\ &= \alpha + \varepsilon \| f \|, \end{aligned}$$

and it now follows that

$$\delta_{(E,E')}(x, C) \geq \inf_{y \in C} \left(\frac{f}{\|f\|}(x) - \frac{f}{\|f\|}(y) \right) \geq \varepsilon,$$

which concludes the proof. □

8.1.11 Corollary (**Mazur in** Top) *The weak closure and the strong closure of convex sets coincide.*

A remarkable fact about the weak distance is that its completeness is equivalent to the completeness of the initial norm.

8.1.12 Theorem *A normed space E is a Banach space if and only if $(E, \delta_{(E,E')})$ is complete.*

Proof Since a uniform approach space is complete if and only if its metric coreflection is complete, this is an immediate consequence of 8.1.1. □

If (E_1, δ_1) and (E_2, δ_2) are approach spaces, $f : E_1 \longrightarrow E_2$ is a function and $k \in \mathbb{R}^+$, we call f *k-Lipschitz* if for every $x \in E_1$ and each subset $A \subseteq E_1$, we have that

$$\delta_2(f(x), f(A)) \leq k \cdot \delta_1(x, A).$$

8.1.13 Proposition *If E_1 and E_2 are normed spaces, and $T : E_1 \longrightarrow E_2$ is a linear function, then the following properties are equivalent.*

1. $T : (E_1, \mathcal{T}_{\| \|_1}) \to (E_2, \mathcal{T}_{\| \|_2})$ *is continuous.*
2. *For some* $k \in \mathbb{R}^+$, $T : (E_1, \delta_{\| \|_1}) \to (E_2, \delta_{\| \|_2})$ *is k-Lipschitz.*
3. *For some* $k \in \mathbb{R}^+$, $T : (E_1, \delta_{(E_1,E_1')}) \to (E_2, \delta_{(E_2,E_2')})$ *is k-Lipschitz.*
4. $T : (E_1, \sigma(E_1, E_1')) \longrightarrow (E_2, \sigma(E_2, E_2'))$ *is continuous.*

In the equivalence of 2 and 3 the Lipschitz constant is preserved in both directions.

Proof 1 ⇔ 2 ⇔ 4. This can be found in e.g. Brezis (1983).

2 ⇒ 3. Suppose that $T : (E_1, \delta_{\|\ \|_1}) \longrightarrow (E_2, \delta_{\|\ \|_2})$ is k-Lipschitz with $k \in \mathbb{R}^+$. Since for any finite subset $F \subseteq B_{E_2'}$

$$F \circ T := \{ f \circ T \mid f \in F \}$$

is a finite subset of $kB_{E_1'}$ we obtain that for every $x \in E_1$ and every $A \subseteq E_1$

$$\begin{aligned}\delta_{(E_2,E_2')}(T(x), T(A)) &= \sup_{F \in 2^{(B_{E_2'})}} \inf_{a\in A} p_{F\circ T}(x-a) \\ &\le \sup_{H \in 2^{(kB_{E_1'})}} \inf_{a\in A} p_H(x-a) \\ &= k\delta_{(E_1,E_1')}(x, A).\end{aligned}$$

3 ⇒ 2. This is analogous to the foregoing and we leave this to the reader. □

Our next result gives an alternative description of the weak distance in a Banach space which in turn will allow us to prove a theorem in the same vain as 3.2.15 and 8.1.10.

We write $\mathscr{S}$ for the set of all closed subspaces F of E with a finite codimension, i.e. such that $\dim(E/F)$ is finite. We denote the canonical quotient map by

$$\pi_F : E \longrightarrow E/F : x \mapsto x + F.$$

We recall that the quotient norm is given by $\|\pi_F(x)\| := \inf_{y\in F} \|x - y\|$. The map

$$\pi_F^\tau : (E/F)' \longrightarrow E' : \varphi \mapsto \varphi \circ \pi_F$$

is well known to be an isometry with image

$$F^\perp := \mathrm{Im}\pi_F^\tau = \{f \in E' \mid f_{|F} = 0\}.$$

8.1.14 Proposition *If E is a Banach space, then for any $x \in E$ and $A \subseteq E$ we have*

$$\delta_{(E,E')}(x, A) = \sup_{F\in\mathscr{S}} \inf_{a\in A} \inf_{z\in F} \|x - a - z\| = \sup_{F\in\mathscr{S}} \delta_{\|\ \|}(x, A + F).$$

Proof Let $x \in E$ and $A \subseteq E$ and for any finite subset $G \subseteq B_{E'}$ put

$$F_G := G^\perp = \{y \in E \mid \forall f \in G : f(y) = 0\}.$$

Then, since the codimension of F_G cannot be larger than the cardinality of G and F_G is closed it follows that $F_G \in \mathscr{S}$. Hence, since clearly $G \subseteq {F_G}^\perp \cap B_{E'}$ and applying the Hahn-Banach theorem we find that

$$\begin{aligned}
\delta_{(E,E')}(x, A) &= \sup_{G\in 2^{(B_{E'})}} \inf_{a\in A} \sup_{f\in G} |f(x-a)| \\
&\leq \sup_{G\in 2^{(B_{E'})}} \inf_{a\in A} \sup_{f\in F_G{}^{\perp}\cap B_{E'}} |f(x-a)| \\
&= \sup_{G\in 2^{(B_{E'})}} \inf_{a\in A} \sup_{\varphi\in B_{(E/F_G)}} |\varphi\circ\pi_{F_G}(x-a)| \\
&= \sup_{G\in 2^{(B_{E'})}} \inf_{a\in A} \|\pi_{F_G}(x-a)\| \\
&= \sup_{G\in 2^{(B_{E'})}} \inf_{a\in A} \inf_{z\in F_G} \|x-a-z\| \\
&\leq \sup_{F\in\mathscr{S}} \inf_{a\in A} \inf_{z\in F} \|x-a-z\|.
\end{aligned}$$

Conversely, let $F \in \mathscr{S}$ be arbitrary. Then $F^{\perp}$ is a finite-dimensional subspace of E' and for any $\varepsilon \in]0, 1[$, by compactness of $B_{F^{\perp}}$ with respect to the topology generated by the dual norm there exists (see Valentine 1965) a finite subset $G_\varepsilon := \{f_1, \ldots, f_n\}$ of $B_{F^{\perp}}$ such that

$$(1-\varepsilon)B_{F^{\perp}} \subseteq \mathrm{conv}(G_\varepsilon),$$

where we recall that conv stands for convex hull. Hence we obtain

$$\begin{aligned}
\inf_{a\in A} \inf_{z\in F} \|x-a-z\| &= \inf_{a\in A} \sup_{f\in B_{F^{\perp}}} |f(x)-f(a)| \\
&\leq (1-\varepsilon)^{-1} \inf_{a\in A} \sup_{f\in \mathrm{conv}(G_\varepsilon)} |f(x)-f(a)| \\
&= (1-\varepsilon)^{-1} \inf_{a\in A} \sup_{f\in G_\varepsilon} |f(x)-f(a)|
\end{aligned}$$

from which, by the arbitrariness of $F \in \mathscr{S}$ and $\varepsilon \in]0, 1[$, the remaining inequality follows. □

The inclusion relation on $\mathscr{S}$ is a partial order which obviously makes $\mathscr{S}$ a directed set. We use the notation $L \preceq M \Leftrightarrow M \subseteq L$.

8.1.15 Definition A subset $A \subseteq E$ is called *weakly complete* if, whenever $(a_L)_{L\in\mathscr{S}'}$ is a subnet of a net $(a_L)_{L\in\mathscr{S}}$ in A such that $(f(a_L))_{L\in\mathscr{S}'}$ converges for every $f \in E'$, then there exists $a \in A$ such that $(a_L)_{L\in\mathscr{S}'}$ converges to a in the weak topology.

8.1.16 Theorem *If E is a Banach space, then for any $A \subseteq E$ which is weakly complete*

$$\delta_{(E,E')}(\cdot, A) = \delta_{\|\ \|}(\cdot, A).$$

Proof If E is finite dimensional there is nothing to show since this is 8.1.3, so we may suppose that E is infinite dimensional. Fix $x \in E$ and put $\alpha := \delta_{(E,E')}(x, A) < \infty$. It follows from 8.1.14 that, for any $L \in \mathscr{S}$ we can find $x_L \in L$ and $a_L \in A$ such that

$$\|x - (a_L + x_L)\| < \alpha + \frac{1}{\operatorname{codim}(L)}.$$

For any $L \in \mathscr{S}$ we have $a_L + x_L \in (\|x\| + \alpha + 1)B_E$, and thus $\widehat{a_L + x_L} \in (\|x\| + \alpha + 1)B_{E''}$. By the theorem of Alaoglu-Bourbaki (see e.g. Brezis 2011), there then exists a subnet $(a_M + x_M)_{M\in\mathscr{S}'}$ and $\varphi \in E''$ such that $(\widehat{a_M + x_M})_{M\in\mathscr{S}'}$ converges to φ with respect to $\sigma(E'', E')$.

Now if $f \in E' \setminus \{0\}$, since $\operatorname{Ker}(f)$ is a subspace of codimension 1, there exists $M_f \in \mathscr{S}$ with $\operatorname{Ker}(f) \preceq M_f$. Clearly, for all $M \in \mathscr{S}'$ such that $M_f \preceq M$ we have $f(a_M) = f(a_M + x_M)$, which proves that for any $f \in E'$ the net $(f(a_M))_{M\in\mathscr{S}'}$ converges to $\varphi(f)$ and hence $(a_M)_{M\in\mathscr{S}'}$ converges to φ in $\sigma(E'', E')$. Since A is weakly complete, this implies that there exists $a \in A$ with $\varphi = \widehat{a}$.

For any $f \in B_{E'}$ and $M \in \mathscr{S}'$ with $M_f \preceq M$ we have

$$\begin{aligned} |f(x - a_M)| &= |f(x - (a_M + x_M))| \\ &\le \|x - (a_M + x_M)\| \\ &< \alpha + \frac{1}{\operatorname{codim}(M)} \end{aligned}$$

and consequently, taking limits, and noting that then $\operatorname{codim}(M) \to \infty$, we obtain that $|f(x - a)| \le \alpha$. This shows that

$$\delta_{\|\ \|}(x, A) \le \|x - a\| \le \alpha$$

which concludes the proof. □

In a similar way as for the weak topology on E, the dual pair E and E' generates the so-called weak* topology $\sigma(E', E)$ on E'. For each finite subset $F \subseteq E$ we write

$$p^F : E' \longrightarrow \mathbb{R} : f \longrightarrow \sup_{x\in F} |\hat{x}(f)| = \sup_{x\in F} |f(x)|$$

and this defines a seminorm on E'. To simplify notations, we put $\mathscr{N}(E', E) := \{p^F \mid F \subseteq E \text{ finite}\}$ and $\mathscr{D}(E', E) := \{d_{p^F} \mid F \subseteq E \text{ finite}\}$. The *weak* topology* $\sigma(E', E)$ is then defined to be the Hausdorff locally convex topology on E' generated by $\mathscr{D}(E', E)$. We now introduce a distance on E' which is a quantification of the weak* topology, and we state a series of results, parallel to those proved in the previous section for the weak topology. We omit all proofs which are similar to the proofs of the analogous statements for the weak distance. $\mathscr{D}(E', E)$ is a collection of metrics on E' which is saturated for the formation of finite suprema. It therefore generates a distance, which we will denote by $\delta_{(E',E)}$ and which we will call the *weak* distance*, given by the following formula:

$$\delta_{(E',E)} : E' \times 2^{E'} \longrightarrow \mathbb{P} : (f, G) \mapsto \sup_{F \in 2^{(B_E)}} \inf_{g \in G} p^F(f - g).$$

8.1.17 Proposition *The topological coreflection of* $(E', \delta_{(E',E)})$ *is* $(E', \sigma(E', E))$ *and the metric coreflection of* $(E', \delta_{(E',E)})$ *is* $(E', d_{\| \|})$.

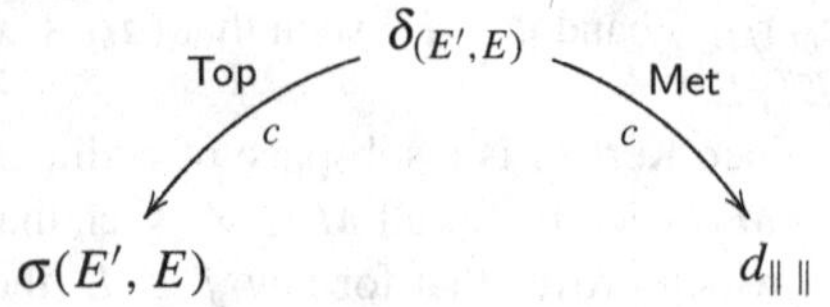

Proof This is analogous to 8.1.1 and we leave this to the reader. □

In this setup, we now have three canonical distances on E': the strong distance $\delta_{\| \|}$, the weak* distance $\delta_{(E',E)}$ and the weak distance $\delta_{(E',E'')}$.

8.1.18 Proposition *The following holds*

$$\delta_{(E',E)} \leq \delta_{(E',E'')} \leq \delta_{\| \|}.$$

Proof The first inequality follows from the fact that $J(B_E) \subseteq B_{E''}$ and that $p^F = p_{J(F)}$ for any finite subset $F \subseteq B_E$. □

As an immediate consequence $\sigma(E', E) \subseteq \sigma(E', E'') \subseteq \mathcal{T}_{\| \|}$.

8.1.19 Theorem *The source* $(\hat{x} : (E', \delta_{(E',E)}) \longrightarrow (\mathbb{R}, \delta_{\mathbb{E}}))_{x \in B_E}$ *is initial.*

Proof This is analogous to 8.1.2 and we leave this to the reader. □

Note that in the following two results we need E to be complete, since their proofs rely on the Banach-Steinhaus theorem.

8.1.20 Theorem *If E is a Banach space and $(f_n)_n$ a sequence in E', then the following properties are equivalent.*

1. *For all* $f \in E'$: $\lambda_{(E',E)}\langle (f_n)_n \rangle (f) < \infty$.
2. *There exists* $f \in E'$ *such that* $\lambda_{(E',E)}\langle (f_n)_n \rangle (f) < \infty$.
3. $\sup_{n \in \mathbb{N}} \|f_n\| < \infty$.

Proof This is analogous to 8.1.4 and we leave this to the reader. □

8.1.21 Corollary (Top) *In the dual of a Banach space every weak* convergent sequence is norm bounded.*

8.1.22 Proposition *If $(f_\kappa)_\kappa$ is a net in E' and $f \in E'$, then*

$$\|f\| \leq \liminf_\kappa \|f_\kappa\| + \lambda_{(E',E)}\langle (f_\kappa)_\kappa\rangle(f).$$

Proof This is analogous to 8.1.6 and we leave this to the reader. □

8.1.23 Corollary (Top) *If $(f_\kappa)_\kappa$ is a net in E' which converges to $f \in E'$ with respect to $\sigma(E', E)$, then $\|f\| \leq \liminf_\kappa \|f_\kappa\|$.*

8.1.24 Proposition *If $(x_\kappa)_\kappa$ and $(f_\kappa)_\kappa)$ are nets in E and E', both based on the same directed set, such that $(f_\kappa)_\kappa$ is norm bounded, $x \in E$ and $f \in E'$, then*

$$\lambda_{d_\mathbb{E}}\langle (f_\kappa(x_\kappa))_\kappa\rangle(f(x)) \leq (\sup_\kappa \|f_\kappa\|) \cdot \lambda_{\|\,\|}\langle (x_\kappa)_\kappa\rangle(x) + \|x\| \cdot \lambda_{(E',E)}\langle (f_\kappa)_\kappa\rangle(f).$$

Proof This is analogous to 8.1.8 and we leave this to the reader. □

8.1.25 Corollary (Top) *If $(x_\kappa)_\kappa$ and $(f_\kappa)_\kappa$ are nets in E and E', both based on the same directed set, $x \in E$ and $f \in E'$, $(x_\kappa)_\kappa$ converges to x and $(f_\kappa)_\kappa$ converges to f with respect to $\sigma(E', E)$, then $(f_\kappa(x_\kappa))_\kappa$ converges to $f(x)$ in $\mathbb{R}$.*

8.1.26 Theorem *The following properties are equivalent.*

1. *E is finite-dimensional.*
2. *The weak distance on E' coincides with the norm distance.*
3. *The weak topology on E' coincides with the norm topology.*

Proof Since E' is finite dimensional if and only if E is, this follows from 8.1.3. □

8.1.27 Theorem *The following properties are equivalent.*

1. *E is reflexive.*
2. *The weak and weak* distances on E' coincide.*
3. *The weak and weak* topologies on E' coincide.*

Proof 1 ⇒ 2. This is obvious since for a reflexive normed space we have $J(B_E) = B_{E''}$ and since for any finite subset $F \subseteq B_E$ we have $p^F = p_{J(F)}$.

2 ⇒ 3. This follows from 8.1.1.

3 ⇒ 1. This is well known, see e.g. Brezis (1983). □

8.1.28 Theorem *The following properties are equivalent.*

1. *E is finite-dimensional.*
2. *The weak* distance and the norm distance coincide.*
3. *The weak* topology and the norm topology coincide.*

Proof 1 ⇒ 2. Since finite dimensional spaces are reflexive, this is a direct consequence of 8.1.26 and 8.1.27.

2 ⇒ 3. This is evident.

3 ⇒ 1. This follows from 8.1.26. □

8.1.29 Theorem *$(E', \delta_{(E',E)})$ is complete.*

Proof This follows directly from 3.5.9 since $(E', \| \ \|)$ is a Banach space. □

8.1.30 Theorem *$J : (E, \delta_{(E,E')}) \to (E'', \delta_{(E'',E')})$ is an embedding.*

Proof It is well known that J is injective. That J is initial, is obvious since for every $x \in E$ and every $A \subseteq E$

$$\delta_{(E'',E')}(J(x), J(A)) = \delta_{(E,E')}(x, A). \qquad \square$$

8.1.31 Corollary (Top) *$J : (E, \sigma(E, E')) \longrightarrow (E'', \sigma(E'', E'))$ is an embedding.*

Proof This follows from 8.1.30 since concrete coreflectors preserve initiality. □

It is well known that a normed space E is reflexive if and only if $(B_E, \sigma(E, E')|_{B_E})$ is compact, see e.g. Brezis (1983). The next proposition extends the mentioned result to the level of approach spaces.

8.1.32 Proposition *The following equality holds.*

$$\chi_c(B_E, \delta_{(E,E')}|_{B_E}) = \inf\{ \varepsilon \in \mathbb{R}_0^+ \mid (J(B_E))^{(\varepsilon)}_{\| \|} \supseteq B_{E''} \}.$$

Proof To prove the "$\leq$"-part, take $\varepsilon \in \mathbb{R}_0^+$ with $(J(B_E))^{(\varepsilon)}_{\| \|} \supseteq B_{E''}$. Let $\mathcal{U} \in \mathsf{U}(B_E)$ be arbitrary. Then $J(\mathcal{U})$ is an ultrafilter basis on $B_{E''}$ and the theorem of Alaoglu-Bourbaki yields that there exists $\varphi \in B_{E''}$ such that $J(\mathcal{U}) \to \varphi$ in $(B_{E''}, \sigma(E'', E')|_{B_{E''}})$. Fix $n \in \mathbb{N}_0$ and take $x \in B_E$ with $\|\hat{x} - \varphi\| \leq \varepsilon + 1/n$. Since

$$J : (B_E, \delta_{(E,E')}|_{B_E}) \to (B_{E''}, \delta_{(E'',E')}|_{B_{E''}})$$

is an embedding, it follows that

$$\begin{aligned}
\lambda_{(E,E')}|_{B_E}(\mathcal{U})(x) &= \lambda_{(E'',E')}|_{B_{E''}}(J(\mathcal{U}))(\hat{x}) \\
&= \sup_{F \in 2^{(B_{E'})}} \inf_{U \in \mathcal{U}} \sup_{y \in U} \sup_{f \in F} |(\hat{x} - \hat{y})(f)| \\
&\leq \sup_{f \in B_{E'}} |(\hat{x} - \varphi)(f)| + \sup_{F \in 2^{(B_{E'})}} \inf_{U \in \mathcal{U}} \sup_{y \in U} \sup_{f \in F} |(\varphi - \hat{y})(f)| \\
&= \|\hat{x} - \varphi\| + \lambda_{(E'',E')}|_{B_{E''}}(J(\mathcal{U}))(\varphi) \\
&\leq \varepsilon + 1/n.
\end{aligned}$$

Because $\mathcal{U} \in \mathsf{U}(B_E)$ and $n \in \mathbb{N}_0$ were taken arbitrarily, it follows that

$$\chi_c(B_E, \delta_{(E,E')}|_{B_E}) = \sup_{\mathcal{U} \in \mathsf{U}(B_E)} \inf_{x \in B_E} \lambda_{(E,E')}|_{B_E}(\mathcal{U})(x) \leq \varepsilon,$$

which completes this part of the proof.

In order to show the "$\geq$"-part of the equality, note that if the right-hand side of the equality equals 0, there is nothing to prove. Hence let $\varepsilon \in \mathbb{R}_0^+$ be such that

$$(J(B_E))^{(\varepsilon)}_{\|\ \|} \not\supseteq B_{E''}$$

and fix $\varphi \in B_{E''} \setminus (J(B_E))^{(\varepsilon)}_{\|\ \|}$. Since for each $x \in B_E$

$$\sup_{f \in B_{E'}} |\varphi(f) - f(x)| = \|\varphi - \hat{x}\| > \varepsilon,$$

we can choose, for every $x \in B_E$, an $f_x \in B_{E'}$ with $|\varphi(f_x) - f_x(x)| > \varepsilon$. Now consider the product space

$$T := \{(F_x)_x \mid \forall x \in E : F_x \subseteq B_{E'} \text{ finite}\}.$$

It then follows that

$$\begin{aligned}\chi_c(B_E, \delta_{(E,E')}|_{B_E}) &= \sup_{(F_x)_x \in T} \inf_{Y \in 2^{(B_E)}} \sup_{x \in B_E} \inf_{y \in Y} \sup_{f \in F_y} |f(x-y)| \\ &\geq \inf_{Y \in 2^{(B_E)}} \sup_{x \in B_E} \inf_{y \in Y} |f_y(x-y)|.\end{aligned}$$

Now fix $n \in \mathbb{N}_0$. Since $J(B_E)$ is dense in $(B_{E''}, \sigma(E'', E')|_{B_{E''}})$ we can fix for every finite subset $Y \subseteq B_E$ an $x_Y \in B_E$ such that

$$\forall y \in Y : |\varphi(f_y) - f_y(x_Y)| < 1/n.$$

Combining the foregoing, we get that

$$\begin{aligned}\chi_c(B_E, \delta_{(E,E')}|_{B_E}) &\geq \inf_{Y \in 2^{(B_E)}} \inf_{y \in Y} |f_y(x_Y) - f_y(y)| \\ &\geq \inf_{Y \in 2^{(B_E)}} \inf_{y \in Y} (|f_y(y) - \varphi(f_y)| - |\varphi(f_y) - f_y(x_Y)|) \geq \varepsilon - 1/n,\end{aligned}$$

which by arbitrariness of $n \in \mathbb{N}_0$ shows that $\chi_c(B_E, \delta_{(E,E')}|_{B_E}) \geq \varepsilon$, completing the proof. □

We recall the following well-known result.

8.1.33 Lemma *If F is a proper closed subspace of E, then for every $\varepsilon \in]0, 1[$ there exists an $x_\varepsilon \in B_E$ with*

$$\delta_{\|\ \|}(x_\varepsilon, F) > 1 - \varepsilon.$$

This enables us to show a "zero-one" law for the index of compactness of B_E in the weak approach structure.

8.1.34 Theorem *The index of compactness of B_E with respect to the weak approach structure is either zero or one.*

Proof Suppose that $\chi_c(B_E, \delta_{(E,E')}|_{B_E}) > 0$. Then it follows from 8.1.32, since J is an isometrical embedding, that $\overline{J(E)}$ is a proper closed subspace of E'' and since

$$0 \leq \chi_c(B_E, \delta_{(E,E')}|_{B_E}) \leq 1,$$

combining 8.1.32 and 8.1.33 yields that in this case $\chi_c(B_E, \delta_{(E,E')}|_{B_E}) = 1$. □

For Banach spaces, this implies the following corollary:

8.1.35 Corollary *If E is a Banach space, then*

$$\chi_c(B_E, \delta_{(E,E')}|_{B_E}) = \begin{cases} 0 & \text{if } E \text{ is reflexive,} \\ 1 & \text{if } E \text{ is not reflexive.} \end{cases}$$

Proof In 3.5.9 it is shown that for an arbitrary uniform approach space Y, we have that the topological coreflection is compact if and only if Y is complete and $\chi_c(Y) = 0$. The statement now follows at once from this, 8.1.34 and the fact that a Banach space is reflexive if and only if its closed unit ball is weakly compact. □

If E is a normed space we will write $\tilde{E} := \overline{J(E)}$. Then

$$J : (E, \delta_{\|\ \|}) \longrightarrow (\tilde{E}, \delta_{\|\ \|}|_{\tilde{E}})$$

is a dense embedding of $(E, \delta_{\|\ \|})$ with respect to the topology underlying the metric coreflection. Consequently this embedding, up to isomorphism, is the App-completion of $(E, \delta_{\|\ \|})$. This implies that the completion of $(E, \delta_{\|\ \|})$ is a cUAp$_2$-object corresponding to a Banach space with a linear embedding.

Since we have proved in 8.1.12 that $(E, \delta_{(E,E')})$ is not complete when E is not a Banach space, and since $(E, \delta_{(E,E')}) \in |\mathsf{UAp}_2|$, it is interesting to study the App-completion of $(E, \delta_{(E,E')})$ in this case.

8.1.36 Theorem *The completion of $(E, \delta_{(E,E')})$ is given by*

$$J : (E, \delta_{(E,E')}) \to (\tilde{E}, \delta_{(E'',E')}|_{\tilde{E}}).$$

Proof This follows from the fact that the metric coreflection of $(\tilde{E}, \delta_{(E'',E')}|_{\tilde{E}})$ is $(\tilde{E}, d_{\|\ \|}|_{\tilde{E}})$, from 3.5.9, and from the fact that $J : (E, \delta_{(E,E')}) \longrightarrow (\tilde{E}, \delta_{(E'',E')}|_{\tilde{E}})$ is dense with respect to the topology underlying the metric coreflection of $\delta_{(E'',E')}|_{\tilde{E}}$. □

8.1.37 Example Let E be a non-reflexive Banach space. Then we know from 8.1.12 that $(E, \delta_{(E,E')})$ is complete in App. However, if E is equipped with the initial

UG-structure Γ for the same source

$$(E \xrightarrow{f} \mathbb{R})_{f \in B_{E'}}$$

but this time with $\mathbb{R}$ being equipped with the usual uniform gauge, then E cannot be UG-complete. Indeed, the uniform coreflection $\mathscr{U}_0$ of Γ is initial for the same source with $\mathbb{R}$ being equipped with the usual uniformity and hence B_E is totally bounded. However, since E is non-reflexive, B_E is not UG-complete and hence neither is (E, Γ).

8.2 Locally Convex Spaces

In the foregoing section we have seen that natural approach structures exist on normed spaces and their duals, which quantify the weak and weak* topologies. These approach structures allowed us to deduce approximations to fundamental theorems of functional analysis but moreover they are easily seen to be natural also from the point of view of the relation with the underlying vector space structure. In this section we prove that this is no coincidence. We isolate the precise conditions required to have approach structures concord with the algebraic operations of a vector space. Not surprisingly we are able to show that topological vector spaces and locally convex spaces fit nicely into our framework, but the conditions for the approach case are more subtle. We also characterize approach vector structures by means of metrics and prenorms, showing relations to notions existing in the literature on topological vector spaces. Finally we prove that categorically all is as it should be, by showing that the categories which are introduced have the right topological and algebraic properties and that the right embeddings, reflections and coreflections, are present.

If X is a group (additive) and $\varphi \in \mathbb{P}^X$ then for any $x \in X$ we write

$$\varphi \ominus x : X \longrightarrow \mathbb{P} : y \mapsto \varphi(y - x) \quad \text{and} \quad \varphi^{(2)} : X \times X \to \mathbb{P} : (x, y) \mapsto \varphi(y - x).$$

For $\Gamma \subseteq \mathbb{P}^X$ we put

$$\Gamma \ominus x := \{\varphi \ominus x \mid \varphi \in \Gamma\} \text{ and } \Gamma^{(2)} = \{\varphi^{(2)} \mid \varphi \in \Gamma\}.$$

Furthermore in the sequel we will say that φ is *sub-additive* if

$$\forall x, y \in X : \varphi(x + y) \leq \varphi(x) + \varphi(y).$$

If X is moreover a vector space (we only consider real vector spaces) we will call φ *balanced* if

$$\forall x \in X, \forall \lambda \in [-1, 1] : \varphi(\lambda x) \leq \varphi(x)$$

and we call φ *absorbing* if

$$\forall x \in X, \forall \varepsilon > 0, \exists \delta > 0, \forall \lambda \in [-\delta, \delta] : \varphi(\lambda x) \leq \varepsilon.$$

This last condition actually implies that, for all $x \in X$, the map $\mathbb{R} \longrightarrow \mathbb{P} : \lambda \mapsto \varphi(\lambda x)$ is continuous in 0. A *prenorm* is a function that is sub-additive, balanced and absorbing. It will follow from several characterizations in the sequel that this terminology is justified. However, note already that if φ is balanced, respectively absorbing, then for any $\varepsilon > 0$ the set $\{\varphi \leq \varepsilon\}$ is balanced, respectively absorbing, in the usual sense.

Approach groups were introduced in Lowen and Windels (2000) and we will first of all make a more detailed investigation of them, in order to pinpoint the right constraints that are needed on an approach system for it to fit nicely with a vector space structure.

8.2.1 Definition If G is at the same time a group and an approach space then it is called an *approach group* if the following properties are fulfilled.

(AG1) $\forall x \in G$: $\mathcal{A}(0) \ominus x = \mathcal{A}(x)$.
(AG2) $\forall \varphi \in \mathcal{A}(0), \forall \varepsilon > 0, \forall \omega < \infty, \exists \psi \in \mathcal{A}(0), \forall x,\ y \in G$

$$\varphi(x+y) \wedge \omega \leq \psi(x) + \psi(y) + \varepsilon.$$

(AG3) $\forall \varphi \in \mathcal{A}(0)$: $\varphi(0) = 0$.
(AG4) $\forall \varphi \in \mathcal{A}(0)$: $\varphi^{(2)}(\cdot, 0) \in \mathcal{A}(0)$.

It can be shown that in (AG2) the ε-part can be dropped and that, by induction, for all $\varphi \in \mathcal{A}(0)$, for all $\omega < \infty$ and $n \in \mathbb{N}$, there exists $\psi \in \mathcal{A}(0)$, such that for all $x_1, \ldots, x_n \in G$

$$\varphi(\sum_{i=1}^{n} x_i) \wedge \omega \leq \sum_{i=1}^{n} \psi(x_i).$$

This "interlinking" between different functions of the approach system sometimes is cumbersome, and it is therefore easier to be able to work with a basis for the approach system consisting of functions where the interlinking condition is trivially fulfilled, because the functions individually have good properties. This is the same as the situation with approach systems and gauges where in the former we have the triangular linking condition (A3) and in the latter we simply have metrics. The following result establishes the analogue in the present context. The proof uses exactly the same technique as in Windels (1997). This technique, in turn, is based on the well-known proof of the Urysohn theorem on the metrizability of a uniformity with a countable basis.

As before, here too, we denote an approach group simply by its underlying set.

8.2.2 Proposition *If G is an approach group, then there exists a basis for $\mathcal{A}(0)$ consisting of sub-additive and symmetric functions.*

Proof For $\varphi_0 \in \mathscr{A}(0)$ and $\omega < \infty$ we shall construct a sub-additive function $\varphi \in \mathscr{A}(0)$ such that $\varphi_0 \wedge \omega \leq \varphi$. So let $(\varphi_m)_{m\in\mathbb{N}_0}$ be an increasing sequence in $\mathscr{A}(0)$ such that for all $x,\ y,\ z \in G$ and $m \in \mathbb{N}$:

$$\varphi_m(x+y+z) \wedge \frac{\omega}{2^m} \leq \varphi_{m+1}(x) + \varphi_{m+1}(y) + \varphi_{m+1}(z).$$

Define

$$\psi := \sup_{m\in\mathbb{N}_0} (\varphi_m \wedge \frac{\omega}{2^{m-1}}),$$

then it follows that $\psi \leq \varphi_m + \frac{\omega}{2^m}$ for all $m \in \mathbb{N}_0$, and so $\psi \in \mathscr{A}(0)$.
Let

$$\varphi(x) := \inf\{\sum_{i=1}^n \psi(x_i) \mid n \in \mathbb{N}_0,\ \sum_{i=1}^n x_i = x\}.$$

Then $\varphi \leq \psi$, so $\varphi \in \mathscr{A}(0)$ and φ is sub-additive by definition. We use induction to prove that for all $n \in \mathbb{N}_0$, the following property $P(n)$ holds:

$$\forall\{x_1,\ldots,x_n\} \subseteq G, \forall m \in \mathbb{N} : \varphi_m(\sum_{i=1}^n x_i) \wedge \frac{\omega}{2^m} \leq \sum_{i=1}^n \psi(x_i).$$

We already have $\varphi_0 \wedge \omega \leq \varphi_1 \wedge \omega \leq \psi$ and for all $m \in \mathbb{N}_0$: $\varphi_m \wedge \frac{\omega}{2^m} \leq \varphi_m \wedge \frac{\omega}{2^{m-1}} \leq \psi$, so $P(1)$ is valid. Fix n and suppose $P(k)$ is valid for $k < n$. Let $\{x_1,\ldots,x_n\} \subseteq G$ and $m \in \mathbb{N}$. We distinguish different cases.

1. Suppose $\sum_{i=1}^n \psi(x_i) \geq \frac{\omega}{2^m}$. Then of course $\varphi_m(\sum_{i=1}^n x_i) \wedge \frac{\omega}{2^m} \leq \sum_{i=1}^n \psi(x_i)$.
2. Suppose $\sum_{i=1}^n \psi(x_i) < \frac{\omega}{2^m}$ and suppose there exists $k \in \{1,\ldots,n-2\}$ such that $\sum_{i=1}^k \psi(x_i) < \frac{\omega}{2^{m+1}}$ and $\sum_{i=k+2}^n \psi(x_i) < \frac{\omega}{2^{m+1}}$. From the induction hypothesis we know that $\varphi_{m+1}(\sum_{i=1}^k x_i) \wedge \frac{\omega}{2^{m+1}} \leq \sum_{i=1}^k \psi(x_i)$ and therefore we have

   $$\varphi_{m+1}(\sum_{i=1}^k x_i) \leq \sum_{i=1}^k \psi(x_i).$$

   In the same way we obtain

   $$\varphi_{m+1}(\sum_{i=k+2}^n x_i) \leq \sum_{i=k+2}^n \psi(x_i).$$

   Since, by definition, $\varphi_{m+1} \wedge \frac{\omega}{2^m} \leq \psi$ and since necessarily $\psi(x_{k+1}) < \frac{\omega}{2^m}$ we also have

   $$\varphi_{m+1}(x_{k+1}) \leq \psi(x_{k+1}).$$

Therefore

$$\begin{aligned}\varphi_m(\sum_{i=1}^{n} x_i) \wedge \frac{\omega}{2^m} &\leq \varphi_{m+1}(\sum_{i=1}^{k} x_i) + \varphi_{m+1}(x_{k+1}) + \varphi_{m+1}(\sum_{i=k+2}^{n} x_i)\\ &\leq \sum_{i=1}^{k} \psi(x_i) + \psi(x_{k+1}) + \sum_{i=k+2}^{n} \psi(x_i)\\ &= \sum_{i=1}^{n} \psi(x_i).\end{aligned}$$

3. Suppose that $\sum_{i=1}^{n} \psi(x_i) < \frac{\omega}{2^m}$ and that $\sum_{i=1}^{n-1} \psi(x_i) < \frac{\omega}{2^{m+1}}$. Then, by induction, it follows that $\varphi_{m+1}(\sum_{i=1}^{n-1} x_i) \leq \sum_{i=1}^{n-1} \psi(x_i)$. We know that $\psi(x_n) < \frac{\omega}{2^m}$, so $\varphi_{m+1}(x_n) \leq \psi(x_n)$, thus

$$\begin{aligned}\varphi_m(\sum_{i=1}^{n} x_i) \wedge \frac{\omega}{2^m} &\leq \varphi_{m+1}(\sum_{i=1}^{n-1} x_i) + \varphi_{m+1}(x_n)\\ &\leq \sum_{i=1}^{n} \psi(x_i).\end{aligned}$$

4. If $\sum_{i=1}^{n} \psi(x_i) < \frac{\omega}{2^m}$ and $\sum_{i=2}^{n} \psi(x_i) < \frac{\omega}{2^{m+1}}$ then applying the same arguments we obtain

$$\varphi_m(\sum_{i=1}^{n} x_i) \wedge \frac{\omega}{2^m} \leq \sum_{i=1}^{n} \psi(x_i).$$

This proves that $P(n)$ is valid for all $n \in \mathbb{N}_0$, from which $\varphi_0 \wedge \omega \leq \varphi$ follows. The symmetric map $x \mapsto \varphi(x) \vee \varphi(-x)$ is then also sub-additive and in $\mathscr{A}(0)$. □

In the following result we use the above to give a simple description of approach groups by means of their gauges.

8.2.3 Proposition *If G is at the same time a group and an approach space, then the following properties are equivalent.*

1. *G is an approach group.*
2. *For all $x \in G$, $\mathscr{A}(x) = \mathscr{A}(0) \ominus x$ and $\mathscr{A}(0)$ has a basis of symmetric and sub-additive functions.*
3. *$\mathscr{G}$ has a basis of translation-invariant metrics.*

Proof 1 ⇒ 2. This is precisely 8.2.2.

2 ⇒ 3. Let $\mathscr{B}$ be a basis for $\mathscr{A}(0)$ consisting of sub-additive functions. Note that for all $\varphi \in \mathscr{B}$: $\varphi^{(2)}(x, \cdot) \in \mathscr{A}(x)$ which yields $\mathscr{B}^{(2)} \subseteq \mathscr{G}$. Let $d \in \mathscr{G}$, fix

$\varepsilon > 0$, $\omega < \infty$ and $x \in G$. Because $\langle \mathscr{B} \ominus x \rangle = \mathscr{A}(x)$, we choose $\psi \in \mathscr{B}$ such that $d(x, y) \wedge \omega \leq \psi(y - x) + \varepsilon$ for all $y \in G$. This yields that $\mathscr{B}^{(2)}$ is a basis for $\mathscr{G}$.

3 ⇒ 1. Let $\mathscr{H}$ be a basis of $\mathscr{G}$ consisting of translation-invariant metrics. Then it is clear that

$$\mathscr{B}(x) := \{d(x, \cdot) \mid d \in \mathscr{H}\}$$

is a basis for $\mathscr{A}(x)$ and hence it is easily verified that the approach system satisfies all the required conditions. □

8.2.4 Definition If E is both a vector space and an approach space, then it is called an *approach vector space* if the following properties are fulfilled.

(AV1) E is an approach group.
(AV2) $\forall \varphi \in \mathscr{A}(0), \forall \varepsilon > 0, \forall \omega < \infty, \exists \psi \in \mathscr{A}(0)$:

$$\forall x \in E, \forall \lambda : |\lambda| \leq 1 \Rightarrow \varphi(\lambda x) \wedge \omega \leq \psi(x) + \varepsilon.$$

(AV3) Every $\varphi \in \mathscr{A}(0)$ is absorbing.

A morphism between approach vector spaces is defined as a linear contraction and the resulting category is denoted ApVec.

In the sequel of this section, unless otherwise stated, E with or without sub- or superscripts, will be a vector space. If necessary to mention the underlying set of E explicitly, we will denote it by $\underline{E}$.

8.2.5 Proposition *If E is at the same time a vector space and an approach space, then the following properties are equivalent.*

1. *E is an approach vector space.*
2. *For all $x \in E$: $\mathscr{A}(x) = \mathscr{A}(0) \ominus x$ and $\mathscr{A}(0)$ has a basis of prenorms.*

Proof 1 ⇒ 2. From 8.2.2 we already know that $\mathscr{A}(0)$ has a basis of sub-additive functions, say $\mathscr{B}$. Let $\varphi \in \mathscr{B}$ and define

$$\varphi^b(x) := \sup_{|\lambda| \leq 1} \varphi(\lambda x).$$

It is easy to see that φ^b is balanced and still sub-additive. In fact φ^b is the smallest balanced function that is larger than φ. We fix $\varepsilon > 0$, $\omega < \infty$ and $\psi \in \mathscr{A}(0)$ such that for all $x \in E$ and for all $\lambda \in [-1, 1]$, $\varphi(\lambda x) \wedge \omega \leq \psi(x) + \varepsilon$. Then for any x we have $\varphi^b(x) \wedge \omega \leq \psi(x) + \varepsilon$ and therefore $\varphi^b \in \mathscr{A}(0)$. In particular φ^b is absorbing and thus a prenorm. Since $\varphi \leq \varphi^b$ and since $\mathscr{B}$ is a basis for $\mathscr{A}(0)$ so is $\{\varphi^b \mid \varphi \in \mathscr{B}\}$.

2 ⇒ 1. This is evident. □

We call $\mathcal{B} \subseteq \Gamma \subseteq \mathbb{P}^E$ *saturated in* Γ if and only if the following condition holds:

$$\forall \varphi \in \Gamma : (\forall \varepsilon > 0,\ \forall \omega < \infty, \exists \psi \in \mathcal{B} : \varphi \wedge \omega \leq \psi + \varepsilon) \Rightarrow \varphi \in \mathcal{B}.$$

If $\Gamma = \mathbb{P}^E$ this is the notion of (local) saturatedness as in 1.1.13.

8.2.6 Definition An ideal $\mathcal{N}$ of prenorms that is saturated in the set of all prenorms (on E) is called a *local prenorm system (on E)*.

For example, if $(E, \mathcal{A})$ is an approach vector space and we define $\mathcal{N}_{\mathcal{A}}$ to be the set of all prenorms in $\mathcal{A}(0)$, then $\mathcal{N}_{\mathcal{A}}$ is a local prenorm system. Every local prenorm system can be obtained in this way. Indeed, let $\mathcal{N}$ be a local prenorm system, then we set, for all $x \in E$, $\mathcal{A}_{\mathcal{N}}(x) := \widehat{\mathcal{N} \ominus x}$. It is not hard to verify that $(E, \mathcal{A}_{\mathcal{N}})$ is an approach vector space and $\mathcal{N}_{\mathcal{A}_{\mathcal{N}}} = \mathcal{N}$. Conversely, if $(E, \mathcal{A})$ is an approach vector space then 8.2.5 yields $\mathcal{A}_{\mathcal{N}_{\mathcal{A}}} = \mathcal{A}$. The extension of this correspondence to the functorial level yields an alternative description of ApVec. A linear map $f : E \longrightarrow F$ between approach vector spaces $(E, \mathcal{A}_E)$ and $(F, \mathcal{A}_F)$ then is a contraction if and only if for all $\varphi \in \mathcal{N}_{\mathcal{A}_F} : \varphi \circ f \in \mathcal{N}_{\mathcal{A}_E}$.

8.2.7 Proposition *A prenorm is finite.*

Proof Let ν be a prenorm and suppose there exists an x such that $\nu(x) = \infty$. Since for all $z \in E : \nu(z) \leq 2\nu(\frac{1}{2}z)$ we can deduce for all $n \in \mathbb{N} : \nu(\frac{1}{2^n}x) = \infty$. This however contradicts the fact that ν is absorbing. □

8.2.8 Definition A function $d : E \times E \longrightarrow \mathbb{P}$ is called a *vector metric* (on E) if there exists a prenorm ν such that $d = \nu^{(2)}$.

8.2.9 Proposition *A function $d : E \times E \longrightarrow \mathbb{P}$ is a vector metric if and only if d is translation-invariant and the map $d(0, \cdot)$ is a prenorm.*

Proof This follows from the definitions. □

8.2.10 Proposition *If E is at the same time a vector space and an approach space, then the following properties are equivalent.*

1. *E is an approach vector space.*
2. *The gauge $\mathcal{G}$ has a basis of vector metrics.*

Proof 1 ⇒ 2. Let $\mathcal{N}$ be the local prenorm system of $\mathcal{A}$. From the proof of 8.2.3 we know that the set of vector metrics $\mathcal{N}^{(2)}$ is a gauge basis for $\mathcal{G}$.

2 ⇒ 1. If $\mathcal{G}$ has a gauge basis of vector metrics, say $\mathcal{H}$, then we know from 8.2.3 that $(E, \mathcal{A})$ is an approach group and that the set of prenorms $\{d(0, \cdot) \mid d \in \mathcal{H}\}$ is a basis for $\mathcal{A}(0)$. □

8.2.11 Theorem *If I is a class, for all $i \in I$, E_i is an approach vector space and $f_i : E \longrightarrow E_i$ a linear map, then the initial approach structure makes E an approach vector space, in other words the forgetful functor* ApVec $\to$ Vec *is topological.*

Proof For each i, let $\mathcal{N}_i$ be the local prenorm system of $\mathcal{A}_i(0)$ and let

$$\mathcal{N} := \{\sup_{j=1}^{n} \nu_{i_j} \circ f_{i_j} \mid \nu_{i_j} \in \mathcal{N}_{i_j},\ n \in \mathbb{N},\ i_j \in I\}.$$

Put $\mathcal{A}_{\text{in}}$ the initial approach system. Then we know that $\mathcal{N}$ is a basis for $\mathcal{A}_{\text{in}}(0)$ and for all $i \in I$ and $x_i \in E_i$ the set $\mathcal{N}_i \ominus x_i$ is a basis for $\mathcal{A}_i(x_i)$. Considering the functions

$$\sup_{j=1}^{n} \nu_{i_j} \circ f_{i_j} \ominus x : E \longrightarrow \mathbb{P} : y \mapsto \sup_{j=1}^{n} \nu_{i_j}(f_{i_j}(y) - f_{i_j}(x))$$

it is easily verified that for all $x \in E$, we have $\mathcal{A}_{\text{in}}(x) = \mathcal{A}_{\text{in}}(0) \ominus x$. Furthermore, using the linearity of f_{i_j}, it is easily verified that the elements of $\mathcal{N}$ are all prenorms. Hence E equipped with $\mathcal{A}_{\text{in}}$ is an approach vector space. □

8.2.12 Definition If d is a vector metric, then E equipped with the associated approach system $\mathcal{A}_d$ is called a *metric vector space*. A morphism between metric vector spaces is a linear contraction and the category thus obtained is denoted MetVec.

Let d be a vector metric and let $\mathcal{G}$ be the gauge of $\mathcal{A}_d$. Since $\{d\}$ is a basis for $\mathcal{G}$, 8.2.10 yields that E equipped with $\mathcal{A}_d$ is an approach vector space.

8.2.13 Theorem MetVec *is a full subcategory of* ApVec*. Moreover, this embedding is an extension of the embedding* Met $\hookrightarrow$ UAp, *in the sense that if an approach vector space is also a metric approach space, then it is a metric vector space.*

Proof In order to prove the second part of the theorem, let $(E, \mathcal{A})$ be an approach vector space where $\mathcal{A}$ is a metric approach structure with gauge $\mathcal{G}$. This means that there exists a metric d such that $\{d\}$ is a basis for $\mathcal{G}$. We know from 8.2.10 that this gauge has a basis of vector metrics, say $\mathcal{H}$. Hence, $d = \sup \mathcal{H}$, which yields $d = d^{(2)}(0, \cdot)$ and $d(0, \cdot)$ is balanced. In order to show that $d(0, \cdot)$ is absorbing, fix $x \in E$ and $\varepsilon > 0$. Let $d' \in \mathcal{H}$ be such that for all $y \in E : d(0, y) \wedge \omega \leq d'(0, y) + \frac{\varepsilon}{2}$, where $\omega > \varepsilon$ is fixed. Let $\delta > 0$ be such that $\forall \lambda \in [-\delta, \delta] : d'(0, \lambda x) \leq \frac{\varepsilon}{2}$. Then it follows that

$$\forall \lambda \in [-\delta, \delta] : d(0, \lambda x) \leq \varepsilon.$$ □

8.2.14 Theorem MetVec *is initially dense in* ApVec.

Proof An object in ApVec is in the initial hull of MetVec if and only if its gauge has a basis of vector metrics. By 8.2.10, this is the case for any object in ApVec. □

8.2.15 Proposition *Let E be a vector space and let $\mathcal{T}$ be a metrizable topology on E. Then $(E, \mathcal{T})$ is a topological vector space if and only if there there exists a vector metric d such that $\mathcal{T}_d = \mathcal{T}$.*

Proof The construction of a vector metric for a metrizable topology of a topological vector space is due to Urysohn and can be found e.g. in Schaefer (1966).

We only prove the if part. Let d be a vector metric on E and let $(x_0, y_0) \in E \times E$. For a neighbourhood V of $x_0 + y_0$ in $\mathcal{T}_d$ we fix $\varepsilon > 0$ such that

$$\{z \mid d(x_0 + y_0, z) \leq \varepsilon\} \subseteq V.$$

Then it follows that

$$\left\{x \mid d(x_0, x) \leq \frac{\varepsilon}{2}\right\} + \left\{y \mid d(y_0, y) \leq \frac{\varepsilon}{2}\right\} \subseteq V.$$

In order to prove continuity of the multiplication, fix $\lambda_0 \in \mathbb{R}$ and $x_0 \in E$. Let V be a neighbourhood of $\lambda_0 x_0$ in $\mathcal{T}_d$ and let $\varepsilon > 0$ be such that

$$\{z \mid d(\lambda_0 x_0, z) \leq \varepsilon\} \subseteq V.$$

Then, using the fact that $d(0, \cdot)$ is absorbing, we can find $\delta > 0$ such that

$$|\lambda - \lambda_0| \leq \delta \Rightarrow d(\lambda_0 x_0, \lambda x_0) \leq \frac{\varepsilon}{2}.$$

Now if $n \in \mathbb{N}_0$ is such that $|\lambda_0| + \delta \leq n$ then it follows that

$$[\lambda_0 - \delta, \lambda_0 + \delta]\left\{z \mid d(x_0, z) \leq \frac{\varepsilon}{2n}\right\} \subseteq V.$$ □

8.2.16 Proposition *The embedding* CReg $\hookrightarrow$ UAp *generates a full embedding of* TopVec *into* ApVec.

Proof Let $(E, \mathcal{T})$ be a topological vector space, let $\mathcal{A}_{\mathcal{T}}$ be the approach system generated by $\mathcal{T}$ and let $\mathcal{V}$ be the neighbourhood system of $\mathcal{T}$.

From the facts that $\mathcal{V}$ is translation invariant and that for all $V \in \mathcal{V}(0)$ there exists $U \in \mathcal{V}(0)$ such that $U + U \subseteq V$ it follows that $(E, \mathcal{A}_{\mathcal{T}})$ is an approach group and from the facts that every $V \in \mathcal{V}(0)$ is absorbing and that $\mathcal{V}(0)$ has a basis of balanced sets it follows that $\mathcal{A}_{\mathcal{T}}$ satisfies (AV2) and (AV3). □

8.2.17 Theorem TopVec *is concretely reflective in* ApVec.

Proof We know that CReg is initially closed in UAp, hence the result follows from 8.2.11. □

8.2.18 Theorem TopVec *is concretely coreflective in* ApVec. *If E is an approach vector space then its* TopVec*-coreflection is determined by the topological coreflection which makes E a topological vector space.*

Proof Let E be an approach vector space with approach system $\mathscr{A}$. By 8.2.14 there is an initial source $((E, \mathscr{A}) \to (E_i, \mathscr{A}_{d_i}))_i$ in MetVec and 8.2.11 implies that $((\underline{E}, \mathscr{A}) \to (\underline{E_i}, \mathscr{A}_{d_i}))_i$ is initial too. Since initial sources are preserved by coreflections,

$$((\underline{E}, \mathscr{T}_{\mathscr{A}}) \to (\underline{E_i}, \mathscr{T}_{d_i}))_i$$

is initial in CReg and, because CReg is initially closed in UAp, this source is initial in UAp. We know from 8.2.15 that $(E_i, \mathscr{T}_{d_i})$ are topological vector spaces and we conclude, by 8.2.17 in connection with 8.2.11 that $(E, \mathscr{T}_{\mathscr{A}})$ is a topological vector space. □

8.2.19 Corollary TopVec *is finally closed in* ApVec.

In the setting of functional analysis the notion of a topological vector space plays a fundamental role. However, interesting results often only work well or are valid in the setting of locally convex spaces. A topological vector space E is locally convex if and only if it can be generated by a collection $\mathscr{P}$ of seminorms, in the sense that

$$\{\{x \in E \mid p(x) < \varepsilon\} \mid p \in \mathscr{P}, \varepsilon > 0\}$$

is a basis for the neighbourhood system of 0.

We now sketch a theory of locally convex approach spaces, leaving a good many details to the reader. Recall that a functional $\varphi \in \mathbb{P}^E$ is called *convex* if

$$\forall x, y \in E, \forall \lambda \in [0, 1] : \varphi(\lambda x + (1 - \lambda)y) \leq \lambda\varphi(x) + (1 - \lambda)\varphi(y),$$

which is equivalent to stating that, whenever we take $x_1, \ldots, x_n \in E$ and real numbers $\lambda_1, \ldots, \lambda_n \in [0, 1]$ with $\sum_{i=1}^n \lambda_i = 1$, we have

$$\varphi(\sum_{i=1}^n \lambda_i x_i) \leq \sum_{i=1}^n \lambda_i \varphi(x_i).$$

8.2.20 Proposition *If E is an approach vector space, then the following properties are equivalent.*

1. *$\mathscr{A}(0)$ has a basis of seminorms.*
2. *$\mathscr{A}(0)$ has a basis of balanced, absorbing and convex functionals.*

Proof This goes along the same lines as foregoing proofs and we leave this to the reader. □

8.2.21 Definition An approach vector space satisfying the properties of 8.2.20 is called a *locally convex approach space*. If we take locally convex approach spaces as objects and linear contractions as morphisms we obtain a full subcategory of ApVec, which we denote lcApVec.

A set $\mathcal{M}$ of seminorms on E is called a *Minkowski system* (on E) if it is a saturated ideal in the lattice of all seminorms on E.

If $(E, \mathcal{A})$ is a locally convex approach space, then

$$\mathcal{M}_{\mathcal{A}} := \{\eta \in \mathcal{A}(0) \mid \eta \text{ is a seminorm}\}$$

is a Minkowski system. Every Minkowski system is obtained in this way. Indeed, let $\mathcal{M}$ be a Minkowski system on E. Then, letting $\mathcal{A}_{\mathcal{M}}$ be the approach system which in each point x is generated by

$$\{\nu^{(2)}(x, \cdot) \mid \nu \in \mathcal{M}\}$$

it follows that $(E, \mathcal{A}_{\mathcal{M}})$ is a locally convex approach space such that $\mathcal{M} = \mathcal{M}_{\mathcal{A}_{\mathcal{M}}}$. Moreover, by 8.2.20, the approach system $\mathcal{A}$ of a locally convex approach space is derived from a Minkowski system as above because we have $\mathcal{A} = \mathcal{A}_{\mathcal{M}_{\mathcal{A}}}$. This shows that, given a vector space E, there is a one-to-one correspondence between locally convex approach structures on E and Minkowski systems on E. Note that if $\mathcal{M}$ is a Minkowski system, then the collection $\{\eta^{(2)} \mid \eta \in \mathcal{M}\}$ is a gauge basis for the gauge of the approach structure derived from $\mathcal{M}$.

It is easily verified that if $(E_1, \mathcal{M}_1)$ and $(E_2, \mathcal{M}_2)$ are locally convex approach spaces given by their Minkowski systems then a linear map $f : E_1 \longrightarrow E_2$ is a morphism in lcApVec if and only if for all $\eta \in \mathcal{M}_2 : \eta \circ f \in \mathcal{M}_1$.

8.2.22 Theorem lcApVec *is initially closed in* ApVec. *Therefore* lcApVec *is topological over* Vec.

Proof Consider a source

$$(f_i : E \longrightarrow (E_i, \mathcal{A}_i))_{i \in I}$$

in lcApVec and let $\mathcal{A}$ be the initial ApVec-structure on E for this source, viewed as a source in ApVec. For each $i \in I$, let $\mathcal{M}_i$ be the Minkowski system of $(E_i, \mathcal{A}_i)$. It was shown in 8.2.11 that initial structures in ApVec are just initial approach structures, and therefore it follows that

$$\{\sup_{j=1}^{n} \eta_{i_j} \circ f_{i_j} \mid n \in \mathbb{N}_0, \ \forall j \in \{1, \ldots, n\} : i_j \in I, \ \eta_{i_j} \in \mathcal{M}_{i_j}\}$$

is a basis for $\mathcal{A}(0)$ which consists of seminorms, yielding that $(E, \mathcal{A})$ is a locally convex approach space. Hence $(f_i : (E, \mathcal{A}) \longrightarrow (E_i, \mathcal{A}_i))_{i \in I}$ is initial in lcApVec. □

8.2.23 Corollary lcApVec *is concretely reflective in* ApVec.

Remark that if $(E, \mathscr{A})$ is an approach vector space then the set of seminorms in $\mathscr{A}(0)$ can be taken as the Minkowski system of an lcApVec structure on E. Clearly this structure determines the reflection of $(E, \mathscr{A})$ in lcApVec.

Now let (E, η) be a seminormed space. Then $\eta^{(2)}$ is a metric and $(E, \mathscr{A}_{\eta^{(2)}})$ is a locally convex approach space. We will let sNorm stand for the category with objects seminormed spaces and morphisms linear nonexpansive maps.

8.2.24 Theorem sNorm *is embedded as a full subcategory of* lcApVec. *Moreover, a locally convex approach space which is at the same time a vector metric space, or for which the approach structure is metric, is a seminormed space.*

Proof If d is a vector metric such that $\mathscr{A}_d(0)$ has a basis of seminorms then the prenorm $d(0, \cdot)$ is a supremum of seminorms and hence is a seminorm. Furthermore, it is easy to see that associating to every seminormed space (E, η) the locally convex approach space $(E, \mathscr{A}_{\eta^{(2)}})$ defines a concrete full embedding of sNorm into lcApVec.

If $(E, \mathscr{A})$ is a locally convex approach space which is at the same time a metric object in App then the approach gauge corresponding to $\mathscr{A}$ contains a largest metric d which can be written as the pointwise supremum of a set of vector metrics. Repeating the proof of 8.2.13 now yields that d itself is a vector metric, hence $d(0, \cdot)$ is a prenorm on E which generates $\mathscr{A}(0)$. □

8.2.25 Theorem sNorm *is initially dense in* lcApVec.

Proof This follows from 8.2.20. If $(E, \mathscr{A})$ is a locally convex approach space, then the source

$$(\mathrm{id}_E : (E, \mathscr{A}) \to (E, \eta))_{\eta \in \mathscr{M}_{\mathscr{A}}}$$

is initial in lcApVec. □

8.2.26 Theorem *The following properties hold.*

1. lcTopVec *is embedded as a full subcategory of* lcApVec.
2. lcTopVec *is concretely coreflective in* lcApVec. *Moreover, the diagram*

$$\begin{array}{ccc} \mathsf{lcTopVec} & \xrightarrow{\;c\;} & \mathsf{lcApVec} \\ \downarrow & & \downarrow \\ \mathsf{CReg} & \xrightarrow{\;c\;} & \mathsf{UAp} \end{array}$$

   where the horizontal arrows are embeddings and the vertical arrows the forgetful functors, is commutative.
3. *A locally convex approach space such that the underlying approach system is topological is a locally convex topological space.*

Proof 1. A locally convex topological space is also an approach vector space and since the characteristic function of a convex set is a convex function we know the approach system of 0 has a basis of convex functions.

2. Let $(E, \mathscr{A})$ be a locally convex approach space and let $(E, \mathscr{T})$ be the topological coreflection of $(E, \mathscr{A})$ in App. From 8.2.18 we know that $(E, \mathscr{T})$ is a topological vector space. Since $\mathscr{M}_{\mathscr{A}}$ is a basis for $\mathscr{A}(0)$, we know that

$$\{\{\nu \leq \varepsilon\} \mid \varepsilon > 0,\ \nu \in \mathscr{M}\}$$

is a basis of convex sets for the neighbourhood system of 0.

3. Let $(E, \mathscr{A})$ be a topological locally convex approach space. Then by 2 above, $(E, \mathscr{A})$ equals its lcTopVec coreflection and is thus a locally convex topological space. □

8.2.27 Proposition *If $\mathscr{M}$ is an ideal of seminorms, then it is the Minkowski system of a locally convex topological space if and only if*

$$\forall \lambda \in \mathbb{R}^+,\ \forall \eta \in \mathscr{M}\ :\ \lambda\eta \in \mathscr{M}.$$

Proof The Minkowski system of a locally convex topological space is just the set of Minkowski functionals of the balanced, convex and absorbing open sets and hence it satisfies the above condition. Conversely, if the condition is satisfied then $\mathscr{M}$ is saturated, and in this case the associated approach system is topological. □

8.2.28 Corollary lcTopVec *is initially and finally closed in* lcApVec.

8.2.29 Corollary *If E is a locally convex approach space with Minkowski system $\mathscr{M}$, then the set*

$$\{\lambda\eta \mid \lambda \in \mathbb{R}^+,\ \eta \in \mathscr{M}\}$$

is the Minkowski system of the topological coreflection of E.

8.3 Comments

1. **Weak and weak* spaces as objects in** lcApVec

In 8.1.2 we have seen that the source $(f : (E, \delta_{(E,E')}) \longrightarrow (\mathbb{R}, \delta_{\mathbb{E}}))_{f \in B_{E'}}$ is initial and analogously in 8.1.19 that the source $(\hat{x} : (E', \delta_{(E',E)}) \longrightarrow (\mathbb{R}, \delta_{\mathbb{E}}))_{x \in B_E}$ is initial. Since $(\mathbb{R}, \delta_{\mathbb{E}})$ clearly is an object in sNorm it follows from our considerations above, in particular 8.2.24 and 8.2.25, that $(E, \delta_{(E,E')})$ and $(E', \delta_{(E',E)})$ are objects in lcApVec. They are moreover clearly objects neither in sNorm nor in lcTopVec, i.e. they are genuine locally convex approach spaces.

2. **Contractivity versus continuity in the definition of** ApVec

Conditions (AV2-3) may seem surprising at first since one might expect some condition on the contractivity of scalar multiplication. However, clearly this would make no sense as multiplication of real numbers is a continuous but not a contractive function in terms of the usual Euclidean metric of $\mathbb{R}$. However, it is the characterization of topological vector spaces by means of neighbourhoods of the origin that suggests the correct way of working, and which shows that conditions (AV2-3) are indeed the correct generalization.

3. **Further results**

For a more extensive study of the concepts introduced in this chapter we refer to Lowen and Verwulgen (2004) where approach vector spaces were introduced, Sioen and Verwulgen (2006) where seminormed spaces are studied, Sioen and Verwulgen (2004) where the particular role of the unit ball is highlighted, Sioen and Verwulgen (2003) where locally convex approach spaces are studied, Van Olmen and Verwulgen (2006) where the concept of reflexivity is treated, Verwulgen (2007a) where the dual of $\mathscr{C}(X, \mathbb{R})$ is studied and Verwulgen (2007b) where links are studied with the notion of absolutely convex modules. Convex modules were extensively studied in Pumplün (2001) and Pumplün and Röhrl (1984, 1985). Further information can be found in the PhD thesis of Verwulgen (2003).

Chapter 9
Approach Theory Meets Probability

The scientific imagination always restrains itself within the limits of probability.

(Thomas Huxley)

It is not certain that everything is uncertain.

(Blaise Pascal)

In this chapter we see that the construction of the weak* approach structure performed in the previous chapter, as in the classical topological setup, when restricted to probability measures allows for a quantification of the weak topology on probability measures. Here it will be important however also to consider other quantifications, depending on the problem at hand. This is not unlike the situation whereby, depending on the situation it is advantageous to be able to have different choices of metrics for a metrizable topological space.

In the first section we consider the general case of probability measures on a Polish space and give a canonical quantification of the weak topology which allows for an indexed version of a Portmanteau theorem both for distances and for limit operators (see 9.1.2 and 9.1.5).

In the second section we consider a quantification of convergence in probability for random variables and again we can give indexed versions of classical theorems relating the weak approach structure and the approach structure of convergence in probability (see 9.2.13).

In the third section we introduce a weak and a strong index of tightness and prove an indexed version of Prokhorov's theorem linking these indices to the index of relative sequential compactness (see 9.3.7).

In the last section we then prove an indexed version of the Lindeberg-Feller central limit theorem (9.4.15) making use of a natural Lindeberg index indicating to what extent the Lindeberg condition is fulfilled.

R. Lowen, *Index Analysis*, Springer Monographs in Mathematics,
DOI 10.1007/978-1-4471-6485-2_9

9.1 Spaces of Probability Measures

Let S be a Polish space with a fixed complete metric d, with topology $\mathcal{T}$, and with set of Borel sets $\mathcal{B}$. $\mathcal{P}(S)$ will denote the set of all probability measures on $\mathcal{B}$. One of the most important and most widely used structures on $\mathcal{P}(S)$ is the so-called *weak topology*, which we denote by $\mathcal{T}_w$, (see Billingsley 1968; Parthasaraty 1967). Although this topology is called the weak topology, from the point of view of functional analysis, and the foregoing chapter, it would better have been called the weak* topology, but we will adhere to the usual terminology. We consider the Banach space of all continuous bounded real-valued functions $\mathcal{C}_b(S)$ equipped with the supremum norm and consider its continuous dual $\mathcal{C}_b(S)'$. $\mathcal{P}(S)$ is embedded in $\mathcal{C}_b(S)'$ by the assignment

$$\mathcal{P}(S) \longrightarrow \mathcal{C}_b(S)' : P \mapsto (f \mapsto \int f dP)$$

and as such is identified with the dual unit sphere. Thus it inherits the weak* topology induced on $\mathcal{C}_b(S)'$ by $\mathcal{C}_b(S)$ via restriction. This weak* topology is a locally convex topology generated by the collection of seminorms $\{p_f \mid f \in \mathcal{C}_b(S), 0 \leq f \leq 1\}$ where

$$p_f(P) := |\int f dP|$$

and the restriction to $\mathcal{P}(S)$ is called the weak topology on probability measures. The above collection of seminorms generates a collection of metrics

$$d_f(P, Q) = |\int f dP - \int f dQ|$$

and it is immediately clear that this collection of metrics is a subbasis for a gauge, precisely

$$\{ \sup_{f \in \mathcal{H}} d_f \mid \mathcal{H} \subseteq \mathcal{C}_b(S) \text{ finite}, \forall f \in \mathcal{H} : 0 \leq f \leq 1\}$$

is an ideal basis which generates a unique and canonical approach structure on $\mathcal{P}(S)$ which we refer to as the *weak approach structure* on probability measures. All associated structures will be denoted by the index w. Thus the gauge generated by the above basis will be denoted $\mathcal{G}_w$.

We have introduced the above entirely internally to the setting, but note that it completely concords with what we did in the previous chapter concerning the weak* distance. If we put $E := \mathcal{C}_b(S)$ then the structure above is nothing else than the restriction of $\delta_{(E',E)}$ to the image of $\mathcal{P}(S)$, i.e. to the unit sphere of E'.

We recall that the weak topology has various different but equivalent bases $\mathcal{B}^i_w$, $i = 1, \ldots, 4$ for the neighbourhoods. Let $P \in \mathcal{P}(S)$. Then $\mathcal{B}^1_w$ consists of the sets

$V^1(P, \mathscr{G}, \varepsilon)$ where $\mathscr{G}$ is a finite collection of open sets, $\varepsilon > 0$ and

$$V^1(P, \mathscr{G}, \varepsilon) := \{Q \in \mathscr{P}(S) | \forall G \in \mathscr{G} : Q(G) > P(G) - \varepsilon\}.$$

$\mathscr{B}_w^2$ consists of the sets $V^2(P, \mathscr{F}, \varepsilon)$ where $\mathscr{F}$ is a finite collection of closed sets, $\varepsilon > 0$ and

$$V^2(P, \mathscr{F}, \varepsilon) := \{Q \in \mathscr{P}(S) \mid \forall F \in \mathscr{F} : Q(F) < P(F) + \varepsilon\}.$$

$\mathscr{B}_w^3$ consists of the sets $V^3(P, \mathscr{H}, \varepsilon)$ where $\mathscr{H}$ is a finite collection of continuous (respectively uniformly continuous or Lipschitz) functions taking values in $[0, 1]$, $\varepsilon > 0$ and

$$V^3(P, \mathscr{H}, \varepsilon) := \left\{Q \in \mathscr{P}(S) | \forall f \in \mathscr{H} : \left|\int f dP - \int f dQ\right| < \varepsilon\right\}.$$

$\mathscr{B}_w^4$ consists of the sets $V^2(P, \mathscr{E}, \varepsilon)$ where $\mathscr{E}$ is a finite collection of P-continuity sets, $\varepsilon > 0$ and

$$V^2(P, \mathscr{E}, \varepsilon) := \{Q \in \mathscr{P}(S) \mid \forall E \in \mathscr{E} : |P(E) - Q(E)| < \varepsilon\}.$$

We also recall that the total variation metric (Parthasaraty and Steerneman 1985) is defined as

$$d_{TV}(P, Q) := \sup_{B \in \mathscr{B}} |P(B) - Q(B)|$$

and is equally well given by various formulas analogous to the various bases for the weak topology given above, and notably by

$$d_{TV}(P, Q) = \sup_{f \in \mathscr{C}_b(S), 0 \leq f \leq 1} \left|\int f dP - \int f dQ\right|.$$

For more information on these metrics we refer to Rachev (1991). The weak approach structure, notably its gauge $\mathscr{G}_w$, also has various different bases. For any finite collection $\mathscr{G}$ of open sets, we let

$$d_1^{\mathscr{G}} : \mathscr{P}(S) \times \mathscr{P}(S) \longrightarrow \mathbb{P} : (P, Q) \mapsto \sup_{G \in \mathscr{G}} P(G) \ominus Q(G)$$

and we put $\mathscr{D}_1 := \left\{d_1^{\mathscr{G}} \mid \mathscr{G} \text{ finite collection of open sets}\right\}$. For any finite collection $\mathscr{F}$ of closed sets, we let

$$d_2^{\mathscr{F}} : \mathscr{P}(S) \times \mathscr{P}(S) \longrightarrow \mathbb{P} : (P, Q) \mapsto \sup_{F \in \mathscr{F}} Q(F) \ominus P(F)$$

and we put $\mathcal{D}_2 := \{d_2^{\mathcal{F}} \mid \mathcal{F} \text{ finite collection of closed sets}\}$. For any finite collection $\mathcal{H}$ of continuous maps, with range [0, 1], we let

$$d_3^{\mathcal{H}} : \mathcal{P}(S) \times \mathcal{P}(S) \longrightarrow \mathbb{P} : (P, Q) \mapsto \sup_{f \in \mathcal{H}} \left| \int f dP - \int f dQ \right|$$

and we put $\mathcal{D}_3 := \{d_3^{\mathcal{H}} \mid \mathcal{H} \subseteq C_b(S) \text{ finite}, \forall f \in \mathcal{H} : 0 \leq f \leq 1\}$. As before, continuous maps may be replaced by uniformly continuous maps or Lipschitz maps. For any finite collection $\mathcal{E}$ of P-continuity sets we put

$$d_4^{\mathcal{E}} : \mathcal{P}(S) \times \mathcal{P}(S) \longrightarrow \mathbb{P} : (P, Q) \mapsto \sup_{E \in \mathcal{E}} |P(E) - Q(E)|$$

and we put $\mathcal{D}_4 := \{d_4^{\mathcal{E}} \mid \mathcal{E} \text{ finite set of } P\text{-continuity sets}\}$. For any $\alpha > 0$ we let

$$d_5^{\alpha} : \mathcal{P}(S) \times \mathcal{P}(S) \longrightarrow \mathbb{P} : (P, Q) \mapsto \sup_{A \in \mathcal{B}} P(A) \ominus Q(A^{(\alpha)})$$

and we put $\mathcal{D}_5 := \{d_5^{\alpha} | \alpha > 0\}$.

The collections $\mathcal{D}_i, i \in \{1, 2\}$ consist of quasi-metrics, whereas the collections $\mathcal{D}_i$, $i \in \{3, 4\}$ consist of metrics. The mappings d_5^{α} do not individually satisfy the triangle inequality, however, as is easily verified, they do satisfy the combined inequality

$$d_5^{\alpha}(P, Q) \leq d_5^{\alpha/2}(P, R) + d_5^{\alpha/2}(R, Q)$$

for any $\alpha > 0$ and any $P, Q, R \in \mathcal{P}(S)$. This last collection is inspired by the so-called Prokhorov metric (Billingsley 1968) which is defined as

$$\rho(P, Q) := \inf \left\{ \alpha > 0 \mid \forall A \in \mathcal{B} : P(A) \leq Q(A^{(\alpha)}) + \alpha \right\}.$$

The collections $\mathcal{D}_w^i$ for $i = 1, \ldots, 4$ are bases for one and the same gauge. The collection $\mathcal{D}_w^5$ allows us to define a basis for an approach system by considering the functions $d_5^{\alpha}(P, \cdot)$. Actually all these collections generate the same structure. In order to prove this we collect the main technical arguments in the following preliminary lemma which will be used several times.

9.1.1 Lemma *The following properties hold.*

1. *For each $P \in \mathcal{P}(S)$, $\varepsilon > 0$ and $\alpha > 0$ there exists a finite collection $\mathcal{G}$ of open sets in S such that for every $Q \in \mathcal{P}(S)$*

$$\sup_{A \in \mathcal{B}_S} (P(A) - Q(A^{(\alpha)})) \leq \sup_{G \in \mathcal{G}} (P(G) - Q(G)) + \varepsilon.$$

2. *For each $P \in \mathscr{P}(S)$, $\varepsilon > 0$ and $F \subseteq S$ closed there exists an $\alpha > 0$ such that for every $Q \in \mathscr{P}(S)$*

$$Q(F) - P(F) \leq \sup_{A \in \mathscr{B}_S} (P(A) - Q(A^{(\alpha)})) + \varepsilon.$$

3. *For each $P \in \mathscr{P}(S)$, $\varepsilon > 0$ and $F \subseteq S$ closed there exists $f \in \mathscr{C}_b(S)$ with $0 \leq f \leq 1$ such that for all $Q \in \mathscr{P}(S)$*

$$Q(F) - P(F) \leq \left| \int f dP - \int f dQ \right| + \varepsilon.$$

4. *For each $f \in \mathscr{C}_b(S)$ such that $0 \leq f \leq 1$ and $\varepsilon > 0$ there exists a finite set of closed sets $\mathscr{F}$ such that for all $P, Q \in \mathscr{P}(S)$*

$$\left| \int f dP - \int f dQ \right| \leq \sup_{F \in \mathscr{F}} (Q(F) - P(F)) + \varepsilon.$$

Proof 1. By separability we can choose a finite collection of open balls $(B_i)_{i=1}^j$ with radii $\alpha/4$ such that $P\left(S \setminus \cup_{i=1}^j B_i\right) \leq \varepsilon$. Then the collection

$$\mathscr{G} := \left\{ \left(B_{i_1} \cup \ldots \cup B_{i_k}\right)^{(\alpha/2)} \mid 1 \leq i_1 < \ldots < i_k \leq j \right\}$$

satisfies the requirement. Indeed, take a probability measure Q in $\mathscr{P}(S)$ and a Borel set A in S. Let I be the set of those natural numbers $1 \leq i \leq j$ for which $B_i \cap A \neq \emptyset$ and put $B := \cup_{i \in I} B_i$. Then we have

$$\begin{aligned} P(A) &\leq P(B) + P(S \setminus \cup_{i=1}^j B_i) \\ &\leq P(B^{(\alpha/2)}) + \varepsilon \\ &\leq \sup_{G \in \mathscr{G}} (P(G) - Q(G)) + Q(B^{(\alpha/2)}) + \varepsilon. \end{aligned}$$

In view of the fact that $B^{(\alpha/2)} \subseteq A^{(\alpha)}$, we conclude that

$$P(A) \leq \sup_{G \in \mathscr{G}} (P(G) - Q(G)) + Q(A^{(\alpha)}) + \varepsilon.$$

2. We can choose $\alpha > 0$ such that $P(F^{(\alpha)}) \leq P(F) + \varepsilon$. For any probability measure Q in $\mathscr{P}(S)$ it then follows that

$$\begin{aligned} Q(F) - P(F) &\leq (Q(F) - P(F^{(\alpha)})) + \varepsilon \\ &\leq \sup_{A \in \mathscr{B}_S} (Q(A) - P(A^{(\alpha)})) + \varepsilon. \end{aligned}$$

3. Again we can choose $\alpha > 0$ such that $P(F^{(\alpha)}) \leq P(F)+\varepsilon$. Then the function f defined by

$$f(x) := 1 \ominus \frac{1}{\alpha} d(x, F)$$

satisfies the requirement. 4. Choose $k \in \mathbb{N}_0$ such that $\frac{1}{k} \leq \varepsilon$ and, for all $i \in \{1, \ldots, k\}$, let $F_i := \{\frac{i}{k} \leq f\}$ and consider the collection $\mathscr{F} := \{F_i \mid i \in \{1, \ldots, k\}\}$. Then for any $P \in \mathscr{P}$,

$$\frac{1}{k} \sum_{F \in \mathscr{F}} P(F) \leq \int f dP \leq \frac{1}{k} + \frac{1}{k} \sum_{F \in \mathscr{F}} P(F)$$

from which it follows that the collection $\mathscr{F}$ satisfies the requirement. □

9.1.2 Theorem (Distance portmanteau theorem) *All collections $\mathscr{D}_i, i \in \{1, \ldots, 5\}$ are bases for $\mathscr{G}_w$ and $(\mathscr{D}_i(P))_P$ is a collection of bases for the associated approach systems. Writing out the associated distance explicitly then gives the following expressions where $P \in \mathscr{P}(S)$ and $\Gamma \subseteq \mathscr{P}(S)$*

$$\begin{aligned}
\delta_w(P, \Gamma) &= \sup_{\mathscr{G}} \inf_{Q \in \Gamma} \sup_{G \in \mathscr{G}} (P(G) \ominus Q(G)) \\
&= \sup_{\mathscr{F}} \inf_{Q \in \Gamma} \sup_{F \in \mathscr{F}} (Q(F) \ominus P(F)) \\
&= \sup_{\mathscr{E}} \inf_{Q \in \Gamma} \sup_{E \in \mathscr{E}} |P(E) - Q(E)| \\
&= \sup_{\mathscr{C}} \inf_{Q \in \Gamma} \sup_{f \in \mathscr{C}} \left| \int f dP - \int f dQ \right| \\
&= \sup_{\alpha > 0} \inf_{Q \in \Gamma} \sup_{A \in \mathscr{B}} (P(A) \ominus Q(A^{(\alpha)}))
\end{aligned}$$

where the first suprema in the first four expressions are respectively taken over all finite collections $\mathscr{G}$ (respectively $\mathscr{F}$) of open (respectively closed) sets, $\mathscr{E}$ of P-continuity sets, $\mathscr{C}$ of continuous (respectively uniformly continuous or Lipschitz) functions with range [0, 1].

Proof In order to see that all collections are equivalent bases for the same gauge it suffices to use lemma 9.1.1 and the saturation condition for gauges and approach systems. The formulas then follow immediately from the characterization of a distance generated by a gauge (see 1.2.6) or by an approach system (see 1.2.34) and again from lemma 9.1.1. □

9.1.3 Proposition *The* Top*-coreflection of $(\mathscr{P}(S), \delta_w)$ is determined by the weak topology and the* Met*-coreflection is determined by the total variation metric.*

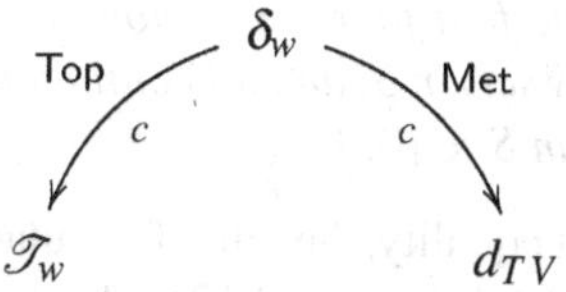

Proof This follows from 8.1.17. □

In Topsøe (1970) the author showed that the weak topology on $\mathscr{P}(S)$ also is initial for the source

$$(\omega_G : \mathscr{P}(S) \longrightarrow (\mathbb{P}, \mathscr{T}_{\mathbb{P}}) : P \mapsto P(G))_{G\in\mathscr{T}}.$$

The next result shows that Topsøe's theorem also holds in our setting.

9.1.4 Theorem *The weak distance δ_w on $\mathscr{P}(S)$ is initial for the source*

$$(\omega_G : \mathscr{P}(S) \longrightarrow \mathbb{P} : P \mapsto P(G))_{G\in\mathscr{T}}.$$

Proof Since the maps ω_G only attain finite values and since a subbasis for the initial gauge is given by

$$\mathscr{B} := \{d \circ (\omega_G \times \omega_G) \mid G \text{ open}, d \in \mathscr{G}_{\mathbb{P}}\}$$

where $\mathscr{G}_{\mathbb{P}}$ stands for the gauge of $\mathbb{P}$, this is an immediate consequence of the fact that a basis for the gauge of $\mathbb{P}$ is given by the quasi-metrics

$$\mathbb{P} \times \mathbb{P} \longrightarrow \mathbb{P} : (x, y) \mapsto (x \wedge a \ominus y \wedge a)$$

where $a < \infty$ (see 1.2.62). Since the values which come into play are bounded by 1 it suffices to consider the quasi-metric $d(x, y) = x \ominus y$. The subbasis $\mathscr{B}$ hence generates the basis $\mathscr{D}_1$. □

9.1.5 Theorem (Convergence portmanteau theorem) *Given a sequence $(P_n)_n$ and P in $\mathscr{P}(S)$ we have*

$$\begin{aligned}
\lambda_w\langle (P_n)_n\rangle(P) &= \sup_G \limsup_n (P(G) \ominus P_n(G)) \\
&= \sup_F \limsup_n (P_n(F) \ominus P(F)) \\
&= \sup_A \limsup_n |P(A) - P_n(A)| \\
&= \sup_f \limsup_n \left| \int f dP - \int f dP_n \right| \\
&= \sup_{\alpha>0} \limsup_n \sup_{A\in\mathscr{B}} (P(A) \ominus P_n(A^{(\alpha)}))
\end{aligned}$$

where the first suprema in the first four expressions respectively run over all open sets, closed sets, P-continuity sets in S, and all continuous (or uniformly continuous, or Lipschitz) functions f from S to [0, 1].

Proof We only prove the first equality, the other formulas then follow from this upon applying once again lemma 9.1.1. From 1.3.12 and 9.1.4 it follows that

$$\lambda_w\langle(P_n)_n\rangle(P) = \sup_{G\in\mathcal{T}} \lambda_{\mathbb{P}}(P_n(G))(P(G)).$$

Now it suffices to remark that all values are finite (less than 1) and hence, as in 9.1.4 the structure of $\mathbb{P}$ which comes into play is only the structure in finite points (the interval [0,1]) and there the structure is simply a quasi-metric ($d(x, y) = (x-y)\vee 0$). Hence it follows that

$$\lambda_{\mathbb{P}}\langle(P_n(G))_n\rangle(P(G)) = \inf_n \sup_{k\geq n}(P(G)\ominus P_k(G))$$

and the formula follows. □

When the expressions in the foregoing result become zero, one obtains all characterizations of weak convergence in the classic portmanteau theorem.

9.1.6 Corollary (Classic portmanteau theorem Billingsley 1968; Parthasarathy 1967) *A sequence $(P_n)_n$ in $\mathcal{P}(S)$ converges weakly to $P\in\mathcal{P}(S)$ if and only if any of the following equivalent properties holds.*

1. $\forall G$ *open* : $P(G)\leq \liminf P_n(G)$.
2. $\forall F$ *closed* : $\limsup_n P_n(F)\leq P(F)$.
3. $\forall P$*-continuity-set* A : $\lim_n P_n(A) = P(A)$.
4. $\forall f\in\mathcal{F}(S,[0,1])$: $\lim_n \int f dP_n = \int f dP$.

where $\mathcal{F}(S,[0,1])$ *stands for all continuous (or uniformly continuous, or Lipschitz) functions from S to* [0, 1].

Another fundamental fact about the weak topology is of a categorical nature. If $f : X \longrightarrow Y$ is a continuous function then its canonical extension $\widehat{f} : \mathcal{P}(X) \longrightarrow \mathcal{P}(Y)$ defined by $\widehat{f}(P)(B) := P\left(f^{-1}(B)\right)$, for all $B\in\mathcal{B}(Y)$, is continuous with respect to the weak topologies. The result here is stronger.

9.1.7 Proposition *If $f : X \longrightarrow Y$ is a continuous function and we equip $\mathcal{P}(X)$ and $\mathcal{P}(Y)$ with the weak approach structures, then $\widehat{f} : \mathcal{P}(X) \longrightarrow \mathcal{P}(Y)$ is a contraction.*

Proof Let $\mathcal{G}$ be a finite collection of open sets in Y, and let $P, Q\in\mathcal{P}(X)$, then

$$\begin{aligned} d_1^{\mathcal{G}}\left(\widehat{f}(P),\widehat{f}(Q)\right) &= \sup_{G\in\mathcal{G}} P\left(f^{-1}(G)\right)\ominus Q\left(f^{-1}(G)\right)\\ &= d_1^{\mathcal{H}}(P,Q), \end{aligned}$$

where $\mathscr{H} := \{f^{-1}(G) \mid G \in \mathscr{G}\}$. This proves that

$$d_1^{\mathscr{G}} \circ \widehat{f} \times \widehat{f} = d_1^{\mathscr{H}} \in \mathscr{G}_w,$$

which proves our claim. □

The following corollary was shown in Topsøe (1970).

9.1.8 Corollary *If $f : X \longrightarrow Y$ is continuous then $\widehat{f} : \mathscr{P}(X) \longrightarrow \mathscr{P}(Y)$ is continuous with respect to the weak topologies.*

9.1.9 Corollary *If $f : X \longrightarrow Y$ is continuous then $\widehat{f} : \mathscr{P}(X) \longrightarrow \mathscr{P}(Y)$ is nonexpansive with respect to the total variation metrics.*

In the following result $\rho \leq d_{TV}$ is a known inequality (see Zolotarev 1983). Moreover, easy examples show that in general all inequalities are strict.

9.1.10 Proposition *The following inequalities hold. $\delta_\rho \leq \delta_w \leq \delta_{d_{TV}}$.*

Proof Fix a probability measure P and a collection of probability measures Γ on S. If $\delta_w(P, \Gamma) < \gamma$ then we can find a measure $Q_\gamma \in \Gamma$ such that

$$\sup_{A \in \mathscr{B}_S} (P(A) - Q_\gamma(A^{(\gamma)})) < \gamma$$

and hence

$$\rho(P, \Gamma) = \inf_{Q \in \Gamma} \inf \left\{\alpha > 0 \mid \forall A \in \mathscr{B}_S : P(A) \leq Q(A^{(\alpha)}) + \alpha\right\} \leq \gamma.$$

This, by the arbitrariness of γ, implies that $\rho(P, \Gamma) \leq \delta_w(P, \Gamma)$. The second inequality is an immediate consequence of the fact that d_{TV} is the metric coreflection of the weak approach structure. □

It is known that the weak topology is completely metrizable. However whereas this requires the choice of a new "external" structure (a complete compatible metric) the weak approach structure does not require this, it is complete itself.

9.1.11 Theorem *The weak approach structure is complete.*

Proof From 3.5.11 it is sufficient to verify that the metric coreflection is complete. This however is a well-known fact (see Jacka and Roberts 1997). □

9.1.12 Theorem *The weak approach structure is locally countable, in particular for each $P \in \mathscr{P}(S)$ the localized gauge*

$$\mathscr{G}_w(P) := \{d(P, \cdot) \mid d \in \mathscr{G}_w\}$$

has a countable basis.

Proof Although by definition the weak approach structure was constructed making use of the (non-separable) Banach space $\mathscr{C}_b(S)$, when considering the various bases for the weak gauge $\mathscr{G}_w$, in particular $\mathscr{D}_3$ we mentioned that we could also restrict ourselves to uniformly continuous maps $S \longrightarrow [0, 1]$. Now since S is a separable metrizable space it can be embedded in a countable product of unit intervals and consequently there exists an equivalent totally bounded metrization of S. The completion $\widehat{S}$ of S under this metric hence is compact and the Banach spaces $\mathscr{C}(\widehat{S})$ and $\mathscr{U}(S)$ of uniformly continuous maps are isomorphic. $\widehat{S}$ being compact, $\mathscr{C}(\widehat{S})$ and hence also $\mathscr{U}(S)$ are separable. Then it follows that also the space of uniformly continuous maps $S \longrightarrow [0, 1]$ is separable. If $\mathscr{E}$ is a countable dense subset then it follows from the definition of $\mathscr{D}_3$ (with uniformly continuous maps) that an alternative equivalent basis for $\mathscr{G}_w$ is given by

$$\mathscr{D}_3' := \left\{ d_3^{\mathscr{H}} \mid \mathscr{H} \subseteq \mathscr{E} \text{ finite}, \forall f \in \mathscr{H} : 0 \leq f \leq 1 \right\}.$$

As this basis is countable, so are the localized bases $\mathscr{D}_3'(P) = \{d(P, \cdot) \mid d \in \mathscr{D}_3^s\}$ and hence $(\mathscr{P}(S), \delta_w)$ is locally countable. □

Theorem 9.3.7 will provide us with an "index-version" of Prokhorov's theorem. For its proof some preparation is required.

For a collection Γ of probability measures on S and $\varepsilon > 0$ we will consider the set

$$\Gamma(\varepsilon) := \{(1 - \varepsilon')P + \varepsilon' Q \mid P \in \Gamma, Q \in \mathscr{P}(X), 0 \leq \varepsilon' \leq \varepsilon\}$$

see e.g. Morgenthaler (2007) and Lo (2000) for the use of these types of "contaminated" sets in robust statistics and game theory. The following result furnishes an estimate for the index of relative sequential compactness of such sets. We denote this index with a superscript w to make clear that it concerns the weak approach structure.

9.1.13 Proposition *For a set $\Gamma \subseteq \mathscr{P}(S)$ and $\varepsilon > 0$ we have*

$$\chi_{rsc}^w(\Gamma(\varepsilon)) \leq \chi_{rsc}^w(\Gamma) + \varepsilon.$$

Proof Take $\Gamma \subseteq \mathscr{P}(S)$ and $\varepsilon > 0$. Then for a sequence $(R_n)_n$ where

$$R_n := (1 - \varepsilon_n)P_n + \varepsilon_n Q_n$$

in $\Gamma(\varepsilon)$ and $\delta > 0$ we can find a subsequence (P_{k_n}) of (P_n) and a probability measure P such that

$$\sup_{\alpha > 0} \limsup_n \sup_{A \in \mathscr{B}_S} (P(A) - P_{k_n}(A^{(\alpha)})) \leq \chi_{rsc}^w(\Gamma) + \delta.$$

Now the result follows from

$$\begin{aligned}
&\sup_{\alpha>0} \limsup_{n} \sup_{A\in\mathscr{B}_S} (P(A) - R_{k_n}(A^{(\alpha)})) \\
&= \sup_{\alpha>0} \limsup_{n} \sup_{A\in\mathscr{B}_S} (P(A) - P_{k_n}(A^{(\alpha)}) + \varepsilon_{k_n}(P_{k_n} - Q_{k_n})(A^{(\alpha)})) \\
&\leq \sup_{\alpha>0} \limsup_{n} \sup_{A\in\mathscr{B}_S} (P(A) - P_{k_n}(A^{(\alpha)})) + \varepsilon \\
&\leq (\chi^{w}_{rsc}(\Gamma) + \delta) + \varepsilon.
\end{aligned}$$

□

9.2 Spaces of Random Variables

Let $(\Omega, \mathscr{A}, P)$ be a fixed probability space and let $\mathscr{R}(S)$ be the set of all S-valued random variables on Ω. An important topology is given by the topology $\mathscr{T}_p$ of convergence in probability and a natural metric is the so-called indicator metric (see Zolotarev 1983) d_I where

$$d_I(\xi, \eta) := P\left(\{d\left(\xi, \eta\right) > 0\}\right) = P(\{\xi \neq \eta\})$$

and where, as usual, $\{d\left(\xi, \eta\right) > 0\}$ stands for the set of points $\omega \in \Omega$ such that $d(\xi(\omega), \eta(\omega)) > 0$. Note that $d_I(\xi, \eta) = 0$ if and only if ξ and η are equal almost everywhere.

We can endow $\mathscr{R}(S)$ in a natural way with an approach structure as follows. Consider the functions φ^a, $a > 0$, defined by

$$\varphi^a\left(\xi, \eta\right) = P\left(\{d\left(\xi, \eta\right) \geq a\}\right) \qquad \xi, \eta \in \mathscr{R}(S).$$

Each function, for a fixed a, gives the probability that the random variables ξ and η lie at a distance larger than or equal to a from each other. Again, as in the case of the basis $\mathscr{D}_5$ for the weak topology on probability measures, these functions do not satisfy the triangle inequality. However, again, they too satisfy an "interlinked" triangle inequality.

9.2.1 Lemma *For any $a, b > 0$ and $\xi, \eta, \zeta \in \mathscr{R}(S)$ we have*

$$\varphi^{a+b}\left(\xi, \zeta\right) \leq \varphi^a\left(\xi, \eta\right) + \varphi^b\left(\eta, \zeta\right).$$

Proof This follows from the additivity of probability measures. □

It follows that the collections

$$\mathscr{B}_P(\xi) := \{\varphi^a(\xi, \cdot) \mid a > 0\} \quad \xi \in \mathscr{R}(S)$$

form a basis for an approach structure. We will denote the generated distance by δ_p and refer to the approach structure as the *c.i.p. approach structure* where c.i.p. stands for *convergence in probability*. All associated structures too will be denoted by the index p.

As was the case for the weak approach structure in the foregoing section, here too we can find alternative bases, one of which is particularly interesting. For any $a > 0$ put

$$K_a(\xi, \eta) := \inf\{\theta \mid \varphi^{\theta a}(\xi, \eta) \le \theta\}.$$

Then it follows from 9.2.1, again using the additivity of probability measures, that the maps K_a are metrics. Actually K_1 is nothing else than the so-called *Ky-Fan metric*

$$K_1(\xi, \eta) := \inf\{\theta \mid P(\{d(\xi, \eta) \ge \theta\})\}$$

(see e.g. Billingsley (1968)). Let us denote $\mathscr{B}_1 := \{\varphi^a \mid a > 0\}$ and $\mathscr{B}_2 := \{K_a \mid a > 0\}$.

9.2.2 Theorem *Both $\mathscr{B}_1$ and $\mathscr{B}_2$ generate the c.i.p. approach structure on random variables.*

Proof By definition $\mathscr{B}_1$ generates δ_p. Let δ stand for the distance generated by $\mathscr{B}_2$. Since for any a and θ we have that $\varphi^{\theta a}(\xi, \eta) < \theta$ implies $K_a(\xi, \eta) \le \theta$ we immediately have $\delta \le \delta_p$. Conversely, if $0 < \theta < \delta_p(\xi, \Sigma)$, then there exists $a > 0$ such that $\theta < \inf_{\eta \in \Sigma} \varphi^a(\xi, \eta)$. Letting $b := a\theta^{-1}$ it follows that $\theta \le \inf_{\eta \in \Sigma} K_b(\xi, \eta)$ which proves that $\delta_p \le \delta$. □

9.2.3 Proposition *The* Top *coreflection of $(\mathscr{R}(S), \delta_p)$ is determined by the topology of convergence in probability and the* Met *coreflection is determined by the indicator metric.*

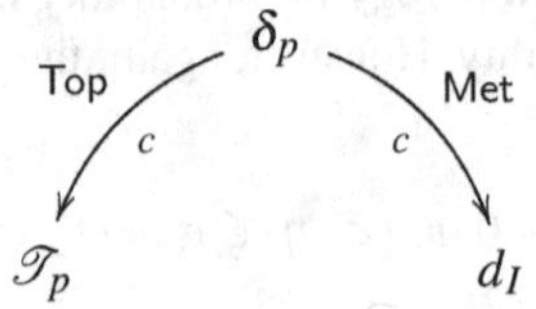

Proof For the Top-coreflection this follows at once from the definitions of both structures involved. As for the metric coreflection it suffices to note that, for any $\xi, \eta \in \mathscr{R}(S)$, we have

$$\sup_{a>0} P(\{d(\xi, \eta) \ge a\}) = P(\{d(\xi, \eta) > 0\}) = d_I(\xi, \eta). \qquad \square$$

9.2.4 Proposition *Given a sequence* $(\xi_n)_n$ *and* ξ *in* $\mathscr{R}(S)$ *we have*

$$\lambda_p \langle (\xi_n)_n \rangle (\xi) = \sup_{a>0} \limsup_{n\to\infty} P\left(\left\{d\left(\xi, \xi_n\right) \geq a\right\}\right).$$

Proof Since the basis for the gauge given by the functions φ^a is increasing with decreasing a this follows at once from the formula for a limit derived from a gauge as given in 1.2.44. □

This formula also gives the usual expression for convergence in probability since the limit operator will produce a zero value exactly if $(\xi_n)_n$ converges to ξ in probability, i.e. if

$$\forall a > 0 : \lim_n P\left(\left\{d\left(\xi, \xi_n\right) \geq a\right\}\right) = 0.$$

In analogy with the results for spaces of probability measures here too our construction is functorial (see also the comments), however of a more metric nature, which is to be expected from the prominent role played by the metric in the definition of the maps φ^a.

9.2.5 Proposition *Suppose* S *and* T *are Polish spaces with fixed metrics* d_S *and* d_T*. If* $f : S \longrightarrow T$ *is a contraction and we equip* $\mathscr{R}(S)$ *and* $\mathscr{R}(T)$ *with the c.i.p. approach structures then* $\widetilde{f} : \mathscr{R}(S) \longrightarrow \mathscr{R}(T)$ *is a contraction where* $\widetilde{f}$ *is defined by* $\widetilde{f}(\xi) := f \circ \xi$.

Proof We denote by φ^a_S and φ^a_T the maps (made by means of the metrics d_S and d_T) which constitute bases for the c.i.p. approach gauges on $\mathscr{R}(S)$ and $\mathscr{R}(T)$ respectively. It suffices now to note that for any $a > 0$ and any $\xi, \eta \in \mathscr{R}(S)$

$$\varphi^a_T(f \circ \xi, f \circ \eta) \leq \varphi^a_S(\xi, \eta).$$ □

In analogy to 9.1.10 we have the following result where again, the inequality $K_1 \leq d_I$ is known (see Zolotarev 1983) and easy examples show that all inequalities are in general strict.

9.2.6 Proposition *The following inequalities hold.* $\delta_{K_1} \leq \delta_p \leq \delta_{d_I}$.

Proof This is an easy consequence of the definitions and the fact that the gauge of the c.i.p. approach structure is generated by the collection $\mathscr{B}_2 = \{K_\gamma \mid \gamma > 0\}$ (see 9.2.2). □

As was the case for the weak approach structure, the c.i.p. approach structure turns out to be complete, and this irrespective of whether the original metric on S was complete or not.

9.2.7 Theorem *The c.i.p. approach structure is complete.*

Proof Again we use 3.5.11. Let $(\xi_n)_n$ be a d_I-Cauchy sequence and choose a subsequence $(\xi_{k_n})_n$ with the property that for each n, $P\left(\{\xi_{k_n} \neq \xi_{k_{n+1}}\}\right) \leq \frac{1}{2^n}$. By the Borel-Cantelli lemma (see e.g. Prokhorov 2001) we now have that if we set

$$A := \bigcap_m \bigcup_{n \geq m} \{\xi_{k_n} \neq \xi_{k_{n+1}}\}$$

then $P(A) = 0$. Observe that for each $\omega \in \Omega \setminus A$ the sequence $\xi_{k_n}(\omega)$ is eventually constant. We denote this constant value by $\xi(\omega)$. Now ξ is an almost everywhere defined random variable and it is obvious that $(\xi_{k_n})_n$ converges almost everywhere, and hence also in probability, to ξ. We claim that even $d_I\left(\xi_{k_n}, \xi\right)$ converges to 0 as n tends to ∞.

In order to prove this, fix $\varepsilon > 0$ and let n_0 be such that for all $m \geq n \geq n_0$ we have $d_I\left(\xi_{k_n}, \xi_{k_m}\right) \leq \varepsilon$. Now since for all $k \in \mathbb{N}_0$ and all $m \geq n \geq n_0$ we have

$$\begin{aligned} P(\{d(\xi_{k_n}, \xi) > \frac{1}{k}\}) &\leq P(\{d(\xi_{k_n}, \xi_{k_m}) > \frac{1}{2k}\}) + P(\{d(\xi_{k_m}, \xi) > \frac{1}{2k}\}) \\ &\leq \varepsilon + P(\{d(\xi_{k_m}, \xi) > \frac{1}{2k}\}), \end{aligned}$$

letting first $m \to \infty$ and then $k \to \infty$, we get that $P(\{d(\xi_{k_n}, \xi) > 0\}) \leq \varepsilon$ for all $n \geq n_0$, and we are finished. □

9.2.8 Theorem *The c.i.p approach structure is locally countable.*

Proof This is an immediate consequence of the definition of either the basis $\mathscr{B}_1$ or the basis $\mathscr{B}_2$ as in both cases the indices of the functions in the basis may be restricted to range over any sequence which decreases to 0. □

There are several interesting relations between the structures which we have introduced on $\mathscr{P}(S)$ and $\mathscr{R}(S)$ respectively. We recall that for a random variable ξ, its *law* is the probability measure $P_\xi \in \mathscr{P}(S)$ defined by $P_\xi(B) := P\left(\xi^{-1}(B)\right)$, for all $B \in \mathscr{B}$. This is the so-called *image measure*. It is well known that convergence in probability of a sequence of random variables implies weak convergence of their laws. Since δ_p has the topology of convergence in probability as topological coreflection and δ_w has the weak topology as topological coreflection it is natural to see what the above property becomes in our setting.

9.2.9 Theorem *The function*

$$L : (\mathscr{R}(S), \delta_p) \longrightarrow (\mathscr{P}(S), \delta_w) : \xi \mapsto P_\xi$$

is a contraction, and consequently for any sequence of random variables $(\xi_n)_n$ *and any random variable* ξ, *we have* $\lambda_w((L(\xi_n))_n)(L(\xi)) \leq \lambda_p((\xi_n)_n)(\xi)$.

Proof Let $\xi \in \mathscr{R}(S)$ and $\mathscr{A} \subseteq \mathscr{R}(S)$. We use the following expressions

$$\delta_w\left(L\left(\xi\right), L\left(\mathscr{A}\right)\right) = \sup_{\mathscr{H}} \inf_{\eta \in \mathscr{A}} d_3^{\mathscr{H}}\left(L\left(\xi\right), L\left(\eta\right)\right)$$

and

$$\delta_p\left(\xi, \mathscr{A}\right) = \sup_{a>0} \inf_{\eta \in \mathscr{A}} \varphi^a\left(\xi, \eta\right)$$

where $\mathscr{H}$ ranges over finite sets of uniformly continuous maps with range [0, 1]. Let $\mathscr{H}$ be such a set and let $\varepsilon > 0$ be fixed. For all $f \in \mathscr{H}$, choose $\theta_f > 0$ such that, for all $x, y \in S$, $d(x, y) \leq \theta_f$ implies that $|f(x) - f(y)| \leq \varepsilon$ and put $\theta := \min_{f \in \mathscr{H}} \theta_f$. Now it suffices to note that, for all $\eta \in \mathscr{A}$, we have

$$\begin{aligned}
d_3^{\mathscr{H}}(L(\xi), L(\eta)) &= \sup_{f \in \mathscr{H}} \left| \int f \circ \xi dP - \int f \circ \eta dP \right| \\
&\leq \sup_{f \in \mathscr{H}} \Big(\int_{\{d(\xi,\eta)<\theta\}} |f \circ \xi - f \circ \eta| dP + \int_{\{d(\xi,\eta)\geq\theta\}} |f \circ \xi - f \circ \eta| dP \Big) \\
&\leq \sup_{f \in \mathscr{H}} (\varepsilon + P(\{\omega \mid d(\xi(\omega), \eta(\omega)) \geq \theta\})) \\
&= \varepsilon + \varphi^\theta(\xi, \eta).
\end{aligned}$$

□

9.2.10 Corollary **(Billingsley 1968)** *If a sequence of random variables $(\xi_n)_n$ converges in probability to a random variable ξ, then it also converges in law to ξ.*

9.2.11 Corollary *If a sequence of random variables $(\xi_n)_n$ converges to a random variable ξ for the indicator metric, then their laws converge to the law of ξ in the total variation metric.*

A converse to 9.2.10 also holds, but only in case the limit random variable is constant (Billingsley 1968). Not only does this result have an appropriate generalization to the context of approach theory, it can also be strengthened because in our context we need not restrict ourselves to constant random variables.

In order to prove this result we require a lemma, the result of which is interesting in its own right. We will calculate the distance between an arbitrary probability measure and the set of all Dirac probability measures, which are of course the laws of constant random variables. It is well known that for the weak topology, S is embedded as a closed subspace of $\mathscr{P}(S)$ by

$$\mathrm{Dir} : S \longrightarrow \mathscr{P}(S) : x \longrightarrow P_x$$

where P_x is the Dirac measure in x (i.e. $P_x(B) = 1$ if $x \in B$ and $P_x(B) = 0$ if $x \notin B$ for all $B \in \mathscr{B}$).

9.2.12 Lemma *For any $P \in \mathscr{P}(S)$, we have*

$$\delta_w(P, \mathrm{Dir}(S)) = 1 - \max_{x \in S} P(\{x\}).$$

Proof We restrict ourselves to the case where P has a countably infinite set of atoms. Let α be the P-measure of an atom $a \in S$ of largest measure. Then we have

$$\begin{aligned}\delta_w(P, \mathrm{Dir}(S)) &\leq \delta_w(P, \{P_a\}) \\ &= \sup_{\mathscr{G} \in 2^{(\mathscr{T})}} \sup_{G \in \mathscr{G}} P(G) \ominus P_a(G) \\ &= \sup_{G \in \mathscr{T}} P(G) \ominus P_a(G) \\ &= P(S \setminus \{a\}) - P_a(S \setminus \{a\}) \\ &= 1 - \alpha.\end{aligned}$$

Conversely,

$$\begin{aligned}\delta_w(P, \mathrm{Dir}(S)) &= \sup_{\mathscr{G} \in 2^{(\mathscr{T})}} \inf_{x \notin \cap \mathscr{G}} \sup_{G \in \mathscr{G}, x \notin G} P(G) \\ &= \sup_{\mathscr{G} \in 2^{(\mathscr{T})}} \inf_{H \in \mathscr{G}} \inf_{x \in S \setminus H} \sup_{G \in \mathscr{G}, x \notin G} P(G) \\ &\geq \sup_{\mathscr{G} \in 2^{(\mathscr{T})}, \cap \mathscr{G} = \emptyset} \inf_{H \in \mathscr{G}} \inf_{x \in S \setminus H} P(H) \\ &\geq \sup_{\mathscr{G} \in 2^{(\mathscr{T})}, \cap \mathscr{G} = \emptyset} \inf_{G \in \mathscr{G}} P(G).\end{aligned}$$

Now let us suppose that the set of P-atoms $\{a_n \mid n \in \mathbb{N}\}$ is ordered in the sense that $P(\{a_{n+1}\}) \leq P(\{a_n\})$, for all $n \in \mathbb{N}$. Then $\alpha = P(\{a_0\})$. Let

$$\beta := \max \{\{P(\{x\}) \mid x \in S\} \setminus \{\alpha\}\},$$

i.e. β is the P-measure of a second largest P-atom. Let $\varepsilon \in \,]0, \alpha - \beta[$ and take $n_0 \in \mathbb{N}$ such that

$$\sum_{n > n_0} P(\{a_n\}) < \frac{\varepsilon}{2}.$$

Next choose $r > 0$ such that, for all $n \in \{0, \ldots, n_0\}$,

$$P(B^*(a_n, r)) \leq P(\{a_n\}) + \varepsilon.$$

Let $Y := S \setminus \{a_n \mid n \in \mathbb{N}\}$ and partition Y into disjoint Borel sets $D_0, \ldots, D_{n_1}$ such that, for all $i \in \{0, \ldots, n_1\}$,

$$P(D_i) < \beta.$$

Next choose $\theta > 0$ such that $n_1\theta \leq \frac{\varepsilon}{2}$. Since P is regular we can find closed sets $K_i \subseteq D_i$, for all $i \in \{0, \ldots, n_1\}$, such that

$$P(K_i) \leq P(D_i) \leq P(K_i) + \theta.$$

Since the sets $K_0, \ldots, K_{n_1}, \{a_n \mid n \leq n_0\}$, are pairwise disjoint and closed it follows from the normality of S and the regularity of P that we can find open sets $O_i \supseteq K_i$ such that the following properties are fulfilled.

1. $O_0, \ldots, O_{n_1}$ are pairwise disjoint,
2. $\forall i \in \{0, \ldots, n_1\} : P(K_i) \leq P(O_i) \leq P(K_i) + \theta$,
3. $\forall i \in \{0, \ldots, n_1\} : O_i \cap \{a_n \mid n \leq n_0\} \neq \emptyset$.

Now consider the following finite collections of open sets in S:

$$\begin{aligned} G_i &:= \bigcup_{j=0, j\neq i}^{n_1} O_j \cup \bigcup_{k=0}^{n_0} B(a_k, \frac{r}{2}) & i \in \{0, \ldots, n_1\}, \\ H_l &:= S \setminus B^*(a_l, r) & l \in \{0, \ldots, n_0\}. \end{aligned}$$

By construction we have

$$(\bigcap_{i=0}^{n_1} G_i) \cap (\bigcap_{l=0}^{n_0} H_l) = \emptyset.$$

For each $i \in \{0, \ldots, n_1\}$, we further have

$$\begin{aligned} P(G_i) &\geq P((\bigcup_{j=0, j\neq i}^{n_1} O_j) \cup \{a_n \mid n \leq n_0\}) \\ &= P(\bigcup_{j=0, j\neq i}^{n_1} O_j) + 1 - P(Y) - \sum_{n_0<n} P(\{a_n\}) \\ &\geq P(\bigcup_{j=0, j\neq i}^{n_1} D_j) - \frac{\varepsilon}{2} + 1 - P(Y) - \frac{\varepsilon}{2} \\ &= 1 - \varepsilon - P(D_i) \\ &\geq 1 - \alpha, \end{aligned}$$

and, for each $l \in \{0, \ldots, n_0\}$, we have

$$P(H_l) = 1 - P(B^*(a_l, r)) \geq 1 - \alpha - \varepsilon.$$

Consequently, if we let

$$\mathscr{G}_\varepsilon := \{G_i \mid i \in \{0, \ldots, n_1\}\} \cup \{H_l \mid l \in \{0, \ldots, n_0\}\},$$

then

$$\inf_{G\in\mathscr{G}_\varepsilon} P(G) \geq 1-\alpha-\varepsilon,$$

and from the arbitrariness of ε it then follows that

$$\begin{aligned}\delta_w(P,\mathrm{Dir}(S)) &\geq \sup_{\mathscr{G}\in 2^{(\mathscr{T})},\cap\mathscr{G}=\emptyset}\ \inf_{G\in\mathscr{G}} P(G)\\ &\geq 1-\alpha,\end{aligned}$$

and we are finished. □

9.2.13 Theorem *If* $(\xi_n)_n$ *and* ξ *are random variables on S, then we have*

$$\lambda_p\langle(\xi_n)_n\rangle(\xi) \leq \lambda_w\langle(L(\xi_n))_n\rangle(L(\xi)) + \delta_w(L(\xi),\mathrm{Dir}(S)).$$

Proof By the 9.2.12, if P_ξ has no atoms, then obviously the result is trivially true. If P_ξ has atoms, then again by the foregoing lemma, we can choose an $x \in S$ such that $P_\xi(\{x\})$ is maximal and such that $\delta(P_\xi,\mathrm{Dir}(S)) = 1 - P_\xi(\{x\})$. Suppose now that $0 < a < \lambda_p\langle(\xi_n)_n\rangle(\xi)$, then it follows from 9.2.4 that we can find $b > 0$ such that

$$\forall n, \exists m \geq n : P\left(\{d(\xi_m,\xi) \geq b\}\right) \geq a.$$

Define

$$f : S \longrightarrow [0,1] : y \mapsto \frac{d(x,y)}{b} \wedge 1.$$

Then f is a continuous map on S with range $[0, 1]$ and

$$\begin{aligned}&\left|\int f dP_{\xi_m} - \int f dP_\xi\right| + (1-P_\xi(\{x\}))\\ &\qquad \geq \left|\int f dP_{\xi_m} - \int f dP_\xi\right| + \left|\int_{S\setminus\{x\}} f dP_\xi\right|\\ &\qquad \geq \left|\int f dP_{\xi_m} - \int f dP_\xi\right| + \left|\int f dP_\xi - \int f dP_x\right|\\ &\qquad \geq \left|\int f dP_{\xi_m} - \int f dP_x\right|\\ &\qquad = \left|\int f dP_{\xi_m}\right|\\ &\qquad \geq \left|\int_{\{d(\xi_m,x)\geq b\}} f\circ\xi_m dP\right|\\ &\qquad \geq a,\end{aligned}$$

and consequently $a \leq \lambda_w\langle(L(\xi_n))_n\rangle\left(L(\xi)\right) + \delta_w(P_\xi, \mathrm{Dir}(S))$, which proves our claim. □

9.2.14 Corollary (Billingsley 1968) *If $x \in S$ and $\left(\xi_n\right)_n$ is a sequence of random variables which converges in law to P_x, then it also converges in probability to the random variable with constant value x.*

9.2.15 Corollary *If $x \in S$ and $\left(\xi_n\right)_n$ is a sequence of random variables which converges for the total variation metric to P_x, then it also converges for the indicator metric to the random variable with constant value x.*

9.3 Prokhorov's Theorem

We recall that a collection Γ of probability measures on S is said to be *tight* if for every $\varepsilon > 0$ there exists a compact set $K \subseteq S$ such that for all $P \in \Gamma$ we have $P(S \setminus K) < \varepsilon$. We generalize this notion in two ways.

9.3.1 Definition (Weak index of tightness) For a collection $\Gamma \subseteq \mathscr{P}(S)$ we define its *weak index of tightness* as the number

$$\boxed{\chi_{wt}(\Gamma) := \sup_{\mathscr{G}} \inf_{\mathscr{G}_0} \sup_{P\in\Gamma} P(X \setminus \cup\mathscr{G}_0)}$$

where $\mathscr{G}$ ranges over all open covers of S and $\mathscr{G}_0$ over all finite subcollections of $\mathscr{G}$.

9.3.2 Proposition *For a metric d metrizing S and $\Gamma \subseteq \mathscr{P}(S)$ we have*

$$\begin{aligned}\chi_{wt}(\Gamma) &= \sup_{\delta_x} \inf_{K} \sup_{P\in\Gamma} P(S \setminus \cup_{x\in K} B_d(x, \delta_x)),\\ &= \sup_{\delta_x} \inf_{Y} \sup_{P\in\Gamma} P(S \setminus \cup_{x\in Y} B_d(x, \delta_x)),\end{aligned}$$

the first supremum on each line ranging over all choices $\delta_x > 0$, $x \in S$, the infimum on the first line over all compact sets K in S and on the second line over all finite sets Y in S.

Proof Let us denote the right hand side of the first line by $j(\Gamma)$ and of the second line by $b(\Gamma)$. To prove that $\chi_{wt}(\Gamma) \leq j(\Gamma)$ fix $\varepsilon > 0$ and an open cover $\mathscr{G}$ of S and assume that $\mathscr{G}$ consists of countably many G_n increasing to S. For each $x \in S$ we let n_x be the smallest number for which $x \in G_{n_x}$. Next we choose $\delta_x > 0$ such that $B_d(x, \delta_x) \subseteq G_{n_x}$. Now pick a compact set K in S such that $P(S \setminus \cup_{x\in K} B(x, \delta_x)) \leq j(\Gamma) + \varepsilon$ for all $P \in \Gamma$. Observe that since K is compact, it must be contained in a set G_{n_0}. Furthermore, for each $x \in K$ we have $B(x, \delta_x) \subseteq G_{n_x} \subseteq G_{n_0}$, by construction of n_x. It follows that $P(S \setminus G_{n_0}) \leq P(S \setminus \cup_{x\in K} B(x, \delta_x)) \leq j(\Gamma) + \varepsilon$ for all $P \in \Gamma$. That $j(\Gamma) \leq b(\Gamma)$ is trivial. Finally, to prove that $b(\Gamma) \leq \chi_{wt}(\Gamma)$

again, fix $\varepsilon > 0$, $\delta_x > 0$ for all $x \in S$ and let $\mathscr{G}$ be the open cover consisting of all balls $B_d(x, \delta_x)$. Since we can pick finitely many x_i such that $P(S \setminus \cup_i B_d(x_i, \delta_{x_i})) \leq \chi_{wt}(\Gamma) + \varepsilon$, it easily follows that $b(\Gamma) \leq \chi_{wt}(\Gamma)$. □

Notice that if (S, d) is a so-called Atsuji or Lebesgue space (Atsuji 1958) then it is possible to replace the choice of radii $(\delta_y)_y$ in the definition of χ_{wt} by a fixed choice for all $y \in S$.

9.3.3 Definition **(Strong index of tightness)** We define the *strong index of tightness* of Γ as the number

$$\boxed{\chi_{st}(\Gamma) := \inf_{K} \sup_{P\in\Gamma} P(S \setminus K)}$$

the infimum being taken over all compact sets $K \subseteq S$.

Observe that the inequality $\chi_{wt}(\Gamma) \leq \chi_{st}(\Gamma)$ always holds true.
The following proposition shows that both indices indeed generalize the classical notion of tightness.

9.3.4 Proposition *For a collection Γ of probability measures the following are equivalent.*

1. *Γ is tight.*
2. $\chi_{wt}(\Gamma) = 0$.
3. $\chi_{st}(\Gamma) = 0$.

Proof The only non trivial assertion is $2 \Rightarrow 1$. Fix $\varepsilon > 0$ and choose a countable dense subset $\{x_i \mid i \in \mathbb{N}\}$. Then for any $m \geq 1$ the family of balls $(B(x_i, 1/m))_i$ is an open cover and thus there exists a finite subset $(B(x_i, 1/m))_{i=0,\ldots,n_m}$ such that

$$\forall P \in \Gamma : P(X \setminus \cup_{i=0}^{n_m} B(x_i, 1/m)) \leq \frac{\varepsilon}{2^n}.$$

Put

$$K := \bigcap_{m=1}^{\infty} \bigcup_{i=0}^{n_m} \overline{B(x_i, 1/m)}$$

then K is compact and for all $P \in \Gamma$, $P(X \setminus K) \leq \varepsilon$. □

That the indices of compactness and tightness also produce meaningful non-zero values is shown by the following simple example.

9.3.5 Example Consider the real line with the usual Borel σ-algebra, fix $\alpha > 0$ and let Γ be the set of all probability measures

$$(1 - \alpha)P_0 + \alpha P_n$$

where P_x stands for the Dirac measure at x and where n is any natural number ≥ 1. Then the weak and strong indices of tightness and the index of relative sequential compactness are all equal to α.

The following lemma is a direct consequence of the classical Prokhorov theorem.

9.3.6 Lemma *A set $\Gamma \subseteq \mathscr{P}(S)$ is weakly relatively sequentially compact if there exists a compact set $K \subseteq S$ containing the support of every probability measure $P \in \Gamma$.*

Proof This follows from Prokhorov's theorem. □

The reason for introducing both a weak and strong index of tightness will become clear in our general form of a Prokhorov theorem for distances, as they turn out to provide respectively a lower and an upper bound for the index of relative sequential weak compactness. The indices of relative sequential compactness have been provided with superscripts to make clear which structure they concern.

9.3.7 Theorem (**Prokhorov for distances**) *For every collection Γ of probability measures on a complete separable metric space S the following inequalities hold.*

$$\chi^{\rho}_{rsc}(\Gamma) \leq \chi_{wt}(\Gamma) \leq \chi^{w}_{rsc}(\Gamma) \leq \chi_{st}(\Gamma) \leq \chi^{d_{TV}}_{rsc}(\Gamma).$$

Proof To prove that $\chi^{\rho}_{rsc}(\Gamma) \leq \chi_{wt}(\Gamma)$ let $\varepsilon > 0$ and let $\mathscr{G}$ be the open cover consisting of all ε-balls. Next let $\mathscr{G}_0$ be an arbitrary finite subcollection of $\mathscr{G}$ and let $\{A_1, \cdots, A_n\}$ be the canonical pairwise disjoint collection generated by $\mathscr{G}_0$ such that $\cup\mathscr{G}_0 = \cup_{i=1}^{n} A_i$. Take arbitrary points $x_i \in A_i$ and, if necessary, $x_{n+1} \in S \setminus \cup\mathscr{G}_0$ and a natural number m for which $n/m \leq \varepsilon$. Consider the finite collection Φ of probability measures of the form $Q = \sum_{i=1}^{n+1}(k_i/m)P_{x_i}$, where $k_i \in \{0, \ldots, m\}$ and $\sum_{i=1}^{n+1} k_i = m$. Fix $P \in \Gamma$ and consider a probability measure $Q = \sum_{i=1}^{n+1}(k_i/m)P_{x_i}$ in Φ such that for all $i \leq n$ we have $P(A_i) \leq k_i/m + 1/m$.

For any Borel set $A \subseteq S$ we denote the set of all numbers $i \leq n$ for which A meets A_i by I. From

$$\begin{aligned}
P(A) &\leq P(\cup_{i \in I} A_i) + P(S \setminus \cup\mathscr{G}_0)\\
&\leq \sum_{i \in I}(k_i/m + 1/m) + P(S \setminus \cup\mathscr{G}_0)\\
&\leq \sum_{i \in I} k_i/m + n/m + P(S \setminus \cup\mathscr{G}_0)\\
&\leq Q(\cup_{i \in I} A_i) + P(S \setminus \cup\mathscr{G}_0) + \varepsilon\\
&\leq Q(A^{(P(S \setminus \cup\mathscr{G}_0)+\varepsilon)}) + P(S \setminus \cup\mathscr{G}_0) + \varepsilon
\end{aligned}$$

it now follows that $\rho(P, Q) \leq P(S \setminus \cup\mathscr{G}_0) + \varepsilon$ and we are finished.

To show that $\chi_{wt}(\Gamma) \leq \chi^w_{rsc}(\Gamma)$ suppose that $\chi^w_{rsc}(\Gamma) < \gamma$ and choose $\varepsilon > 0$ such that $\chi^w_{rsc}(\Gamma) < \gamma - \varepsilon$. Take a countable open cover $\mathcal{G} := \{G_n \mid n \in \mathbb{N}\}$ and suppose that for all $n \in \mathbb{N}$ there exists $Q_n \in \Gamma$ such that

$$Q_n(\cup_{i=0}^n G_i) < 1 - \gamma.$$

Since $\chi^w_{rsc}(\Gamma) < \gamma - \varepsilon$ there exists a subsequence $(Q_{k_n})_n$ and a $P \in \mathcal{P}(X)$ such that

$$\lambda_w \langle (Q_{k_n})_n \rangle (P) < \gamma - \varepsilon.$$

This implies that for all n

$$\begin{aligned} P(\cup_{i=0}^n G_i) &\leq \sup_m \inf_{l \geq m} Q_{k_l}(\cup_{i=0}^n G_i) + \gamma - \varepsilon \\ &\leq \sup_{m, k_m \geq n} \inf_{l \geq m} Q_{k_l}(\cup_{i=0}^{k_l} G_i) + \gamma - \varepsilon \\ &\leq 1 - \gamma + \gamma - \varepsilon = 1 - \varepsilon. \end{aligned}$$

However, since $\cup_{i=0}^n G_i \uparrow X$ this is impossible. Hence there exists a finite subset $\mathcal{G}_0 \subseteq \mathcal{G}$ such that for all $P \in \Gamma$ we have $P(X \setminus \cup \mathcal{G}_0) \leq \gamma$, and thus $\chi_{wt}(\Gamma) \leq \gamma$.

To show that $\chi^w_{rsc}(\Gamma) \leq \chi_{st}(\Gamma)$ fix $\varepsilon > 0$ and take a compact set $K \subseteq S$ such that the inequality $P(S \setminus K) \leq \chi_{st}(\Gamma) + \varepsilon$ is valid for every probability measure $P \in \Gamma$. If we put $\Gamma(\cdot \mid K) := \{P(\cdot \mid K) \mid P \in \Gamma\}$, then the relation $P = P(K)P(\cdot \mid K) + P(S \setminus K)P(\cdot \mid S \setminus K)$ shows that $\Gamma \subseteq \Gamma(\cdot \mid K)(t_s(\Gamma) + \varepsilon)$. Applying 9.1.13, 9.3.6 and 4.3.58, we conclude that

$$\begin{aligned} \chi^w_{rsc}(\Gamma) &\leq \chi^w_{rsc}(\Gamma(. \mid K)(\chi_{st}(\Gamma) + \varepsilon)) \\ &\leq \chi^w_{rsc}(\Gamma(. \mid K)) + \chi_{st}(\Gamma) + \varepsilon \\ &= \chi_{st}(\Gamma) + \varepsilon \end{aligned}$$

and the result follows.

Finally, to show that $\chi_{st}(\Gamma) \leq \chi^{d_{TV}}_{rsc}(\Gamma)$ take $\varepsilon > 0$ and consider a finite set $\Phi \subseteq \mathcal{P}(S)$ such that for each $P \in \Gamma$ there exists a probability measure $Q \in \Phi$ for which $d_{TV}(P, Q) \leq \chi^{d_{TV}}_{rsc}(\Gamma) + \varepsilon/2$. The completeness of S implies the tightness of Φ, and thus we can choose a compact set $K \subseteq S$ such that $Q(S \setminus K) \leq \varepsilon/2$ for all $Q \in \Phi$. Since for $P \in \Gamma$ we have

$$P(S \setminus K) \leq Q(S \setminus K) + \chi^{d_{TV}}_{rsc}(\Gamma) + \varepsilon/2 \leq \chi^{d_{TV}}_{rsc}(\Gamma) + \varepsilon$$

again the result follows. □

9.3.8 Corollary **(Prokhorov's theorem)** *Let Γ be a collection of probability measures on a complete separable metric space S. Then Γ is weakly relatively sequentially compact if and only if it is tight.*

Proof Let Γ be weakly relatively sequentially compact, then by 4.3.58, $\chi^w_{rsc}(\Gamma) = 0$, and by 9.3.7 $\chi_{wt}(\Gamma) = 0$. Now from 9.3.4 it follows that Γ is tight.

Conversely, if Γ is tight then by 9.3.4, $\chi_{wt}(\Gamma) = 0$, and by 9.3.7, $\chi^\rho_{rsc}(\Gamma) = 0$. Now 4.3.58 and the completeness of ρ imply that Γ is weakly relatively sequentially compact. □

9.3.9 Theorem *If there exists a sequence $(U_n)_n$ of relatively compact open sets which increases to S then for $\Gamma \subseteq \mathscr{P}(S)$ we have*

$$\chi_{wt}(\Gamma) = \chi^w_{rsc}(\Gamma) = \chi_{st}(\Gamma).$$

Proof It suffices to show that in this case $\chi_{st}(\Gamma) \leq \chi_{wt}(\Gamma)$. Let $\varepsilon > 0$. Now it is possible to find a U_n such that $\sup_{P\in\Gamma} P(S \setminus U_n) \leq \chi_{wt}(\Gamma) + \varepsilon$. Let K be the compact set $\overline{U_n}$ and observe that, since $U_n \subseteq K$, we have $\sup_{P\in\Gamma} P(S \setminus K) \leq \sup_{P\in\Gamma} P(S \setminus U_n) \leq \chi_{wt}(\Gamma) + \varepsilon$. We conclude that $\chi_{st}(\Gamma) \leq \chi_{wt}(\Gamma)$. □

Theorem 9.3.9 has the following obvious corollary for Euclidean spaces.

9.3.10 Corollary *For $\Gamma \subseteq \mathscr{P}(\mathbb{R}^d)$ we have $\chi_{wt}(\Gamma) = \chi^w_{rsc}(\Gamma) = \chi_{st}(\Gamma)$.*

9.4 An Indexed Central Limit Theorem in One Dimension

In the foregoing section we have seen approach structures which have as underlying topologies various well-known classical structures, such as e.g. the weak topology. However, just as in classical analysis where often different choices of metrics generating a given topology are chosen depending on the problem at hand, it is often advantageous here too, to be able to choose various approach structures overlying a given topology.

We recall that, as usual, $\mathbb{E}$ stands for the expected value function.

By a *standard triangular array* we mean a triangular array of real square integrable random variables

$$\begin{array}{lll} \xi_{1,1} & & \\ \xi_{2,1} & \xi_{2,2} & \\ \xi_{3,1} & \xi_{3,2} & \xi_{3,3} \\ & \vdots & \end{array}$$

satisfying the following properties.

(a) $\forall n : \xi_{n,1}, \ldots, \xi_{n,n}$ are independent.

(b) $\forall n, k : \mathbb{E}\left[\xi_{n,k}\right] = 0$.

(c) $\forall n : \sum_{k=1}^{n} \sigma^2_{n,k} = 1$, where $\sigma^2_{n,k} = \mathbb{E}\left[\xi^2_{n,k}\right]$.

We recall that *Feller's negligibility condition* states that

$$\max_{k=1}^{n} \sigma_{n,k}^2 \to 0.$$

9.4.1 Theorem (Lindeberg-Feller CLT) *If* $\{\xi_{n,k}\}$ *is a standard triangular array which satisfies Feller's negligibility condition, and* ξ *is a standard normally distributed random variable, then the following properties are equivalent.*

$$(a) \quad \sum_{k=1}^{n} \xi_{n,k} \xrightarrow{w} \xi.$$

$$(b)\ \forall \varepsilon > 0 : \sum_{k=1}^{n} \mathbb{E}\left[\xi_{n,k}^2 \mid |\xi_{n,k}| \geq \varepsilon\right] \to 0.$$

Here (b) is usually referred to as *Lindeberg's condition*. Throughout the remainder of this chapter, ξ will be a standard normally distributed random variable and $\{\xi_{n,k}\}$ will be a standard triangular array satisfying Feller's negligibility condition.

We will be defining and using various structures on the set of probability measures of the real line (or equivalently on the set of distribution functions on the real line, which in the context of $\mathbb{R}$ is often advantageous to work with). These structures will then be transported to random variables in the usual way by considering the map, which was already introduced in the foregoing section

$$L : \mathscr{R}(\mathbb{R}) \longrightarrow \mathscr{P}(\mathbb{R}) : \xi \mapsto P_\xi$$

or, in the present context

$$L : \mathscr{R}(\mathbb{R}) \longrightarrow \mathscr{F}(\mathbb{R}) : \xi \mapsto F_\xi$$

where F_ξ stands for the distribution function of P_ξ (or of ξ), and where $\mathscr{F}(\mathbb{R})$ stands for the set of all such probability distribution functions. We will use the same notations for the structures whether they are functioning on random variables or on distributions. Furthermore, the random variables are supposed to be defined on a fixed probability space $(\Omega, \mathscr{A}, P)$ as in Sect. 9.2. Since in this section we will always be working with real random variables we omit reference to $\mathbb{R}$ and simply write $\mathscr{F}$ for $\mathscr{F}(\mathbb{R})$ and further we put $\mathscr{F}_c$ for the continuous distributions in $\mathscr{F}$. Similarly the set of all real random variables will simply be denoted $\mathscr{R}$ and the set of continuously distributed real random variables will be denoted $\mathscr{R}_c$.

We also recall that the *Kolmogorov metric* between distribution functions on the real line is defined as

$$K(F, G) := \sup_{x \in \mathbb{R}} |F(x) - G(x)|.$$

This then is transported to real random variables η and ξ by

$$K(\eta, \xi) := K(F_\eta, F_\xi).$$

As usual in this context we also transport the weak topology on $\mathscr{F}$ to $\mathscr{R}$ and we say for instance that a sequence of random variables $(\eta_n)_n$ converges weakly to a random variable η, (denoted as $\eta_n \stackrel{w}{\to} \eta$), if the sequence $(F_{\eta_n})_n$ converges weakly to F_η.

In general, K is too strong to metrize weak convergence, but it is well known that if η is continuously distributed, then the following are equivalent for any sequence of random variables $(\eta_n)_n$:

$$(a)\ \eta_n \stackrel{w}{\to} \eta.$$
$$(b)\ \limsup_{n\to\infty} K(\eta, \eta_n) = 0.$$

Now we define the approach structure wherein we will be working. We denote convolution by $\circledast$. From Bergström (1949) and Råde (1997) we know that weak convergence in $\mathscr{F}$ of a sequence $(F_n)_n$ to F is equivalent with uniform convergence of the sequence $(F_n \circledast G)_n$ to $F \circledast G$ for every continuous $G \in \mathscr{F}_c$. In other words, if we let $\mathscr{T}_w$ stand for the topology of weak convergence on $\mathscr{F}$ and $\mathscr{T}_K$ for the topology of uniform convergence (i.e. generated by the Kolmogorov metric) on $\mathscr{F}_c$, then $\mathscr{T}_w$ is the weakest topology on $\mathscr{F}$ making all mappings

$$(\mathscr{F} \longrightarrow (\mathscr{F}_c, \mathscr{T}_K) : F \mapsto F \circledast G)_{G\in\mathscr{F}_c}$$

continuous.

If we replace the uniform topology $\mathscr{T}_K$ by its generating metric K, then we end up with the mappings

$$(\mathscr{F} \longrightarrow (\mathscr{F}_c, K) : F \mapsto F \circledast G)_{G\in\mathscr{F}_c}.$$

As in previous examples, again we are not able to construct a weakest metric on $\mathscr{F}$, metrizing the weak topology and making all mappings contractive, since it simply does not exist. But we are able to construct a weakest such approach structure. We call it the *continuity approach structure*, which hence is a quantification of $(\mathscr{F}, \mathscr{T}_w)$.

It follows from the general theory that a basis for the initial approach system in $F \in \mathscr{F}$ is given by

$$\mathscr{B}_c(F) := \{K(F \circledast G, \cdot \circledast G) \mid G \in \mathscr{F}_c\}$$

and that, for any $F \in \mathscr{F}$ and $\mathscr{D} \subseteq \mathscr{F}$ the initial distance (see 1.2.34) is given by

$$\delta_c(F, \mathscr{D}) = \sup_{\mathscr{H}\in 2_0^{(\mathscr{F}_c)}} \inf_{H\in\mathscr{D}} \sup_{G\in\mathscr{H}} K(F \circledast G, H \circledast G)$$

and that the limit operator of a sequence $(F_n)_n$ in $\mathscr{F}$ evaluated at $F \in \mathscr{F}$ (see 1.2.43) is given by

$$\lambda_c\langle (F_n)_n\rangle(F) = \sup_{G\in\mathscr{F}_c} \limsup_{n\to\infty} K\,(F \circledast G,\, F_n \circledast G)\,.$$

As before, in what follows we freely transport this structure from $\mathscr{F}$ to $\mathscr{R}$. This means that for any real random variable ξ, any sequence of random variables $(\xi_n)_n$ and any collection of random variables $\mathscr{D}$ we have

$$\delta_c\,(\xi, \mathscr{D}) = \delta_c(F_\xi, \{F_\zeta \mid \zeta \in \mathscr{D}\}),$$

$$\lambda_c\langle(\xi_n)_n\rangle(\xi) = \lambda_c\langle(F_{\xi_n})_n\rangle(F_\xi).$$

9.4.2 Proposition *The* Top *coreflection of* $(\mathscr{F}, \delta_c)$ *is determined by the weak topology and the* Met *coreflection is determined by the Kolmogorov metric.*

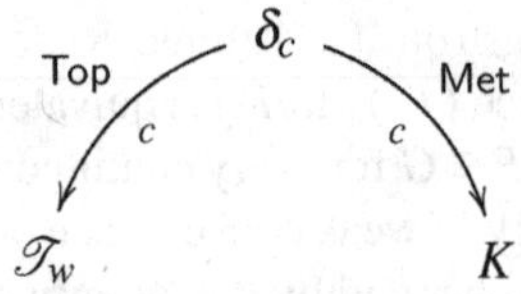

Proof Both for the Top coreflection and the Met coreflection this follows at once from the definition. □

For each probability distribution F and each $\alpha > 0$ we now consider the map

$$\varphi_{F,\alpha} : \mathscr{F} \longrightarrow [0,1] : H \mapsto \sup_{x\in\mathbb{R}} (F(x-\alpha) - H(x)) \vee (H(x) - F(x+\alpha))$$

and we define $j(F)$ to be the largest discontinuity-jump of F.

9.4.3 Lemma *For any* $F \in \mathscr{F}$ *and* $\varepsilon > 0$ *there exists* $\alpha > 0$ *such that*

$$K(F,\cdot) \le \varphi_{F,\alpha} + j(F) + \varepsilon$$

Proof Fix F and $\varepsilon > 0$. Choose points $x_0 < x_1 < \cdots < x_{n-1} < x_n$ such that

$$F(x_0) \le \varepsilon/2 \text{ and } F(x_n) \ge 1 - \varepsilon/2$$

and such that for all $i \in \{0, \ldots, n\}$ and all $x, y \in [\,x_i, x_{i+1}\,[$ the inequality

$$|F(x) - F(y)| \le \varepsilon/2$$

holds. Now if $0 < \alpha < \min_{i=0}^{n-1} |x_{i+1} - x_i|$ then, for any $x \in \mathbb{R}$, distinguishing cases as to whether $x \in [\,x_i, x_{i+1}\,[$ for some $i \in \{0, \ldots, n-1\}$, $x < x_0$ or $x \ge x_n$ one easily verifies that for any $H \in \mathcal{F}$

$$|F(x) - H(x)| \leq \varphi_{F,\alpha}(H) + j(F) + \varepsilon. \qquad \square$$

9.4.4 Proposition *A basis for the continuity approach structure in $F \in \mathcal{F}$ is given by*

$$\mathcal{C}_c(F) := \{\varphi_{F,\alpha} \mid \alpha > 0\}.$$

Proof First, it suffices to note that for any $F, H, G \in \mathcal{F}$ and $\alpha > 0$ we have

$$\varphi_{F,\alpha}(H) \leq \varphi_{F,\frac{\alpha}{2}}(G) + \varphi_{G,\frac{\alpha}{2}}(H)$$

to see that $(\mathcal{C}_c(F))_{F\in\mathcal{F}}$ is indeed a basis for an approach system.

Second, fix $F \in \mathcal{F}$ and $G \in \mathcal{F}_c$, then on the one hand, from 9.4.3, it follows that for any $\varepsilon > 0$ there exists α such that for any $H \in \mathcal{F}$

$$K(F \circledast G, H \circledast G) \leq \varphi_{F\circledast G,\alpha}(H \circledast G) + \varepsilon$$

and on the other hand we have

$$\begin{aligned}
\varphi_{F\circledast G,\alpha}(H \circledast G) &= \sup_{x\in\mathbb{R}} \max\{F \circledast G(x-\alpha) - H \circledast G(x), H \circledast G(x) - F \circledast G(x+\alpha)\} \\
&\leq \int_{-\infty}^{\infty} \sup_{x\in\mathbb{R}} \max\{F(x-\alpha-y) - H(x-y), H(x-y) - F(x+\alpha-y)\}dG(y) \\
&= \int_{-\infty}^{\infty} \sup_{z\in\mathbb{R}} \max\{F(z-\alpha) - H(z), H(z) - F(z+\alpha)\}dG(y) \\
&= \varphi_{F,\alpha}(H).
\end{aligned}$$

This proves that $\mathcal{B}_c(F) \subseteq \widehat{\mathcal{C}_c(F)}$.

To prove that $\mathcal{C}_c(F) \subseteq \widehat{\mathcal{B}_c(F)}$ fix $F \in \mathcal{F}$, $\alpha > 0$ and $\varepsilon > 0$. Now choose $G \in \mathcal{F}_c$ such that $G(-\frac{\alpha}{2}) \leq \frac{\varepsilon}{2}$ and $G(\frac{\alpha}{2}) \geq 1 - \frac{\varepsilon}{2}$. Then it follows that for any $H \in \mathcal{F}$ and $x \in \mathbb{R}$

$$\begin{aligned}
F(x-\alpha) - H(x) &= \int_{-\infty}^{\infty} 1_{]-\infty,x-\alpha]}(y)dF(y) - \int_{-\infty}^{\infty} 1_{]-\infty,x]}(y)dH(y) \\
&\leq \int_{-\infty}^{\infty} G(x - \frac{\alpha}{2} - y)dF(y) - \int_{-\infty}^{\infty} G(x - \frac{\alpha}{2} - y)dH(y) + \varepsilon \\
&= (F \circledast G)(x - \frac{\alpha}{2}) - (H \circledast G)(x - \frac{\alpha}{2}) + \varepsilon
\end{aligned}$$

from which it follows that $F(x-\alpha) - H(x) \leq K(F \circledast G, H \circledast G) + \varepsilon$. Analogously one finds that $H(x) - F(x+\alpha) \leq K(F \circledast G, H \circledast G) + \varepsilon$ which finally gives that

$$\varphi_{F,\alpha} \leq K(F \circledast G, \cdot \circledast G) + \varepsilon$$

showing that $\mathcal{C}_c(F) \subseteq \widehat{\mathcal{B}_c(F)}$. $\square$

9.4.5 Theorem *For $F \in \mathscr{F}$ and $\mathscr{D} \subseteq \mathscr{F}$*

$$\delta_c(F, \mathscr{D}) \leq \delta_K(F, \mathscr{D}) \leq \delta_c(F, \mathscr{D}) + j(F).$$

Proof The first inequality is immediate from the fact that the Kolmogorov metric determines the Met coreflection of δ_c and the second inequality is an immediate consequence of 9.4.3 and 9.4.4 and the formula for a distance in terms of the approach system as in 1.2.34. □

9.4.6 Corollary *For $F \in \mathscr{F}_c$ and $\mathscr{D} \subseteq \mathscr{F}$*

$$\delta_c(F, \mathscr{D}) = \delta_K(F, \mathscr{D}).$$

9.4.7 Corollary *For $F \in \mathscr{F}_c$ and any sequence $(F_n)_n$ in $\mathscr{F}$*

$$\lambda_c \langle (F_n)_n \rangle (F) = \limsup_{n\to\infty} K(F, F_n).$$

9.4.8 Theorem *$(\mathscr{F}, \delta_c)$ is complete and locally countable.*

Proof Completeness is an immediate consequence of the completeness of K and 3.5.11 and the local countability follows at once from the fact that the functions $\varphi_{F,\frac{1}{n}}$ for $n \geq 1$ constitute a basis for the approach system in F (see 9.4.4). □

Before we can prove the promised result concerning the central limit theorem we need to prove some technical facts.

Let $\mathscr{H}$ stand for the collection of all strictly decreasing functions $h : \mathbb{R} \longrightarrow \mathbb{R}$, with a bounded first and second derivative and a bounded and piecewise continuous third derivative, and for which $\lim_{x\to-\infty} h(x) = 1$ and $\lim_{x\to\infty} h(x) = 0$. The conditions on the set $\mathscr{H}$ are required, among other things, for the application of Stein's method (see 9.4.11).

9.4.9 Lemma *If $\eta \in \mathscr{R}_c$ and $(\eta_n)_n$ is a sequence in $\mathscr{R}$ then*

$$\lambda_c \langle (\eta_n)_n \rangle (\eta) = \sup_{h\in\mathscr{H}} \limsup_{n\to\infty} \left| \mathbb{E}\left[h(\eta) - h(\eta_n)\right] \right|.$$

Proof We make use of 9.4.7. Let $\varepsilon > 0$ be arbitrary. The continuity of F_η allows us to construct points $x_1 < \cdots < x_m$ such that for each n

$$K(\eta, \eta_n) \leq \max_{k=1}^{m} \left| F_\eta(x_k) - F_{\eta_n}(x_k) \right| + \varepsilon.$$

Indeed, suppose that $\varepsilon < 1/2$ and choose points $0 = y_0 < y_1 < \cdots < y_m < y_{m+1} = 1$ such that $\max_{k=0}^{m} |y_{k+1} - y_k| < \varepsilon/2$. Then the continuity of F_η, combined with the fact that $\lim_{x\to-\infty} F_\eta(x) = 0$ and $\lim_{x\to\infty} F_\eta(x) = 1$, allows us to choose points $-\infty = x_0 < x_1 < \cdots < x_m < x_{m+1} = \infty$ such that $F_\eta(x_k) = y_k$ for all $k \in \{0, \ldots, m+1\}$. Hence we have

$$\max_{k=0}^{m} \left|F_\eta(x_{k+1}) - F_\eta(x_k)\right| < \varepsilon/2.$$

Also, for each n and $k \in \{0, \ldots, m\}$, it follows from the monotonicity of F_η and F_{η_n} that

$$\begin{aligned}\sup_{x_k \le x \le x_{k+1}} \left|F_\eta(x) - F_{\eta_n}(x)\right| &= \sup_{x_k \le x \le x_{k+1}} \max\left\{F_\eta(x) - F_{\eta_n}(x), F_{\eta_n}(x) - F_\eta(x)\right\} \\ &\le \max\{F_\eta(x_{k+1}) - F_{\eta_n}(x_k), F_{\eta_n}(x_{k+1}) - F_\eta(x_k)\}. \\ &\le \max\{F_\eta(x_k) - F_{\eta_n}(x_k), F_{\eta_n}(x_{k+1}) - F_\eta(x_{k+1})\} + \varepsilon,\end{aligned}$$

and thus we can conclude that for each n

$$K(\eta, \eta_n) \le \max_{k=1}^{m} \left|F_\eta(x_k) - F_{\eta_n}(x_k)\right| + \varepsilon.$$

Again from the continuity of F_η, it follows that for each $x \in \mathbb{R}$ there exists $\delta > 0$ such that for each n

$$|F_\eta(x) - F_{\eta_n}(x)| \le \max\{F_\eta(x - \delta) - F_{\eta_n}(x), F_{\eta_n}(x) - F_\eta(x + \delta)\} + \varepsilon/2.$$

Now, we can find $h \in \mathscr{H}$ such that $h([-\infty, x - \delta]) \subseteq [1 - \frac{\varepsilon}{4}, 1]$ and $h([x, \infty[) \subseteq [0, \frac{\varepsilon}{4}]$, and then it follows that

$$\begin{aligned}F_\eta(x - \delta) - F_{\eta_n}(x) &= \int 1_{]-\infty, x-\delta]} dP_\eta - \int 1_{]-\infty, x]} dP_{\eta_n} \\ &\le (\int h dP_\eta + \varepsilon/4) - (\int h dP_{\eta_n} - \varepsilon/4) \\ &= \mathbb{E}\left[h(\eta) - h(\eta_n)\right] + \varepsilon/2.\end{aligned}$$

Since an analogous reasoning holds for the second term in the maximum above, we can conclude that we can find functions $h, h' \in \mathscr{H}$ such that

$$\left|F_\eta(x) - F_{\eta_n}(x)\right| \le \max\left\{\mathbb{E}\left[h(\eta) - h(\eta_n)\right], \mathbb{E}[h'(\eta_n) - h'(\eta)]\right\} + \varepsilon.$$

Hence there exist functions $h_1, \ldots, h_{2m} \in \mathscr{H}$ such that

$$\begin{aligned}\limsup_{n\to\infty} K(\eta, \eta_n) &\le \limsup_{n\to\infty} \max_{k=1}^{2m} |\mathbb{E}[h_k(\eta) - h_k(\eta_n)]| + 2\varepsilon \\ &= \max_{k=1}^{2m} \limsup_{n\to\infty} |\mathbb{E}[h_k(\eta) - h_k(\eta_n)]| + 2\varepsilon \\ &\le \sup_{h \in \mathscr{H}} \limsup_{n\to\infty} |\mathbb{E}[h(\eta) - h(\eta_n)]| + 2\varepsilon,\end{aligned}$$

which by the arbitrariness of ε proves one inequality.

For the converse inequality, let $h \in \mathcal{H}$. The layer cake representation (see Lieb and Loss 2001) states that, for any positive random variable ξ,

$$\mathbb{E}[\xi] = \int_0^{\infty} P[\xi \geq t]\, dt.$$

Taking into account that h is a strictly decreasing map taking values between 0 and 1 we find that

$$\mathbb{E}[h(\eta)] = \int_0^{\infty} P[h(\eta) \geq t]\, dt = \int_0^{1} P\left[\eta \leq h^{-1}t\right] dt,$$

which allows us to conclude that

$$|\mathbb{E}[h(\eta) - h(\eta_n)]| \leq \int_0^1 \left|F_\eta(h^{-1}t) - F_{\eta_n}(h^{-1}t)\right| dt \leq K(\eta, \eta_n). \qquad \square$$

9.4.10 Definition **(Lindeberg index)** Given a standard triangular array $\{\xi_{n,k}\}$ we define

$$\boxed{\chi_{Lin}(\xi_{n,k}) = \sup_{\varepsilon>0} \limsup_{n\to\infty} \sum_{k=1}^{n} \mathbb{E}\left[\xi_{n,k}^2 \mid |\xi_{n,k}| \geq \varepsilon\right]}$$

which we call the *Lindeberg index*. It is clear that $\left\{\xi_{n,k}\right\}$ satisfies Lindeberg's condition if and only if $\chi_{Lin}(\xi_{n,k}) = 0$.

In what follows we will use Stein's method Stein (1972, 1986), to prove a generalization of the Lindeberg-Feller central limit theorem providing an upper bound for

$$\lambda_c\big(\big(\sum_{k=1}^{n} \xi_{n,k}\big)_n\big)(\xi) = \limsup_{n\to\infty} K(\xi, \sum_{k=1}^{n} \xi_{n,k})$$

where ξ is a normally distributed random variable and $\{\xi_{n,k}\}$ is a standard triangular array which is asymptotically negligible in the sense of Feller.

The basics of Stein's method which we will be using are contained in the following lemma. The proofs can be found in e.g. Barbour and Chen (2005).

9.4.11 Lemma **(Stein's method)** *Let $h : \mathbb{R} \longrightarrow \mathbb{R}$ be measurable and bounded. Put*

$$f_h(x) = e^{x^2/2} \int_{-\infty}^{x} (h(t) - \mathbb{E}[h(\xi)])\, e^{-t^2/2} dt.$$

Then for any $x \in \mathbb{R}$

$$\mathbb{E}\left[h(\xi)\right] - h(x) = xf_h(x) - f_h'(x).$$

Moreover, if h is absolutely continuous, then

$$\left\| f_h'' \right\|_\infty \leq 2 \left\| h' \right\|_\infty ,$$

and if $h_z = 1_{]-\infty,z]}$ *for* $z \in \mathbb{R}$*, then for all* $x, y \in \mathbb{R}$

$$\left| f_{h_z}'(x) - f_{h_z}'(y) \right| \leq 1.$$

Stein's method was used by Barbour and Hall to derive Berry-Esseen type bounds in Barbour and Hall (1984). The following lemma is inspired by their paper.

9.4.12 Lemma *Let* $h \in \mathscr{H}$ *and put*

$$\delta_{n,k} = f_h\big(\sum_{i\neq k} \xi_{n,i} + \xi_{n,k}\big) - f_h\big(\sum_{i\neq k} \xi_{n,i}\big) - \xi_{n,k} f_h'\big(\sum_{i\neq k} \xi_{n,i}\big)$$

and

$$\varepsilon_{n,k} = f_h'\big(\sum_{i\neq k} \xi_{n,i} + \xi_{n,k}\big) - f_h'\big(\sum_{i\neq k} \xi_{n,i}\big) - \xi_{n,k} f_h''\big(\sum_{i\neq k} \xi_{n,i}\big).$$

Then the following equality holds:

$$\mathbb{E}\big[\big(\sum_{k=1}^n \xi_{n,k}\big) f_h\big(\sum_{k=1}^n \xi_{n,k}\big) - f_h'\big(\sum_{k=1}^n \xi_{n,k}\big)\big] = \sum_{k=1}^n \mathbb{E}\left[\xi_{n,k}\delta_{n,k}\right] - \sum_{k=1}^n \sigma_{n,k}^2 \mathbb{E}\left[\varepsilon_{n,k}\right].$$

Proof Taking into account that $\xi_{n,k}$ and $\sum_{i\neq k} \xi_{n,i}$ are independent, $\mathbb{E}[\xi_{n,k}] = 0$ and $\sum_{k=1}^n \sigma_{n,k}^2 = 1$ we obtain

$$\begin{aligned}
&\sum_{k=1}^n \mathbb{E}[\xi_{n,k}\delta_{n,k}] - \sum_{k=1}^n \sigma_{n,k}^2 \mathbb{E}[\varepsilon_{n,k}] \\
&\quad = \sum_{k=1}^n \mathbb{E}[\xi_{n,k} f_h(\sum_{k=1}^n \xi_{n,k})] - \sum_{k=1}^n \mathbb{E}[\xi_{n,k} f_h(\sum_{i\neq k} \xi_{n,i})] \\
&\quad - \sum_{k=1}^n \mathbb{E}[\xi_{n,k}^2 f_h'(\sum_{i\neq k} \xi_{n,i})] - \sum_{k=1}^n \sigma_{n,k}^2 \mathbb{E}[f_h'(\sum_{k=1}^n \xi_{n,k})] \\
&\quad + \sum_{k=1}^n \mathbb{E}[\xi_{n,k}^2]\mathbb{E}[f_h'(\sum_{i\neq k} \xi_{n,i})] + \sum_{k=1}^n \sigma_{n,k}^2 \mathbb{E}[\xi_{n,k} f_h''(\sum_{i\neq k} \xi_{n,i})]
\end{aligned}$$

$$
\begin{aligned}
&= \mathbb{E}[(\sum_{k=1}^{n} \xi_{n,k}) f_h(\sum_{k=1}^{n} \xi_{n,k})] - \sum_{k=1}^{n} \mathbb{E}[\xi_{n,k}]\mathbb{E}[f_h(\sum_{i\neq k} \xi_{n,i})] \\
&- \sum_{k=1}^{n} \mathbb{E}[\xi_{n,k}^2 f_h'(\sum_{i\neq k} \xi_{n,i})] - \mathbb{E}[f_h'(\sum_{k=1}^{n} \xi_{n,k})] \\
&+ \sum_{k=1}^{n} \mathbb{E}[\xi_{n,k}^2 f_h'(\sum_{i\neq k} \xi_{n,i})] + \sum_{k=1}^{n} \sigma_{n,k}^2 \mathbb{E}[\xi_{n,k}]\mathbb{E}[f_h''(\sum_{i\neq k} \xi_{n,i})] \\
&= \mathbb{E}[(\sum_{k=1}^{n} \xi_{n,k}) f_h(\sum_{k=1}^{n} \xi_{n,k}) - f_h'(\sum_{k=1}^{n} \xi_{n,k})]. \qquad \square
\end{aligned}
$$

9.4.13 Lemma *Let $f : \mathbb{R} \longrightarrow \mathbb{R}$ have a bounded derivative and a bounded and piecewise continuous second derivative. Then for any $a, x \in \mathbb{R}$*

$$
|f(a+x) - f(a) - f'(a)x| \leq \min\{(\sup_{x_1,x_2\in\mathbb{R}} |f'(x_1) - f'(x_2)|)|x|, \frac{1}{2}\|f''\|_\infty x^2\}.
$$

Proof Put $\varphi(t) = f(a+tx)$. Then we get

$$
\begin{aligned}
f(a+x) - f(a) - xf'(a) &= \varphi(1) - \varphi(0) - \varphi'(0) \\
&= \int_0^1 \varphi'(t)dt - \varphi'(0) \\
&= \int_0^1 x(f'(a+tx) - f'(a))dt,
\end{aligned}
$$

and thus

$$
\left|f(a+x) - f(a) - xf'(a)\right| \leq |x| \sup_{x_1,x_2\in\mathbb{R}} \left|f'(x_1) - f'(x_2)\right|.
$$

Analogously, performing an integration by parts on the right hand side of $\varphi(1) - \varphi(0) = \int_0^1 \varphi'(t)dt$, rearranging terms, and writing $\varphi'(1) - \varphi'(0) = \int_0^1 \varphi''(t)dt$ gives

$$
\varphi(1) - \varphi(0) - \varphi'(0) = \int_0^1 (1-t)\varphi''(t)dt,
$$

which finally gives

$$
f(a+x) - f(a) - xf'(a) = \int_0^1 (1-t)x^2 f''(a+tx)dt,
$$

and

$$\left|f(a+x)-f(a)-xf'(a)\right| \le x^2\|f''\|_\infty \int_0^1 (1-t)dt = \frac{1}{2}\|f''\|_\infty x^2. \qquad \square$$

9.4.14 Lemma *Let $h \in \mathscr{H}$. Then for all $x, y \in \mathbb{R}$*

$$\left|f_h'(x)-f_h'(y)\right| \le 1.$$

Proof From 9.4.11 we derive that

$$f_h'(x) = xe^{x^2/2}\int_{-\infty}^{x} \left(h(t)-\mathbb{E}[h(\xi)]\right)e^{-t^2/2}dt + h(x)-\mathbb{E}[h(\xi)].$$

Furthermore, for all $h \in \mathscr{H}$, applying the layer cake representation to the constant random variable ζ with value $h(x)$ on the probability space $([0,1], \mathscr{B}_{[0,1]}, \lambda)$, with $\mathscr{B}_{[0,1]}$ the Borel σ-field and λ Lebesgue measure, we get

$$h(x) = \int_0^1 \lambda\left(\{t \in [0,1] : \zeta(t) \ge s\}\right) ds = \int_0^1 \Phi(x,s)ds$$

with

$$\Phi(x,s) = \begin{cases} 0 & \text{if } h(x) < s \\ 1 & \text{if } h(x) \ge s, \end{cases}$$

and hence

$$h(x) = \int_0^1 h_{h^{-1}(s)}(x)ds.$$

Combining the foregoing and applying Fubini yields

$$\begin{aligned} f_h'(x)-f_h'(y) &= xe^{x^2/2}\int_{-\infty}^{x} \left(h(t)-\mathbb{E}[h(\xi)]\right)e^{-t^2/2}dt + h(x) \\ &\quad - ye^{y^2/2}\int_{-\infty}^{y} \left(h(t)-\mathbb{E}[h(\xi)]\right)e^{-t^2/2}dt - h(y) \\ &= xe^{x^2/2}\int_{-\infty}^{x}\int_0^1 (h_{h^{-1}(s)}(t)-\mathbb{E}[h_{h^{-1}(s)}(\xi)])e^{-t^2/2}dsdt + \int_0^1 h_{h^{-1}(s)}(x)ds \\ &\quad - ye^{y^2/2}\int_{-\infty}^{y}\int_0^1 (h_{h^{-1}(s)}(t)-\mathbb{E}[h_{h^{-1}(s)}(\xi)])e^{-t^2/2}dsdt - \int_0^1 h_{h^{-1}(s)}(y)ds \end{aligned}$$

$$= \int_0^1 \Big[x e^{x^2/2} \int_{-\infty}^{x} (h_{h^{-1}(s)}(t) - \mathbb{E}[h_{h^{-1}(s)}(\xi)]) e^{-t^2/2} dt + h_{h^{-1}(s)}(x)$$

$$- y e^{y^2/2} \int_{-\infty}^{y} (h_{h^{-1}(s)}(t) - \mathbb{E}[h_{h^{-1}(s)}(\xi)]) e^{-t^2/2} dt + h_{h^{-1}(s)}(y) \Big] ds$$

$$= \int_0^1 [f'_{h_{h^{-1}(s)}}(x) - f'_{h_{h^{-1}(s)}}(y)] ds$$

which again by 9.4.11 proves the lemma. □

Now we obtain the following indexed version of the sufficiency of Lindeberg's condition in the Lindeberg-Feller CLT.

9.4.15 Theorem (Indexed CLT) *If $\{\xi_{n,k}\}$ is a standard triangular array which satisfies Feller's negligibility condition, then the following inequality holds.*

$$\lambda_c\langle (\sum_{k=1}^{n} \xi_{n,k})_n \rangle (\xi) \leq \chi_{Lin}((\xi_{n,k})_{n,k}).$$

Proof Combining the foregoing gives that for any $h \in \mathscr{H}$ and $\theta > 0$

$$\begin{aligned}
&\Big|\mathbb{E}[h(\xi) - h(\sum_{k=1}^{n} \xi_{n,k})]\Big| \\
&= \Big|\mathbb{E}[(\sum_{k=1}^{n} \xi_{n,k}) f_h(\sum_{k=1}^{n} \xi_{n,k}) - f'_h(\sum_{k=1}^{n} \xi_{n,k})]\Big| \\
&\leq \sum_{k=1}^{n} \mathbb{E}[|\xi_{n,k}\delta_{n,k}|] + \sum_{k=1}^{n} \sigma_{n,k}^2 \mathbb{E}[|\varepsilon_{n,k}|] \\
&\leq \frac{1}{2} \|f''_h\|_\infty \sum_{k=1}^{n} \mathbb{E}\Big[|\xi_{n,k}|^3 \mid |\xi_{n,k}| < \theta\Big] \\
&\quad + \big(\sup_{x_1, x_2 \in \mathbb{R}} |f'_h(x_1) - f'_h(x_2)|\big) \sum_{k=1}^{n} \mathbb{E}\Big[|\xi_{n,k}|^2 \mid |\xi_{n,k}| \geq \theta\Big] \\
&\quad + \big(\sup_{x_1, x_2 \in \mathbb{R}} |f''_h(x_1) - f''_h(x_2)|\big) \sum_{k=1}^{n} \sigma_{n,k}^2 \mathbb{E}[|\xi_{n,k}|] \\
&\leq \frac{1}{2} \|f''_h\|_\infty \theta + \sum_{k=1}^{n} \mathbb{E}\Big[|\xi_{n,k}|^2 \mid |\xi_{n,k}| \geq \theta\Big] \\
&\quad + \big(\sup_{x_1, x_2 \in \mathbb{R}} |f''_h(x_1) - f''_h(x_2)|\big) \max_{k=1}^{n} \sigma_{n,k}.
\end{aligned}$$

This implies, making use of Feller's negligibility condition, that for all h and θ

$$\limsup_{n\to\infty} \left|\mathbb{E}[h(\xi) - h(\sum_{k=1}^{n} \xi_{n,k})]\right| \leq \frac{1}{2} \| f_h'' \|_\infty \theta + \chi_{Lin}(\xi_{n,k})$$

and the result now follows from 9.4.9 and the arbitrariness of h and θ. □

9.5 Comments

1. **Further results**
Most results in this chapter come from Berckmoes et al. (2011a, b, 2013) and we refer to those papers for more information. Many more interesting results and details can be found in the PhD thesis of Berckmoes (2014).

2. **Categorical aspect of the weak approach structure**
If we put Pol for the category of completely metrizable separable topological spaces (Polish spaces) and continuous maps then it follows from 9.1.7 that

$$\mathsf{Pol} \longrightarrow \mathsf{App} : \begin{cases} S \longrightarrow (\mathscr{P}(S), \delta_w) \\ f \longrightarrow \widehat{f} \end{cases}$$

is functorial.

3. **Categorical aspect of the approach structure of convergence in probability**
If we put Pol_m for the category of complete separable metric spaces and contractions then it follows from 9.2.5 that

$$\mathsf{Pol}_m \longrightarrow \mathsf{App} : \begin{cases} S \longrightarrow (\mathscr{R}(S), \delta_p) \\ f \longrightarrow \widetilde{f} \end{cases}$$

is functorial.

4. **Categorical relation between the structures**
The combined results of 9.1.7, 9.2.5 and 9.2.9 show that if $f : S \longrightarrow T$ is a contraction then the following is a commutative diagram of contractions for the weak and c.i.p. approach structures. As an immediate consequence, applying the topological and metric coreflections, the diagram is also a commutative diagram of continuous maps for respectively the weak topologies and the topologies of convergence in probability and a commutative diagram of contractions for respectively the total variation metrics and the indicator metrics.

$$\begin{array}{ccc} \mathscr{R}(S) & \xrightarrow{\tilde{f}} & \mathscr{R}(T) \\ {\scriptstyle L}\downarrow & & \downarrow{\scriptstyle L} \\ \mathscr{P}(S) & \xrightarrow[\hat{f}]{} & \mathscr{P}(T) \end{array}$$

5. **Concerning the Lindeberg index**

The Lindeberg index which we introduced is not trivial, by which we mean that it can attain any possible value besides the obvious ones 0 and 1. In order to see this an example suffices. Therefore, fix $0 < \alpha < 1$, let $\beta = \frac{\alpha}{1-\alpha}$ and put

$$s_n^2 = (1+\beta)n - \beta \sum_{k=1}^{n} k^{-1} = n + \beta \sum_{k=1}^{n} \left(1 - k^{-1}\right).$$

Notice that $s_n^2 \to \infty$. Now consider the standard triangular array $\{\eta_{\alpha,n,k}\}$ such that

$$\mathbb{P}\left[\eta_{\alpha,n,k} = -1/s_n\right] = \mathbb{P}\left[\eta_{\alpha,n,k} = 1/s_n\right] = \frac{1}{2}\left(1 - \beta k^{-1}\right)$$

and

$$\mathbb{P}\left[\eta_{\alpha,n,k} = -\sqrt{k}/s_n\right] = \mathbb{P}\left[\eta_{\alpha,n,k} = \sqrt{k}/s_n\right] = \frac{1}{2}\beta k^{-1}.$$

It can then be verified that $\{\eta_{\alpha,n,k}\}$ satisfies Feller's negligibility condition and that

$$\mathrm{Lin}\left(\{\eta_{\alpha,n,k}\}\right) = \alpha.$$

We refer to Berckmoes et al. (2013) for details.

6. **Underbound in the indexed CLT**

In Berckmoes et al. (submitted for publication) the theory introduced in this chapter is taken further. In Berckmoes et al. (2013) also an underbound is given for the limit in 9.4.15. Hereto a relaxed Lindeberg index is introduced, namely

$$\chi^*_{Lin}\left(\{\xi_{n,k}\}\right) = \limsup_{n\to\infty} \sum_{k=1}^{n} \mathbb{E}\left[\xi_{n,k}^2 \varphi(|\xi_{n,k}|)\right]$$

where

$$\varphi(x) = 1 - e^{-\frac{1}{2}x^2}.$$

It can then be verified that

$$\chi_{Lin}^{*}\left(\left\{\xi_{n,k}\right\}\right) \leq \chi_{Lin}\left(\left\{\xi_{n,k}\right\}\right)$$

and that this relaxed version too will be zero if and only if the Lindeberg condition is fulfilled. Then it is possible to prove that there exists a constant $C > 0$, not depending on $\left\{\xi_{n,k}\right\}$, such that

$$\chi_{Lin}^{*}\left(\left\{\xi_{n,k}\right\}\right) \leq C\lambda_c\langle(\sum_{k=1}^{n}\xi_{n,k})_n\rangle(\xi).$$

Moreover, this formula can be shown to hold with $C \leq 30.3$. Together with 9.4.15 this then provides us with a full indexed version of the usual CLT.

7. **Further research in higher dimensions**

In Berckmoes et al. (submitted for publication) the theory is taken one step further to the realm of finite-dimensional random vectors by making use of a multivariate version of Stein's method to obtain a finite-dimensional quantitative Lindeberg central limit theorem in the same vain as 9.4.15.

Chapter 10
Approach Theory Meets Hyperspaces

Traveling through hyperspace ain't like dusting crops, boy! Without precise calculations we could fly right through a star or bounce too close to a supernova, and that'd end your trip real quick, wouldn't it?

(Han Solo, to Luke Skywalker)

In this chapter we are mainly interested in approach structures (or uniform gauge structures) on hyperspaces of closed sets of metric spaces. In the first section we study a natural quantification of the Wijsman topology. This has several advantages over the Wijsman topology. For instance, the Wijsman topology is metrizable only if the original metric space is separable. For quantification with an approach structure no condition is required, the Wijsman topology is always quantifiable (see 10.1.1). We also compare the Wijsman approach structure with the Hausdorff metric. For instance it will turn out that the indices of compactness of all three spaces, the original metric space, the Wijsman approach hyperspace and the hyperspace with the Hausdorff metric, coincide, from which several classical results can be deduced (see 10.1.6).

In the second section we study the proximal topologies where we see the same behaviour as for the Wijsman topology, they are always quantifiable by canonical approach structures (see 10.2.5). We extend some results which can be found in (Beer et al. 1992; Beer and Lucchetti 1993), characterizing some well-known hyperspace topologies as suprema of collections of other hyperspace topologies. Further, here too we see that the index of compactness of the hyperspace equipped with a proximal structure coincides with the index of compactness of the original metric space. For the proximal structures we also give a description of the completion showing that the completion of the hyperspace is isomorphic to the hyperspace of the completion (see 10.2.12).

In the last example, we study a quantified version of the Vietoris structure in the more general setup of closed sets in an arbitrary T_1 approach space. This line of attack has several advantages over the topological one, not in the least because the Vietoris construction can now also be considered intrinsically for metric spaces. Further we mainly pay attention to properties involving compactness. In the first place we prove that, again, the indices of compactness of the original approach space and of the Vietoris hyperspace coincide (see 10.3.7, 10.3.12 and 10.3.13). In the second place

R. Lowen, *Index Analysis*, Springer Monographs in Mathematics,
DOI 10.1007/978-1-4471-6485-2_10

the well-known result (see e.g. Naimpally 2003) which says that if the original space is compact metric then the Vietoris topology is metrizable by the Hausdorff metric gets strengthened in the sense that in the approach setting under the same conditions the Vietoris approach structure actually coincides with the Hausdorff metric. Classic results follow as easy corollaries. Besides these main results we also draw attention to the good functorial relationship between the Vietoris approach structures and the associated topologies.

10.1 The Wijsman Structure

In the literature a large variety of topologies, uniformities, and metrics have been considered on hyperspaces of topological, uniform, or metric spaces. See e.g. Beer and Luchetti (1993), Beer (1989, 1991). In particular in the case of metric spaces, the so-called Wijsman topology on the hyperspace of all closed sets has been extensively studied.

Let (X, d) be a metric space. We denote by $CL(X)$ the set of all nonempty closed subsets of X.

We begin by recalling the definitions of the Hausdorff metric and of the Wijsman topology.

The *Hausdorff metric* on $CL(X)$ can be defined in several equivalent ways, one of which is important for us. We denote this metric by H_d. It is given by the following formula (see e.g. Beer 1991). For any $A, B \in CL(X)$:

$$H_d(A, B) = \sup_{x \in X} \left| \delta_d(x, A) - \delta_d(x, B) \right|.$$

The *Wijsman topology* too can be introduced in several equivalent ways. Two of these are important for our considerations.

First, it is the initial topology on $CL(X)$ for the source

$$(CL(X) \longrightarrow \mathbb{R}^+ : A \mapsto \delta_d(x, A))_{x \in X},$$

where $\mathbb{R}^+$ is equipped with the usual Euclidean topology.

Second, it can also be characterized as being the initial topology for the source

$$CL(X) \longrightarrow (\mathbb{R}^+)^X : A \mapsto \delta_d(\cdot, A),$$

where $\mathbb{R}^+$ is again equipped with the usual Euclidean topology and $(\mathbb{R}^+)^X$ is equipped with the product topology. In both cases the usual topology on the real line plays a crucial role and the Wijsman topology owes its existence to the fact that we are able to construct initial topologies. In the first method we are simply constructing the Wijsman topology as an initial topology of the usual topology on $\mathbb{R}^+$ for a collection of maps, and in the second method we are actually identifying

$CL(X)$ with the set of all distance functionals $\{\delta_d(\cdot, A) \mid A \in CL(X)\}$ and we construct the Wijsman topology as a subspace topology of a product (or pointwise) topology.

If, however, we start not with the usual topology on $\mathbb{R}^+$ but with the usual metric, then classically we are unable to perform either of these constructions. Again, using approach spaces we can overcome these obstacles.

For each finite subset F of X we define

$$d_F : CL(X) \times CL(X) \longrightarrow \mathbb{P}$$

by

$$d_F(A, B) := \sup_{x \in F} \left|\delta_d(x, A) - \delta_d(x, B)\right|.$$

Now consider the collection

$$\mathscr{D}_{W_d} := \left\{d_F \mid F \in 2^{(X)}\right\}.$$

The set $\mathscr{D}_{W_d}$ of metrics is closed under the formation of finite suprema and hence is a symmetric basis for a symmetric gauge. The distance generated by this gauge is given by

$$\delta_{W_d} : CL(X) \times 2^{CL(X)} \longrightarrow \mathbb{P} : (A, \mathscr{A}) \longrightarrow \sup_{F \in 2^{(X)}} \inf_{B \in \mathscr{A}} d_F(A, B).$$

We will refer to this approach structure as the *Wijsman (approach) structure*, and our first task of course is to justify this terminology. We will denote the space $CL(X)$ equipped with this approach structure by $CL_{W_d}(X)$ and analogously we will denote $CL(X)$ equipped with the Hausdorff metric by $CL_{H_d}(X)$.

10.1.1 Proposition *The topological coreflection of $CL_{W_d}(X)$ is $CL(X)$ equipped with the Wijsman topology $\mathscr{T}_{W_d}$ and the metric coreflection is $CL_{H_d}(X)$.*

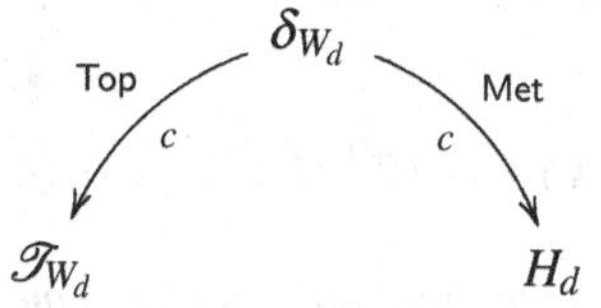

Proof For the first claim, it follows from 3.1.11 that a basis for the neighbourhoods of $A \in CL(X)$ in the underlying topology of $CL_{W_d}(X)$ is given by the collection

$$\{B \in CL(X) \mid d_F(A, B) < \varepsilon\} \qquad F \in 2^{(X)}, \varepsilon > 0.$$

This, however, is precisely a basis for the neighbourhoods of A in the Wijsman topology $\mathscr{T}_{W_d}$.

For the second claim, it follows from 3.1.11 that, for all $A, B \in CL(X)$, we have

$$\begin{aligned} d_{\delta_{W_d}}(A, B) &= \sup_{F \in 2^{(X)}} d_F(A, B) \\ &= \sup_{x \in X} d_{\{x\}}(A, B) \\ &= H_d(A, B). \end{aligned}$$

□

In Lechicki and Levi (1987) it was shown that the Wijsman topology on $CL(X)$ is metrizable if and only if $(X, \mathcal{T}_d)$ is separable. Such a condition is not required in our case. $(CL(X), \mathcal{T}_{W_d})$ is always canonically quantifiable by δ_{W_d}. The canonicity is enhanced by the following results which are to be compared with the analogous results for the Wijsman topology in the setting of topology.

10.1.2 Theorem *The Wijsman approach structure is the initial structure on $CL(X)$ for the source*

$$(CL(X) \longrightarrow \mathbb{R}^+ : A \mapsto \delta_d(x, A))_{x \in X},$$

where $\mathbb{R}^+$ is equipped with the usual Euclidean metric.

Proof According to 1.3.11 the initial approach structure is given by the basis for the gauge

$$\left\{ \sup_{x \in F} d_{\mathbb{E}} \circ (d(x, \cdot) \times d(x, \cdot)) \mid F \in 2^{(X)} \right\}.$$

The result then follows from the fact that, for all $F \in 2^{(X)}$ and all $A, B \in CL(X)$, we have

$$\sup_{x \in F} d_{\mathbb{E}}(\delta_d(x, A), \delta_d(x, B)) = d_F(A, B).$$

□

10.1.3 Theorem *The Wijsman approach structure is the initial structure on $CL(X)$ for the source*

$$CL(X) \longrightarrow (\mathbb{R}^+)^X : A \mapsto \delta_d(\cdot, A),$$

where $\mathbb{R}^+$ is again equipped with the usual Euclidean metric and $(\mathbb{R}^+)^X$ is equipped with the product distance.

Proof The product distance on $(\mathbb{R}^+)^X$ is given by

$$\delta(f, \mathcal{F}) := \sup_{F \in 2^{(X)}} \inf_{g \in \mathcal{F}} \sup_{x \in F} |f(x) - g(x)|.$$

Consequently, the initial distance is given by

$$\begin{aligned}\delta'(A, \mathscr{A}) &= \sup_{F\in 2^{(X)}} \inf_{B\in\mathscr{A}} \sup_{x\in F} \left|\delta_d(x, A) - \delta_d(x, B)\right| \\ &= \sup_{F\in 2^{(X)}} \inf_{B\in\mathscr{A}} d_F(A, B) \\ &= \delta_{W_d}(A, \mathscr{A}).\end{aligned}$$

□

A topological structure on a hyperspace is called *admissible* (Michael 1951), if the map $X \longrightarrow CL(X) : x \mapsto \{x\}$ is well defined and an embedding and we use exactly the same terminology in our case where of course the embedding is considered in App.

10.1.4 Proposition *Let (X, d) be a metric space. The Wijsman distance on $CL(X)$ is admissible.*

Proof Since $(X, \mathscr{T}_d)$ is Hausdorff the function

$$\psi : X \longrightarrow CL_{W_d}(X) : x \mapsto \{x\}$$

is well defined. Now if $x \in X$ and $A \subseteq X$, then on the one hand we have

$$\begin{aligned}\delta_{W_d}(\psi(x), \psi(A)) &\geq \inf_{a\in A} |d(x, x) - d(x, a)| \\ &= \delta_d(x, A),\end{aligned}$$

whereas on the other hand we have

$$\begin{aligned}\delta_{W_d}(\psi(x), \psi(A)) &\leq \sup_{F\in 2^{(X)}} \inf_{a\in A} \sup_{y\in F} d(x, a) \\ &= \delta_d(x, A).\end{aligned}$$

Consequently, ψ is an embedding. □

Identifying a set with the set of its singletons and a metric space with its "copy" in App, the foregoing result says that (X, δ_d) is a metric subspace of $CL_{W_d}(X)$. Hence, the link between metric and Wijsman distance is much stronger than between metric and Wijsman topology, since under certain conditions, described in Costantini et al. (1993), different metrics on X may generate the same Wijsman topology. The foregoing result actually shows that $\mathsf{Met}(X) \longrightarrow \mathsf{App}(X) : d \mapsto \delta_{W_d}$ is an injection.

10.1.5 Proposition *If (X, d) is separable then $CL_{W_d}(X)$ is gauge-countable. In particular, if Y is a countable dense subset of X, then the countable set of metrics*

$$\mathscr{D}_0 := \left\{d_F \mid F \in 2^{(Y)}\right\}$$

is a gauge basis.

Proof Evidently, for any $A \in CL(X)$ and $\mathscr{A} \subseteq CL(X)$, we have

$$\delta_{W_d}(A, \mathscr{A}) \geq \sup_{F \in 2^{(Y)}} \inf_{B \in \mathscr{B}} d_F(A, B).$$

Conversely, let $\varepsilon > 0$ and let $F \in 2^{(X)}$. Then, for each $x \in F$, there exists $y_x \in Y$ such that $d(x, y_x) \leq \frac{\varepsilon}{2}$. It follows that if we let $F_\varepsilon := \{y_x \mid x \in F\}$, then

$$\begin{aligned} \delta_{W_d}(A, \mathscr{A}) &= \sup_{F \in 2^{(X)}} \inf_{B \in \mathscr{B}} d_F(A, B) \\ &\leq \sup_{F \in 2^{(X)}} \inf_{B \in \mathscr{B}} d_{F_\varepsilon}(A, B) + \varepsilon \\ &\leq \sup_{F \in 2^{(Y)}} \inf_{B \in \mathscr{B}} d_F(A, B) + \varepsilon, \end{aligned}$$

which by the arbitrariness of ε proves our claim. □

It is well known that if the metric space (X, d) is totally bounded then the Wijsman topology is metrized by the Hausdorff metric. A last interesting result concerning the relationship between the Hausdorff metric and the Wijsman distance considerably strengthens this property.

10.1.6 Proposition *In any metric space (X, d) the following inequalities hold.*

$$\delta_{W_d} \leq \delta_{H_d} \leq \delta_{W_d} + 2\chi_c(X).$$

Proof By 10.1.1 it suffices to prove the second inequality. Let $A \in CL(X)$ and let $\mathscr{A} \subseteq CL(X)$. Further let $\chi_c(X) < \alpha$, then there exist $Y \subseteq X$ finite such that

$$\bigcup_{x \in Y} B(x, \alpha) = X.$$

Then, for any $x \in X$, there exists $y \in Y$ such that, for all $B \in \mathscr{A}$,

$$|\delta_d(x, A) - \delta_d(x, B)| \leq |\delta_d(y, A) - \delta_d(y, B)| + 2\alpha,$$

and consequently

$$\begin{aligned} \delta_{H_d}(A, \mathscr{A}) &= \inf_{B \in \mathscr{A}} \sup_{x \in X} |\delta_d(x, A) - \delta_d(x, B)| \\ &\leq \sup_{Y \in 2^{(X)}} \inf_{B \in \mathscr{A}} \sup_{y \in Y} |\delta_d(y, A) - \delta_d(y, B)| + 2\alpha \\ &= \delta_{W_d}(A, \mathscr{A}) + 2\alpha, \end{aligned}$$

which proves our claim. □

10.1.7 Corollary *If (X, d) is totally bounded then $\delta_{W_d} = \delta_{H_d}$.*

10.1.8 Theorem *For any metric space (X, d) the following properties are equivalent.*

1. *$CL_{W_d}(X)$ is complete.*
2. *$CL_{H_d}(X)$ is complete.*
3. *X is complete.*

Proof 1 ⇔ 2 This follows from 3.5.11.
2 ⇔ 3 This can be found, for example, in Kuratowski (1996). □

10.1.9 Theorem *For any metric space (X, d) we have*

$$\chi_c(CL_{W_d}(X)) = \chi_c(CL_{H_d}(X)) = \chi_c(X).$$

Proof For any $A \in CL(X)$ choose a point $x_A \in A$ and put $G_A := \{x_A\}$. Then, if $\mathscr{B} \in 2^{(CL(X))}$ and we put $Y_{\mathscr{B}} := \{x_A \mid A \in \mathscr{B}\}$ we obtain

$$\begin{aligned}\sup_{z\in X}\inf_{x\in Y_{\mathscr{B}}} d(x,z) &= \sup_{z\in X}\inf_{A\in\mathscr{B}} \delta_d(x_A,\{z\})\\ &= \sup_{z\in X}\inf_{A\in\mathscr{B}} |\delta_d(x_A,A) - \delta_d(x_A,\{z\})|\\ &= \sup_{z\in X}\inf_{A\in\mathscr{B}} d_{G_A}(A,\{z\})\\ &\le \sup_{C\in CL(X)}\inf_{A\in\mathscr{B}} d_{G_A}(A,C)\end{aligned}$$

from which it follows that

$$\chi_c(X) \le \chi_c(CL_{W_d}(X)).$$

That $\chi_c(CL_{W_d}(X)) \le \chi_c(CL_{H_d}(X))$ follows from 10.1.1.

To show the remaining required inequality suppose that $X = \cup_{i=1}^k B(x_i, \varepsilon)$ and let $\mathscr{B}$ stand for the set of all nonempty subsets of $\{x_1, \ldots, x_k\}$. Take $A \in CL(X)$ arbitrary then obviously

$$I_A := \{i \in \{1, \ldots, k\} \mid A \cap B(x_i, \varepsilon) \neq \emptyset\} \neq \emptyset$$

and

$$A \subseteq \bigcup_{i\in I_A} B(x_i, \varepsilon).$$

If we now put $B_A := \{x_i \mid i \in I_A\}$ then $B_A \in \mathscr{B}$ and

$$H_d(A, B_A) = \sup_{a\in A} \delta_d(a, B_A) \vee \sup_{i\in I_A} \delta_d(x_i, A) \le \varepsilon.$$

Consequently, for any $n \in \mathbb{N}_0$ we have

$$CL(X) = \bigcup_{B \in \mathscr{B}} B_{H_d}(B, \varepsilon + \frac{1}{n})$$

from which it follows that $\chi_c(CL_{H_d}(X)) \leq \varepsilon + \frac{1}{n}$ which by the arbitrariness of ε shows that $\chi_c(CL_{H_d}(X)) \leq \chi_c(X)$. □

Some immediate corollaries of the foregoing result are

10.1.10 Corollary (Kuratowski) *A metric space X is totally bounded if and only if $CL_{H_d}(X)$ is totally bounded.*

10.1.11 Corollary *A metric space X is bounded if and only if $CL_{H_d}(X)$ is bounded.*

10.1.12 Corollary (Kuratowski, Lechicki, Levi) *A metric space is compact if and only if $(CL(X), \mathscr{T}_{W_d})$ is compact.*

10.2 The Proximal Structures

Besides the set of all nonempty closed subsets we will in this section also consider the set $CLB(X)$ of all nonempty closed and bounded subsets of X. For a subset A of X and for $\varepsilon \in \mathbb{R}_0^+$, $S_\varepsilon(A)$ will stand for the set $\{x \in X \mid d(x, A) < \varepsilon\}$. When studying $CL(X)$, the *gap functional* D_d and the *Hausdorff excess functional* e_d play an important role. They are defined as follows:

$$\begin{aligned} D_d(A, B) &:= \inf\{d(x, y) \mid x \in A,\ y \in B\}, \\ e_d(A, B) &:= \sup_{x \in A} \delta_d(x, B). \end{aligned}$$

If $A \subseteq X$, we also use the following standard notations:

$$\begin{aligned} A^- &:= \{B \in CL(X) \mid B \cap A \neq \emptyset\}, \\ A^+ &:= \{B \in CL(X) \mid B \subseteq A\}, \\ A^{++} &:= \{B \in CL(X) \mid \exists \varepsilon \in \mathbb{R}_0^+ : S_\varepsilon(B) \subseteq A\}, \\ &= \{B \in CL(X) \mid D_d(B, A^c) > 0\}. \end{aligned}$$

In fact, A^-, A^+ and A^{++} consist of all closed subsets of X that hit A, respectively miss A^c, respectively miss A^c in a metric-detectable way.

The study of proximal hit-and-miss-topologies on hyperspaces of metric spaces began with the study of the so-called *proximal topology* (see Beer et al. 1992), defined as the topology on $CL(X)$ having $\{V^- \mid V \in \mathscr{T}_d\} \cup \{V^{++} \mid V \in \mathscr{T}_d\}$ as a subbasis. In the sequel we will write $\mathscr{T}_{prox(d)}$ to denote the proximal topology determined by d. That the proximal topology indeed depends on the metric of X follows from a remark given in Beer et al. (1992), stating that two metrics d and d' on X determine the same proximal topology if and only if d and d' determine the same metric proximity. For information on the proximal topology we refer to Beer et al. (1992), Beer and Lucchetti (1993) and Beer (1993). In Beer (1993), Beer and Lucchetti (1993) and Di Maio and Holà (1997), a broader class of proximal hit-and-miss topologies was considered: for an arbitrary nonempty subset Δ of $CL(X)$, the Δ-proximal topology on $CL(X)$, which we will denote by $\mathscr{T}_{prox(\Delta,d)}$, is defined to be the topology on $CL(X)$ having $\{V^- \mid V \in \mathscr{T}_d\} \cup \{(D^c)^{++} \mid D \in \Delta\}$ as a subbasis. Note that $\mathscr{T}_{prox(CL(X),d)}$ coincides with the proximal topology. In the special case when $\Delta = CLB(X)$ the associated proximal hypertopology is called the bounded proximal topology and will be denoted by $\mathscr{T}_{bprox(d)}$.

10.2.1 Definition Let (X, d) be a metric space. A nonempty subset Δ of $CL(X)$ is called *stable under enlargements* if

$$\forall D \in \Delta, \forall \varepsilon \in \mathbb{R}^+ : D^{(\varepsilon)} \in \Delta,$$

and it is called a *p-cover of* X if it satisfies the following properties.

(P1) $\{\{x\} \mid x \in X\} \subseteq \Delta$.
(P2) Δ is stable under enlargements.

Note that $(X^c)^{++} = \emptyset$, which yields that $\mathscr{T}_{prox(\Delta,d)} = \mathscr{T}_{prox(\Delta\cup\{X\},d)}$ for every nonempty subset Δ of $CL(X)$. Also in the setting of 10.2.4, we see that adding X to Δ has no effect since $D_d(\cdot, X)$ is the constant zero-functional on $CL(X)$, which is continuous with respect to any topology on $CL(X)$.

We denote the set of all finite non-empty subset of Δ by $2_0^{(\Delta)}$, and then for each $\Gamma \in 2_0^{(\Delta)}$, we define

$$d^\Gamma : CL(X) \times CL(X) \longrightarrow \mathbb{R}^+ : (A, B) \mapsto \sup_{D\in\Gamma} \mid D_d(A, D) - D_d(B, D) \mid,$$

and

$$\mathscr{D}^{\Delta,d} := \{ d^\Gamma \mid \Gamma \in 2_0^{(\Delta)}\}.$$

We now have the following proposition:

10.2.2 Proposition *Let (X, d) be a metric space and let Δ be a p-cover of X. Then $\mathscr{D}^{\Delta,d}$ is a collection of metrics on $CL(X)$, closed for the formation of finite suprema. Therefore, $\mathscr{D}^{\Delta,d}$ generates a uniform approach structure on $CL(X)$ with distance*

$$\delta_{prox(\Delta,d)} : CL(X) \times 2^{CL(X)} \longrightarrow \mathbb{P} : (A, \mathscr{A}) \mapsto \sup_{\Gamma\in 2_0^{(\Delta)}} \inf_{B\in\mathscr{A}} d^\Gamma(A, B).$$

Proof This is straightforward and we leave this to the reader. □

The distance $\delta_{prox(\Delta,d)}$ will be called the *Δ-proximal distance* and the next two results will prove this terminology to be plausible. $CL(X)$ equipped with this structure will be denoted $CL_{prox(\Delta,d)}(X)$.

10.2.3 Theorem *Let (X, d) be a metric space and let Δ be a p-cover of X. Then $\delta_{prox(\Delta,d)}$ is the initial distance on $CL(X)$ for the source*

$$\left(D_d(\cdot, D) : CL(X) \longrightarrow (\mathbb{R}^+, \delta_{d_{\mathbb{E}}}) : A \mapsto D_d(A, D)\right)_{D\in\Delta}.$$

Proof If we denote the initial distance for the source above by δ, it follows that for every $A \in CL(X)$ and every $\mathscr{A} \in 2^{CL(X)}$

$$\begin{aligned}\delta(A, \mathscr{A}) &= \sup_{\Gamma\in 2_0^{(\Delta)}} \inf_{B\in\mathscr{A}} \sup_{D\in\Gamma} d_{\mathbb{E}}(D_d(A, D), D_d(B, D))\\ &= \delta_{prox(\Delta,d)}(A, \mathscr{A}).\end{aligned}$$

□

10.2.4 Corollary **(G. Beer, R. Lucchetti)** *Let (X, d) be a metric space and let Δ be a p-cover, then $\mathscr{T}_{prox(\Delta,d)}$ is the initial topology on $CL(X)$ for the source*

$$\left(D_d(\cdot, D) : CL(X) \longrightarrow (\mathbb{R}^+, \mathscr{T}_{\mathbb{E}}) : A \mapsto D_d(A, D)\right)_{D\in\Delta}.$$

10.2.5 Proposition *Let (X, d) be a metric space and let Δ be a p-cover of X. Then the topological coreflection of $CL_{prox(\Delta,d)}(X)$ is $(CL(X), \mathscr{T}_{prox(\Delta,d)})$ and the metric coreflection is $CL_{H_d}(X)$.*

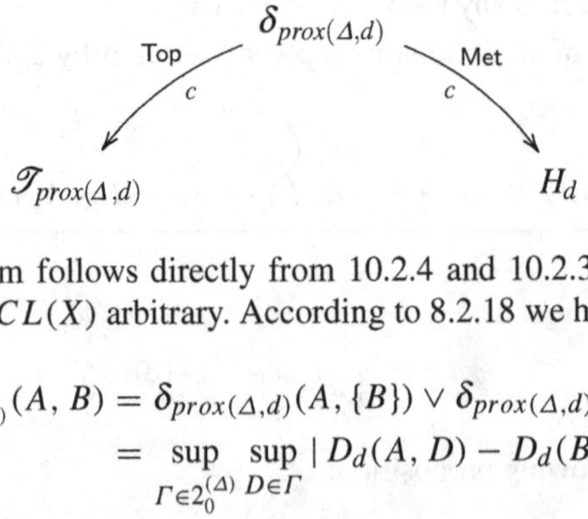

Proof The first claim follows directly from 10.2.4 and 10.2.3 and for the second claim take $A, B \in CL(X)$ arbitrary. According to 8.2.18 we have that

$$\begin{aligned}d_{\delta_{prox(\Delta,d)}}(A, B) &= \delta_{prox(\Delta,d)}(A, \{B\}) \vee \delta_{prox(\Delta,d)}(B, \{A\})\\ &= \sup_{\Gamma\in 2_0^{(\Delta)}} \sup_{D\in\Gamma} | D_d(A, D) - D_d(B, D) |\\ &= \sup_{D\in\Delta} | D_d(A, D) - D_d(B, D) |.\end{aligned}$$

On the one hand, it now follows from (P1) that

$$\begin{aligned}d_{\delta_{prox(\Delta,d)}}(A, B) &\geq \sup_{x\in X} | D_d(A, \{x\}) - D_d(B, \{x\}) |\\ &= H_d(A, B).\end{aligned}$$

On the other hand, we have for every $D \in \Delta$ that $D_d(A, D) \leq e_d(B, A)+D_d(B, D)$ and $D_d(B, D) \leq e_d(A, B)+D_d(A, D)$. From the last two inequalities, we obtain that $|\, D_d(A, D) - D_d(B, D)\,| \leq e_d(A, B) \vee e_d(B, A) = H_d(A, B)$, which implies that

$$d_{\delta_{prox(\Delta,d)}}(A, B) \leq H_d(A, B). \qquad \square$$

10.2.6 Proposition *Let (X, d) be a metric space and let Δ be a p-cover of X. The Δ-proximal distance is admissible on $CL(X)$.*

Proof This is analogous to 10.1.4 and we leave this to the reader.

In the following proposition we extend a result from Di Maio and Holà (1997), which provides a necessary and sufficient condition for the comparability of two proximal hypertopologies determined by the same metric and different p-covers.

10.2.7 Proposition *Let (X, d) be a metric space, and let Δ, Δ' both be p-covers of X. Then the following properties are equivalent.*

1. $\delta_{prox(\Delta,d)} \leq \delta_{prox(\Delta',d)}$.
2. $\mathscr{T}_{prox(\Delta,d)} \subseteq \mathscr{T}_{prox(\Delta',d)}$.
3. $\forall D \in \Delta \setminus \{X\}, \forall \varepsilon \in \mathbb{R}_0^+ : \exists\, D_1', \ldots, D_n' \in \Delta' : D \subseteq \bigcup_{k=1}^{n} D_k' \subseteq D^{(\varepsilon)}$.

Proof 1 $\Rightarrow$ 2 This follows from 10.2.5.

2 $\Rightarrow$ 3. This can be found in Di Maio and Holà (1997).

3 $\Rightarrow$ 1. Take $A \in CL(X)$ and $\mathscr{A} \in 2^{CL(X)}$ arbitrary. Further let $\Gamma \in 2_0^{(\Delta)}$ and fix $\varepsilon \in \mathbb{R}_0^+$. First note that since $D_d(\cdot, X)$ equals the constant zero-functional on $CL(X)$, we may assume without loss of generality that $X \notin \Gamma$. To simplify notations, we write $\Gamma = \{\, D_1, \ldots, D_n \,\}$. By condition 3, for every $j \in \{\, 1, \cdots, n \,\}$ we can find, $D'_{j,1}, \ldots, D'_{j,m(j)} \in \Delta'$ such that

$$D_j \subseteq \bigcup_{k=1}^{m(j)} D'_{j,k} \subseteq D_j^{(\varepsilon)}.$$

Let $\Gamma' := \{\, D'_{j,k} \mid j \in \{1, \ldots, n\},\ k \in \{1, \ldots, m(j)\,\}\}$. Then obviously $\Gamma' \in 2_0^{(\Delta')}$ and it can easily be verified that

$$d^{\Gamma} \leq d^{\Gamma'} + \varepsilon.$$

By the arbitrariness of $\Gamma \in 2_0^{(\Delta)}$ and $\varepsilon \in \mathbb{R}_0^+$, this shows that

$$\sup_{\Gamma \in 2_0^{(\Delta)}} \inf_{B \in \mathscr{A}} d^{\Gamma}(A, B) \leq \sup_{\Gamma' \in 2_0^{(\Delta')}} \inf_{B \in \mathscr{A}} d^{\Gamma'}(A, B)$$

and we are finished. $\square$

For a given metric space (X, d), we will use $\mathscr{E}(d)$, respectively $\mathscr{E}_u(d)$ and $\mathscr{E}_u^b(d)$ to denote the set of all metrics on X which are equivalent to d, respectively uniformly equivalent to d, respectively uniformly equivalent to and determine the same bounded subsets as d.

For any metric space (X, d), $CL(X)$ and $CLB(X)$ obviously are p-covers of X. As justified by the foregoing results, we will call the corresponding distances on $CL(X)$ the *proximal distance*, respectively the *bounded proximal distance* and we will denote them by $\delta_{prox(d)}$, respectively $\delta_{bprox(d)}$.

10.2.8 Theorem *Let (X, d) be a metric space. Then we have*

$$\delta_{\mathscr{T}_{prox(d)}} = \bigvee_{e \in \mathscr{E}_u(d)} \delta_{W_e}.$$

Proof To simplify notations, we will put $\delta := \bigvee_{e \in \mathscr{E}_u(d)} \delta_{W_e}$. Now take $A \in CL(X)$ and $\mathscr{A} \in 2^{CL(X)}$ arbitrary. It then follows that

$$\delta(A, \mathscr{A}) = \sup_{\mathscr{D} \in 2^{(\mathscr{E}_u(d))}} \sup_{(F_e)_{e \in \mathscr{D}} \in \left(2_0^{(X)}\right)^{\mathscr{D}}} \inf_{B \in \mathscr{A}} \sup_{e \in \mathscr{D}} \sup_{x \in F_e} |\, \delta_e(x, A) - \delta_e(x, B)\,|.$$

Suppose that $\delta(A, \mathscr{A}) > 0$. Then there exist $\mathscr{D}_0 \in 2_0^{(\mathscr{E}_u(d))}$ and $(F_e^0)_{e \in \mathscr{D}_0} \in \left(2_0^{(X)}\right)^{\mathscr{D}_0}$ such that

$$\alpha := \inf_{B \in \mathscr{A}} \sup_{e \in \mathscr{D}_0} \sup_{x \in F_e^0} |\, \delta_e(x, A) - \delta_e(x, B)\,| > 0.$$

Moreover, since $n \cdot e \in \mathscr{E}_u(d)$ for every $e \in \mathscr{D}_0$ and every $n \in \mathbb{N}_0$, we see that

$$\begin{aligned}\delta(A, \mathscr{A}) &\geq \sup_{n \in \mathbb{N}_0} \inf_{B \in \mathscr{A}} \sup_{e \in \mathscr{D}_0} \sup_{x \in F_e^0} |\, n \cdot \delta_e(x, A) - n \cdot \delta_e(x, B)\,| \\ &= \sup_{n \in \mathbb{N}_0} (n \cdot \alpha) = \infty.\end{aligned}$$

Hence $\delta(CL(X) \times 2^{CL(X)}) = \{0, \infty\}$, which implies that $\delta = \delta_{\mathscr{T}_\delta}$. On the other hand, it follows from the fact that concrete coreflectors preserve initiality that

$$\mathscr{T}_\delta = \bigvee_{e \in \mathscr{E}_u(d)} \mathscr{T}_{W_e}$$

and applying a result proved in Beer et al. (1992), we obtain that $\mathscr{T}_\delta = \mathscr{T}_{prox(d)}$, which completes the proof. □

10.2.9 Theorem *Let (X, d) be a metric space. Then we have*

$$\delta_{\mathscr{T}_{bprox(d)}} = \bigvee_{e \in \mathscr{E}_u^b(d)} \delta_{W_e}.$$

Proof This proof is completely analogous to the one of 10.2.8 and now uses the fact that

$$\mathscr{T}_{bprox(d)} = \bigvee_{e \in \mathscr{E}_u^b(d)} \mathscr{T}_{W_e},$$

which can be found in Beer and Lucchetti (1993). □

We will now show that the index of compactness, as well as completeness are preserved when going from the original metric space to its hyperspace and vice versa.

10.2.10 Theorem *Let (X, d) be a metric space and let Δ be a p-cover of X. Then we have*

$$\chi_c(CL_{prox(\Delta,d)}(X)) = \chi_c(X).$$

Proof First note that, since Δ satisfies (P1):

$$\begin{aligned}\chi_c(CL_{W_d}(X)) &\leq \chi_c(CL_{prox(\Delta,d)}(X)) \\ &\leq \inf_{\mathscr{B} \in 2^{(CL(X))}} \sup_{C \in CL(X)} \inf_{A \in \mathscr{B}} H_d(A, C) \\ &= \chi_c(CL_{H_d}(X)).\end{aligned}$$

The result now follows from 10.1.9. □

10.2.11 Theorem *Let (X, d) be a metric space and let Δ be a p-cover of X. Then the following properties are equivalent.*

1. *$CL_{prox(\Delta,d)}(X)$ is complete.*
2. *$CL_{H_d}(X)$ is complete.*
3. *X is complete.*

Proof It is well known that 2 and 3 are equivalent, and we refer the reader to Kuratowski (1966) for a proof of this equivalence. The equivalence of 1 and 2 follows directly from 3.5.9. □

We now give a full description of the completion of the hyperspace endowed with the Δ-proximal distance. We will start by proving the following theorem, which indicates that the completion of the hyperspace $CL(X)$ of a given metric space (X, d) is in fact nothing but the hyperspace $CL(\hat{X})$ of the usual metric completion $(\hat{X}, \hat{d})$ of (X, d), endowed with a suitable approach structure. First of all, note that

$(CL(X), \delta_{prox(\Delta,d)}) \in$ UAp by definition. On the other hand the fact that Δ satisfies (P1) implies that $\delta_{prox(\Delta,d)} \geq \delta_{W_d}$ and hence $(CL(X), \delta_{prox(\Delta,d)})$ is also Hausdorff, so we may apply the completion theory of 6.1.

10.2.12 Theorem *Let (X, d) be a metric space, let Δ be a p-cover of X and let*

$$e_X : (X, d) \longrightarrow (\hat{X}, \hat{d}) : x \mapsto \mathcal{V}_{\mathcal{T}_{\hat{d}}}(x)$$

stand for the usual metric completion of (X, d). If we define

$$\Delta^* := \{cl_{\mathcal{T}_{\hat{d}}}(e_X(D)) \mid D \in \Delta\},$$

and

$$\mathcal{D}^* := \{\hat{d}^{\Phi} \mid \Phi \in 2_0^{(\Delta^*)}\},$$

then $\mathcal{D}^*$ is a collection of metrics on $CL(\hat{X})$ which is closed for the formation of finite suprema, so it determines a uniform approach structure on $CL(\hat{X})$, the distance of which we will denote by δ^*. Then the completion of $(CL(X), \delta_{prox(\Delta,d)})$ is isomorphic to $(CL(\hat{X}), \delta^*)$.

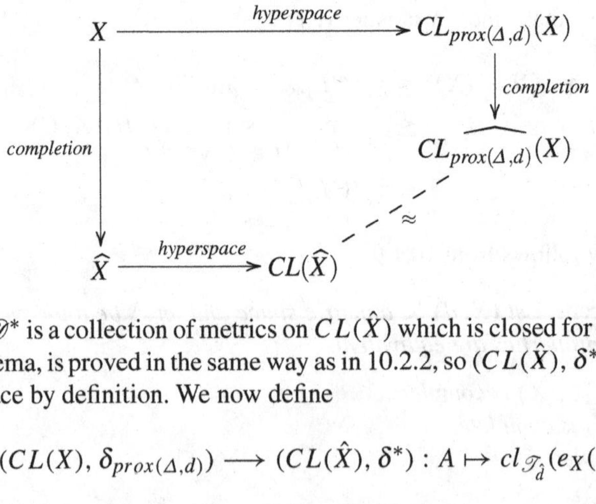

Proof That $\mathcal{D}^*$ is a collection of metrics on $CL(\hat{X})$ which is closed for the formation of finite suprema, is proved in the same way as in 10.2.2, so $(CL(\hat{X}), \delta^*)$ is a uniform approach space by definition. We now define

$$\theta : (CL(X), \delta_{prox(\Delta,d)}) \longrightarrow (CL(\hat{X}), \delta^*) : A \mapsto cl_{\mathcal{T}_{\hat{d}}}(e_X(A)).$$

It suffices to show that $(CL(\hat{X}), \delta^*)$ is Hausdorff and complete, that θ is an embedding and that $\theta(CL(X))$ is dense in $(CL(\hat{X}), \mathcal{T}_{d_{\delta^*}})$.

Fix $\mathfrak{A} \in CL(\hat{X})$ and $\aleph \in 2^{CL(\hat{X})}$. Then since Δ satisfies (P1), we obtain that

$$\begin{aligned}\delta^*(\mathfrak{A}, \aleph) &\geq \sup_{F \in 2_0^{(X)}} \inf_{\mathfrak{B} \in \aleph} \hat{d}^{e_X(F)}(\mathfrak{A}, \mathfrak{B}) \\ &= \sup_{F \in 2_0^{(X)}} \inf_{\mathfrak{B} \in \aleph} \sup_{x \in F} |\, \delta_{\hat{d}}(e_X(x), \mathfrak{A}) - \delta_{\hat{d}}(e_X(x), \mathfrak{B}) \,|\end{aligned}$$

$$= \sup_{\mathfrak{F}\in 2_0^{(\hat{X})}} \inf_{\mathfrak{B}\in \aleph} \sup_{F\in\mathfrak{F}} |\, \delta_{\hat{d}}(F, \mathfrak{A}) - \delta_{\hat{d}}(F, \mathfrak{B})\, |$$
$$= \delta_{W_{\hat{d}}}(\mathfrak{A}, \aleph),$$

where only the "$\geq$"-part of the last but one equality needs some explanation. Therefore, fix $\mathfrak{F} \in 2_0^{(\hat{X})}$ and $\varepsilon \in \mathbb{R}_0^+$. Again by denseness of $e_X(X)$ in $(\hat{X}, \mathscr{T}_{\hat{d}})$ we have that for every $F \in \mathfrak{F}$ there exists $x_F \in X$ such that $\hat{d}(e_X(x_F), F) \leq \varepsilon/2$. It now follows that for every $\mathfrak{B} \in CL(\hat{X})$

$$\begin{aligned}\sup_{F\in\mathfrak{F}} |\, \delta_{\hat{d}}(F, \mathfrak{A}) - \delta_{\hat{d}}(F, \mathfrak{B})\, | &\leq \sup_{F\in\mathfrak{F}} \Big(|\, \delta_{\hat{d}}(F, \mathfrak{A}) - \delta_{\hat{d}}(e_X(x_F), \mathfrak{A})\, | \\ &\quad + |\, \delta_{\hat{d}}(e_X(x_F), \mathfrak{A}) - \delta_{\hat{d}}(e_X(x_F), \mathfrak{B})\, | \\ &\quad + |\, \delta_{\hat{d}}(e_X(x_F), \mathfrak{B}) - \delta_{\hat{d}}(F, \mathfrak{B})\, | \Big) \\ &\leq \sup_{F\in\mathfrak{F}} \Big(\hat{d}(F, e_X(x_F)) + |\, \delta_{\hat{d}}(e_X(x_F), \mathfrak{A}) - \delta_{\hat{d}}(e_X(x_F), \mathfrak{B})\, | \\ &\quad + \hat{d}(e_X(x_F), F) \Big) \\ &\leq \sup_{F\in\mathfrak{F}} |\, \delta_{\hat{d}}(e_X(x_F), \mathfrak{A}) - \delta_{\hat{d}}(e_X(x_F), \mathfrak{B})\, | + \varepsilon.\end{aligned}$$

Since $\mathscr{T}_{W_{\hat{d}}}$ is Hausdorff and since $\mathscr{T}_{\delta_{W_{\hat{d}}}} = \mathscr{T}_{W_{\hat{d}}}$, this shows that δ^* is Hausdorff. Our next step is to show that the metric coreflection of $(CL(\hat{X}), \delta^*)$ is $(CL(\hat{X}), H_{\hat{d}})$. Therefore, fix $\mathfrak{B}, \mathfrak{B}' \in CL(\hat{X})$. That $d_{\delta^*}(\mathfrak{B}, \mathfrak{B}') \leq H_{\hat{d}}(\mathfrak{B}, \mathfrak{B}')$ is easily seen. On the other hand, since Δ satisfies (P1), we have that

$$\begin{aligned}d_{\delta^*}(\mathfrak{B}, \mathfrak{B}') &\geq \sup_{x\in X} |\, \delta_{\hat{d}}(e_X(x), \mathfrak{B}) - \delta_{\hat{d}}(e_X(x), \mathfrak{B}')\, | \\ &= \sup_{F\in\hat{X}} |\, \delta_{\hat{d}}(F, \mathfrak{B}) - \delta_{\hat{d}}(F, \mathfrak{B}')\, | \\ &= H_{\hat{d}}(\mathfrak{B}, \mathfrak{B}'),\end{aligned}$$

where the last but one equality follows from the denseness of $e_X(X)$ in $(\hat{X}, \mathscr{T}_{\hat{d}})$ by an argument completely similar to above. It now follows from 10.2.11 that $(CL(\hat{X}), \delta^*)$ is complete. The initiality of

$$\theta : (CL(X), \delta_{prox(\Delta,d)}) \longrightarrow (CL(\hat{X}), \delta^*) : A \mapsto cl_{\mathscr{T}_{\hat{d}}}(e_X(A))$$

follows from the fact that for every $A \in CL(X)$ and every $\mathscr{A} \in 2^{CL(X)}$

$$\begin{aligned}&\sup_{\Phi\in 2_0^{(\Delta^*)}} \inf_{B\in\mathscr{A}} \hat{d}^{\Phi}(\theta(A), \theta(B)) \\ &\qquad = \sup_{\Gamma\in 2_0^{(\Delta)}} \inf_{B\in\mathscr{A}} \sup_{D\in\Gamma} |\, D_{\hat{d}}(e_X(A), e_X(D)) - D_{\hat{d}}(e_X(B), e_X(D))\, |\end{aligned}$$

$$= \sup_{\Gamma \in 2_0^{(\Delta)}} \inf_{B \in \mathscr{A}} \sup_{D \in \Gamma} | D_d(A, D) - D_d(B, D) |$$
$$= \delta_{prox(\Delta,d)}(A, \mathscr{A}).$$

From this it also immediately follows that θ is injective.

Finally to prove the denseness of $\theta(CL(X))$ in $(CL(\hat{X}), \mathscr{T}_{d_{\delta*}} = \mathscr{T}_{H_{\hat{d}}})$, fix $\mathfrak{A} \in CL(\hat{X})$ and $\varepsilon \in \mathbb{R}_0^+$. By denseness of $e_X(X)$ in $(\hat{X}, \mathscr{T}_{\hat{d}})$, for every $F \in \mathfrak{A}$ there exists an $x_{\mathrm{F}} \in X$ with $\hat{d}(\mathrm{F}, e_X(x_{\mathrm{F}})) < \varepsilon/3$. Now we define

$$A := cl_{\mathscr{T}_d}(\{x_{\mathrm{F}} \mid \mathrm{F} \in \mathfrak{A}\}),$$

and take $\mathscr{G} \in \theta(A)$ arbitrary. Then there exists $x \in A$ such that $\hat{d}(\mathscr{G}, e_X(x)) < \varepsilon/3$, and hence there also exists $\mathrm{F} \in \mathfrak{A}$ such that $d(x_{\mathrm{F}}, x) < \varepsilon/3$. We then have that

$$\begin{aligned}\hat{d}(\mathscr{G}, \mathfrak{A}) &\leq \hat{d}(\mathscr{G}, \mathrm{F}) \\ &\leq \hat{d}(\mathscr{G}, e_X(x)) + \hat{d}(e_X(x), e_X(x_{\mathrm{F}})) + \hat{d}(e_X(x_{\mathrm{F}}), \mathrm{F}) < \varepsilon,\end{aligned}$$

which by arbitrariness of $\mathscr{G} \in \theta(A)$ yields that $e_{\hat{d}}(\theta(A), \mathfrak{A}) < \varepsilon$. On the other hand, we have that

$$e_{\hat{d}}(\mathfrak{A}, \theta(A)) = \sup_{\mathrm{F} \in \mathfrak{A}} \hat{d}(\mathrm{F}, \theta(A)) \leq \sup_{\mathrm{F} \in \mathfrak{A}} \hat{d}(\mathrm{F}, e_X(x_{\mathrm{F}})) \leq \varepsilon/3,$$

so it follows that $H_{\hat{d}}(\mathfrak{A}, \theta(A)) < \varepsilon$, and by the arbitrariness of ε we are finished. □

10.3 The Vietoris Structure

In this section we start not from a metric or normed space but from an arbitrary approach space and see how we can, in a natural way, extend the Vietoris hyperspace construction from topology to approach theory.

Starting from an approach space $(X, \mathfrak{L})$, we consider the hyperspaces $CL(X)$ of all closed and nonempty subsets and $K(X)$ of all compact and nonempty subsets (in the underlying topology).

If X is a set and $A \subseteq X$, then A^+ and A^- have the same meaning as in the foregoing section. If $\mu \in [0, \infty]^X$, we let

$$\mu^\wedge : CL(X) \longrightarrow [0, \infty] : A \mapsto \inf_{x \in A} \mu(x) = \inf \mu(A)$$

and

$$\mu^\vee : CL(X) \longrightarrow [0, \infty] : A \mapsto \sup_{x \in A} \mu(x) = \sup \mu(A).$$

We will make use of the obvious formulas $\theta_{\cap\mathscr{A}} = \sup_{A\in\mathscr{A}} \theta_A$, $\theta_{\cup\mathscr{A}} = \inf_{A\in\mathscr{A}} \theta_A$ for $\mathscr{A} \subseteq CL(X)$, and $\theta_A^{\wedge} = \theta_{A^-}$, $\theta_A^{\vee} = \theta_{A^+}$ for $A \in CL(X)$.

The hyperspace topologies which we are about to use were introduced in Vietoris (1922, 1923). We recall that if $(X, \mathscr{T})$ is a topological space, the *upper Vietoris topology* $\mathscr{T}_v^+$ on $CL(X)$ is the topology with basis for the closed sets $\{F^- \mid F \in CL(X)\}$, while the *lower Vietoris topology* $\mathscr{T}_v^-$ on $CL(X)$ is the topology with subbasis for the closed sets $\{F^+ \mid F \in CL(X)\}$. The Vietoris topology $\mathscr{T}_v$ then is simply $\mathscr{T}_v^- \vee \mathscr{T}_v^+$. Note that these topological structures are usually introduced via their open sets but in our case it is more convenient to do it via their closed sets.

Given the approach space $(X, \mathfrak{L})$, we define, what we will call the Vietoris approach structures on $CL(X)$ in the following way:

1. *The Vietoris $\wedge$-structure*: the collection $\mathfrak{L}^{\wedge} := \{\mu^{\wedge} \mid \mu \in \mathfrak{L}\}$ is translation-invariant and stable for finite infima, hence it is a basis for a lower regular function frame
$$\mathfrak{L}_v^{\wedge} := [\mathfrak{L}^{\wedge}] = \{\sup_{j\in J} \mu_j^{\wedge} \mid \forall j \in J : \mu_j \in \mathfrak{L}\}.$$
2. *The Vietoris $\vee$-structure*: the collection $\mathfrak{L}^{\vee} := \{\mu^{\vee} \mid \mu \in \mathfrak{L}\}$ is translation-invariant and hence is a subbasis for a lower regular function frame
$$\mathfrak{L}_v^{\vee} := [\mathfrak{L}^{\vee}] = \{\sup_{j\in J} \inf_{k\in K_j} \mu_{j,k}^{\vee} \mid \forall j, k : K_j \text{ finite }, \mu_{j,k} \in \mathfrak{L}\}.$$
3. *The Vietoris structure*: the translation-invariant collection $\mathfrak{L}^{\wedge} \cup \mathfrak{L}^{\vee}$ is a subbasis for a lower regular function frame
$$\mathfrak{L}_v = \{\sup_{j\in J}(\mu_j^{\wedge} \wedge \inf_{k\in K_j} \mu_{j,k}^{\vee}) \mid \forall j, k : K_j \text{ finite}, \mu_j, \mu_{j,k} \in \mathfrak{L}\}.$$

In the sequel we will need a more appropriate basis for $\mathfrak{L}_v$. For that purpose we introduce a notation. If $\{\mu_1, \mu_2, ..., \mu_n\} \subseteq \mathfrak{L}$ we will write $\langle \mu_1, \mu_2, ..., \mu_n\rangle := \mu_0^{\wedge} \wedge \inf_{1\leq k\leq n} \mu_k^{\vee}$ where $\mu_0 := \sup_{1\leq k\leq n} \mu_k$.

10.3.1 Lemma *$\{\langle \mu_1, \mu_2, \ldots, \mu_n\rangle \mid \{\mu_1, \mu_2, \ldots, \mu_n\} \subseteq \mathfrak{L}, n \in \mathbb{N}_0\}$ is a basis for $\mathfrak{L}_v$.*

Proof Considering cases one verifies that $\mu^{\wedge} \wedge (\lambda \wedge \mu)^{\vee} = \mu^{\wedge} \wedge \lambda^{\vee}$ for any $\lambda, \mu \in \mathfrak{L}$. To complete the proof it is then sufficient to observe that
$$\mu_0^{\wedge} \wedge \inf_{1\leq k\leq n} \mu_k^{\vee} = \mu_0^{\wedge} \wedge (\mu_0^{\vee} \wedge \inf_{1\leq k\leq n} (\mu_0 \wedge \mu_k)^{\vee}). \qquad \square$$

The following result shows that our definitions are appropriate when compared to the existing Vietoris topologies. Since we will be dealing with various topologies

$\mathcal{T}$, in the present section it will be useful to use the notation $\mathcal{T}^c$ for the collection $\{X \setminus G \mid G \in \mathcal{T}\}$ of closed sets.

10.3.2 Proposition *If $(X, \mathfrak{L})$ is topological, with underlying topology $\mathcal{T}$, then the following properties hold.*

1. *$(CL(X), \mathfrak{L}_v^\wedge)$ is topological and $\mathcal{T}_{\mathfrak{L}_v^\wedge} = \mathcal{T}_v^+$.*
2. *$(CL(X), \mathfrak{L}_v^\vee)$ is topological and $\mathcal{T}_{\mathfrak{L}_v^\vee} = \mathcal{T}_v^-$.*
3. *$(CL(X), \mathfrak{L}_v)$ is topological and $\mathcal{T}_{\mathfrak{L}_v} = \mathcal{T}_v$.*

Proof We only prove 1 leaving the remaining cases for the reader. If $\mu \in \mathfrak{L}$ and $\varepsilon \geq 0$, we put $F_\varepsilon := \{\mu \leq \varepsilon\}$ and first observe that $\theta_{F_\varepsilon} \in \mathfrak{L}$. Then it is immediately verified that $\mu^\wedge(A) = 0$ if and only if $\theta_{F_\varepsilon}^\wedge(A) = 0$ for all $\varepsilon > 0$ and therefore $\theta_{Z\mu^\wedge} = \sup_{\varepsilon>0} \theta_{F_\varepsilon}^\wedge \in \mathfrak{L}^\wedge$.

If $\rho = \sup_{j\in J} \mu_j^\wedge$ with $\mu_j \in \mathfrak{L}$ then $Z\rho = \bigcap_{j\in J} Z\mu_j$, and thus

$$\theta_{Z\rho} = \sup_{j\in J} \theta_{Z\mu_j} \in \mathfrak{L}_v^\wedge.$$

For the second claim, let $\mu \in \mathfrak{L}$ and $\alpha \in \mathbb{R}^+$. Since $\{\mu^\wedge \leq \alpha\} = \bigcap_{\varepsilon>0} F_{\alpha+\varepsilon}^-$ and $F_{\alpha+\varepsilon} \in \mathcal{T}^c$, we have $\{\mu^\wedge \leq \alpha\} \in (\mathcal{T}_v^+)^c$. Consequently $\mathcal{T}_{\mathfrak{L}_v^\wedge} \subseteq \mathcal{T}_v^+$.

As for the reverse inclusion, it suffices to note that for any $F \in \mathcal{T}^c$ we have $\theta_F^\wedge \in \mathfrak{L}^\wedge$ and $F^- = Z\theta_F^\wedge$. □

Starting from a given approach space $(X, \mathfrak{L})$, there are now two ways to arrive at associated hyperspace topologies as depicted in the diagram below. In this diagram Δ stands for respectively $(\wedge, \vee, \text{void})$ and $\square$ stands for respectively $(+, -, \text{void})$.

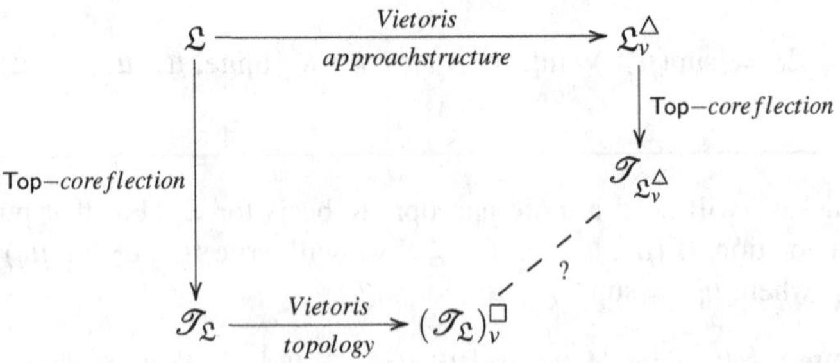

We can first perform the topological coreflection obtaining $\mathcal{T}_\mathfrak{L}$ and then consider the associated Vietoris topologies $(\mathcal{T}_\mathfrak{L})_v^+$, respectively $(\mathcal{T}_\mathfrak{L})_v^-$ and $(\mathcal{T}_\mathfrak{L})_v$. We can also first consider the associated Vietoris approach structures $\mathfrak{L}_v^\wedge$, respectively $\mathfrak{L}_v^\vee$ and $\mathfrak{L}_v$ and then consider their topological coreflections obtaining $\mathcal{T}_{\mathfrak{L}_v^\wedge}$, respectively $\mathcal{T}_{\mathfrak{L}_v^\vee}$, and $\mathcal{T}_{\mathfrak{L}_v}$. It will be seen that on $CL(X)$, $(\mathcal{T}_\mathfrak{L})_v^- = \mathcal{T}_{\mathfrak{L}_v^\vee}$, but there is in general no relation between $(\mathcal{T}_\mathfrak{L})_v^+$ and $\mathcal{T}_{\mathfrak{L}_v^\wedge}$, or between $(\mathcal{T}_\mathfrak{L})_v$ and $\mathcal{T}_{\mathfrak{L}_v}$ on the whole of $CL(X)$. The following theorem shows that if we add compactness to the picture the situation changes.

10.3.3 Theorem *Given an approach space $(X, \mathfrak{L})$, then the following properties hold.*

1. *$(\mathscr{T}_{\mathfrak{L}})_v^-$ and $\mathscr{T}_{\mathfrak{L}_v^\vee}$ coincide.*
2. *The restrictions of $(\mathscr{T}_{\mathfrak{L}})_v^+$ and $\mathscr{T}_{\mathfrak{L}_v^\wedge}$ to $K(X)$ coincide.*
3. *The restrictions of $(\mathscr{T}_{\mathfrak{L}})_v$ and $\mathscr{T}_{\mathfrak{L}_v}$ to $K(X)$ coincide.*

Proof 1. This follows from the fact that $(Z\mu)^+ = Z\mu^\vee$ (for any $\mu \in \mathfrak{L}$).

2. It is clear that always $(Z\mu)^- \subseteq Z\mu^\wedge$ and for $\mu \in \mathfrak{L}$ the converse inclusion follows from the lower semicontinuity of μ and the compactness of the sets in $\mathscr{K}(X)$.

3. This follows from $(\mathscr{T}_{\mathfrak{L}})_v = (\mathscr{T}_{\mathfrak{L}})_v^- \vee (\mathscr{T}_{\mathfrak{L}})_v^+$ and the coreflectivity of Top in App. □

The foregoing results shed some light on the functorial relationship between the various constructions. It is well known that in the topological case, and for $K(X)$, the association of any of the Vietoris hyperspaces to a given topological space is functorial, and it can easily be verified that this remains true in the approach case. More precisely in the diagram below, V_a and V_t stand for any of the functors giving respectively the Vietoris approach structures and the corresponding Vietoris topologies. Note that the action on morphisms in all cases is determined by

$$f : X \longrightarrow Y \mapsto [K(X) \longrightarrow K(Y) : A \mapsto f(A)].$$

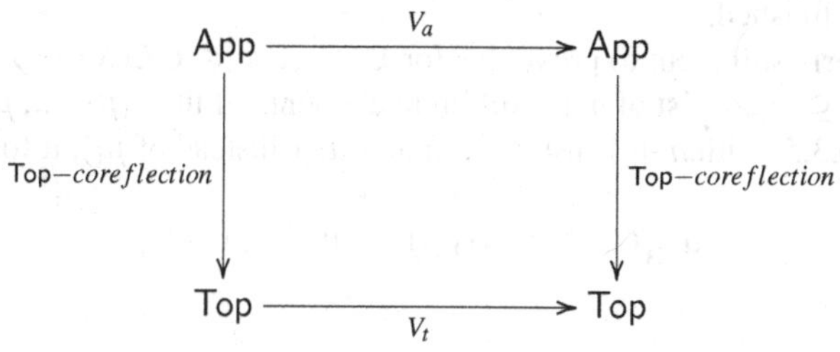

What the foregoing results show is that the restriction of V_a to Top always coincides with V_t, and that in all cases the diagram commutes.

10.3.4 Proposition *If $(X, \mathfrak{L})$ is a T_1 approach space, then all structures $\mathfrak{L}_v^\wedge$, $\mathfrak{L}_v^\vee$ and $\mathfrak{L}_v$ are admissible.*

Proof This is analogous to 10.1.4 and we leave this to the reader. □

10.3.5 Lemma *If $\{x_1, x_2, ..., x_n\}$ is a set of n different points in a T_2 approach space $(X, \mathfrak{L})$, there exist elements $\mu_1, \mu_2, ..., \mu_n$ in $\mathfrak{L}$ with the following properties.*

1. *$\mu_k(x_k) > 0$ for $1 \leq k \leq n$.*
2. *$\mu_j \wedge \mu_k = 0$ for $1 \leq j \leq n, 1 \leq k \leq n, j \neq k$.*

Proof Since $(X, \mathfrak{L})$ is (T_2), there exist $\mu_{jk} \in \mathfrak{L}$ for $1 \leq j \leq n, 1 \leq k \leq n, j \neq k$ such that $\mu_{jk}(x_j) > 0$, $\mu_{jk} \wedge \mu_{kj} = 0$. Then $\mu_k = \bigwedge_{j \neq k} \mu_{kj}$, $1 \leq k \leq n$ has the announced properties. □

In the following result we put $\mathscr{J}_n(X)$ (respectively $\mathscr{J}(X)$) for the subsets of X which have no more than n points (respectively which are finite).

10.3.6 Proposition *Given an approach space* $(X, \mathfrak{L})$, *then for all* $n \in \mathbb{N}_0$ *the following properties hold.*

1. $\mathscr{J}_n(X)$ *and* $\mathscr{J}(X)$ *are dense in* $(CL(X), \mathfrak{L}_v^{\wedge})$.
2. $\mathscr{J}(X)$ *is dense in* $(CL(X), \mathfrak{L}_v^{\vee})$ *and in* $(CL(X), \mathfrak{L}_v)$.
3. *If* $(X, \mathfrak{L})$ *is* T_2, *then* $\mathscr{J}_n(X)$ *is closed in* $(CL(X), \mathfrak{L}_v^{\vee})$ *and in* $(CL(X), \mathfrak{L}_v)$.
4. $p_n : (X, \mathfrak{L})^n \longrightarrow (\mathscr{J}_n(X), \mathscr{S}) : (x_1, ..., x_n) \mapsto \{x_1, ..., x_n\}$ *is a contraction if* $\mathscr{S} \in \{\mathfrak{L}_v^{\wedge}, \mathfrak{L}_v^{\vee}, \mathfrak{L}_v\}$.

Proof 1. This follow from the fact that if $\mu \in \mathfrak{L}$ and $\mu^{\wedge}|_{i(X)} = 0$, then $\mu^{\wedge} = 0$.

2. It is clearly sufficient to prove this for $\mathfrak{L}_v$. So let

$$\lambda := \mu^{\wedge} \wedge \inf_{k \in K} \mu_k^{\vee}$$

in the basis of $\mathfrak{L}_v$ be such that $\lambda|_{\mathscr{J}(X)} = 0$ and take $A \in CL(X) \setminus \mathscr{J}(X)$. If $\mu_k^{\vee}(A) = 0$ for some $k \in K$, then $\lambda(A) = 0$. If on the contrary $\mu_k^{\vee}(A) \neq 0$ for all $k \in K$, there is for each $k \in K$ an $x_k \in A$ with $\mu_k(x_k) \neq 0$. Then $B := \{x_k \mid k \in K\} \in \mathscr{J}(X)$ and therefore clearly $\inf_{k \in K} \mu_k^{\vee}(B) \neq 0$ which implies $\mu^{\wedge}(B) = 0$. Since $B \subseteq A$, it follows that also $\mu^{\wedge}(A) = 0$, hence $\lambda(A) = 0$, and so we are finished.

3. It is clearly sufficient to prove this for $\mathfrak{L}_v^{\vee}$. Let $A \in CL(X) \setminus \mathscr{J}_n(X)$ and let $\{x_1, ..., x_{n+1}\} \subseteq A$ consist of $n + 1$ different elements. If then $\{\mu_1, ..., \mu_{n+1}\} \subseteq \mathfrak{L}$ is taken as in 10.3.5 (with $n + 1$ instead of n and $\mu_{i,A}$ instead of μ_i), it follows that

$$\mu_{i,A}^{\vee}(A) \geq \mu_{i,A}^{\vee}(\{x_i\}) = \mu_{i,A}(x_i) > 0$$

and so $\inf_{i=1}^{n+1} \mu_{i,A}^{\vee}(A) > 0$.

On the other hand, if $\{y_1, ..., y_p\} \in \mathscr{J}_n(X)$, then it follows from (2) in 10.3.5 that $\mu_{i,A}(y_j) > 0$ for at most one $i \in \{1..., n+1\}$. It follows that $\mu_{i,A}(y_j) = 0$ for some $i \in \{1, ..., n+1\}$ and all $j \in \{1..., p\}$, so $\mu_{i,A}^{\vee}(\{y_1, ..., y_p\}) = 0$. This means that $\inf_{i=1}^{n+1} \mu_{i,A}^{\vee}(B) = 0$ for all $B \in \mathscr{J}_n(X)$, and if

$$\mu := \sup_{A \in CL(X) \setminus \mathscr{J}_n(X)} \inf_{i=1}^{n+1} \mu_{i,A}^{\vee}$$

we obtain $\mathscr{J}_n(X) = Z\mu$ and so we are finished.

4. This follows from the definitions. $\square$

We put $CL^{\Delta}(X)$ the hyperspace equipped with the structure $\mathfrak{L}_v^{\Delta}$ for $\Delta \in \{\wedge, \vee, \text{void}\}$. We use the notations and results from 4.3.4 to 4.3.9.

10.3.7 Theorem *For any approach space X we have*

$$\chi_c(CL^\wedge(X)) = 0.$$

Proof Consider the subbasis $\mathfrak{L}^\wedge$ for $\mathfrak{L}_v^\wedge$. If $\mathscr{I} \in B_s(\mathfrak{L}^\wedge)$ then $\inf_{A\in CL(X)}\mu^\wedge(A) = 0$ for any $\mu^\wedge \in \mathscr{I}$ and hence $\mu^\wedge(X) = 0$, therefore $\sup \mathscr{I}(X) = 0$, and then it follows that $\inf_{A\in CL(X)} \sup \mathscr{I}(A) = 0$. The result now follows from 4.3.9. □

In order to study the relation between $\chi_c(CL^\vee(X))$ and $\chi_c(X)$, we need some lemmas concerning the relations between ideals in $\mathfrak{L}$ and ideals in $\mathfrak{L}_v^\vee$. If $\mathscr{I} \in I(\mathfrak{L})$, then $\mathscr{I}^\vee := \{\mu^\vee \mid \mu \in \mathscr{I}\}$ is an ideal basis in $\mathfrak{L}_v^\vee$ and we denote the generated ideal by $\mathscr{I}_v^\vee$.

10.3.8 Lemma *If X is T_1, $\mu \in \mathfrak{L}_v^\vee$ and $\mathscr{I} \in I(\mathfrak{L})$ then the following properties hold.*

1. $(\mu_{|X})^\vee \leq \mu$.
2. $\sup \mathscr{I}_v^\vee = \sup_{\mu\in\mathscr{I}} \mu^\vee$.
3. $\inf_{x\in X}(\sup \mathscr{I})(x) = \inf_{A\in CL(X)}(\sup \mathscr{I}_v^\vee)(A)$.
4. $\inf_{x\in X}\mu_{|X}(x) = \inf_{A\in CL(X)}\mu(A)$.

Proof 1. This follows from the fact that liminf $\leq$ lim sup.

2. This follows from the definition of $\mathscr{I}_v^\vee$.

3. For every $A \in CL(X)$ and $a \in A$ we have

$$\inf_{x\in X}(\sup \mathscr{I})(x) \leq \sup_{\mu\in\mathscr{I}} \mu(a) \leq \sup_{\mu\in\mathscr{I}} \mu^\vee(A)$$

and therefore

$$\begin{aligned}\inf_{x\in X}(\sup \mathscr{I})(x) &\leq \inf_{A\in CL(X)} (\sup_{\mu\in\mathscr{I}} \mu^\vee)(A) = \inf_{A\in CL(X)} (\sup \mathscr{I}_v^\vee)(A)\\ &\leq \inf_{x\in X}(\sup \mathscr{I}_v^\vee)(\{x\}) = \inf_{x\in X}(\sup \mathscr{I})(x).\end{aligned}$$

4. This is analogous to 3. □

10.3.9 Lemma *If X is T_1 then $\mathscr{I} \in I_s(\mathfrak{L})$ if and only if $\mathscr{I}_v^\vee \in I_s(\mathfrak{L}_v^\vee)$.*

Proof This follows from the definition of $\mathscr{I}_v^\vee$ and 10.3.8. □

10.3.10 Lemma *If X is T_1 and $\mathscr{I} \in I(\mathfrak{L}_v^\vee)$ then the following properties hold.*

1. $\mathscr{I}_{|X} \in I(\mathfrak{L})$.
2. $\inf_{x\in X}(\sup \mathscr{I}_{|X})(\{x\}) = \inf_{A\in CL(X)}(\sup \mathscr{I})(A)$.

Proof 1. This follows from the definition.

2. This is analogous to 3 and 4 of 10.3.8. □

10.3.11 Lemma *If X is T_1 then $\mathscr{I} \in I_s(\mathfrak{L}_v^\vee)$ if and only if $\mathscr{I}_{|X} \in I_s(\mathfrak{L})$.*

Proof This follows from 10.3.8 and 10.3.10. □

Note that again, mutatis mutandis, all conclusions in the foregoing lemmas can be reformulated in terms of sets with the finite-sup property and we will freely use this fact.

10.3.12 Theorem *For any T_1 approach space X we have*

$$\chi_c(CL^\vee(X)) = \chi_c(X).$$

Proof If $\mathscr{I} \in I_s(\mathfrak{L})$, then $\mathscr{I}_v^\vee \in I_s(\mathfrak{L}_v^\vee)$ by 10.3.9, and therefore by 4.3.9 and 10.3.8

$$\begin{aligned}\inf_{x\in X}(\sup \mathscr{I})(x) &= \inf_{A\in CL(X)}(\sup \mathscr{I}_v^\vee)(A)\\ &\leq \chi_c(CL^\vee(X))\end{aligned}$$

which implies that $\chi_c(X) \leq \chi_c(CL^\vee(X))$.

If $\mathscr{I} \in I_s(\mathfrak{L}_v^\vee)$, then $\mathscr{I}_{|X} \in I_s(\mathfrak{L})$ by 10.3.10, and therefore also by 4.3.9

$$\begin{aligned}\inf_{A\in CL(X)}(\sup \mathscr{I})(A) &= \inf_{x\in X}(\sup \mathscr{I}_{|X})(x)\\ &\leq \chi_c(X)\end{aligned}$$

which implies that $\chi_c(CL^\vee(X)) \leq \chi_c(X)$. □

10.3.13 Theorem *For any T_1 approach space X we have*

$$\chi_c(CL(X)) = \chi_c(X).$$

Proof That $\chi_c(X) \leq \chi_c(CL(X)$ follows at once from the foregoing theorem.

In order to prove the other inequality we use 4.3.9. Let $\varepsilon > 0$ and take

$$\mathscr{B} := \{\mu_k^\vee \mid k \in K\} \cup \{\mu_l^\wedge \mid l \in L\} \in B_s(\mathfrak{L}^\vee \cup \mathfrak{L}^\wedge)$$

and suppose that for all $CL \in B_s(\mathfrak{L}) : \inf_{x\in X} \sup CL(x) \leq b$.

For $l \in L$, define $\mathscr{B}_l := \{\mu_k^\vee \mid k \in K\} \cup \{\mu_l^\wedge\}$. Since for any $K_0 \subseteq K$ finite, $\inf_{x\in X}((\sup_{k\in K_0} \mu_k)(x) \vee \mu_l(x)) = 0$, it follows that

$$\{\mu_k \mid k \in K\} \cup \{\mu_l \mid l \in L\} \in B_s(\mathfrak{L})$$

and therefore $\inf_{x\in X}((\sup_{k\in K} \mu_k) \vee \mu_l)(x) \leq b$. Consequently there exists $x_l \in X$ such that

$$(\sup_{k\in K} \mu_k) \vee \mu_l(x_l) \leq b + \varepsilon.$$

If we now put $M := \{x_l \mid l \in L\}$, then for all $l \in L$

$$(\sup_{k\in K} \mu_k)^{\vee}(M) \vee \mu_l^{\wedge}(M) \leq (\sup_{k\in K} \mu_k)^{\vee}(M) \vee \mu_l(x_l) \leq b + \varepsilon$$

and therefore

$$\inf_{A\in CL(X)} \sup \mathscr{B}(A) \leq (\sup_{k\in K} \mu_k)^{\vee}(M) \vee \sup_{l\in L} \mu_l^{\wedge}(M) \leq b + \varepsilon$$

and by arbitrariness of ε and b we are finished. □

We now consider the particular case of a compact metric space (X, d). It is a well-known result (see e.g. Beer 1993; Charalambos 2007; Naimpally 2003) that in this case the Vietoris topology is metrizable by the Hausdorff metric. Again, as in the previous section concerning the index of compactness, we will considerably strengthen this result by proving that, always under the same conditions, actually the Vietoris approach structure coincides with the Hausdorff metric. We recall that for a metric space (X, d), quasi-metrics are defined on $CL(X)$ by

$$d_H^-(A, B) = \sup_{a\in A} \inf_{b\in B} d(a, b) \text{ and } d_H^+(A, B) = \sup_{b\in B} \inf_{a\in A} d(a, b).$$

The Hausdorff-metric d_H on $CL(X)$ then is given by $d_H = d_H^- \vee d_H^+$.

10.3.14 Theorem *If (X, d) is compact, then $\delta_d^{\wedge} = \delta_{d_H^+}$.*

Proof Since $\{\mu^{\wedge} \mid \mu \in \mathfrak{L}\}$ is a basis for $\mathfrak{L}_v^{\wedge}$, it follows that for any $A \in CL(X)$ and $\mathscr{B} \subseteq CL(X)$

$$\begin{aligned}
\delta_d^{\wedge}(A, \mathscr{B}) &= \sup\{\mu^{\wedge}(A) \mid \forall B \in \mathscr{B} : \mu^{\wedge}(B) = 0, \mu \in \mathfrak{L}\} \\
&= \sup\{\mu^{\wedge}(A) \mid \forall B \in \mathscr{B}, \exists b \in B : \mu(b) = 0, \mu \in \mathfrak{L}\} \\
&= \sup\{(\delta_d)^{\wedge}_{\varphi(\mathscr{B})}(A) \mid \varphi \in \prod_{B\in\mathscr{B}} B\} \\
&= \sup_{\varphi\in\prod_{B\in\mathscr{B}} B} \inf_{a\in A} \inf_{B\in\mathscr{B}} d(a, \varphi(B)) \\
&= \inf_{B\in\mathscr{B}} \sup_{b\in B} \inf_{a\in A} d(a, b) \\
&= \delta_{d_H^+}(A, \mathscr{B})
\end{aligned}$$

where the second equality follows from the compactness of the sets in $\mathscr{B}$ and the lower semicontinuity of μ. □

10.3.15 Theorem *If (X, d) is compact then $\delta_d^{\vee} = \delta_{d_H^-}$.*

Proof Since in this case $\{\inf_{k\in K} \mu_k^{\vee} \mid K \text{ finite}, \forall k : \mu_k \in \mathfrak{L}\}$ is a basis for $\mathfrak{L}_v^{\vee}$, we now have

$$\begin{aligned}\delta_d^{\vee}(A, \mathscr{B}) &= \sup\{\inf_{k\in K} \mu_k^{\vee}(A) \mid K \text{ finite}, \forall k : \mu_k \in \mathfrak{L}, \inf_{k\in K} (\mu_k^{\vee})_{|\mathscr{B}} = 0\} \\ &= \sup\{\inf_{j\in J} (\delta_d)^{\vee}_{\cup\mathscr{B}_j}(A) \mid (\mathscr{B}_j)_{j\in J} \in \mathsf{R}(\mathscr{B})\}\end{aligned}$$

where $\mathsf{R}(\mathscr{B})$ stands for the set of finite partitions of $\mathscr{B}$. If in particular $\mathscr{B} = \{B\}$, $B \in CL(X)$, this clearly gives

$$\delta_d^{\vee}(A, \{B\}) = \sup_{a\in A} \inf_{b\in B} d(a, b) = \delta_{d_H}^{-}(A, \mathscr{B}),$$

and from this it follows that

$$\delta_d^{\vee}(A, \mathscr{B}) \le \inf_{B\in\mathscr{B}} \sup_{a\in A} \inf_{b\in B} d(a, b) = \delta_{d_H^-}(A, \mathscr{B}).$$

To show the converse inequality, let $\varepsilon > 0$ and consider a finite covering $\mathfrak{S}$ of X by open balls with diameter ε. For each nonempty subset $\mathscr{M}$ of $\mathfrak{S}$ put

$$\mathscr{B}_{\mathscr{M}} = \{B \in \mathscr{B} \mid \forall S \in \mathscr{M} : B \cap S \neq \emptyset, \forall S \in \mathfrak{S} \setminus \mathscr{M} : B \cap S = \emptyset\}$$

and let $\mathfrak{M}$ be the set of nonempty subsets $\mathscr{M}$ of $\mathfrak{S}$ for which $\mathscr{B}_{\mathscr{M}} \neq \emptyset$. Clearly then $(\mathscr{B}_{\mathscr{M}})_{\mathscr{M}\in\mathfrak{M}} \in \mathsf{R}(\mathscr{B})$. Consider an arbitrary function $\varphi : \mathfrak{M} \longrightarrow \mathscr{B}$ such that for each $\mathscr{M} \in \mathfrak{M}$: $\varphi(\mathscr{M}) \in \mathscr{B}_{\mathscr{M}}$. Then we have

$$\begin{aligned}\delta_d^{\vee}(A, \mathscr{B}) &= \sup\{\inf_{j\in J} (\delta_d)^{\vee}_{\cup\mathscr{B}_j}(A) \mid (\mathscr{B}_j)_{j\in J} \in \mathsf{R}(\mathscr{B})\} \\ &\ge \inf_{\mathscr{M}\in\mathfrak{M}} \sup_{a\in A} \inf_{b\in\cup\mathscr{B}_{\mathscr{M}}} d(a, b) \\ &\ge \inf_{\mathscr{M}\in\mathfrak{M}} \sup_{a\in A} \inf_{b\in\varphi(\mathscr{M})} d(a, b) - \varepsilon \\ &\ge \inf_{B\in\mathscr{B}} \sup_{a\in A} \inf_{b\in B} d(a, b) - \varepsilon\end{aligned}$$

which by arbitrariness of ε proves the remaining inequality. □

In our last theorem it will be convenient to work with gauges. The foregoing theorems have shown that the gauges associated with the $\wedge$- and $\vee$-Vietoris approach structures are respectively the principal gauges (i.e. generated by a unique element) $\mathscr{G}^{\wedge} = \{d \mid d \le d_H^+\}$ and $\mathscr{G}^{\vee} = \{d \mid d \le d_H^-\}$.

10.3.16 Theorem *If* (X, d) *is compact then* $(\delta_d)_v = \delta_{d_H}$.

Proof Since the source

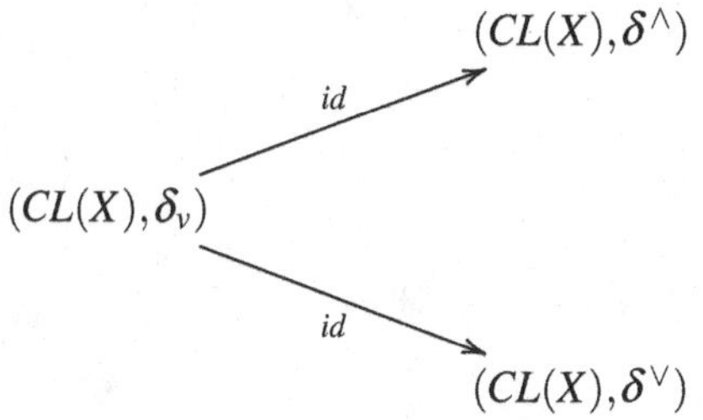

is initial, using 1.3.11, it follows from 10.3.14 and 10.3.15 that the initial gauge is given by

$$\mathcal{G}_v := \{d \mid d \leq d_H^+ \vee d_H^-\}$$

and since $d_H = d_H^+ \vee d_H^-$ we are finished. □

10.4 Comments

1. **Further results**

The material covered in Sects. 10.1 and 10.2 of this chapter, mainly comes from a series of papers by Lowen and Sioen dealing with various aspects of hyperspaces in a metric setting (Lowen and Sioen 1996, 1998, 2000a). More information can be found in those papers.

2. **Metric spaces versus approach spaces as starting point**

There is a major difference in the hyperspaces considered in Sects. 10.1 and 10.2 compared to the Vietoris structures considered in 10.3. Whereas in 10.1 and 10.2 we always started from a metric space, in 10.3 we start from an arbitrary approach space. In the first two sections this was motivated by the desire to lay the link with the extensive body of results on the associated classical hyperspaces on metric spaces (see e.g. Beer 1993 for a comprehensive bibliography). In 10.3, noting that the Vietoris topologies are defined in a topological rather than in a metric setting, it was natural to take general approach spaces as starting point making use of lower regular function frames. However, an interested reader will be able to see that many things done in Sects. 10.1 and 10.2 with metric spaces as starting point can actually be generalized to arbitrary (or at least uniform) approach spaces. For more information on the Vietoris approach structure see the paper by Lowen and Wuyts (2013).

Chapter 11
Approach Theory Meets DCPO's and Domains

The real danger is not that computers will begin to think like men, but that men will begin to think like computers.
(H. Eves in Return to Mathematical Circles)

Motivated by central problems in theoretical computer science, mathematical structures have been created to model semantics of programming languages. These are the so called semantic domains and they are mainly intended to define the meaning of a computer program. The models that are useful in this respect are directed complete partial orders, called dcpo's, or continuous dcpo's, called domains (Gierz et al. 2003). This theory originated with the work of Scott (1972) who endowed a given dcpo $(X, \leq)$ with the Scott topology $\sigma(X)$ which is then used as a tool to study convergence phenomena in X and to describe the Scott continuous functions. In Scott's model the latter represent the computable functions.

For a Scott continuous map on a dcpo with a bottom element this setting provides the "Scott least fixed point theorem" where the least fixed point is obtained by iterating the function on the bottom element (Gierz et al. 2003). In Scott's model fixed point theorems are extremely important since they represent the "meaning" of the algorithm. As is known from the work of Edalat (1998), the Scott least fixed point theorem implies the classical "Banach fixed point theorem" for Lipschitz functions with Lipschitz factor strictly smaller than 1, on a complete metric space.

However domains alone are insufficient in a more refined quantitative reasoning. The mathematical structures that have been used to capture quantitative data are dcpo's endowed with a weightable quasi-metric structure or equivalently a partial quasi-metric structure inducing the Scott topology (Matthews 1994). This has led to the study of quantifiability of domains, see Schellekens (2003) and Waszkiewicz (2003), who independently showed that all domains with a countable basis are quantifiable. In this context weightable quasi-metrics are constructed by taking some infinite sum $\Sigma \frac{1}{2^n}$ over some suitable subset of $\mathbb{N}$. The role of $(\frac{1}{2^n})_n$ could be replaced by any other suitable sequence, which means that, although the existence of such a quasi-metric is important, numerical values computed with it are not canonically determined. Moreover the procedure heavily depends on countability properties.

R. Lowen, *Index Analysis*, Springer Monographs in Mathematics,
DOI 10.1007/978-1-4471-6485-2_11

Quantified domain theory is applied for instance in complexity analysis (Schellekens 1995) where fixed point theorems are again extremely important since they are the clue for estimating the complexity of an algorithm. As a generalization of the classical Banach theorem, existence of fixed points was proved for Lipschitz functions with Lipschitz factor strictly smaller than 1 on a bicomplete quasi-metric space by Oltra and Valero (2004). Martin considers a measurement on a domain, which can be seen as an alternative for a weightable quasi-metric in quantitative domain theory, and he developed fixed point theorems for non-monotone maps in that setting (Martin 2000a, b).

In Sect. 11.1 we propose an intrinsic solution for the problem of quantifiability. We propose to use an approach structure rather than a quasi-metric. The approach structure on a domain $(X, \leq)$ is supposed to quantify $\sigma(X)$. As was shown in Colebunders et al. (2011), such an approach structure can be intrinsically defined, regardless of cardinality conditions on bases. We show that every domain X is quantifiable in this sense. We get weightability for free and in the case of an algebraic domain satisfying the Lawson condition (Lawson 1997), a quantifying approach space can be obtained with a weight satisfying the so-called kernel condition. This allows to extract the set of maximal elements of the domain. We also prove that there are important structural advantages of working in the category of approach spaces. With respect to contractions, in 11.4 we study fixed point theorems. Given a function $f : X \longrightarrow X$, with domain and codomain structured as approach spaces, and given an arbitrary point $a \in X$ the use of the limit operator will allow us to estimate "how far" a is from being a fixed point. An upper bound for the distance between $f(a)$ and a in the quasi-metric coreflection of the approach space will be given. For monotone as well as for non-monotone maps we establish new fixed point theorems and we recover some existing ones, as the ones of Scott (Gierz et al. 2003) and Martin (2000a, b) mentioned above.

11.1 Basic Structures

In this section we will consider approach structures on a given dcpo. We refer the reader to Gierz et al. (2003) for terminology and basic results. To fix notations recall that for a partially ordered set (poset) $(X, \leq)$ and elements x and y we write $x\#y$ if x and y have no common upper bound. A subset $D \subseteq X$ is *directed* if it is nonempty and any pair of elements of D has an upper bound in D. A poset in which every directed subset D has a supremum ($\sup D$) is called a *directed complete poset (dcpo)*. We use the notations $\uparrow x := \{y \mid x \leq y\}$ and $\downarrow x := \{y \mid y \leq x\}$.

We say that x is *way below* y if for all directed subsets $D \subseteq X$, $y \leq \sup D$ implies $x \leq a$ for some $a \in D$. We denote this by $x \ll y$. We say that x is a *compact element* if $x \ll x$. We also use the notations $\twoheaduparrow x := \{y \mid x \ll y\}$ and $\twoheaddownarrow x := \{y \mid y \ll x\}$. A subset $B \subseteq X$ is said to be a *basis* for X if for every element $x \in X$ the set $B \cap \twoheaddownarrow x$ is directed with supremum x. A poset is called a *domain* if it is a dcpo having a basis.

If the class $K(X)$ of all compact elements is a domain basis then we call the domain *algebraic*. If a domain has a countable domain basis it is called an *ω-domain*.

On a dcpo $(X, \leq)$ there are some intrinsic topologies. The *Scott topology* denoted by $\sigma(X)$, is the topology for which the open sets are upper-sets inaccessible for directed suprema. If X is a domain then $\sigma(X)$ has a basis $\{ \twoheaduparrow x \mid x \in X \}$. The specialisation order of $\sigma(X)$ coincides with the original order. We denote by lX the *lower-topology* generated by $\{\downarrow x \mid x \in X\}$ and by wX the topology generated by $\{X \setminus \uparrow x \mid x \in X\}$. Other intrinsic topologies on $(X, \leq)$ are the *Lawson topology* defined as $L(X) := \sigma(X) \vee w(X)$ and the *Martin topology* defined as $M(X) := \sigma(X) \vee l(X)$.

11.1.1 Definition A quasi-metric space (X, q) is called *weightable* if there exists a function $w : X \longrightarrow \mathbb{P}$ (called a *weight*), not identically ∞, such that

$$q(x, y) + w(x) = q(y, x) + w(y)$$

whenever $x, y \in X$. A weight w is called *forcing* for q if $x \in X$ and $w(x) = \infty$ imply that the function $q(x, .)$ is identically zero on X. We let wqMet be the category consisting of weightable quasi-metric spaces with non-expansive maps. For a weightable quasi-metric q we denote by $\mathscr{W}_q$ the collection of all its weights.

The following example is well known and will appear to be crucial.

11.1.2 Example Consider $\mathbb{P}$ endowed with $d_{\mathbb{P}}^-$ (see 1.2.62, 2.4.13). The function $w_{\mathbb{P}} : \mathbb{P} \longrightarrow \mathbb{P}$ defined as $w_{\mathbb{P}}(x) = x$ is a weight for $d_{\mathbb{P}}^-$. Note that it is forcing since the only point in which the weight is infinite is ∞ and $d_{\mathbb{P}}^-(\infty, y) = 0$ for all y. The underlying topology $\mathscr{T}_{d_{\mathbb{P}}^-}$ on $[0, \infty]$ is

$$\{[0, b[\mid b \leq \infty\} \cup \{[0, \infty]\} \cup \{\emptyset\}.$$

When we endow $\mathbb{P}$ with the opposite order $x \preceq y \Leftrightarrow y \leq x$ the Scott topology $\sigma(\mathbb{P}^{op})$ associated to the dcpo $\mathbb{P}^{op} = (\mathbb{P}, \preceq)$ is exactly the topology $\mathscr{T}_{d_{\mathbb{P}}^-}$.

We establish some categorical properties of the category wqMet.

11.1.3 Proposition *Let $f : X \longrightarrow Y$ be an initial morphism in* qMet *with Y a weightable quasi-metric space, then the initial structure on X is weightable.*

Proof Let $Y = (Y, q)$ with q some weightable quasi-metric, and let w be a weight for q. Then clearly the initial structure $q \circ f \times f$ has the weight function $w \circ f$. □

A similar result does not hold for arbitrary sources. Even a pointwise supremum of two weightable quasi-metrics need not be weightable.

11.1.4 Example We consider $X = \{x, y, z\}$ and the following quasi-metrics on X.

$$q(x, y) = p(x, y) = 1;\ q(y, x) = p(y, x) = 2$$
$$q(x, z) = p(x, z) = 1;\ q(z, x) = 3;\ p(z, x) = 2$$
$$q(z, y) = p(z, y) = 1;\ q(y, z) = 0;\ p(y, z) = 1$$

Both quasi-metrics q and p are weightable whereas their pointwise supremum is not.

11.1.5 Theorem wqMet *is finally dense in* qMet.

Proof Let (X, d) be a quasi-metric space with more than one point. For a fixed $x \in X$ define $X_x = (X, d_x)$ as follows:

$$\begin{cases} d_x(x, y) = d(x, y) \text{ for every } y \in X, \\ d_x(z, y) = \infty \text{ whenever } z \neq x \text{ and } y \in X, z \in X. \end{cases}$$

Clearly X_x is weightable by the weight $w : X \longrightarrow \mathbb{P}$ defined by $w(x) = \infty$ and $w(y) = 0$ for every $y \neq x$.

Next form the coproduct $\Sigma_{x \in X} X_x$, which is clearly weightable, and consider the identification

$$\varphi : \Sigma_{x \in X} X_x \longrightarrow X : (z, x) \mapsto z$$

Clearly for the final quasi-metric structure d_{fin} on X we have

$$d_{\text{fin}}(z, y) = \inf_{x \in X} d_x(z, y) = d_z(z, y) = d(z, y)$$

and hence $\varphi : \Sigma_{x \in X} X_x \longrightarrow (X, d)$ is final in qMet. □

Since wqMet is finally dense in qMet and by 11.1.4 not concretely reflective in it, the category wqMet is not topological.

11.1.6 Proposition *For a weightable quasi-metric q on X with weight w the following properties hold.*

1. $d_{\mathbb{P}}^{-} \circ w \times w \leq q$ *and hence* $w : (X, q) \longrightarrow (\mathbb{P}, d_{\mathbb{P}}^{-})$ *is non expansive.*
2. $w : (X, \leq_q) \longrightarrow \mathbb{P}^{op}$ *is monotone.*
3. *If w is forcing for q then* $q \leq q^{-1} + d_{\mathbb{P}}^{-} \circ w \times w$.
4. *If w is forcing for q then for $x \in X$ we have*

$$q(x, .) \leq d_{\mathbb{P}}^{-} \circ w \times w(x, .) \vee \theta_{\{y | y \leq_q x\}}.$$

Proof 1. Let $x, y \in X$. The only nontrivial case to be considered is with $w(x) < \infty$, $w(x) < w(y)$ and $q(x, y) < \infty$. It follows that $q(y, x) < \infty$ and $w(y) < \infty$. Hence $(w(y) - w(x)) \vee 0 \leq q(x, y)$.

2. Let $x \leq_q y$ then by assumption $q(x, y) = 0$ and applying 1 we have $d_{\mathbb{P}}^{-}(w(x), w(y)) = 0$. Hence $w(x) \preceq w(y)$.

3. Since w is assumed to be forcing for q, in case $w(x) = \infty$ the inequality is trivially fulfilled. In case $w(x) < \infty$ we have

$$q(x, y) = q(y, x) + w(y) - w(x) \leq q(y, x) + (w(y) - w(x)) \vee 0$$
$$= q^{-1}(x, y) + d_{\mathbb{P}}^{-} \circ w \times w(x, y).$$

4. When evaluating both sides in $y \in X$, the only nontrivial case is with $q(x, y) \neq 0$ and $\theta_{\{y \leq_q x\}} \neq \infty$. So we may assume $w(x) < \infty$, $y \leq_q x$ and $q(y, x) = 0$. Then

$$q(x, y) = w(y) - w(x) \leq (w(y) - w(x)) \vee 0 = d_{\mathbb{P}}^{-} \circ w \times w(x, y) \vee \theta_{\{y \leq_q x\}}(y).$$

□

We now investigate how wqMet is embedded in App.

11.1.7 Proposition *Let $f : X \longrightarrow Y$ be an initial morphism in* App *with Y a weightable quasi-metric space, then the approach space X too is a weightable quasi-metric space.*

Proof Suppose the gauge $\mathcal{G}_Y$ has a gauge basis $\{q\}$, with q some weightable quasi-metric. Then clearly the initial gauge $\mathcal{G}_X$ has a gauge basis $\{q \circ f \times f\}$. Moreover it follows from 11.1.3 that the quasi-metric $q \circ f \times f$ is weightable. □

11.1.8 Theorem *The following properties hold and are equivalent.*

1. App *is the epireflective hull of* wqMet.
2. App *is the concretely reflective hull of* wqMet *(or equivalently of $\{d_{\mathbb{P}}^{-}\}$) in* App.
3. *An approach space has a gauge subbasis consisting of weightable quasi-metrics.*
4. *An approach space is the supremum in* App *of all weightable quasi-metric spaces that are coarser.*

Proof Since by 2.4.13 clearly 2 holds, if suffices to prove the equivalence of all the assertions.

1 ⇒ 2. Since indiscrete quasi-metric spaces are weightable, the epireflective hull coincides with the concrete reflective hull.

2 ⇒ 3. Let the source $(f_i : X \longrightarrow Y_i)_{i \in I}$ be initial in App with each Y_i a weightable quasi-metric space. Suppose each $\mathcal{G}_i$ has a gauge basis $\{q_i\}$ with q_i weightable. As pointed out in 11.1.7 the quasi-metric $q_i \circ f_i \times f_i$ is weightable. Moreover

$$\{q_i \circ f_i \times f_i | i \in I\}$$

is a subbasis for the gauge $\mathcal{G}_X$.

3 ⇒ 4. Suppose X has a gauge $\mathcal{G}_X$ with a gauge subbasis $\mathcal{H}_X$ consisting of weightable quasi-metrics. For each $q \in \mathcal{H}_X$ let X_q be the quasi-metric space on X

with gauge basis $\{q\}$. Then the source $(1_X : X \longrightarrow X_q)_{q \in \mathcal{H}}$ is initial. It follows that X is the supremum of all quasi-metric spaces coarser than X.

4 $\Rightarrow$ 1. Since a supremum can be seen as being initial for a point separating source this source can be decomposed as a subspace of a product. □

11.1.9 Definition If (X, δ) is an approach space and $\mathcal{H}$ is a gauge subbasis consisting of weightable quasi-metrics, then an element in $\prod_{q \in \mathcal{H}} \mathcal{W}_q$ is called a *weight* associated with $\mathcal{H}$ and $\bigcap_{q \in \mathcal{H}} w_q^{-1}(0)$ is called its *kernel*. The weight $(w_q)_{q \in \mathcal{H}}$ is called *forcing* for $\mathcal{H}$ if every w_q with $q \in \mathcal{H}$ is forcing for q.

11.1.10 Definition Given an approach space (X, δ) with gauge $\mathcal{G}$, one defines the *specialization preorder* as follows

$$x \leq y \Leftrightarrow (q(x, y) = 0 \text{ whenever } q \in \mathcal{G}).$$

Remark that as the quasi-metric coreflection (X, q_δ) of an approach space (X, δ) is given by $q_\delta(x, y) = \delta(x, \{y\}) = \sup_{q \in \mathcal{G}} q(x, y)$ the following expressions are equivalent

$$(q(x, y) = 0 \text{ whenever } q \in \mathcal{G}) \Leftrightarrow q_\delta(x, y) = 0.$$

Moreover, since the topology $\mathcal{T}_\delta$ of the topological coreflection of (X, δ) is the supremum of the topologies $\{\mathcal{T}_q \mid q \in \mathcal{G}\}$ we also have

$$(q(x, y) = 0 \text{ whenever } q \in \mathcal{G}) \Leftrightarrow x \in cl_{\mathcal{T}_\delta}\{y\}.$$

So the specialization preorder of (X, δ) defined in 11.1.10 coincides with the specialization preorders determined by the quasi-metric or topological coreflections.

11.1.11 Proposition *The following properties hold.*

1. *An approach space X is T_0 if and only if the specialization preorder $\leq_X$ is a partial order.*
2. *Let X be an approach space. The open sets in the topological coreflection $(X, \mathcal{T}_X)$ are upsets in the preorder $\leq_X$.*
3. *Every contraction between approach spaces $f : X \longrightarrow Y$ is monotone as map $f : (X, \leq_X) \longrightarrow (Y, \leq_Y)$.*

Proof 1. This follows from the definitions.

2. Let U be $\mathcal{T}_X$-open, $x \in U$ and $x \leq_X y$. Since $x \in \mathrm{cl}_{\mathcal{T}_X}\{y\}$ we clearly have $y \in U$.

3. Suppose $f(x) \not\leq f(y)$. Since $\downarrow f(y) = \mathrm{cl}_{\mathcal{T}_Y}\{f(y)\}$ this set is $\mathcal{T}_Y$-closed. It follows that the set $f^{-1}(X \setminus \downarrow f(y))$ is $\mathcal{T}_X$-open. Since this set is an upset for $\leq_X$ and contains x but not y, it follows that $x \not\leq y$. □

We now investigate the special situation where X carries a dcpo structure and its associated Scott topology, in particular we study the impact of having an approach structure quantifying the Scott topology.

11.1.12 Definition Let $(X, \leq)$ be a dcpo, $\sigma(X)$ the associated Scott topology and δ an approach structure on X. The approach structure is called *compatible* with the dcpo if the specialization preorder coincides with the given dcpo partial order.

We say that (X, δ) *quantifies the dcpo* if its topological coreflection coincides with the Scott topology, i.e. if $\mathcal{T}_\delta = \sigma(X)$. Clearly if (X, δ) quantifies the dcpo then it is compatible with the dcpo.

If moreover (X, δ) has a gauge subbasis $\mathcal{H}$ of weightable quasi-metrics and a weight $(w_q)_{q\in\mathcal{H}}$ such that its kernel $\bigcap_{q\in\mathcal{H}} w_q^{-1}(0) = \mathrm{Max}(X)$ (the maximal elements of the dcpo) then we say that (X, δ) quantifies the dcpo and *satisfies the kernel condition.*

Let $(X, \leq)$ be a preordered space and $\mathcal{W} \subseteq \mathbb{P}^X$.

11.1.13 Definition $\mathcal{W}$ is called *monotone for* $\leq$ if all functions

$$w : (X, \leq) \longrightarrow \mathbb{P}^{op}$$

are monotone. $\mathcal{W}$ is called *strictly monotone for* $\leq$ if $\mathcal{W}$ is monotone and

$$(y \leq x \text{ and } \forall w \in \mathcal{W} : w(x) = w(y)\,) \Rightarrow x = y.$$

The source $(w : X \longrightarrow (\mathbb{P}, d_{\mathbb{P}}^-))_{w\in\mathcal{W}}$ has an initial lift $X_{\mathcal{W}}^{in}$ in App and we denote $\mathcal{A}_{\mathcal{W}}^{in}$ its approach system. Further we denote by $\mathcal{T}_{\mathcal{W}}^{in}$ the initial topology determined by the source $(w : X \longrightarrow (\mathbb{P}, \mathcal{T}_{d_{\mathbb{P}}^-}))_{w\in\mathcal{W}}$ in Top.

11.1.14 Theorem *Let $(X, \leq)$ be a dcpo and suppose $\mathcal{H} \subseteq \mathbb{P}^{X\times X}$ is a collection of weightable quasi-metrics, with weight $(w_q)_{q\in\mathcal{H}} \in \prod_{q\in\mathcal{H}} \mathcal{W}_q$, and consider the approach space X for which $\mathcal{Q} := \mathcal{H}^\vee$ is a gauge basis. With $\mathcal{W} = \{w_q | q \in \mathcal{H}\}$ the following properties hold.*

1. *$w : X \longrightarrow (\mathbb{P}, d_{\mathbb{P}}^-)$ is a contraction, $w : (X, \mathcal{T}_X) \longrightarrow (\mathbb{P}^{op}, \sigma(\mathbb{P}^{op}))$ is continuous and $w : (X, \leq_X) \longrightarrow \mathbb{P}^{op}$ is monotone, for every $w \in \mathcal{W}$.*
2. *If X is a compatible approach space for the dcpo then $\bigcap_{q\in\mathcal{H}} w_q^{-1}(0) \subseteq$ Max(X).*
3. *If the weight is forcing then:*
   a. *If the approach space is T_0 (in particular when $\leq_X = \leq$) we have that $\mathcal{W}$ is strictly monotone for $\leq_X$.*
   b. *If the specialization preorder satisfies $\leq \subseteq \leq_X$ then*
      i. *$X \leq X_{\mathcal{W}}^{in} \vee X_{l(X)}$ with the supremum taken in* App.
      ii. *$X \vee X_{l(X)} = X_{\mathcal{W}}^{in} \vee X_{l(X)}$ with the supremum taken in* App.
      iii. *$\mathcal{T}_X \subseteq \mathcal{T}_{\mathcal{W}}^{in} \vee l(X)$ with the supremum taken in* Top.

c. *If X is quantifying then*
 i. $\sigma(X) \subseteq \mathcal{T}^{in}_{\mathcal{W}} \vee l(X)$ *with the supremum taken in* Top.
 ii. $M(X) = \sigma(X) \vee l(X) = \mathcal{T}^{in}_{\mathcal{W}} \vee l(X)$ *with the supremum taken in* Top.

Proof 1. From 11.1.6 we know that $d_{\mathbb{P}}^{-} \circ w \times w \leq q$ for every $q \in \mathcal{H}$. Since q belongs to the gauge the map w is a contraction. The rest follows by application of 11.1.2 and 11.1.11.

2. Let $x \in \bigcap_{q \in \mathcal{H}} w_q^{-1}(0)$ and $x \leq y$. Applying 1 we have that $w_q(y) = 0$ whenever $q \in \mathcal{H}$ and thus $q(x, y) = q(y, x)$ for all $q \in \mathcal{H}$ and hence $q(y, x) = 0$ for all $q \in \mathcal{H}$ and we can conclude that $x = y$.

3.a. First observe that by 1, $\mathcal{W}$ is monotone for $\leq_X$. By 11.1.6 (3) under the assumptions $y \leq_X x$ and $w(x) = w(y)$ whenever $w \in \mathcal{W}$ we have

$$q(x, y) \leq d_{\mathbb{P}}^{-} \circ w \times w(x, y) = 0$$

for every $q \in \mathcal{H}$. Hence $x \leq_X y$ and in view of the T_0 property we have $x = y$.

3.b.i. Under the extra assumption we have $\leq \subseteq \leq_q$ for every $q \in \mathcal{H}$. By application of 11.1.6 for $x \in X$ we have

$$q(x, .) \leq d_{\mathbb{P}}^{-}(w(x), w(.)) \vee \theta_{\{y | y \leq_q x\}} \leq d_{\mathbb{P}}^{-}(w(x), w(.)) \vee \theta_{\downarrow x}.$$

Clearly $d_{\mathbb{P}}^{-}(w(x), w(.))$ belongs to $\mathcal{A}^{in}_{\mathcal{W}}(x)$ and $\theta_{\downarrow x}$ to $\mathcal{A}_{l(X)}(x)$. Hence $q(x, .)$ is dominated by a finite sup of functions belonging to $(\mathcal{A}^{in}_{\mathcal{W}} \cup \mathcal{A}_{l(X)})(x)$ for every $q \in \mathcal{H}$. From this we can conclude that $X \leq X^{in}_{\mathcal{W}} \vee X_{l(X)}$.

3.b.ii. In view of 1 the weights w are contractions $w : X \longrightarrow (\mathbb{P}, d_{\mathbb{P}}^{-})$ so $X^{in}_{\mathcal{W}} \leq X$. From 3.b.i we now have

$$X^{in}_{\mathcal{W}} \vee X_{l(X)} \leq X \vee X_{l(X)} \leq X^{in}_{\mathcal{W}} \vee X_{l(X)}$$

from which the equality follows.

3.b.iii. Apply the coreflector from App to Top to the inequality in 3.b.i. Since the coreflector preserves initial sources we immediately get the result, where this time the suprema are taken in Top.

3.c.i This follows from 3.b.iii.

3.c.ii. This follows by applying of the coreflector from App to Top to the equality in 3.b.ii. □

11.2 Quantification of Algebraic Domains

In this section we prove that on any algebraic domain there are intrinsic quantifying approach spaces. We start by an example of one particular algebraic domain which will later be shown to be universal for all algebraic ones. The example generalizes the well-known construction of Plotkin (1978) which he developed in the setting of ω-domains.

11.2.1 Example The domain $\mathbb{T}^\gamma$ is determined as follows. For any cardinal γ,

$$\mathbb{T}^\gamma := \{u = (u_0, u_1) \mid u_0 \subseteq \gamma,\ u_1 \subseteq \gamma \text{ and } u_0 \cap u_1 = \emptyset\}$$

with the order defined by

$$u \leq v \iff u_0 \subseteq v_0 \text{ and } u_1 \subseteq v_1.$$

There is a least element $(\emptyset, \emptyset)$. The maximal elements are of the form $u = (u_0, u_1)$ with

$$u_0 \cup u_1 = \gamma.$$

The way-below relation is defined by

$$u \ll v \iff u_0 \text{ and } u_1 \text{ finite, and } u \leq v.$$

The compact elements are the finite elements, i.e. the elements $u = (u_0, u_1) \in \mathbb{T}^\gamma$ such that u_0 and u_1 are finite sets. The compact elements form a domain basis for $\mathbb{T}^\gamma$. Hence it is an algebraic domain.

11.2.2 Theorem *The domain $\mathbb{T}^\gamma$ has a quantifying approach structure $\delta_{\mathbb{T}^\gamma}$ satisfying the kernel condition.*

Proof The approach space we are looking for will be defined by means of a suitable gauge basis. For any finite subset $K \subseteq \gamma$, let $q_K : \mathbb{T}^\gamma \times \mathbb{T}^\gamma \longrightarrow \mathbb{P}$ be defined as follows

$$q_K(x, y) = |\, K \cap [(x_0 \setminus y_0) \cup (x_1 \setminus y_1)]\, |,$$

where $|.|$ stands for the cardinality of the set. Clearly for points $x, y, z \in \mathbb{T}^\gamma$ we have $x_i \setminus y_i \subseteq x_i \setminus z_i \cup z_i \setminus y_i$ for $i = 0$ and $i = 1$, so q_K is a quasi-metric. Furthermore it is easily seen to be weightable by the function

$$w_K : \mathbb{T}^\gamma \longrightarrow \mathbb{P} \text{ with } w_K(x) = |K \setminus (x_0 \cup x_1)|.$$

Consider $\mathscr{H} = \{q_K | K \subseteq \gamma, \text{finite}\}$. Observe that $\mathscr{H}$ is an ideal basis since $q_K \vee q_{K'} \leq q_{K \cup K'}$. The gauge $\mathscr{G}_{\mathbb{T}^\gamma}$ generated by $\mathscr{H}$ via saturation, defines an approach space.

In order to compare the topological coreflection $\sup\{\mathscr{T}_{q_K} | K \subseteq \mathbb{T}^\gamma, \text{finite}\}$ of the approach space with the Scott topology, we prove the following equality with respect to open balls with $\varepsilon < 1$:

$$B_{q_K}(x, \varepsilon) = \bigcap \{\, \Uparrow t \mid t \ll x,\ t_0 \cup t_1 \subseteq K\}.$$

One inclusion follows from the fact that $y \in B_{q_K}(x, \varepsilon)$ implies that points of K that belong to x_i also are in y_i. So whenever $t \ll x$, $t_0 \cup t_1 \subseteq K$ clearly $t \ll y$. For the other inclusion assume that $t \ll y$, whenever $t \ll x$ and $t_0 \cup t_1 \subseteq K$. Let $a \in K \cap x_0$ and put $t_0 = \{a\}$ and $t_1 = x_1 \cap (K \setminus \{a\})$. Then we have $t \ll x$ and $t_0 \cup t_1 \subseteq K$, and hence $t \ll y$. It follows that $a \in y_0$. Similarly we deduce that points in K that belong to x_1 also are in y_1.

Applying 11.1.14 we already know that $\bigcap\{w_{q_K}^{-1}(0) | K \subseteq \gamma, \text{finite}\} \subseteq \operatorname{Max}(\mathbb{T}^\gamma)$. Conversely if x is maximal and K is an arbitrary finite subset of γ then the fact that $x_0 \cup x_1 = \gamma$ implies that $w_{q_K}(x) = 0$. □

As we will see next the approach space constructed in the previous theorem is neither topological nor quasi-metric. We start by calculating its quasi-metric coreflection.

11.2.3 Proposition *With the notations of the previous theorem, the quasi-metric coreflection of the approach space* $(\mathbb{T}^\gamma, \delta_{\mathbb{T}^\gamma})$ *is given by*

$$q(x, y) = \begin{cases} |(x_0 \setminus y_0) \cup (x_1 \setminus y_1)| & \text{if the set involved is finite,} \\ \infty & \text{otherwise.} \end{cases}$$

Proof Let x and y be fixed elements in the domain. We have

$$\begin{aligned} q(x, y) = \delta_{\mathbb{T}^\gamma}(x, \{y\}) &= \sup\{q_K(x, y) | K \subseteq \gamma, \text{finite}\} \\ &= \sup\{| K \cap [(x_0 \setminus y_0) \cup (x_1 \setminus y_1)] \mid | K \subseteq \gamma, \text{finite} \} \end{aligned}$$

which proves the assertion. □

Hence we have the following situation for the topological and quasi-metric coreflections of $\delta_{\mathbb{T}^\gamma}$.

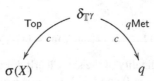

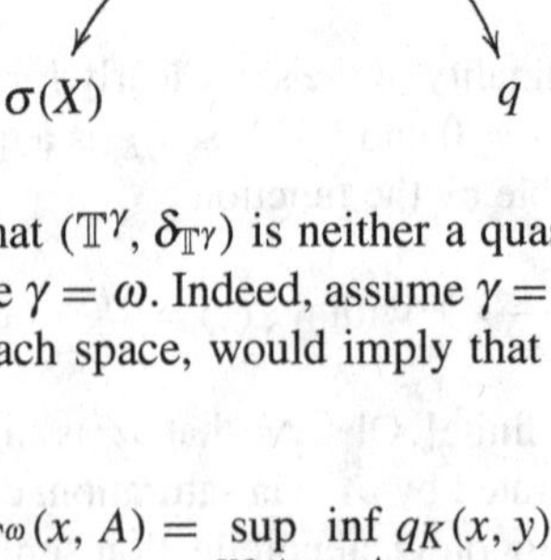

11.2.4 Example We note that $(\mathbb{T}^\gamma, \delta_{\mathbb{T}^\gamma})$ is neither a quasi-metric nor a topological approach space, even in case $\gamma = \omega$. Indeed, assume $\gamma = \omega$. Remark that $(\mathbb{T}^\omega, \delta_{\mathbb{T}^\omega})$ being a quasi-metric approach space, would imply that for all $x \in \mathbb{T}^\omega$ and for all $A \subseteq \mathbb{T}^\omega$ we would have

$$\begin{aligned} \delta_{\mathbb{T}^\omega}(x, A) &= \sup_{K \text{finite}} \inf_{y \in A} q_K(x, y) \\ &= \inf_{y \in A} \delta(x, \{y\}) \\ &= \inf_{y \in A} \sup_{K \text{finite}} q_K(x, y). \end{aligned}$$

Put $x = (2\mathbb{N}, 2\mathbb{N}+1)$ and put $A = \{y \mid y_0 \text{ and } y_1 \text{ finite}\}$. We first compute the first expression. Let $K \subseteq \omega$ be a finite set and let $y \in \mathbb{T}^\omega$ be a finite element, then

$$q_K(x, y) = |\, K \cap [(x_0 \setminus y_0) \cup (x_1 \setminus y_1)]\,|$$

and thus

$$\inf_{y \in A} q_K(x, y) \leq \inf_{\{y \in A \,\mid\, y_0 \subseteq 2\mathbb{N} \,\&\, y_1 \subseteq 2\mathbb{N}+1\}} |\, K \cap [(x_0 \setminus y_0) \cup (x_1 \setminus y_1)]\,| = 0.$$

However, computing the third expression, for any y finite we get $\sup_{K \text{finite}} q_K(x, y) = \infty$ which yields a contradiction, showing that the space is not quasi-metric.

To show that the space also is not topological, consider $x = (\{0, \ldots, n\}, \{n+1\})$ and $y = (\{n+2\}, \{n+1\})$, then

$$q(x, y) = |(x_0 \setminus y_0) \cup (x_1 \setminus y_1)| = n$$

and thus all values $n \in \mathbb{N}$ are obtained.

The coreflection described in the previous proposition actually gives a bicomplete quasi-metric. The symmetrization of the quasi-metric q defined in 11.2.3 is

$$q^*(x, y) = \begin{cases} \max(|(x_0 \setminus y_0) \cup (x_1 \setminus y_1)|, |(y_0 \setminus x_0) \cup (y_1 \setminus x_1)| & \text{if both sets are finite,} \\ q^*(x, y) = \infty & \text{otherwise.} \end{cases}$$

It is clear that a Cauchy sequence in $\mathbb{T}^\gamma$ eventually becomes constant.

We proceed in proving that every algebraic domain has a quantifying approach structure and that under the Lawson condition the quantifying approach structure also satisfies the kernel condition. So in the sequel $(X, \leq)$ is an algebraic domain with Scott topology $\sigma(X)$. We use a technique inspired by the one developed by Waszkiewicz (2003) for ω-algebraic domains based on a theorem of Plotkin (1978). The proof of the next result is quite similar to the ω case given in Waszkiewicz (2003).

11.2.5 Theorem *For every algebraic domain $(X, \leq)$ the space $(X, \sigma(X))$ can be topologically embedded in some space $(\mathbb{T}^\gamma, \sigma(\mathbb{T}^\gamma))$, for a suitable cardinal γ.*

Proof Suppose the cardinality of the set of compact elements in the given domain is $|K(X)| = \gamma$, so this set can be labeled as $K(X) = \{b_\alpha | \alpha < \gamma\}$. Consider the map

$$\eta : X \longrightarrow \mathbb{T}^\gamma : x \mapsto (\{\alpha \mid b_\alpha \ll x\}, \{\alpha \mid \exists \rho \in \gamma : \ b_\rho \ll x \ \text{and} \ b_\rho \# b_\alpha\})$$

We make the following observations:

1. η is properly defined: for all $\alpha, \rho \in \gamma$ with $b_\alpha \ll x$ and $b_\rho \ll x$ we have that b_α and b_ρ are compatible.

2. η is injective: Let $\eta(x) = \eta(y)$, then $\Downarrow x \cap K(X) = \Downarrow y \cap K(X)$ and thus by continuity of $(X, \leq)$ we have that $x = y$.

3. η is order preserving: Let $x, y \in (X, \leq)$ such that $x \leq y$, then clearly $\eta(x) \leq \eta(y)$ in $\mathbb{T}^\gamma$.

4. η is Scott continuous: It suffices to observe that for $x \in X$ we have

$$\eta(x) = \mathrm{Sup}\{\eta(b_\beta) | b_\beta \ll x\}.$$

5. $\eta : (X, \sigma(X)) \longrightarrow (\mathbb{T}^\gamma, \sigma(\mathbb{T}^\gamma))$ is initial in Top and therefore also an order embedding.

It is sufficient to prove that $\eta : (X, \sigma(X)) \longrightarrow (\eta(X), \sigma(\mathbb{T}^\gamma)|_{\eta(X)})$ is open. Let $U = \Uparrow b_\rho = \uparrow b_\rho$, with $\rho \in \gamma$ be basic open in the Scott topology. Define

$$T = \{u = (u_0, u_1) \in \mathbb{T}^\gamma \mid \exists \alpha \in \gamma : \ b_\rho \leq b_\alpha \text{ and } \alpha \in u_0\}.$$

This is clearly an upper set which is inaccessible for directed suprema and hence Scott open in $\mathbb{T}^\gamma$. We prove that $\eta(U) = \eta(X) \cap T$. Since for any $x \in X$ we have that $b_\rho \leq x$ implies $\rho \in (\eta(x))_0$, the inclusion $\eta(U) \subseteq T$ is clear.

For the converse let $x \in X$ such that $\eta(x) \in T$. Then there is some $\alpha \in \gamma$ with $b_\rho \leq b_\alpha$ and $\alpha \in (\eta(x))_0$. This implies $b_\alpha \ll x$ and so $x \in U$.

That η is also an order embedding follows from the following standard argument. If $x \not\leq y$ then $X \setminus \downarrow y$ is Scott open for the initial structure, so there exists a Scott open subset U in $\mathbb{T}^\gamma$ such that $x \in \eta^{-1}(U) \subseteq X \setminus \downarrow y$. Since U is an upper-set we can conclude that $\eta(x) \not\leq \eta(y)$. □

11.2.6 Theorem *Every algebraic domain has a quantifying approach structure.*

Proof The proof is based on the embedding described in 11.2.5. So with the same notations as before consider

$$\eta : (X, \sigma(X)) \longrightarrow (\mathbb{T}^\gamma, \sigma(\mathbb{T}^\gamma) : x \mapsto (\{\alpha \mid b_\alpha \ll x\}, \{\alpha \mid \exists \rho \in \gamma : \ b_\rho \ll x \,\&\, b_\rho \# b_\alpha\})$$

which is initial in Top. We endow $\mathbb{T}^\gamma$, with the quantifying approach structure $\delta_{\mathbb{T}^\gamma}$ described in 11.2.2 with gauge basis

$$\mathscr{H} = \{q_K | K \subseteq \gamma, \text{finite}\}.$$

Let δ_X be the approach structure on X which makes the source

$$\eta : (X, \delta_X) \longrightarrow (\mathbb{T}^\gamma, \delta_{\mathbb{T}^\gamma})$$

initial in App. By initiality the approach structure δ_X has a gauge basis

$$\mathscr{H} \circ \eta \times \eta = \{q_K \circ \eta \times \eta | K \subseteq \gamma, \text{finite}\}$$

and a weight

$$\mathscr{W} \circ \eta = (w_K \circ \eta)_{K \subseteq \gamma, \text{finite}}.$$

Applying the coreflector to Top, we obtain that the source

$$\eta : (X, \mathscr{T}_{\delta_X}) \longrightarrow (\mathbb{T}^\gamma, \mathscr{T}_{\delta_{\mathbb{T}^\gamma}})$$

is initial in Top. As was shown in 11.2.2 $\mathscr{T}_{\delta_{\mathbb{T}^\gamma}} = \sigma(\mathbb{T}^\gamma)$ and so in view of 11.2.5 we can conclude that $\mathscr{T}_{\delta_X} = \sigma(X)$. □

Next we investigate whether the quantifying approach structure on X satisfies the kernel condition. In this respect we need the Lawson (1997) condition which makes use of the Lawson topology.

11.2.7 Definition A domain satisfies the *Lawson condition* (L) if the Lawson and Scott topologies agree on the set of maximal elements, i.e.

$$L(X)|_{\text{Max}(X)} = \sigma(X)|_{\text{Max}(X)}.$$

We will use the following characterization of maximal elements given in Waszkiewicz (2003).

11.2.8 Proposition *In an algebraic domain that satisfies* (L)*, the following properties are equivalent.*

1. $x \in MaxX$.
2. $\forall b \in K(X) : b \ll x$ *or* $\exists c \in K(X)$ *such that* $c \ll x$ *and* $c\#b$.

11.2.9 Theorem *An algebraic domain satisfying* (L) *has a quantifying approach structure satisfying the kernel condition.*

Proof With the same notations as before consider again

$$\eta : X \longrightarrow \mathbb{T}^\gamma : x \mapsto (\{\alpha \mid b_\alpha \ll x\}, \{\alpha \mid \exists \rho \in \gamma : b_\rho \ll x \,\&\, b_\rho \# b_\alpha\}).$$

The condition (L) ensures that

$$x \in \text{Max}(X) \Leftrightarrow \eta(x) \in \text{Max}(\mathbb{T}^\gamma).$$

So we have

$$x \in \text{Max}(X) \Leftrightarrow \forall K \subseteq \gamma, \text{ finite}, \ w_K \circ \eta(x) = 0.$$

As we already know from 11.2.6 the collection $\mathscr{W} \circ \eta = \{w_K \circ \eta | K \subseteq \gamma, \text{ finite}\}$ is a weight for (X, δ), we are finished. □

As in 11.2.3 we can calculate the quasi-metric coreflection (X, q_X) of the approach space (X, δ_X) defined in 11.2.6. Using the gauge basis $\mathscr{H} \circ \eta \times \eta$ we get that

$$q_X(x, y) = \sup\{q_K(\eta(x), \eta(y)) | K \subseteq \gamma, \text{ finite}\}$$
$$= \begin{cases} |((\eta(x))_0 \setminus (\eta(y))_0) \cup ((\eta(x))_1 \setminus (\eta(y))_1)| & \text{if the set involved is finite,} \\ q_X(x, y) = \infty & \text{otherwise.} \end{cases}$$

It is clear that for a Cauchy sequence $(x_n)_n$ for the symmetrization q_X^* the sequence $(\eta(x_n))_n$ eventually becomes constant. Since η is injective the sequence $(x_n)_n$ too becomes constant. Hence q_X is bicomplete.

11.3 Quantification of Arbitrary Domains

We now turn to arbitrary domains and we develop another technique for the construction of a quantifying approach structure.

11.3.1 Proposition *Let $(X, \leq)$ be a domain, $B \subseteq X$ a domain basis and $\sigma(X)$ the Scott topology. Then the collection*

$$\mathscr{H}^B := \{q_K^B \mid K \subseteq B, \textit{ finite }\}$$

with, for all $x, y \in X$

$$q_K^B(x, y) := |\{\, b \in K \mid x \in \Uparrow b,\ y \notin \Uparrow b \,\}|$$

is a gauge basis for a quantifying approach space (X, δ^B) with weight $(w_K^B)_{K \subseteq B, \textit{ finite}}$ where $w_K^B(x) = | \{b \in K \mid x \notin \Uparrow b \} |$.

Proof It is clear that each of the functions $q_K^B : X \times X \longrightarrow \mathbb{P}$ defined above satisfies the triangular inequality and has weight w_K^B. Moreover for K and K' finite subsets of B we have $q_K^B \vee q_{K'}^B \leq q_{K \cup K'}^B$, so $\mathscr{H}^B = \{q_K^B \mid K \subseteq B, \text{ finite }\}$ is a gauge basis. Let (X, δ^B) be the approach space generated by $\mathscr{H}^B$.

For $x \in X$ we have

$$B_{q_K^B}(x, 1) = \bigcap\{\Uparrow b \mid b \in K,\ x \in \Uparrow b\,\},$$

so these sets are Scott open. Conversely, given a Scott basic open set $\Uparrow b$ we let $K = \{b\}$. Then clearly $B_{q_{\{b\}}^B}(x, 1) = \Uparrow b$ and consequently $\sup\{\mathscr{T}_{q_K^B} \mid K \subseteq B,$ finite $\} = \sigma(X)$. □

In general the approach space constructed in 11.3.1 is neither quasi-metric nor topological.

11.3.2 Example On the domain $X = \mathbb{N} \cup \{\infty\}$ with the usual order, the domain basis X generates an approach space δ^X with gauge basis $\mathcal{H}^X = \{ q_K \mid K \subseteq X, \text{ finite } \}$ with $q_K(m, n) = | \{b \in K \mid b \leq m, n < b \} |$. The quasi-metric coreflection is given by $q(m, n) = (m - n) \vee 0$ for m and n in $\mathbb{N} \cup \{\infty\}$. And since it is clearly not $\{0, \infty\}$-valued, neither is δ^X, and so the approach space is not topological. In order to show that it is also not quasi-metric, as in 11.2.3, we show that for some $x \in X$ and $A \subseteq X$ we have

$$\sup_{K \text{finite}} \inf_{y \in A} q_K(x, y) < \inf_{y \in A} \sup_{K \text{finite}} q_K(x, y).$$

For $x = \infty$ and $A = \mathbb{N}$ the left hand side equals 0 whereas on the right hand side we obtain $\inf_{n \in A}(\infty - n) \vee 0 = \infty$.

Remark that in general, even assuming the Lawson condition, the gauge basis constructed in 11.3.1 does not satisfy the kernel condition.

11.3.3 Example We consider the domain of partial functions on the naturals. A partial function $f : X \longrightarrow Y$ between sets X and Y is a function $f : A \longrightarrow Y$ defined on a subset $A \subseteq X$. We write $\text{dom}(f) = A$ for the domain of a partial map $f : X \longrightarrow Y$. The set X of partial mappings from $\mathbb{N}$ to $\mathbb{N}$ is ordered by the extension order $f \leq g \iff \text{dom}(f) \subseteq \text{dom}(g)$ and $g|_{\text{dom}(f)} = f$. X is a domain with the way below relation characterized by $f \ll g \iff f \leq g$ and $\text{dom}(f)$ finite. The set of compact elements $K(X) = \{f \in X \mid \text{dom}(f) \text{ finite}\}$ forms a countable domain basis. Applying our construction 11.3.1 to this example yields an approach space $(X, \delta^{K(X)})$ with gauge basis $\mathcal{H}^{K(X)} = \{ q_K \mid K \subseteq K(X) \}$ where q_K is defined by $q_K(f, g) = | \{b \in K \mid b \leq f, b \not\leq g\} |$ and has weight $w_K(f) = | \{b \in K \mid b \not\leq f \} |$. For this domain $\text{Max}(X)$ consists of those functions $f \in X$ with $\text{dom}(f) = \mathbb{N}$. It can be seen that X satisfies the Lawson condition but the gauge basis $\mathcal{H}^{K(X)}$ does not satisfy the kernel condition.

Next we investigate whether the approach spaces constructed in 11.3.1 via different domain bases coincide. In order to obtain refined results this study has to be pursued locally. So we use approach systems rather than gauges. Using the notations of 11.3.1, for a given domain basis B let $(\mathcal{B}^B(x))_{x \in X}$ be the basis for the approach system associated with the gauge basis $\mathcal{H}^B$. More explicitly we have

$$\mathcal{B}^B(x) = \{ q(x, .) \mid q \in \mathcal{H}^B \} = \{ q_K^B(x, .) \mid K \subseteq B, \text{ finite}\}$$

and let $(\mathcal{A}^B(x))_{x \in X}$ be the generated approach system.

For $x \in X$ and $K \subseteq X$ finite, we denote $K_x = K \cap \downarrow x$. Using these notations we have the following results.

11.3.4 Proposition *If K_x contains only compact elements then we have $q_K^X(x, .) \in \mathcal{A}^B(x)$.*

Proof If K_x contains only compact elements then also $K_x \subseteq B$. With $L = K_x$ we then have $q_K^X(x, .) = q_L^B(x, .)$ □

It follows from 11.3.4 that in domains like the powerset of $\mathbb{N}$, $\mathbb{T}^\gamma$ (11.2.1) or the domain of partial functions on the naturals (11.3.3), where the relation $z \ll y$ implies that z is compact, the constructed approach space will not depend on the domain basis.

11.3.5 Proposition *Let $x \in X$, $K \subseteq X$ finite, and for every $k \in K_x$, put $B_k = \{b \in B \mid k \ll b \ll x\}$. If $|\varphi(K_x)| = |K_x|$ for some choice $\varphi \in \prod_{k\in K_x} B_k$, then we have $q_K^X(x,.) \in \mathscr{A}^B(x)$.*

Proof Let $y \in X$ be arbitrary. For φ chosen with $|\varphi(K_x)| = |K_x|$ we have that $\varphi : K_x \longrightarrow \varphi(K_x)$ is bijective, so we also have a bijection between

$$\{k \mid k \in K,\ k \in \Downarrow x, k \notin \Downarrow y\} \text{ and } \{\varphi(k) \mid k \in K,\ k \in \Downarrow x, k \notin \Downarrow y\}.$$

Moreover since $k \ll \varphi(k) \ll x$ whenever $k \in K_x$, for $L = \varphi(K_x)$ we have

$$\{\varphi(k) \mid k \in K,\ k \in \Downarrow x, k \notin \Downarrow y\} \subseteq \{l \mid l \in L,\ l \in \Downarrow x, l \notin \Downarrow y\}.$$

So the conclusion $q_K^X(x,y) \leq q_L^B(x,y)$ follows. □

11.3.6 Theorem *If $x \in X$ is not compact then for two domain bases B and D the approach systems $(\mathscr{A}^B(x))_{x\in X}$ and $(\mathscr{A}^D(x))_{x\in X}$ coincide.*

Proof Let $x \in X$ and suppose there is a domain basis B such that $\mathscr{A}^B(x) \neq \mathscr{A}^X(x)$. Since we clearly have $\mathscr{A}^B(x) \subseteq \mathscr{A}^X(x)$, this means that there exists $K \subseteq X$ with $q_K(x,.) \notin \mathscr{A}^B(x)$. So for every $L \subseteq B$ there exists $y \in X$ with

$$q_L^B(x,y) < q_K^X(x,y).$$

From 11.3.5 and using the same notations as before, we have $|\varphi(K_x)| < |K_x|$ for every $\varphi \in \prod_{k\in K_x} B_k$. Let

$$m = \max\{|\varphi(K_x)| \mid \varphi \in \prod_{k\in K_x} B_k\}$$

and take φ_0 such that $|\varphi_0(K_x)| = m$. Since $B \cap \Downarrow x$ is directed there exists $b_0 \in B$, $b_0 \ll x$, such that $\varphi_0(k) \leq b_0$ whenever $k \in K_x$.

Either $b_0 = x$, and then the conclusion $x \ll x$ follows, or $b_0 \neq x$, and then in view of the particular choice for φ_0 there exists $k_0 \in K_x$ such that $b_0 = \varphi_0(k_0)$. In this case, for $b \in B$ with $b \ll x$ arbitrary, again using the fact that $B \cap \Downarrow x$ is directed we choose $b' \in B$ satisfying $\varphi_0(k_0) \leq b' \ll x$ and $b \leq b' \ll x$. Either one of the chosen b' equals x, and then the conclusion $x \ll x$ follows. Or, in view of the special choice of φ_0, for all chosen b' we have $b' = \varphi_0(k_0)$. This in particular implies $b \leq \varphi_0(k_0)$. Taking into account that $\bigvee \Downarrow x \cap B = x$ we have that $\varphi_0(k_0) = x$, and then again the conclusion $x \ll x$ follows. □

As a corollary we obtain that in all domains where the set of compact elements is empty, as is the case in the interval domain on $\mathbb{R}$ or in the formal ball model constructed from a complete metric space (Edalat 1998), the approach spaces constructed from different domain bases coincide. Also in example 11.1.2 on $\mathbb{P}$ with the reverse order, where ∞ is the only compact element, the previous results imply that the approach spaces constructed from different domain bases coincide.

However in compact elements the approach systems associated with different bases can differ, as shown by the following example.

11.3.7 Example Let $X = \mathbb{P} \cup \{c, d\}$ where $c \neq d$ are new points added to $\mathbb{P}$. We take the usual order on $\mathbb{P}$ and we define $\infty < c < d$. This is a complete chain and therefore a domain. The only compact elements are 0, c and d. Clearly $B = X \setminus \{\infty\}$ is a domain basis for X. We investigate the approach structures in the point d associated with the domain basis X and B respectively. For $K = \{\, \infty, c, d \,\}$ we have

$$q_K^X(d, \infty) = |\, \{k \in K \mid k \in \mathop{\downarrow\!\!\!\!\downarrow} d,\ k \notin \mathop{\downarrow\!\!\!\!\downarrow} \infty \}\, | = 3.$$

For every finite subset $L \subseteq B$ we have $\infty \notin L$ and so

$$q_L^B(d, \infty) = |\, \{l \in L \mid l \in \mathop{\downarrow\!\!\!\!\downarrow} d,\ l \notin \mathop{\downarrow\!\!\!\!\downarrow} \infty \}\, | = 2.$$

So it is clear that $q_K^X(d, .)$ does not belong to $\mathscr{A}^B(d)$.

11.4 Fixed Points for Contractive Functions

In the first part of this section we consider a function $f : X \longrightarrow X$ and for an arbitrary point $a \in X$ we will estimate "how far" a is from being a fixed point. This will be done by comparing $f(a)$ to a. An estimation for the "distance" between $f(a)$ and a will be obtained by structuring X as an approach space and using its limit operator.

As in 11.1.13, given an arbitrary $\mathscr{W} \subseteq \mathbb{P}^X$ we consider the initial lift $X_{\mathscr{W}}^{in}$ in App of the source $(w : X \to ([0, \infty]^{op}, d_{\mathbb{P}}^{-}))_{w \in \mathscr{W}}$ with $d_{\mathbb{P}}^{-}$.

11.4.1 Theorem *Let X be an approach space and let $\mathscr{W} \subseteq \mathbb{P}^X$ be arbitrary. For a contraction*

$$f : X \to X_{\mathscr{W}}^{in}$$

fix $x \in X$ and let

$$a_n = f^n(x)$$

be the values obtained by iterating f on x. For $w \in \mathscr{W}$ let

$$l_w = \limsup_{n\to\infty} w(a_n).$$

For $a \in X$ arbitrary we have the following estimation

$$\sup_{w\in\mathscr{W}} d_{\mathbb{P}}^{-}(w(f(a)), w(a)) \leq \lambda_X \langle (a_n)_n \rangle (a) + \sup_{w\in\mathscr{W}} d_{\mathbb{P}}^{-}(l_w, w(a)).$$

Proof For $w \in \mathscr{W}$ we have

$$\begin{aligned} d_{\mathbb{P}}^{-}(w(f(a)), w(a)) &\leq d_{\mathbb{P}}^{-}(w(f(a)), l_w) + d_{\mathbb{P}}^{-}(l_w, w(a)) \\ &= (\limsup_{n\to\infty} w(a_n) - w(f(a))) \vee 0 + d_{\mathbb{P}}^{-}(l_w, w(a)) \\ &= \limsup_{n\to\infty}(d_{\mathbb{P}}^{-}(w(f(a)), w(a_n))) + d_{\mathbb{P}}^{-}(l_w, w(a)) \\ &= \lambda_{d_{\mathbb{P}}^{-}} \langle (w(a_n))_n \rangle (w(f(a))) + d_{\mathbb{P}}^{-}(l_w, w(a)). \end{aligned}$$

By taking the supremum on both sides it follows that

$$\begin{aligned} \sup_{w\in\mathscr{W}} d_{\mathbb{P}}^{-}(w(f(a)), w(a)) &\leq \sup_{w\in\mathscr{W}} \lambda_{d_{\mathbb{P}}^{-}} \langle (w(a_n))_n \rangle (w(f(a))) + \sup_{w\in\mathscr{W}} d_{\mathbb{P}}^{-}(l_w, w(a)) \\ &= \lambda_{X_{\mathscr{W}}^{in}} \langle (a_n)_n \rangle (f(a)) + \sup_{w\in\mathscr{W}} d_{\mathbb{P}}^{-}(l_w, w(a)) \\ &\leq \lambda_X \langle (a_n)_n \rangle (a) + \sup_{w\in\mathscr{W}} d_{\mathbb{P}}^{-}(l_w, w(a)) \end{aligned}$$

where, in the last equality we use the fact that f is a contraction. □

Given a dcpo $(X, \leq)$ with compatible approach space X we can use the previous theorem to evaluate the distance between a and $f(a)$ in the metric coreflection of X.

11.4.2 Theorem *Let $(X, \leq)$ be a dcpo with compatible approach space X having a forcing weight $\mathscr{W}$. For a contraction*

$$f : X \to X_{\mathscr{W}}^{in}$$

and for an element $a \in X$ with $a \leq f(a)$, with the same notations as the previous theorem we get

$$d_X(f(a), a) \leq \lambda_X \langle (a_n)_n \rangle (a) + \sup_{w\in\mathscr{W}} d_{\mathbb{P}}^{-}(l_w, w(a)).$$

Proof Let $\mathscr{H}$ be the gauge subbasis of weightable quasi-metrics where $q \in \mathscr{H}$ has associated weight w_q and $\mathscr{W} := \{w_q \mid q \in \mathscr{H}\}$. Since X is compatible with $(X, \leq)$ we have that $a \leq f(a)$ implies $q(a, f(a)) = 0$ for all $q \in \mathscr{H}$. Since w_q is forcing for

q it follows from 11.1.6 that

$$\begin{aligned} q(f(a), a) &\leq q(a, f(a)) + d_{\mathbb{P}}^{-} \circ w_q \times w_q(f(a), a) \\ &= d_{\mathbb{P}}^{-}(w_q(f(a)), w_q(a)). \end{aligned}$$

Taking suprema of both sides we get

$$d_X(f(a), a) = \sup_{q \in \mathscr{H}} q(f(a), a) \leq \sup_{q \in \mathscr{H}} d_{\mathbb{P}}^{-}(w_q(f(a)), w_q(a)).$$

The property now follows from 11.4.1. □

Next we concentrate on both terms on the right-hand side of the previous inequality.

With the same notations as in 11.4.1 suppose moreover that X carries a dcpo structure $\leq$.

11.4.3 Theorem *If $\mathscr{W}$ is monotone for $(X, \leq)$ and if there is a subsequence $(a_{k_n})_n$ such that $a_{k_n} \leq a$ for all $n \in \mathbb{N}$, then we have*

$$\sup_{w \in \mathscr{W}} d_{\mathbb{P}}^{-}(l_w, w(a)) = 0.$$

Proof Let $w \in \mathscr{W}$ be arbitrary. Applying monotonicity we have

$$a_{k_n} \leq a \Rightarrow w(a_{k_n}) \geq w(a)$$

for the given subsequence. Hence we get

$$\limsup_{n \to \infty} w(a_n) \geq w(a)$$

and thus $d_{\mathbb{P}}^{-}(l_w, w(a)) = 0$. □

11.4.4 Theorem *Let $(X, \leq)$ be a dcpo endowed with an approach structure X and let $\mathscr{W} \subseteq \mathbb{P}^X$ be strictly monotone for $\leq$. For a contraction*

$$f : X \to X_{\mathscr{W}}^{in}$$

fix $x \in X$ and let

$$a_n = f^n(x)$$

be the values obtained by iterating f on x. We make the following assumptions on some $a \in X$.

1. *There is a subsequence* $(a_{k_n})_n$ *such that* $a_{k_n} \leq a \ \forall n \in \mathbb{N}$.
2. $(a_n)_n$ *converges to* a *in the topological coreflection* $\mathscr{T}_X$.
3. $a \leq f(a)$.

Then the point a *is a fixed point of* f.

Proof For $w \in \mathscr{W}$ let

$$l_w = \limsup_{n\to\infty} w(a_n).$$

By 11.4.3 condition 1 implies that $d_{\mathbb{P}}^{-}(l_w, w(a)) = 0$ and by condition 2 also $\lambda_X\langle(a_n)_n\rangle(a) = 0$. Applying 11.4.1 we can conclude that

$$\sup_{w\in\mathscr{W}} d_{\mathbb{P}}^{-}(w(f(a)), w(a)) = 0.$$

It follows that $w(a) = w(f(a))$ for every $w \in \mathscr{W}$ and by strict monotonicity we have $a = f(a)$. □

Next we list some situations implying the conditions in 11.4.4.

11.4.5 Theorem *Let* $(X, \leq)$ *be a dcpo endowed with an approach structure* X *with* $\mathscr{T}_X \subseteq M(X)$ *and let* $\mathscr{W} \subseteq \mathbb{P}^X$ *be strictly monotone for* $\leq$. *For a contraction*

$$f : X \to X_{\mathscr{W}}^{in}$$

fix $x \in X$ *and let*

$$a_n = f^n(x)$$

be the values obtained by iterating f *on* x. *If* $(a_n)_n$ *is monotone and*

$$\{z \in X \mid z \leq f(z)\}$$

is closed under directed suprema then $a = \bigvee_n a_n$ *is a fixed point.*

Proof Since $(a_n)_n$ is monotone the supremum $a = \bigvee_n a_n$ is well defined and $(a_n)_n$ converges to a in the Scott topology. Since $a_n \leq a$ for every n, the sequence $(a_n)_n$ also converges to a in the Martin topology $M(X)$ (Martin 2000a, b). In view of the assumption $\mathscr{T}_X \subseteq M(X)$ the convergence also holds in $\mathscr{T}_X$. Clearly for all n the term a_n satisfies

$$a_n \leq a_{n+1} = f(a_n).$$

Therefore the directed supremum a satisfies $a \leq f(a)$. By 11.4.4 the conclusion follows. □

11.4.6 Theorem *Let $(X, \leq)$ be a dcpo endowed with an approach structure X with $\mathcal{T}_X \subseteq M(X)$ and let $\mathcal{W} \subseteq \mathbb{P}^X$ be strictly monotone for $\leq$. For a contraction*

$$f : X \to X^{in}_{\mathcal{W}}$$

the following properties hold.

1. *If $I \subseteq X$ is nonempty and closed under directed suprema and $f : I \longrightarrow I$ is splitting in the terminology of Martin (2000a) (inflationary (Gierz et al. 2003)) in the sense that $z \leq f(z)$ whenever $z \in I$, then f has a fixed point.*
2. *If f is splitting on X then f has a fixed point.*

Proof Clearly we only have to prove the first assertion. Choose $x \in I$ then the sequence $(a_n)_n$ obtained by iterating f on x is a monotone sequence in I. So the supremum $a = \bigvee_n a_n$ belongs to I. Since $a_n \leq a$ for every n, the sequence $(a_n)_n$ also converges to a in $M(X)$ and hence also in $\mathcal{T}_X$. Again 11.4.4 can be applied. □

Next we turn to situations where apparently no specific class $\mathcal{W}$ is given.

11.4.7 Theorem *Let $(X, \leq)$ be a dcpo endowed with approach spaces $(X, \mathcal{G})$ and $(X, \mathcal{G}')$ defined by their gauges, with $\mathcal{T}_{\mathcal{G}} \subseteq M(X)$ and $\leq = \leq_{\mathcal{G}'}$. For a contraction*

$$f : (X, \mathcal{G}) \to (X, \mathcal{G}')$$

fix $x \in X$ and let

$$a_n = f^n(x)$$

be the values obtained by iterating f on x. If $(a_n)_n$ is monotone and

$$\{z \in X \mid z \leq f(z)\}$$

is closed under directed suprema then $a = \bigvee_n a_n$ is a fixed point.

Proof We consider the collection of quasi-metrics

$$\mathcal{H} = \{d^-_{\mathbb{P}} \circ g \times g \mid g : (X, \mathcal{G}') \longrightarrow (\mathbb{P}, d^-_{\mathbb{P}}) \text{ contraction}\}$$

and $\mathcal{Q} = \mathcal{H}^{\vee}$ as in 11.1.14. Clearly $\mathcal{Q}$ is a gauge basis for the approach space $(X, \mathcal{G}')$.

Using the notations of 11.1.2 every quasi-metric $d^-_{\mathbb{P}} \circ g \times g$ is weighted by the function $w_{d^-_{\mathbb{P}}} \circ g = g$. If $g(x) = \infty$ then clearly $d^-_{\mathbb{P}}(g(x), g(y)) = 0$ hence g is forcing for $d^-_{\mathbb{P}} \circ g \times g$.

The collection $\mathcal{W}$ of all the weights coincides with the collection of all contractions $g : (X, \mathcal{G}') \longrightarrow (\mathbb{P}, d^-_{\mathbb{P}})$, hence we have $X^{in}_{\mathcal{W}} = (X, \mathcal{G}')$. Moreover by 11.1.14, the collection $\mathcal{W}$ is strictly monotone for $\leq$. The rest follows at once from 11.4.5. □

The following example should be compared to Theorem 3.2.1 in Martin (2000a).

11.4.8 Example Taking for $\mathcal{G}$ the gauge of the Martin topology $M(X)$ (embedded as an approach space) and $\mathcal{G}'$ the gauge of the Scott topology, on a dcpo $(X, \leq)$, then both conditions $\mathcal{T}_{\mathcal{G}} \leq M(X)$ and $\leq = \leq_{\mathcal{G}'}$ are fulfilled. This means that for a continuous map

$$f : (X, M(X)) \longrightarrow (X, \sigma(X)),$$

assuming that $(a_n)_n$ is monotone and $\{z \in X \mid z \leq f(z)\}$ is closed under directed suprema, we have that $a = \bigvee_n a_n$ is a fixed point.

Finally we include some applications to monotone functions.

11.4.9 Theorem *Let $(X, \leq)$ be a dcpo endowed with an approach structure on X such that $\mathcal{T}_X \leq M(X)$ and $\leq = \leq_X$. For a contractive map*

$$f : X \to X$$

and a fixed $x \in X$ satisfying $x \leq f(x)$, the sequence $(a_n)_n$ obtained by iterating f on x has a supremum $a = \bigvee_n a_n$ which is a fixed point for f. If moreover $(X, \leq)$ has a bottom element and x is taken as $x = \bot$ then a is the least fixed point of f.

Proof As in 11.4.7, we consider the collection of quasi-metrics

$$\mathcal{H} = \{d_{\mathbb{P}}^{-} \circ g \times g \mid g : X \longrightarrow (\mathbb{P}, d_{\mathbb{P}}^{-}) \text{ contraction}\}.$$

Using the notations of 11.1.2 every quasi-metric $d_{\mathbb{P}}^{-} \circ g \times g$ is weighted by the forcing weight $w_{d_{\mathbb{P}}^{-}} \circ g = g$. The collection $\mathcal{W}$ of all the weights coincides with the collection of all contractions $g : X \longrightarrow (\mathbb{P}, d_{\mathbb{P}}^{-})$, hence we have $X_{\mathcal{W}}^{in} = X$ and by 11.1.14 $\mathcal{W}$ is strictly monotone for $\leq$.

Next we consider the given contraction f. By 11.1.11 it is monotone as a map $f : (X, \leq) \to (X, \leq)$. Since the fixed element x satisfies $x \leq f(x)$ it follows that the sequence $(a_n)_n$ is monotone and therefore the supremum $a = \bigvee_n a_n$ is well defined. Moreover condition 1. in 11.4.4 is trivially fulfilled. Since the sequence converges in $M(X)$ it also converges in $\mathcal{T}_X$, so condition 2 too is satisfied. Finally in order to prove condition 3 observe that $a_{n-1} \leq a$ and hence $a_n \leq f(a)$ for every $n \in \mathbb{N}$. Finally also $a \leq f(a)$ is fulfilled. So we can apply 11.4.4 in order to conclude that a is a fixed point for f.

Assuming that $x = \bot$ and applying the monotonicity of f we get that $a_n \leq b$ for every fixed point b. So clearly a is the least fixed point. □

As an application of 11.4.3 we recover Scott's least fixed point theorem (Gierz et al. 2003) which, as was shown by Edalat and Heckmann (1998), implies the classical Banach fixed point theorem for Lipschitz functions with Lipschitz factor strictly smaller than 1, on a complete metric space.

11.4.10 Example On a dcpo $(X, \leq)$, by taking for the approach structure the Scott topology on X (embedded in App), both conditions $\mathscr{T}_X \leq M(X)$ and $\leq = \leq_X$ are fulfilled and we obtain the following.

For a continuous map

$$f : (X, \sigma(X)) \longrightarrow (X, \sigma(X)),$$

and a fixed $x \in X$ satisfying $x \leq f(x)$ the sequence $(a_n)_n$ with $a_n = f^n(x)$, obtained by iterating f on x, has a supremum $a = \bigvee_n a_n$ which is a fixed point for f. In case $(X, \leq)$ has a bottom element $\perp$, and choosing $x = \perp$, the supremum a is the least fixed point of f.

11.5 Comments

1. **Weightability in the literature**

In 11.1.1 we adapted the definition of a weightable quasi-metric given by Matthews (1994) and Künzi (2001) to our setting of extended quasi-metrics. Our notion *forcing weight* takes care of the situation with weight values infinity. Matthews (1994) showed that weightable quasi-metric spaces are in one to one correspondence with partial metric spaces. Künzi and Vajner (1994) studied topological spaces that can be induced by a weightable quasi-metric and formulated necessary as well as sufficient conditions on the topology to ensure quasi-metrizability by some weightable quasi-metric. There is a vast literature on the applications of weightable quasi-metrics to domain theory. See for instance Waszkiewicz (2003) and Schellekens (2003). Whenever q is a weightable quasi-metric on a domain X, with weight w_q and inducing the Scott topology then (X, δ_q) with weight $\{w_q\}$ quantifies the domain in our sense.

2. **Complexity**

The complexity quasi-metric is an example of a weightable quasi-metric space that was introduced by Schellekens (2003). Consider $\mathscr{C} \subseteq]0, \infty]^{\mathbb{N}_0}$, such that for all $f \in \mathscr{C}$ the series $\sum_{n \in \mathbb{N}_0} \frac{1}{2^n} \frac{1}{f(n)}$ converges. The *complexity quasi-metric*, $d_{\mathscr{C}}$ on $\mathscr{C}$ is defined as follows: Let $f, g \in \mathscr{C}$

$$d_{\mathscr{C}}(f, g) = \sum_{n \in \mathbb{N}_0} \frac{1}{2^n} (\frac{1}{g(n)} - \frac{1}{f(n)}) \vee 0.$$

This quasi-metric is weightable by the following weight function:

$$w_{\mathscr{C}} : \mathscr{C} \to [0, \infty] : f \mapsto \sum_{n \in \mathbb{N}_0} \frac{1}{2^n} \frac{1}{f(n)}.$$

The space $(\mathscr{C}, d_{\mathscr{C}})$ is called the complexity quasi-metric space. Schellekens et al. use this quasi-metric space to model the complexity of recursive algorithms, such as for example Divide & Conquer algorithms as illustrated in the example below. The basic idea is that the complexity of a recursive algorithm typically is the solution to a recurrence equation based on its recursive structure. The complexity C of a Divide & Conquer algorithm is modelled as the solution to a recurrence equation of the form

$$\begin{cases} C(1) = c \\ C(n) = a.C(\frac{n}{b}) + h(n) & \text{whenever } n \neq 1 \end{cases}$$

with given numbers a, b, c and a function $h : \mathbb{N}_0 \to]0, \infty[$ such that $h(1) = c$. There is an associated self-map Φ on the complexity quasi-metric space $(\mathscr{C}, d_{\mathscr{C}})$ such that the problem of solving the recurrence equation is reduced to finding a fixed point for Φ. The associated self-map Φ on the complexity quasi-metric space is defined by

$$\Phi(g) = \begin{cases} c & \text{if } n = 1 \\ ag(\frac{n}{b}) + h(n) & \text{otherwise} \end{cases}$$

where $g \in \mathscr{C}$. The existence of a fixed point for Φ is proved using a generalization of the Banach fixed point theorem to quasi-metric spaces and by showing that Φ is a $d_{\mathscr{C}}$-Lipschitz function with Lipschitz factor strictly smaller than 1. This holds on condition that $a > 1$.

3. A fixed-point example

Next we consider the same example and we fit the fixed point problem into 11.4.9. Let $X =]0, \infty]^{\mathbb{N}_0}$ and $\Phi : X \to X$ defined as in 2. On $]0, \infty]$ we consider the quasi-metric $p :]0, \infty] \times]0, \infty] \to [0, \infty]$ defined by

$$p(x, y) = (\frac{1}{y} - \frac{1}{x}) \vee 0, \ \ p(0, 0) = 0 \tag{11.1}$$

where we assume $\frac{1}{0} = \infty$. This quasi-metric is weightable by $w_p :]0, \infty] \to [0, \infty]$ defined as $w_p(x) = \frac{1}{x}$ which is clearly forcing for p.

Consider the product approach space on X which is generated by the following gauge subbasis:

$$\{p \circ pr_n \times pr_n \mid n \in \mathbb{N}_0\}$$

with $pr_n :]0, \infty]^{\mathbb{N}_0} \to]0, \infty]$ the projections. A weight for the resulting approach space is

$$\mathscr{W} = \{w_p \circ pr_n = pr_n \mid n \in \mathbb{N}_0\},$$

The approach space X is compatible with $(]0,\infty]^{\mathbb{N}_0}, \leq)$ for the pointwise order and its topological coreflection is coarser than the Martin topology. It can be shown that Φ is a contraction $X \longrightarrow X$ on condition that $a \geq 1$. Thus we can apply 11.4.9 to conclude that Φ has a fixed point on X. Remark that this result is better than the one obtained under 2. This result can still be improved by using a more suitable approach setting as was done in Colebunders et al. (2014) where it was shown that even $a > 0$ is sufficient.

4. **Another fixed-point example**

Next we consider an application of 11.4.6. By making the right choices for the approach space X and for the collection $\mathscr{W}$ we can recover a fixed point theorem proved by Martin (2000a).

Let $(X, \leq)$ be a domain with a measurement $\mu : X \longrightarrow \mathbb{P}^{op}$ such that $\sigma(X) \subseteq \mathscr{T}_{\mu}^{in} \vee \downarrow\mathscr{T}$. If $f : (X, \leq) \longrightarrow (X, \leq)$ is splitting and

$$\mu \circ f : (X, M(X)) \longrightarrow (\mathbb{P}^{op}, \sigma(\mathbb{P}^{op}))$$

is continuous, then for every $x \in X$ the sequence $(a_n)_n$ obtained by iterating f on x has a supremum which is a fixed point for f.

Indeed, let $\mathscr{W} = \{\mu\}$ and observe that as was proved by Martin, $\mathscr{W}$ is strictly monotone. Consider $M(X)$ embedded as an approach space X, then $\mathscr{T}_X = M(X)$. The condition that $\mu \circ f$ is continuous from $M(X)$ to $\mathbb{P}^{op}$ is equivalent to saying that $f : M(X) \longrightarrow (X, \mathscr{T}_{\mathscr{W}}^{in})$ is continuous. Since Top is concretely coreflective in App this in turn is equivalent to $f : X \longrightarrow X_{\mathscr{W}}^{in}$ being contractive.

5. **And yet another fixed-point example**

By making other choices for the approach spaces on X we recover another fixed point theorem of Martin (2000b).

Let $(X, \leq)$ be a domain with a measurement $\mu : X \longrightarrow \mathbb{P}^{op}$ such that $\sigma(X) \subseteq \mathscr{T}_{\mu}^{in} \vee \downarrow\mathscr{T}$. If $I \subseteq X$ is nonempty and closed under directed suprema, $f : I \longrightarrow I$ is splitting and

$$\mu \circ f : I \longrightarrow \mathbb{P}^{op}$$

is Scott continuous, then choosing $x \in I$ the sequence $(a_n)_n$ obtained by iterating f on x has a supremum which is a fixed point for f.

Indeed, again let $\mathscr{W} = \{\mu\}$, but this time the approach space X is the embedding in App of the Scott topology on $(X, \leq)$. The condition that $\mu \circ f$ is continuous from $\sigma(X)$ to $\mathbb{P}^{op}$ is equivalent to $f : X \longrightarrow X_{\mathscr{W}}^{in}$ being contractive.

Chapter 12
Categorical Considerations

Although extensions of constructs follow the rule "bigger is better", i.e. stronger convenience stipulations require bigger extensions, they also follow the rule "smaller is better", i.e. smaller extensions generally preserve more structure of the original construct.

(Horst Herrlich)

It is better to have a good category with bad objects than a bad category with good objects.

(Alexander Grothendieck)

In this chapter we treat several different categorical aspects of the theory of approach spaces.

First, we have seen that Top is a simultaneously concretely reflective and coreflective subcategory of App. We call subcategories having this behaviour *stable*. This situation is not common, it is indeed well known that in many familiar topological categories, such as e.g. Top, PrTop, Unif, Bor, there do not exist non-trivial stable subcategories (see e.g. Giuli 1983; Hušek 1973; Kannan 1972). The situation is quite different in App, qMet and in Met where in all cases an infinite collection of stable subcategories exists as we will see in the first section of this chapter.

Second, App is a topological category which is neither cartesian closed nor extensional and hence also not a quasi-topos. It is well known that categories, under mild conditions which are fulfilled in the case of App, can have "smallest" cartesian closed, extensional and quasi-topos hulls wherein they are fully embedded. For the case of Top we refer to Antoine (1966c), Bourdaud (1976), Herrlich (1988a, b), Machado (1973) and Wyler (1976, 1991). In the second section we begin by constructing a large quasi-topos supercategory of App which will serve as a starting-point for the construction of all the hulls. In the third section we first construct the extensional topological hull, PrAp. In Sects. 12.4 and 12.5 we then construct the quasi-topos hull of App making use of the fact that this can be done in a two-step process, first making the extensional topological hull PrAp and then taking the cartesian closed topological hull of PrAp (see Schwarz 1989). In Sect. 12.6 we then construct the cartesian closed topological hull of App. We also show the close relationship between the hulls

R. Lowen, *Index Analysis*, Springer Monographs in Mathematics,
DOI 10.1007/978-1-4471-6485-2_12

of App and those of Top which actually can all be described as stable subcategories of the corresponding hulls of App.

Third, App can also be described as the category of lax algebras for a particular extension of the ultrafilter monad. See Clementino and Hofmann (2003) and Clementino et al. (2004). We give a new proof of this result in the last section.

12.1 Stable Subcategories of App, qMet and Met

A central role in our investigations will be played by certain subsemigroups and certain subadditive functions. We shall therefore at first collect the properties of these concepts which are needed in the sequel. In what follows Ω will stand for a complete lattice ordered commutative semigroup with a bottom element which is at the same time the neutral element for the semigroup.

For functions $\gamma \in \Omega^\Omega$ we consider the following properties.

(S1) γ is increasing.
(S2) $\gamma \geq 1_\Omega$.
(S3) γ is sup-preserving.
(S4) γ is subadditive.
(S5) $\gamma \circ \gamma = \gamma$.

A function in Ω^Ω satisfying (S1)-(S5) will be called a *suitable subadditive function* (or shortly suitable), and we denote by Φ_Ω the set of all suitable subadditive functions in Ω^Ω.

Likewise a subset of Ω which is at the same time a closed sublattice and a subsemigroup and which contains both the top and bottom elements of Ω is called a *suitable subsemigroup* (or shortly suitable), and we denote the set of all such suitable subsemigroups of Ω by S_Ω.

For $\gamma \in \Phi_\Omega$ we define

$$\Gamma_\gamma := \mathrm{Fix}\gamma = \{x \in \Omega \mid \gamma(x) = x\}$$

and for $\Gamma \in S_\Omega$ we define the function γ_Γ as

$$\gamma_\Gamma : \Omega \longrightarrow \Omega : x \mapsto \inf\{y \in \Gamma \mid x \leq y\}.$$

Φ_Ω and S_Ω are partially ordered by respectively the pointwise order and the inclusion order.

12.1.1 Lemma *The following properties hold for $\Gamma \in S_\Omega$ and $\gamma \in \Phi_\Omega$.*

1. $\Gamma_\gamma = \mathrm{Im}\gamma$.
2. $x \in \Gamma \Leftrightarrow \gamma_\Gamma(x) = x$.
3. $\mathrm{Im}\gamma_\Gamma = \Gamma$.

Proof This follows from the definitions. □

12.1.2 Proposition *The maps*

$$S_\Omega \longrightarrow \Phi_\Omega : \Gamma \mapsto \gamma_\Gamma \text{ and } \Phi_\Omega \longrightarrow S_\Omega : \gamma \mapsto \Gamma_\gamma$$

are mutually inverse and order-preserving.

Proof This follows by applying 12.1.1. □

In this section we will encounter two types of lattice ordered semigroups Ω, a one-dimensional one which will provide the basis for the cases of approach spaces and of metric spaces, and a two-dimensional one which will be the basis for the case of quasi-metric spaces.

For the time being, Ω will be $\mathbb{P}$ with its usual additive semigroup, order and Euclidean topology which we have been using throughout. Therefore, in this case, $\Gamma \subseteq \mathbb{P}$ is closed in the lattice-theoretical sense, if and only if it is closed in the topological sense.

12.1.3 Lemma *If $\Gamma \subseteq \mathbb{P}$ is a suitable subsemigroup, then either $\Gamma = \mathbb{P}$ and then $\gamma_\Gamma = 1_\Gamma$, or $m = inf(\Gamma \setminus \{0\}) > 0$ and then for $x \in \Gamma$ we have*

$$\gamma_\Gamma(x) < m \Leftrightarrow \gamma_\Gamma(x) = 0 \Leftrightarrow x = 0.$$

Proof This is straightforward and we leave this to the reader. □

12.1.4 Lemma *If $\Gamma \subseteq \mathbb{P}$ is a suitable subsemigroup and if δ is a distance on X, then $\delta_\Gamma := \gamma_\Gamma \circ \delta$ too is a distance on X.*

Proof (D1) and (D2) are trivial, and (D3) follows immediately from (S1). As to (D4), this follows from (S3) and (S4) making use of 1.1.2. □

12.1.5 Proposition *If $\Gamma \subseteq \mathbb{P}$ is a suitable subsemigroup then for any approach space X the following properties are equivalent.*

1. $\mathrm{Im}\delta \subseteq \Gamma$.
2. $\forall \mu \in \mathfrak{L} : \gamma_\Gamma \circ \mu \in \mathfrak{L}$.

Proof 1 ⇒ 2. We know from 1.3.5 that $\mu \in \mathfrak{L}$ if and only if $\mu : X \to \mathbb{P}$ is a contraction, and so we have to verify that

$$(\gamma_\Gamma \circ \mu(x) - \sup \gamma_\Gamma \circ \mu(A)) \vee 0 \leq \delta(x, A)$$

for all $x \in X$, $A \subseteq X$. Since $\mu(x) \leq \sup \mu(A) + \delta(x, A)$, it follows from the fact that $\delta(x, A) \in \Gamma$ that

$$\gamma_\Gamma \circ \mu(x) \leq \gamma_\Gamma(\sup \mu(A)) + \delta(x, A) \leq \sup \gamma_\Gamma \circ \mu(A) + \delta(x, A).$$

2 $\Rightarrow$ 1. We know from 1.2.45 that $\delta(x, A) = \sup\{\rho(x) | \rho \in \mathfrak{L}, \rho|_A = 0\}$. It follows from 12.1.3 that $\rho|_A = 0 \Leftrightarrow \gamma_\Gamma \circ \rho|_A = 0$. Hence, since $\rho \leq \gamma_\Gamma \circ \rho$ we have $\delta(x, A) = \sup\{\gamma_\Gamma \circ \rho(x) | \rho \in \mathfrak{L}, \gamma_\Gamma \circ \rho|_A = 0\}$, and therefore $\delta(x, A) \in \Gamma$ by closedness of Γ. □

If Γ is a suitable subsemigroup, we write App_Γ for the full subcategory of App with objects those (X, δ) for which $\mathrm{Im}\,\delta \subseteq \Gamma$.

12.1.6 Theorem *If $\Gamma \subseteq \mathbb{P}$ is a suitable subsemigroup then* App_Γ *is a stable subcategory of* App.

Proof It is sufficient to prove that App_Γ is closed under the formation of initial and final structures.

In order to prove that App_Γ is closed under the formation of initial structures we recall from 1.3.17 that if $(f_j : X \longrightarrow (X_j, \delta_j))_{j \in J}$ is a source in App then the initial distance is given by the formula

$$\delta_{in}(x,A) := \sup_{\mathscr{P} \in \mathsf{P}(A)} \min_{P \in \mathscr{P}} \sup_{j \in J} \delta_j(f_j(x), f_j(P))$$

where $\mathsf{P}(A)$ is the set of finite covers of A by means of subsets of A.

If the source is taken in App_Γ, which means that all distances have their images contained in Γ, it follows from the formula for δ_{in} and by closedness of Γ, that the image of the initial distance too is contained in Γ.

In order to prove that App_Γ is closed under the formation of final structures, we use the characterization of approach spaces by means of lower regular function frames. So let

$$(f_j : (X_j, \mathscr{R}_j) \longrightarrow X)_{j \in J}$$

be a sink in App_Γ. From 1.3.13 it follows that the final lower regular function frame $\mathscr{R}_{fin}$ on X is given by

$$\mathscr{R}_{fin} := \{\mu \in \mathbb{P}^X \mid \forall j \in J : \mu \circ f_j \in \mathscr{R}_j\},$$

and hence the result follows immediately from 12.1.5. □

We now proceed to the converse of the foregoing result.

12.1.7 Theorem *If $\mathfrak{S}$ is a stable subcategory of* App *then it is of type* App_Γ *for some suitable subsemigroup $\Gamma \subseteq \mathbb{P}$.*

Proof First, note that since a concretely coreflective subcategory necessarily contains the subcategory Dis of discrete spaces and since Top has no stable subcategories, it follows that Top is a subcategory of $\mathfrak{S}$.

Second, we define

$$\Gamma_{\mathfrak{S}} := \bigcup_{(X,\delta) \in |\mathfrak{S}|} \mathrm{Im}\,\delta.$$

Third, in what follows we will denote by (x,y,α,β) the quasi-metric space with underlying set $\{x,y\}$ and distance from x to y equal to α and from y to x equal to β.

Fourth we can apply 12.1.4 to $\mathbb{P}$ and we write $\mathbb{P}_\Gamma$ for the approach space $(\mathbb{P}, \delta_{\mathbb{P},\Gamma})$ where we recall that $\delta_{\mathbb{P},\Gamma} := \gamma_\Gamma \circ \delta_{\mathbb{P}}$.

The proof of the theorem now requires a number of lemmas.

12.1.8 Lemma *If $a \in \Gamma_{\mathfrak{S}}$, there exists an object $(\{x,y\},\delta)$ in $\mathfrak{S}$ for which*

$$\delta(x,\{y\}) = a.$$

Proof Indeed, since $a \in \Gamma_{\mathfrak{S}}$, there exist $(X,\delta') \in |\mathfrak{S}|$, $A \subseteq X$, $z \in X$ such that $\delta'(z,A) = a$. Put $Y := A \cup \{z\}$, then taking first the subspace $(Y,\delta'|_Y)$, and then the quotient $f : (Y,\delta'|_Y) \to (\{x,y\},\delta)$ with $f(z) = x$, $f(A) = \{y\}$ gives the desired object. □

12.1.9 Lemma *If $a \in \Gamma_{\mathfrak{S}}$, then both $(x,y,a,0)$ and (x,y,a,∞) are in $\mathfrak{S}$.*

Proof Since $(x,y,0,\infty)$ and $(x,y,\infty,0)$ are in Top, the result follows by taking the supremum respectively infimum of each of these structures with the one obtained in 12.1.8. □

12.1.10 Lemma *$\Gamma_{\mathfrak{S}}$ is a suitable subsemigroup.*

Proof That $\{0,\infty\} \subseteq \Gamma_{\mathfrak{S}}$ follows from the fact that Top is a subcategory of $\mathfrak{S}$.

If $a_1 \in \Gamma_{\mathfrak{S}}$, $a_2 \in \Gamma_{\mathfrak{S}}$, and both (x,y,a_1,b_1) and (x,y,a_2,b_2) are in $\mathfrak{S}$, it is readily verified that the sink

$$(f_k : (x,y,a_k,b_k) \to (\{u,v,w\},\delta))_{k\in\{1,2\}}$$

with $f_1(x) = u$, $f_1(y) = f_2(x) = v$, $f_2(y) = w$ and $\delta(u,\{v\}) = a_1$, $\delta(v,\{w\}) = a_2$, $\delta(u,\{w\}) = a_1 + a_2$, $\delta(v,\{u\}) = b_1$, $\delta(w,\{v\}) = b_2$, $\delta(w,\{u\}) = b_1 + b_2$, is final. Hence $a_1 + a_2 \in \Gamma_{\mathfrak{S}}$ which proves that $\Gamma_{\mathfrak{S}}$ is a subsemigroup.

If $\emptyset \neq A \subseteq \Gamma_{\mathfrak{S}}$, then from 12.1.8 it follows that we can consider the initial source

$$(1_{\{x,y\}} : (\{x,y\},\delta') \longrightarrow (\{x,y\},\delta_a))_{a\in A}$$

and the final sink

$$(1_{\{x,y\}} : (\{x,y\},\delta_a) \longrightarrow (\{x,y\},\delta''))_{a\in A},$$

with $\delta_a(x,\{y\}) = a$ for all $a \in A$. It is clear that $\delta'(x,\{y\}) = \sup A$ and $\delta''(x,\{y\}) = \inf A$, and so $\Gamma_{\mathfrak{S}}$ is a closed sublattice. □

12.1.11 Lemma *If (X,δ) is in App_Γ, then $\delta_{A,\Gamma} : X \longrightarrow \mathbb{P}_\Gamma : x \mapsto \delta_{\mathbb{P},\Gamma}(x,A)$ is a contraction for all $A \subseteq X$.*

Proof Take $x \in X$, $B \subseteq X$. If $B = \emptyset$, there is nothing to show. If $B \neq \emptyset$, then, from the supposition and making use of 12.1.4

$$\begin{aligned}\delta_{\mathbb{P},\Gamma}(\delta_{A,\Gamma}(x), \delta_{A,\Gamma}(B)) &= \gamma_\Gamma((\gamma_\Gamma \circ \delta(x,A) - \sup_{b\in B} \gamma_\Gamma \circ \delta(b,A)) \vee 0)\\ &= \gamma_\Gamma((\delta(x,A) - \sup_{b\in B} \delta(b,A)) \vee 0)\\ &\leq \gamma_\Gamma(\delta(x,B) \vee 0) = \gamma_\Gamma(\delta(x,B)) = \delta(x,B).\end{aligned}$$

□

12.1.12 Lemma *$\mathbb{P}_\Gamma$ is initially dense in* App_Γ.

Proof The proof is analogous to the one of 1.3.19 and therefore we omit it. Note however that it actually also follows from that same result upon observing that the App_Γ-coreflection of an approach space (X, δ) is given by

$$1_X : (X, \gamma_\Gamma \circ \delta) \longrightarrow (X, \delta).$$

□

12.1.13 Lemma *If $\mathfrak{S}$ is a stable subcategory of* App *then $\mathbb{P}_{\Gamma_\mathfrak{S}}$ is in $\mathfrak{S}$.*

Proof We will use the function $\delta_\infty : \mathbb{P} \times 2^\mathbb{P} \longrightarrow \mathbb{P}$, defined by

$$\delta_\infty(x,A) = \begin{cases} 0 & \text{if } x \leq \sup A \text{ and } A \neq \emptyset,\\ \infty & \text{if } x > \sup A \text{ or } A = \emptyset,\end{cases}$$

which is a topological distance. We denote the space $(\mathbb{P}, \delta_\infty)$ by $\mathbb{P}_\infty$.

From 12.1.9 it follows that we can consider the sink consisting of all maps

$$(f_{a,b} : (a,b,0,\gamma_{\Gamma_\mathfrak{S}}(b-a)) \longrightarrow \mathbb{P})_{0\leq a<b\leq\infty}$$

where $f_{a,b}(a) = a$, $f_{a,b}(b) = b$, together with

$$1_\mathbb{P} : \mathbb{P}_\infty \longrightarrow \mathbb{P}.$$

Let δ_{fin} be the final distance. We will show that this distance coincides with $\delta_{\mathbb{P},\Gamma_\mathfrak{S}}$. Hereto take $z \in \mathbb{P}$, $\emptyset \neq A \subseteq \mathbb{P}$. Four cases have to be considered.

(i) $z < \infty$, $z < \sup A$. Take $t \in A$, $z < t$. Since $f_{z,t}$ is a contraction, we have $\delta_{fin}(z,\{t\}) = 0$, and thus $\delta_{fin}(z,A) \leq \inf_{a\in A} \delta_{fin}(z,\{a\}) = 0$, hence $\delta_{fin}(z,A) = 0 = \delta_{\mathbb{P},\Gamma_\mathfrak{S}}(z,A)$.

(ii) $\sup A \leq z < \infty$. If $A_{fin}^{(0)} = \{t \mid \delta_{fin}(t,A) = 0\}$ then $\delta_{fin}(z,A) = \delta_{fin}(z,A_{fin}^{(0)})$. If $t \leq \sup A$ (i.e. $\delta_\infty(t,A) = 0$), we have $\delta_{fin}(t,A) = 0$ since the identity is a contraction, and therefore $B := [0, \sup A] \subseteq A_{fin}^{(0)}$. Since $\sup B = \sup A$, we now obtain that

$$\begin{aligned}\delta_{fin}(z,A) &= \delta_{fin}(z,A^{(0)}_{fin}) \leq \delta_{fin}(z,B)\\ &\leq \inf_{t\in B} \delta_{fin}(z,\{t\}) \leq \inf_{t\in B} \gamma_{\Gamma_{\mathfrak{S}}}(z-t)\\ &= \delta_{\mathbb{P},\Gamma_{\mathfrak{S}}}(z,B) = \gamma_{\Gamma_{\mathfrak{S}}}(z-\sup B) = \gamma_{\Gamma_{\mathfrak{S}}}(z-\sup A)\\ &= \delta_{\mathbb{P},\Gamma_{\mathfrak{S}}}(z,A)\end{aligned}$$

(iii) $z = \infty = \sup A$. Then $\delta_\infty(z,A) = 0$ and so $\delta_{fin}(z,A) = 0 = \gamma_{\Gamma_{\mathfrak{S}}}(z,A)$.
(iv) $\sup A < \infty = z$. Then $\delta_{fin}(z,A) \leq \infty = \delta_{\mathbb{P},\Gamma_{\mathfrak{S}}}(z,A)$.

Hence $\delta_{fin} \leq \delta_{\mathbb{P},\Gamma}$ and since $\delta_{\mathbb{P},\Gamma_{\mathfrak{S}}}$ is a distance on $\mathbb{P}$ for which all the functions in the sink are contractions, it necessarily is the final distance. □

To conclude the proof of the theorem, on the one hand note that from the definition of $\Gamma_{\mathfrak{S}}$, obviously $\mathfrak{S}$ is a subcategory of $\mathsf{App}_{\Gamma_{\mathfrak{S}}}$. On the other hand by 12.1.12 and 12.1.13 the object $\mathbb{P}_{\Gamma_{\mathfrak{S}}}$ is initially dense in $\mathsf{App}_{\Gamma_{\mathfrak{S}}}$ and belongs to $\mathfrak{S}$. Hence, by stability, it follows that $\mathsf{App}_{\Gamma_{\mathfrak{S}}}$ is a subcategory of $\mathfrak{S}$. □

12.1.14 Theorem *The collection of all stable subcategories of* App *is the set* $\{\mathsf{App}_\Gamma \mid \Gamma \subseteq \mathbb{P}$ suitable subsemigroup$\}$.

Proof This follows from 12.1.6 and 12.1.7. □

12.1.15 Proposition *If Γ and Γ' are suitable subsemigroups, the subcategories App_Γ and $\mathsf{App}_{\Gamma'}$ are concretely isomorphic if and only if there exists an $m \in]0,\infty[$ such that $\Gamma' = m\Gamma$, or equivalently such that $\gamma_{\Gamma'}(x) = m\gamma_\Gamma(x/m)$ for all x.*

Proof That the condition is sufficient is evident.

Let $\mathsf{F} : \mathsf{App}_\Gamma \longrightarrow \mathsf{App}_{\Gamma'}$ be a concrete isomorphism. Given a two-point set $\{x,y\}$ it then defines a bijection between the approach structures on $\{x,y\}$ respectively in App_Γ and $\mathsf{App}_{\Gamma'}$. Thus it induces a bijection

$$\lambda : \Gamma^2 \longrightarrow \Gamma'^2 : (a,b) \mapsto \lambda(a,b) = (\lambda_1(a,b), \lambda_2(a,b))$$

such that

$$\mathsf{F}((x,y,a,b)) = (x,y,\lambda_1(a,b), \lambda_2(a,b)).$$

Since the isomorphism $(x,y,a,b) \longrightarrow (x,y,b,a) : x \mapsto y,\ y \mapsto x$ is mapped by F onto an isomorphism, it follows immediately that $\lambda_1(a,b) = \lambda_2(b,a)$, and therefore $\lambda_1(a,a) = \lambda_2(a,a)$. This means that F also induces a map

$$\mu : \Gamma \longrightarrow \Gamma' : a \mapsto \mu(a) \text{ such that } \lambda(a,a) = (\mu(a), \mu(a)).$$

If now $a < b$, then the map $1_{\{x,y\}} : (x,y,a,a) \longrightarrow (x,y,b,b)$ is not a morphism, and so neither is its image, and therefore $\mu(a) < \mu(b)$, which implies that μ is a bijection.

Since $\Gamma = \{0, \infty\}$ is the only suitable subsemigroup with a finite number of elements, it follows that in that case also $\Gamma' = \{0, \infty\}$, in which case the condition is trivially satisfied and hence we can, from now on, suppose $\Gamma \neq \{0, \infty\}$ (and therefore also $\Gamma' \neq \{0, \infty\}$).

The final sink in part (2) of the proof of 12.1.10 is mapped by F onto a final sink, and therefore $\lambda(a_1 + a_2, b_1 + b_2) = \lambda(a_1, b_1) + \lambda(a_2, b_2)$.

From this it follows that $\mu(a_1 + a_2) = \mu(a_1) + \mu(a_2)$, and therefore $\mu(0) = 0$, $\mu(\infty) = \infty$. Moreover we have $\mu(a) \neq 0$ for all $a \in \Gamma \setminus \{0, \infty\}$ and $\mu(pa) = p\mu(a)$ for all $a \in \Gamma$, $p \in \mathbb{N}$.

Finally now, let $a \in \Gamma \setminus \{0, \infty\}$, $b \in \Gamma \setminus \{0, \infty\}$ with $\mu(a) = ma$, $\mu(b) = nb$ such that $m \neq n$, e.g. $m < n$. Then $\frac{ma}{nb} < \frac{a}{b}$, and so there exist $p \in \mathbb{N}, q \in \mathbb{N}$ such that $\frac{ma}{nb} < \frac{p}{q} < \frac{a}{b}$. From this it follows that $qma < pnb$ and $pb < qa$ so that $\mu(qa) = qma < pnb = \mu(pb)$ and $pb < qa$. However, then

$$1_{\{x,y\}} : (x,y,qa,qa) \longrightarrow (x,y,pb,pb)$$

is a morphism whilst

$$1_{\{x,y\}} : (x,y,\mu(qa),\mu(qa)) \longrightarrow (x,y,\mu(pb),\mu(pb))$$

is not, and this contradiction shows that $m = n$. □

12.1.16 Proposition *The set of stable subcategories of* App *has cardinality* $2^{\aleph_0}$*, and there is a set of the same cardinality of non-concretely isomorphic such subcategories.*

Proof The set of suitable subsemigroups is easily seen to have cardinality $2^{\aleph_0}$. Since each $\Gamma_a := \{p + qa \mid p \in \mathbb{N}, q \in \mathbb{N}, q \geq 1\}$ with $a \in]0, 1[\cap(\mathbb{R} \setminus \mathbb{Q})$ is a suitable subsemigroup and since no two different such subsemigroups satisfy the condition in the foregoing proposition, the result follows. □

Note that the only subcategory in the family $(\mathsf{App}_\Gamma)_\Gamma$, which is a supercategory of both Top and qMet, is App. It is not even necessary to make the requirement of being a supercategory of the whole of qMet, just Met, or the unique object $(\mathbb{R}, \delta_{d_\mathbb{E}})$, already suffice to force the conclusion. In this sense, App is the smallest reasonable extension of Top and qMet.

We now turn our attention to qMet. Now Ω will be $\mathbb{P}^2$ with the product semigroup structure and the product lattice structure. We will denote the partial order on Ω by $\preceq$ so that $(a,b) \preceq (c,d)$ if and only if $a \leq c$ and $b \leq d$. A suitable subsemigroup Γ of Ω will be called *symmetric* if $(a,b) \in \Gamma \Leftrightarrow (b,a) \in \Gamma$.

If $\Gamma \subseteq \Omega$ is symmetric, then $\mathrm{pr}_1(\Gamma) = \mathrm{pr}_2(\Gamma)$ and we will simply write $\mathrm{pr}(\Gamma)$ for both.

If $\mathfrak{S}$ is a stable subcategory of qMet, we define a function $\varphi_\mathfrak{S} \in \Omega^\Omega$ as follows. Let $\{x, y\}$ be an arbitrary two-point set, then obviously for any $a,b \in \mathbb{P}$ the space (x,y,a,b) is in qMet. Consider its $\mathfrak{S}$-coreflection

$$1_{\{x,y\}} : (x,y,\varphi_1(a,b),\varphi_2(a,b)) \longrightarrow (x,y,a,b)$$

and then define

$$\varphi_{\mathfrak{S}} : \Omega \longrightarrow \Omega : (a,b) \mapsto (\varphi_1(a,b), \varphi_2(a,b)).$$

12.1.17 Lemma *If (X,d) is in qMet, $x,y \in X$ and*

$$1_{\{x,y\}} : (\{x,y\}, d_{\mathfrak{S}}) \to (\{x,y\}, d)$$

is the $\mathfrak{S}$-coreflection then

$$d_{\mathfrak{S}}(x,y) = \varphi_1(d(x,y), d(y,x)) \text{ and } d_{\mathfrak{S}}(y,x) = \varphi_2(d(x,y), d(y,x)).$$

Proof Put $a := d(x,y)$ and $b := d(y,x)$. By coreflection the initial source $j : (x,y,a,b) \to (X,d)$ gives the initial source

$$j : (x,y, \varphi_1(a,b), \varphi_2(a,b)) \to (X, d_{\mathfrak{S}}).$$

□

12.1.18 Lemma *For all $(a,b) \in \Omega$, we have $\varphi_1(b,a) = \varphi_2(a,b)$.*

Proof Consider the isomorphism $(x,y,a,b) \longrightarrow (x,y,b,a) : x \mapsto y, y \mapsto x$ and its image by the $\mathfrak{S}$-coreflection. □

12.1.19 Lemma *For any stable subcategory $\mathfrak{S}$ of qMet, $\varphi_{\mathfrak{S}}$ is a suitable subadditive function and hence*

$$\Gamma_{\varphi_{\mathfrak{S}}} = \mathrm{Fix}\varphi_{\mathfrak{S}} = \{(a,b) \mid (x,y,a,b) \in \mathfrak{S}\}$$

is a symmetric suitable subsemigroup.

Proof (S1), given $(a,b) \prec (c,d)$, follows by considering the morphism

$$1_{\{x,y\}} : (x,y,c,d) \longrightarrow (x,y,a,b)$$

and its image by the $\mathfrak{S}$-coreflection and (S2) follows immediately from the definition.

To prove (S3) note first that if $A = \emptyset$ we have $\sup A = (0,0) = \varphi_{\mathfrak{S}}(0,0)$ since $\mathfrak{S}$ contains all indiscrete spaces. Second, if $A \neq \emptyset$, it suffices to consider the initial source

$$(1_{\{x,y\}} : (x,y, \sup_{(a',b)\in A} a', \sup_{(a,b')\in A} b') \longrightarrow (x,y,a,b))_{(a,b)\in A}$$

and its image by coreflection.

To prove (S4), if $(a_1,b_1) \in \Gamma$, $(a_2,b_2) \in \Gamma$, consider the final sink

$$(f_k : (x,y,a_k,b_k) \longrightarrow (\{u,v,w\}, \delta))_{k\in\{1,2\}}$$

with $f_1(x) = u$, $f_1(y) = f_2(x) = v$, $f_2(y) = w$. It is readily verified (see also 12.1.10), that $\delta(u,\{v\}) = a_1$, $\delta(v,\{w\}) = a_2$, $\delta(u,\{w\}) = a_1 + a_2$, $\delta(v,\{u\}) = b_1$,

$\delta(w, \{v\}) = b_2$, $\delta(w, \{u\}) = b_1 + b_2$. The result then follows from the fact that the coreflection of a final sink is final too.

To prove (S5) note that by definition the fact that $(x, y, \varphi_1(a,b), \varphi_2(a,b))$ and (x,y,a,b) are in $\mathfrak{S}$ is equivalent to $\varphi_{\mathfrak{S}}(a,b) = (a,b)$.

The final claim is an immediate consequence of 12.1.2. □

12.1.20 Lemma *If $(a,b) \in \Gamma$, then also $(a,a) \in \Gamma$.*

Proof By symmetry and closedness of Γ. □

Note that if we define $\Delta_\Gamma := \{(a \in \mathbb{P} \mid (a,a) \in \Gamma\}$, it follows that

$$\Delta_\Gamma = \bigcup_{(X,d)\in\mathfrak{S}} \operatorname{Im} d$$

but that it is not true that $a \in \Delta_\Gamma$, $b \in \Delta_\Gamma \Rightarrow (a,b) \in \Gamma$.

12.1.21 Definition If $\Gamma \subseteq \Omega$ is a suitable subsemigroup, we define $q\mathsf{Met}_\Gamma$ to be the full subcategory of $q\mathsf{Met}$ containing all (X,d) such that for all $(x,y) \in X \times X$: $(d(x,y), d(y,x)) \in \Gamma$.

If, for any space (X,d) we define

$$\hat{d} : X \times X \longrightarrow \mathbb{P} \times \mathbb{P} : (x,y) \mapsto (d(x,y), d(y,x))$$

then (X,d) will be in $q\mathsf{Met}_\Gamma$ if and only if $\operatorname{Im}\hat{d} \subseteq \Gamma$.

12.1.22 Theorem *For any symmetric suitable subsemigroup Γ, $q\mathsf{Met}_\Gamma$ is a stable subcategory of $q\mathsf{Met}$.*

Proof For any object (X,d) in $q\mathsf{Met}$ we define $d_\Gamma : X \times X \longrightarrow \Gamma$ by

$$(d_\Gamma(x,y), d_\Gamma(y,x)) = \gamma_\Gamma(d(x,y), d(y,x)).$$

From the subadditivity of γ_Γ it follows that d_Γ in turn is a quasi-metric, and it is then easily verified that $1_X : (X,d_\Gamma) \longrightarrow (X,d)$ is the $q\mathsf{Met}_\Gamma$-coreflection.

Since the initial quasi-metric for the source $(f_j : X \longrightarrow (X_j,d_j))_{j\in J}$ is given by $d(x,y) = \sup_{j\in J} d_j(f_j(x), f_j(y))$, reflectivity follows from the definition of $q\mathsf{Met}_\Gamma$. □

12.1.23 Theorem *If $\mathfrak{S}$ is a stable subcategory of $q\mathsf{Met}$ then it is of type $q\mathsf{Met}_\Gamma$ for some symmetric suitable subsemigroup Γ.*

Proof If $\mathfrak{S}$ is a stable subcategory of $q\mathsf{Met}$, we know that $\Gamma_{\varphi_\mathfrak{S}}$ is a symmetric suitable subsemigroup, and so it defines the stable subcategory $q\mathsf{Met}_{\Gamma_{\varphi_\mathfrak{S}}}$. Since $\gamma_{\Gamma_{\varphi_\mathfrak{S}}} = \varphi_\mathfrak{S}$, it follows that the objects which are invariant by coreflection are the same in either case, and so $\mathfrak{S} = q\mathsf{Met}_{\Gamma_{\varphi_\mathfrak{S}}}$. □

12.1.24 Theorem *The collection of all stable subcategories of qMet is the set* $\{q\mathsf{Met}_\Gamma \mid \Gamma \subseteq \mathbb{P}^2$ suitable symmetric subsemigroup$\}$.

Proof This follows from 12.1.22 and 12.1.23. □

In the proof of the next result we will make use of the following additional properties of the function $\gamma = \gamma_\Gamma$:

1. For all $a \in \mathbb{P} : \gamma_1(\infty, a) = \gamma_2(a, \infty) = \infty$.
2. If $a < b$, then $\gamma_1(a,b) \leq \gamma_2(a,b)$.

To see (2) it is sufficient to consider the diagram

$$\begin{array}{ccc} (x,y,\gamma_1(a,b),\gamma_2(a,b)) & \xrightarrow{\;c\;} & (x,y,a,b) \\ {\scriptstyle 1_{\{x,y\}}}\big\downarrow & \nearrow{\scriptstyle 1_{\{x,y\}}} & \\ (x,y,\gamma_2(a,b),\gamma_1(a,b)) & & \end{array}$$

where c stands for the coreflection.

12.1.25 Proposition *If Γ and Γ' are symmetric suitable subsemigroups, the subcategories $q\mathsf{Met}_\Gamma$ and $q\mathsf{Met}_{\Gamma'}$ are concretely isomorphic if and only if there exists an $m \in]0, \infty[$ such that $\Gamma' = m\Gamma$.*

Proof That the condition is sufficient is evident. In order to prove the necessity, we proceed step by step.

1. Let $\mathsf{F} : q\mathsf{Met}_\Gamma \longrightarrow q\mathsf{Met}_{\Gamma'}$ be a concrete isomorphism. As in the proof of 12.1.15 it defines a bijection between the quasi-metrics in $q\mathsf{Met}_\Gamma$ and $q\mathsf{Met}_{\Gamma'}$ on a two-point set $\{x,y\}$. Hence it induces a bijection

$$\lambda = \lambda_{\mathsf{F}} : \Gamma \longrightarrow \Gamma' : (a,b) \mapsto \lambda(a,b) = (\lambda_1(a,b), \lambda_2(a,b))$$

such that

$$\mathsf{F}((x,y,a,b)) = (x,y,\lambda_1(a,b),\lambda_2(a,b)).$$

As in 12.1.15 it follows that $\lambda_1(a,b) = \lambda_2(b,a)$, and therefore $\lambda_1(a,a) = \lambda_2(a,a)$, and so F also induces a map

$$\mu = \mu_{\mathsf{F}} : \mathrm{pr}(\Gamma) \longrightarrow \mathrm{pr}(\Gamma') : a \mapsto \mu(a)$$

such that $\lambda(a,a) = (\mu(a), \mu(a))$, and which once more is a bijection.

2. If (X,d) is in $q\mathsf{Met}_\Gamma$, $\{x,y\} \subseteq X$, $d(x,y) = a$, $d(y,x) = b$, the initial source $j : (x,y,a,b) \to (X,d)$ is mapped by F onto the initial source

$$j : (x,y,\lambda_1(a,b),\lambda_2(a,b)) \to \mathsf{F}(X,d) = (X, \mathsf{F}(d)),$$

and so $\mathsf{F}(d)(x,y) = \lambda_1(a,b)$, $\mathsf{F}(d)(y,x) = \lambda_2(a,b)$.

3. Since $\Gamma_2 = \{(0,0), (\infty,\infty)\}$ and $\Gamma_4 = \{(0,0), (\infty,\infty), (0,\infty), (\infty,0)\}$ are the only finite symmetric suitable subsemigroups of Ω, it follows from (1) that $\Gamma = \Gamma_2 \Rightarrow \Gamma' = \Gamma_2$, $\Gamma = \Gamma_4 \Rightarrow \Gamma' = \Gamma_4$, and in both cases the result is obtained with $m = 1$. From now on we can suppose that $\Gamma \notin \{\Gamma_2, \Gamma_4\}$ and therefore that $\mathrm{pr}(\Gamma) \neq \{0,\infty\}$.

4. Once more, the same proof as in 12.1.15 shows that

(a) λ and μ are additive.
(b) $\mu(a) \in]0,\infty[\cap \mathrm{pr}(\Gamma)$ for $a \in]0,\infty[\cap \mathrm{pr}(\Gamma)$.
(c) $\mu(pa) = p\mu(a)$ for all $a \in \mathrm{pr}(\Gamma)$, $p \in \mathbb{N}$.
(d) There exists an $m \in]0,\infty[$ such that $\mu(a) = ma$ for all $a \in \mathrm{pr}(\Gamma)$.

5. For all $(a,b) \in \Gamma$: $\lambda(a,b) \in \{(\mu(a),\mu(b)), (\mu(b),\mu(a))\}$. Indeed, the initial source

$$\begin{array}{ccc} & & (x,y,a,b) \\ & \overset{1_{\{x,y\}}}{\nearrow} & \\ (x,y,a \vee b, a \vee b) & & \\ & \underset{1_{\{x,y\}}}{\searrow} & \\ & & (x,y,b,a) \end{array}$$

is mapped by F onto an initial source. The initial structure on $\{x,y\}$ is given on one side by

$$(x,y,\lambda_1(a,b) \vee \lambda_2(a,b), \lambda_1(a,b) \vee \lambda_2(a,b)),$$

on the other side by

$$(x,y,\mu(a \vee b), \mu(a \vee b)),$$

and therefore $\lambda_1(a,b) \vee \lambda_2(a,b) = (\mu(a \vee b), \mu(a \vee b))$.

Considering in the same way a final sink, we also obtain $\lambda_1(a,b) \wedge \lambda_2(a,b) = (\mu(a \wedge b), \mu(a \wedge b))$. The result now immediately follows by considering the cases $a < b$ and $a > b$.

6. Finally we prove that either for all $(a,b) \in \Gamma : \lambda(a,b) = (\mu(a),\mu(b))$ or for all $(a,b) \in \Gamma : \lambda(a,b) = (\mu(b),\mu(a))$. To do this we only have to consider the case $a \neq b$. So suppose there exist (a,b) and (c,d) such that $\lambda(a,b) = (ma,mb)$ and $\lambda(c,d) = (md,mc)$. Since $\lambda_1(r,s) = \lambda_2(s,r)$ we may suppose $a < b, c < d$ and so $b \neq 0$, $d \neq 0$. Four cases then are to be considered.

(i) $(a,b) \preceq (c,d)$. Then $a<b\leq d$, $a\leq c<d$. If $c\neq 0$, we have $a/c<b/c$, $b/d<b/c$, so $(a/c)\vee(b/d)<b/c$ and there exist $p \in \mathbb{N}$, $q \in \mathbb{N}$ such that $(a/c)\vee(b/d)<q/p<b/c$ and therefore $pa<qc$, $pb<qd$, $qc<pb$.
It follows that $1_{\{x,y\}} : (x,y,qc,qd) \longrightarrow (x,y,pa,pb)$ is a morphism and that its image $1_{\{x,y\}} : (x,y,mqd,mqc) \rightarrow (x,y,mpa,mpb)$ by F is not. If $c = 0$, then

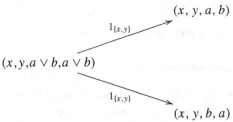

$a = 0$, so $b < d$. Taking $p \in \mathbb{N}, q \in \mathbb{N}$ such that $b/d < q/p$ we obtain the same result.

(ii) $(c,d) \preceq (a,b)$. Exchange the roles of (a,b) and (c,d) in (i).

(iii) $a < c, d < b$. Now we have $a < c < d < b$. The sink

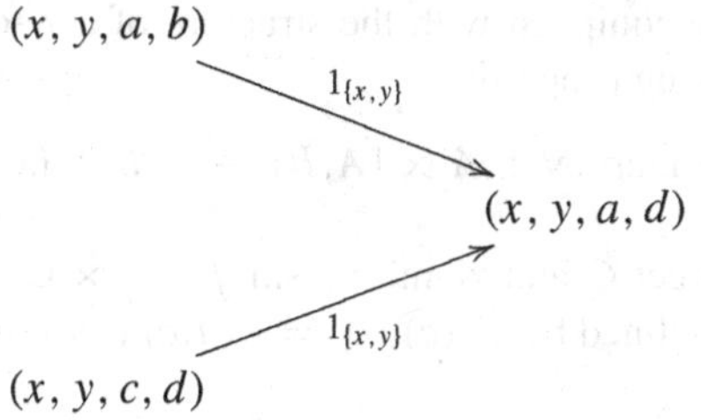

is final since $a \wedge c = a, b \wedge d = d$, so its image by F is final too. Since in this image the final structure is given by either (x,y,ma,md) or (x,y,md,ma), the first alternative is not acceptable because of $mc < md$, and the second not because of $ma < md$.

(iv) $c < a, b < d$. Again exchange the roles of (a,b) and (c,d) in (iii). □

12.1.26 Example Some interesting particular cases of the foregoing result arise.

1. If $\Gamma = \Gamma_2 = \{(0,0), (\infty,\infty)\}$, then $q\mathsf{Met}_\Gamma$ is concretely isomorphic to the subcategory of Top consisting of all coproducts of indiscrete spaces.
2. If $\Gamma = \Gamma_4 = \{(0,0), (\infty,\infty), (0,\infty), (\infty,0)\}$, then $q\mathsf{Met}_\Gamma$ is concretely isomorphic to the subcategory Fin of Top consisting of all finitely generated spaces.
3. If $\Gamma = \{(a,a) \mid a \in \mathbb{P}\}$, then $q\mathsf{Met}_\Gamma = \mathsf{Met}$.
4. If Γ is a suitable subsemigroup of $\mathbb{P}$, then $q\mathsf{Met}_{\Gamma\times\Gamma} = q\mathsf{Met} \cap \mathsf{App}_\Gamma$.

The third observation, together with the foregoing result immediately give the following result for the case of Met. Let Met_Γ be defined as the full subcategory of Met with objects all (X,d) for which $\mathrm{Im}d \subseteq \Gamma$.

12.1.27 Theorem *The collection of all stable subcategories of* Met *is the set* $\{\mathsf{Met}_\Gamma \mid \Gamma \subseteq \mathbb{P}$ suitable subsemigroup$\}$.

Proof This follows from the foregoing remarks. □

12.2 A Quasi-topos Supercategory of App

A topological category $\mathscr{C}$ is a *quasi-topos* (see Herrlich 1987; Wyler 1976, 1991 and the references therein) if it is at the same time cartesian closed (see Bentley et al. 1987; Brandenburg and Hušek 1982; Booth and Tillotson 1980; Lee 1976 and the references therein) and extensional (see Herrlich 1988a, b and the references therein). As a general reference see the books of Preuss (1987, 2002).

A topological category $\mathscr{C}$ is *cartesian closed* if for each $\mathscr{C}$-object A the functor $A \times -$ has a right adjoint. However, it is often more informative to describe a topological category as being cartesian closed if it has nice function spaces in the sense of the following definition.

A topological category $\mathscr{C}$ is cartesian closed if for every pair A, B of $\mathscr{C}$-objects the set $\mathscr{C}(A,B)$ can de equipped with the structure of a $\mathscr{C}$-object, denoted $[A,B]$ which fulfils the following properties.

(CC1) The evaluation map $\mathrm{ev} : A \times [A,B] \longrightarrow B : (x,f) \mapsto f(x)$ is a $\mathscr{C}$-morphism.

(CC2) For each $\mathscr{C}$-object C and $\mathscr{C}$-morphism $f : A \times C \longrightarrow B$, the map $f^* : C \longrightarrow [A,B]$ defined by $f^*(c)(a) := f(a,c)$ is a $\mathscr{C}$-morphism.

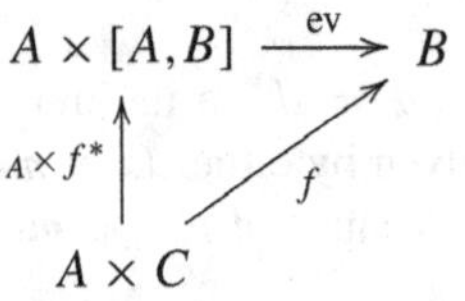

A topological category $\mathscr{C}$ is *extensional* if partial morphisms are representable, precisely, if it fulfils the following property.

(ET) Every object (B,ξ) can be embedded in a so-called one-point extension $(B^\#, \xi^\#)$, where $B^\# = B \cup \{\infty_B\}$, $\infty_B \notin B$, such that for every object A, for every subobject C of A, and for every morphism $f : C \longrightarrow B$, the extension, $f^\# : A \longrightarrow B^\#$, defined by $f^\#(A \backslash C) := \{\infty_B\}$ is a morphism.

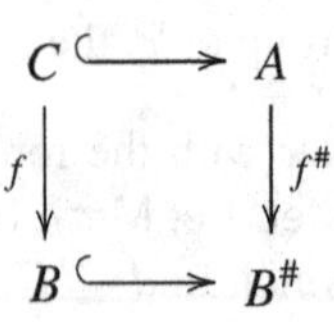

We begin by constructing a quasi-topos which contains App as a fully embedded subcategory. We will not be interested in this category in itself but it does serve our purpose for constructing hulls since it will be a supercategory of App having all the properties of the various hulls. Let X be a set and consider a map

$$\lambda : \mathsf{F}(X) \longrightarrow \mathbb{P}^X$$

which fulfils the following properties.

(L1) $\forall x \in X : \lambda \dot{x}(x) = 0$.
(CAL) $\forall \mathscr{F}, \mathscr{G} \in \mathsf{F}(X) : \lambda(\mathscr{F} \cap \mathscr{G}) = \lambda\mathscr{F} \vee \lambda\mathscr{G}$.

We call this a *convergence-approach limit operator* or shortly CA-limit operator and the space (X, λ) a *convergence-approach space* or shortly a CA-space. Given CA-spaces (X, λ) and (X', λ') a map $f : X \longrightarrow X'$ is said to be a *contraction* if for all $\mathscr{F} \in \mathsf{F}(X) : \lambda'(f(\mathscr{F})) \circ f \leq \lambda\mathscr{F}$. The category consisting of all CA-spaces with contractions is denoted CAp.

In order to prove our next theorem we need some preparation.

1. *First we need the definition of hom-sets in* CAp. Suppose that X and Y are CA-spaces and consider the set $\mathsf{CAp}(X,Y)$ of all contractions between X and Y. For a filter $\Psi \in \mathsf{F}(\mathsf{CAp}(X,Y))$ and a filter $\mathscr{F} \in \mathsf{F}(X)$ we write

$$\Psi(\mathscr{F}) := \{\psi(F) \mid \psi \in \Psi, F \in \mathscr{F}\}$$

where

$$\psi(F) := \{g(y) \mid g \in \psi, y \in F\}.$$

Clearly $\Psi(\mathscr{F}) \in \mathsf{F}(Y)$. Further, if $f \in \mathsf{CAp}(X,Y)$ then we define

$$L(\Psi, f) := \{\alpha \in \mathbb{P} \mid \forall \mathscr{F} \in \mathsf{F}(X) : \lambda_Y(\Psi(\mathscr{F})) \circ f \leq \lambda_X\mathscr{F} \vee \alpha\}.$$

Since $L(\Psi, f)$ is a nonempty subinterval of $\mathbb{P}$ the following function is well-defined

$$\lambda : \mathsf{F}(\mathsf{CAp}(X,Y)) \longrightarrow \mathbb{P}^{\mathsf{CAp}(X,Y)} : \Psi \mapsto \inf L(\Psi, \cdot).$$

We leave the verification that $(\mathsf{CAp}(X,Y), \lambda)$ is a CA-space to the reader but we note that (CAL) follows from the fact that for any $f \in \mathsf{CAp}(X,Y)$ and $\Psi, \Phi \in \mathsf{F}(\mathsf{CAp}(X,Y))$ we have $L(\Psi \cap \Phi, f) = L(\Psi, f) \cap L(\Phi, f)$.

2. *Second we need the definition of* $\cdot^{\#}$*-extensions in* CAp. Given a CA-space Y we put

$$Y^{\#} := Y \cup \{\infty_Y\}$$

where $\infty_Y \notin Y$ and define

$$\lambda^{\#} : \mathsf{F}(Y^{\#}) \longrightarrow \mathbb{P}^{Y^{\#}}$$

as follows

$$\lambda^{\#}\mathscr{F}(y) = \begin{cases} \lambda\mathscr{F}_{|Y}(y) & y \in Y \text{and } \mathscr{F} \neq \text{stack}\,\infty_Y, \\ 0 & \text{otherwise.} \end{cases}$$

Again we leave the rather lengthy but straightforward verification that $(Y^{\#}, \lambda^{\#})$ is a CA-space to the reader. Moreover it is also straightforward to see that Y is a subspace of $Y^{\#}$.

12.2.1 Theorem CAp *is a quasi-topos.*

Proof 1. CAp *is a topological category.* Consider the source

$$(f_j : X \longrightarrow (X_j, \lambda_j))_{j\in J}$$

then the initial CA-limit is given by

$$\lambda\mathscr{F} = \sup_{j\in J} \lambda_j(f_j\mathscr{F}) \circ f_j.$$

We leave the verification of the details to the reader.

2. CAp *is cartesian closed.* First, take two CA-spaces X and Y and consider the evaluation

$$\mathrm{ev} : X \times \mathsf{CAp}(X,Y) \longrightarrow Y : (x,f) \mapsto f(x).$$

Let $\mathfrak{G} \in \mathsf{F}(X \times \mathsf{CAp}(X,Y))$ and put

$$\mathscr{F} := \mathrm{pr}_X\, \mathfrak{G} \text{ and } \Psi := \mathrm{pr}_{\mathsf{CAp}(X,Y)}\, \mathfrak{G}.$$

Now fix $(x, f) \in X \times \mathsf{CAp}(X,Y)$, then it follows from the definition of λ and the description of initial structures in CAp that

$$\begin{aligned}(\lambda_X \times \lambda)(\mathfrak{G})(x,f) &= \lambda_X\mathscr{F}(x) \vee \lambda\Psi(f)\\ &= \inf\{\lambda_X\mathscr{F}(x) \vee \alpha \mid \alpha \in L(\Psi, f)\}.\end{aligned}$$

From this and the definition of $L(\Psi, f)$ we then obtain

$$\begin{aligned}\lambda_Y(\text{stack}\,\mathrm{ev}(\mathfrak{G})(f(x)) &\le \lambda_Y(\text{stack}\,\mathrm{ev}(\mathscr{F} \times \Psi))(f(x))\\ &= \lambda_Y(\text{stack}\,\Psi(\mathscr{F}))(f(x))\\ &\le \lambda\mathscr{F}(x) \vee \lambda\Psi(f)\\ &= (\lambda_X \times \lambda)(\mathfrak{G})(x,f)\end{aligned}$$

which proves that the evaluation map is a contraction.

Second, take three CA-spaces X, Y and Z, a contraction

$$f : X \times Z \longrightarrow Y$$

and consider the transpose

$$f^* : Z \longrightarrow \mathsf{CAp}(X,Y) : z \mapsto [x \mapsto f(x,z)].$$

If $\mathcal{G} \in \mathsf{F}(Z)$, $\mathcal{F} \in \mathsf{F}(X)$, $z \in Z$ and $x \in X$ then

$$\begin{aligned}\lambda_Y(\text{stack}\, f^*(\mathcal{G})(\mathcal{F}))(f^*(z)(x)) &= \lambda_Y(\text{stack}\, f(\mathcal{F} \times \mathcal{G}))(f(x,z)) \\ &\leq \lambda_X\mathcal{F}(x) \vee \lambda_Z\mathcal{G}(z)\end{aligned}$$

which implies that $\lambda_Z\mathcal{G}(z) \in Lf^*(\mathcal{G}), f^*(z))$ and thus

$$\lambda(f^*\mathcal{G})(f^*(z)) \leq \lambda_Z\mathcal{G}(z).$$

This proves that f^* is a contraction.

3. CAp *is extensional.* Let X and Y be CA-spaces and let $Z \subseteq X$. The subobject Z has as CA-limit operator $\lambda_Z\mathcal{F} = \lambda_X \,\text{stack}\, \mathcal{F}$ for any $\mathcal{F} \in \mathsf{F}(Z)$. Now let $f : Z \longrightarrow Y$ be a contraction, i.e. a so-called partial morphism from X to Y and define

$$f^\# : X \longrightarrow Y^\# : x \mapsto \begin{cases} f(x) & x \in Z, \\ \infty_Y & x \in X \setminus Z. \end{cases}$$

Let $\mathcal{F} \in \mathsf{F}(X)$ and $x \in X$. If $\mathcal{F}$ has a trace on Z and $x \in Z$ then it follows that

$$\begin{aligned}\lambda^\# \,\text{stack}_{Y^\#}\, f^\#(\mathcal{F})(f^\#(x)) &= \lambda \,\text{stack}_Y\, f(\mathcal{F}_{|Z})(f(x)) \\ &\leq \lambda_Z\mathcal{F}_{|Z}(x) \\ &= \lambda_X \,\text{stack}\, \mathcal{F}_{|Z}(x) \\ &\leq \lambda_X\mathcal{F}(x).\end{aligned}$$

If $x \in X \setminus Z$ or if $\mathcal{F}$ does not have a trace on Z then again the same inequality holds by definition of $\lambda^\#$. Hence $f^\#$ is a contraction, which proves that partial morphisms are representable and hence that CAp is extensional.

Since a quasi-topos is by definition a cartesian closed extensional topological category we are finished. □

12.2.2 Theorem App *is finally dense in* CAp.

Proof It suffices to prove that any CA-space (X, λ) can be obtained via a final sink from a particular set of filter spaces (see 1.2.63). For each $\mathcal{F} \in \mathsf{F}(X)$ consider the filter space $(X, \lambda_{(\mathcal{F},\lambda\mathcal{F})})$ then it is easily verified that

$$(1_X : (X, \lambda_{(\mathcal{F},\lambda\mathcal{F})}) \longrightarrow (X, \lambda))_{\mathcal{F}\in\mathsf{F}(X)}$$

is a final sink in CAp. □

12.2.3 Corollary App *is a concretely reflective subcategory of* CAp.

We recall that Conv is the category of convergence spaces and continuous maps as defined e.g. in Colebunders and Lowen (2001). In the literature this category is sometimes also referred to as the category of limit spaces (sometimes with a slight change in the definition).

12.2.4 Theorem Conv *is stable in* CAp.

Proof Given a CA-space (X, λ) it is easily seen that on the one hand

$$\lambda_*(\mathscr{F})(x) := \begin{cases} 0 & \lambda(\mathscr{F})(x) < \infty, \\ \infty & \lambda(\mathscr{F})(x) = \infty, \end{cases}$$

is a CA-limit which determines the concrete reflection and on the other hand

$$\lambda^*(\mathscr{F})(x) := \begin{cases} 0 & \lambda(\mathscr{F})(x) = 0, \\ \infty & \lambda(\mathscr{F})(x) > 0, \end{cases}$$

too is a CA-limit which determines the concrete coreflection. □

12.3 The Extensional Topological Hull of App

The extensional topological hull of a category $\mathscr{C}$(shortly denoted by ETH($\mathscr{C}$)), if it exists, is defined as the smallest extensional topological category $\mathscr{B}$ in which $\mathscr{C}$ is finally dense. Given an extensional topological category $\mathscr{A}$ in which $\mathscr{C}$ is finally dense, the extensional topological hull of $\mathscr{C}$ is the full subcategory of $\mathscr{A}$ with those objects C for which there exists an initial source $(f_i : C \longrightarrow A_i^\#,)_{i\in I}$, such that $\forall i \in I : A_i \in \mathscr{C}$. In short, the extensional topological hull of $\mathscr{C}$ is the initial (or bireflective) hull in $\mathscr{A}$ of the one-point extensions of $\mathscr{C}$-objects. See Herrlich (1987,1988a, b).

12.3.1 Definition Given a set X a map $\lambda : \mathsf{F}(X) \longrightarrow \mathbb{P}^X$ which fulfils (L1) and (L2) is called a *pre-limit* and the pair (X, λ) is called a *pre-approach space*. We denote PrAp the full subcategory of CAp with objects all pre-approach spaces.

A quick inspection of the theorems in the first chapter showing the equivalence of limit operators and distances in App shows that a pre-approach limit is equivalent to what we call a *pre-distance*, i.e. a function $\delta : X \times 2^X \longrightarrow \mathbb{P}$ satisfying properties (D1), (D2) and (D3) and that the transition from one structure to another goes via precisely the same formulas. Therefore objects in PrAp carry two equivalent structures and we will of course at each instance choose whichever is most convenient.

12.3.2 Theorem **(pL $\rightrightarrows$ pD)** *If* $\lambda : \mathsf{F}(X) \longrightarrow \mathbb{P}^X$ *is a pre-limit operator on* X*, then the function*

$$\delta : X \times 2^X \longrightarrow \mathbb{P} : (x, A) \mapsto \inf_{\mathscr{U} \in \mathsf{U}(A)} \lambda\mathscr{U}(x)$$

is a pre-distance on X. Moreover, for any $\mathcal{F} \in \mathsf{F}(X)$ *and* $x \in X$, *we have*

$$\lambda\mathcal{F}(x) = \sup_{U \in \sec \mathcal{F}} \delta(x, U).$$

Proof This is contained in the proof of 1.2.2. □

12.3.3 Theorem (**pD** $\rightrightarrows$ **pL**) *If* $\delta : X \times 2^X \longrightarrow \mathbb{P}$ *is a pre-distance on X, then the function*

$$\lambda : \mathsf{F}(X) \longrightarrow \mathbb{P}^X : \mathcal{F} \mapsto \sup_{U \in \sec \mathcal{F}} \delta_U$$

is a pre-limit operator on X. Moreover, for any $x \in X$ *and* $A \in 2^X$, *we have*

$$\delta(x, A) = \mathit{inf}_{\mathcal{U} \in \mathsf{U}(A)} \lambda\mathcal{U}(x).$$

Proof This is contained in the proof of 1.2.1. □

12.3.4 Theorem PrAp *is an extensional concretely reflective subcategory of* CAp *containing* App *as a concretely reflective subcategory.*

Proof That PrAp is a concretely reflective subcategory of CAp is easily seen. The reflection of a CA-space (X, λ) is determined by the pre-distance

$$\delta(x, A) := \inf_{\mathcal{U} \in \mathsf{U}(A)} \lambda\mathcal{U}(x).$$

That PrAp contains App as a concretely reflective subcategory follows at once from 12.2.3. That PrAp is extensional finally goes as in 12.2.1. □

In the following result we require $\mathbb{P}^{\#}$ and in particular its pre-distance

$$\delta_{\mathbb{P}}^{\#} : \mathbb{P}^{\#} \times 2^{\mathbb{P}^{\#}} \longrightarrow \mathbb{P}$$

which, as the reader can easily verify is the unique distance which extends $\delta_{\mathbb{P}}$ and further satisfies

$$\delta_{\mathbb{P}}^{\#}(x, A) = 0 \text{ if } \infty_{\mathbb{P}} \in \{x\} \cup A \text{ and } A \neq \emptyset.$$

The reader will indeed easily verify that $\delta_{\mathbb{P}}^{\#}$ and $\lambda_{\mathbb{P}}^{\#}$ are equivalent in the sense of 12.3.1.

12.3.5 Theorem $\mathbb{P}^{\#}$ *is initially dense in* PrAp.

Proof Let X and Y be objects in PrAp and suppose that $f : X \longrightarrow Y$ is not a contraction. This implies that there exist $x \in X$ and $A \subseteq X$ such that $\delta_X(x, A) < \delta_Y(f(x), f(A))$. We now define the following function

$$g : Y \longrightarrow \mathbb{P}^{\#} : y \mapsto \begin{cases} \delta_Y(f(x), f(A)) & y = f(x), \\ \infty_{\mathbb{P}} & y \notin f(A), y \neq f(x), \\ 0 & y \in f(A). \end{cases}$$

To see that g is a contraction note that for $y \in Y$ and B a nonempty subset of Y the value $\delta^{\#}_{\mathbb{P}}(g(y), g(B))$ is non zero only if

$$y = f(x), \infty_{\mathbb{P}} \notin g(B) \text{ and } \delta_Y(f(x), f(A)) \notin g(B).$$

In that case however

$$\begin{aligned} \delta^{\#}_{\mathbb{P}}(g(y), g(B)) &= \delta_{\mathbb{P}}(\delta_Y(f(x), f(A)), \{0\}) \\ &= \delta_Y(f(x), f(A)) \\ &\leq \delta_Y(y, B). \end{aligned}$$

Finally, that $g \circ f$ is not a contraction follows from

$$\begin{aligned} \delta_X(x, A) &< \delta_Y(f(x), f(A)) \\ &= \delta_{\mathbb{P}}(\delta_Y(f(x), f(A)), \{0\}) \\ &= \delta^{\#}_{\mathbb{P}}(g(f(x)), g(f(A))) \end{aligned}$$

and this completes the proof. □

12.3.6 Theorem PrAp *is the extensional topological hull of* App.

Proof By 12.3.4 PrAp is extensional, and by 12.2.2 App is finally dense in PrAp. Finally by 12.3.5 the class of objects

$$\{(X^{\#}, \lambda^{\#}) \mid (X, \lambda) \in \mathsf{App}\}$$

is initially dense in PrAp and hence the result follows from Herrlich (1987). □

12.3.7 Corollary App *is finally dense in* PrAp.

12.3.8 Theorem PrTop *is stable in* PrAp.

Proof Given a pre-approach space (X, λ) it is easily seen that λ_* and λ^* as defined in 12.2.4 are pre-approach limits and determine respectively the concrete reflection and coreflection. □

12.4 The Cartesian Closed Topological Hull of PrAp

The CCT hull of a category $\mathscr{C}$ (shortly denoted by CCTH($\mathscr{C}$)) (if it exists) is defined as the smallest CCT category $\mathscr{B}$ in which $\mathscr{C}$ is closed under finite products (see Herrlich and Nel 1977; Nel 1977; Weck-Schwarz 1991 and the references therein). Also from Herrlich and Nel (1977), we recall that given a CCT category $\mathscr{D}$ in which $\mathscr{C}$ is finally dense, the CCT hull of $\mathscr{C}$ is the full subcategory of $\mathscr{D}$ determined by

$$\mathrm{CCTH}(\mathscr{C}) := \{C \text{ a } \mathscr{D}\text{-object } | \text{there exists an initial source } (f_i : C \longrightarrow [A_i, B_i])_{i \in I} \text{ where } \forall i \in I : A_i \text{ and } B_i \text{ are in } \mathscr{C}\}.$$

In short, the CCT hull of $\mathscr{C}$ is the initial hull in $\mathscr{D}$ of the power-objects of $\mathscr{C}$-objects.

A more recent survey of such properties and hull concepts can be found in Herrlich (1987) and Schwarz and Weck-Schwarz (1991).

We use the notations and general construction from Bourdaud (1975, 1976). Let $C(\mathbb{P}^{\#})$ stand for the full subcategory of CAp with objects those spaces X which carry the initial structure for the source

$$j : X \longrightarrow \mathsf{CAp}(\mathsf{CAp}(X, \mathbb{P}^{\#}), \mathbb{P}^{\#}) : x \mapsto [f \mapsto f(x)]$$

12.4.1 Theorem *$C(\mathbb{P}^{\#})$ is the cartesian closed topological hull of* PrAp.

Proof By Bourdaud (1976), $C(\mathbb{P}^{\#})$ is cartesian closed topological with hom-objects formed as in CAp, $C(\mathbb{P}^{\#})$ is concretely reflective in CAp and $\mathbb{P}^{\#}$ is in $C(\mathbb{P}^{\#})$. By 12.3.4 PrAp is concretely reflective in CAp. Hence it follows from 12.3.5 that PrAp is a subcategory of $C(\mathbb{P}^{\#})$ which is closed under the formation of finite products in $C(\mathbb{P}^{\#})$. That PrAp is finally dense in $C(\mathbb{P}^{\#})$ follows at once from 12.2.2. Moreover, since the functor

$$\mathsf{CAp}(\cdot, \mathbb{P}^{\#}) : \mathsf{CAp} \longrightarrow \mathsf{CAp}$$

transforms final epi-sinks into initial sources, 12.2.2 also implies that powers of objects in PrAp are initially dense in $C(\mathbb{P}^{\#})$.

Consequently, following Herrlich and Nel (1977), $C(\mathbb{P}^{\#})$ is indeed the cartesian closed topological hull of PrAp. □

Of course we need to give an internal characterization of the objects in $C(\mathbb{P}^{\#})$ and therefore we need an explicit formulation of the initial limit determined by the source

$$j : X \longrightarrow \mathsf{CAp}(\mathsf{CAp}(X, \mathbb{P}^{\#}), \mathbb{P}^{\#}) : x \mapsto [f \mapsto f(x)]$$

where X is a CA-space.

We will use the following notations:

1. The limit on $\mathsf{CAp}(X, \mathbb{P}^{\#})$ is denoted by $\lambda_H^{\mathbb{P}}$.
2. The limit on $\mathsf{CAp}(\mathsf{CAp}(X, \mathbb{P}^{\#}), \mathbb{P}^{\#})$ is denoted by $\lambda_{HH}^{\mathbb{P}}$.
3. For $A \subseteq X$ and $\beta \in \mathbb{P}$ we put

$$A^{<\beta>} := \{k \in \mathsf{CAp}(X, \mathbb{P}^{\#}) \mid k^{-1}[0, \beta] \cap A \neq \emptyset\}.$$

In the following result we use the function l as defined in 1.2.62.

12.4.2 Proposition *If μ stands for the initial limit operator on X, determined by the source $j : X \longrightarrow \mathsf{CAp}(\mathsf{CAp}(X, \mathbb{P}^{\#}), \mathbb{P}^{\#})$, $\mathscr{H} \in \mathsf{F}(X)$, $a \in X$ and $\alpha \in \mathbb{P}$ then the following properties are equivalent.*

1. $\mu\mathscr{H}(a) \leq \alpha$.
2. $\forall \Psi \in \mathsf{F}(\mathsf{CAp}(X, \mathbb{P}^{\#})), \forall f \in \mathsf{CAp}(X, \mathbb{P}^{\#})$ *such that* $f(a) \neq \infty_{\mathbb{P}}$, $\text{stack}\,\Psi(\mathscr{H}) \neq \text{stack}\,\infty_{\mathbb{P}}$ *and* $\lambda_H^{\mathbb{P}}\Psi(f) < \infty$ *and* $\forall \beta < f(a) - \lambda_H^{\mathbb{P}}\Psi(f) \vee \alpha$ *there exists* $H \in \mathscr{H}$ *such that* $H^{<\beta>} \in \Psi$.

Proof For simplicity in notation we put D for the set of all those pairs $(\Psi, f) \in \mathsf{F}(\mathsf{CAp}(X, \mathbb{P}^{\#})) \times \mathsf{CAp}(X, \mathbb{P}^{\#})$ such that

$$f(a) \neq \infty_{\mathbb{P}},\ \text{stack}\,\Psi(\mathscr{H}) \neq \text{stack}\,\infty_{\mathbb{P}} \text{ and } \lambda_H^{\mathbb{P}}\Psi(f) < \infty.$$

Then we have

$$\begin{aligned}
&\mu\mathscr{H}(a) = \lambda_{HH}^{\mathbb{P}}\,\text{stack}\,j(\mathscr{H})(j(a)) \leq \alpha \\
&\Leftrightarrow \forall \Psi \in \mathsf{F}(\mathsf{CAp}(X, \mathbb{P}^{\#})), \forall f \in \mathsf{CAp}(X, \mathbb{P}^{\#}) : \lambda_{\mathbb{P}}^{\#}\,\text{stack}\,\Psi(\mathscr{H})(f(a)) \leq \lambda_H^{\mathbb{P}}\Psi(f) \vee \alpha \\
&\Leftrightarrow \forall (\Psi, f) \in D : \lambda_{\mathbb{P}}\,\text{stack}\,\Psi(\mathscr{H})_{|\mathbb{P}}(f(a)) \leq \lambda_H^{\mathbb{P}}\Psi(f) \vee \alpha \\
&\Leftrightarrow \forall (\Psi, f) \in D : f(a) \ominus l(\text{stack}\,\Psi(\mathscr{H})_{|\mathbb{P}}) \leq \lambda_H^{\mathbb{P}}\Psi(f) \vee \alpha \\
&\Leftrightarrow \forall (\Psi, f) \in D, \forall \beta \in [0, f(a) - \lambda_H^{\mathbb{P}}\Psi(f) \vee \alpha[:]\beta, \infty] \in \text{stack}\,\Psi(\mathscr{H})_{|\mathbb{P}} \\
&\Leftrightarrow \forall \beta \in [0, f(a) - \lambda_H^{\mathbb{P}}\Psi(f) \vee \alpha[\exists H \in \mathscr{H} : H^{<\beta>} \in \Psi.
\end{aligned}$$

□

12.4.3 Definition We define and denote by PsAp the full subcategory of CAp with objects those CA-spaces which moreover fulfil the following property.

(PSAL) $\forall \mathscr{F} \in \mathsf{F}(X) : \lambda\mathscr{F} = \sup_{\mathscr{U} \in \mathsf{U}(\mathscr{F})} \lambda\mathscr{U}$.

An object in PsAp is called a *pseudo-approach space* and its limit operator is called a *pseudo-limit operator*.

12.4.4 Lemma *Suppose X and Y are CA-spaces, $\Psi \in \mathsf{F}(\mathsf{CAp}(X, Y))$, $\mathscr{F} \in \mathsf{F}(X)$ and $\mathscr{W} \in \mathsf{U}(\text{stack}\,\Psi(\mathscr{F}))$ then the following properties hold.*

1. *There exists* $\mathscr{U} \in \mathsf{U}(\mathscr{F}) : \text{stack}\,\Psi(\mathscr{U}) \subseteq \mathscr{W}$.
2. *There exists* $\Phi \in \mathsf{U}(\Psi) : \text{stack}\,\Phi(\mathscr{F}) \subseteq \mathscr{W}$.

Proof The proof of 2 is perfectly analogous to that of 1 so we only prove 1. Suppose that for every $\mathcal{U} \in \mathsf{U}(\mathcal{F})$ there exists $\psi \in \Psi$ and $U \in \mathcal{U}$ such that $\psi(U) \notin \mathcal{W}$. Then we apply 1.1.4 to select a finite collection $\mathcal{U}_1, \ldots, \mathcal{U}_n \in \mathsf{U}(\mathcal{F})$ and corresponding sets $\psi_1, \ldots, \psi_n \in \Psi$ and $U_i \in \mathcal{U}_i$ for $i \in \{1, \ldots, n\}$ such that

$$\bigcup_{i=1}^{n} U_i \in \mathcal{F} \text{ and } \psi_i(U_i) \notin \mathcal{W} \text{ for all } i \in \{1, \ldots, n\}.$$

Then however

$$(\bigcap_{i=1}^{n} \psi_i)(\bigcup_{i=1}^{n} U_i) \in \mathcal{W}$$

which is a contradiction. □

12.4.5 Proposition *Suppose that X is a CA-space and Y a pseudo-approach space. For $\Psi \in \mathsf{F}(\mathsf{CAp}(X,Y))$, $f \in \mathsf{CAp}(X,Y)$ and $\alpha \in \mathbb{P}$ the following properties are equivalent.*

1. $\forall \mathcal{F} \in \mathsf{F}(X) : \lambda_Y \operatorname{stack} \Psi(\mathcal{F}) \circ f \leq \lambda_X \mathcal{F} \vee \alpha$.
2. $\forall \mathcal{U} \in \mathsf{U}(X) : \lambda_Y \operatorname{stack} \Psi(\mathcal{U}) \circ f \leq \lambda_X \mathcal{U} \vee \alpha$.
3. $\forall \mathcal{F} \in \mathsf{F}(X), \forall \Phi \in \mathsf{U}(\Psi) : \lambda_Y \operatorname{stack} \Phi(\mathcal{F}) \circ f \leq \lambda_X \mathcal{F} \vee \alpha$.
4. $\forall \mathcal{U} \in \mathsf{U}(X), \forall \Phi \in \mathsf{U}(\Psi) : \lambda_Y \operatorname{stack} \Phi(\mathcal{U}) \circ f \leq \lambda_X \mathcal{U} \vee \alpha$.

Proof The implications $1 \Rightarrow 2 \Rightarrow 4$ and $1 \Rightarrow 3 \Rightarrow 4$ are evident.
$4 \Rightarrow 1$. This follows from the foregoing lemma and

$$\begin{aligned}
\lambda_Y \operatorname{stack} \Psi(\mathcal{F}) \circ f &= \sup_{\mathcal{W} \in \mathsf{U}(\Psi(\mathcal{F}))} \lambda_Y \mathcal{W} \circ f \\
&\leq \sup_{\mathcal{U} \in \mathsf{U}(\mathcal{F}))} \sup_{\Phi \in \mathsf{U}(\Psi)} \lambda_Y \operatorname{stack} \Phi(\mathcal{U}) \circ f \\
&\leq \sup_{\mathcal{U} \in \mathsf{U}(\mathcal{F}))} \lambda_X \mathcal{U} \vee \alpha \\
&\leq \lambda_X \mathcal{F} \vee \alpha.
\end{aligned}$$

□

12.4.6 Corollary *Suppose that X is a CA-space and Y a pseudo-approach space. For a map $f : X \longrightarrow Y$ the following properties are equivalent.*

1. *f is a contraction.*
2. $\forall \mathcal{U} \in \mathsf{U}(X) : \lambda_Y \operatorname{stack} f(\mathcal{U}) \circ f \leq \lambda_X \mathcal{U}$.

12.4.7 Theorem PsAp *is a concretely reflective subcategory of* CAp.

Proof Making use of the foregoing corollary it is immediately verified that for a given CA-space (X, λ) its PsAp-reflection is determined by $\widetilde{\lambda}$ where

$$\widetilde{\lambda}\mathscr{F} = \sup_{\mathscr{U} \in \mathsf{U}(\mathscr{F})} \lambda\mathscr{U}.$$

□

12.4.8 Proposition *If X is a CA-space and Y is a pseudo-approach space then* $\mathsf{CAp}(X,Y)$ *is a pseudo-approach space.*

Proof Let λ stand for the limit operator on $\mathsf{CAp}(X,Y)$, let $\Psi \in \mathsf{F}(\mathsf{CAp}(X,Y))$ and let $f \in \mathsf{CAp}(X,Y)$. Put

$$\alpha := \sup_{\Phi \in \mathsf{U}(\Psi)} \lambda\Phi(f)$$

then for all $\Phi \in \mathsf{U}(\Psi)$ and $\mathscr{F} \in \mathsf{F}(X)$ we have

$$\lambda_Y \Phi(\mathscr{F}) \circ f \leq \lambda_X \mathscr{F} \vee \alpha$$

which by 12.4.5 implies that for all $\mathscr{F} \in \mathsf{F}(X)$

$$\lambda_Y \Psi(\mathscr{F}) \circ f \leq \lambda_X \mathscr{F} \vee \alpha$$

and thus that $\lambda\Psi(\mathscr{F}) \leq \alpha$. □

We are now in a position to prove the main result of this section, namely the internal characterization of the objects in the cartesian closed topological hull of PrAp.

12.4.9 Theorem PsAp *is the cartesian closed topological hull of* PrAp.

Proof If X is an object in $C(\mathbb{P}^{\#})$ then it follows from 12.4.8 that

$$\mathsf{CAp}(\mathsf{CAp}(X, \mathbb{P}^{\#}), \mathbb{P}^{\#})$$

is a pseudo-approach space. By 12.4.7 it then follows that also X is a pseudo-approach space.

Conversely, let X be a pseudo-approach space and let μ stand for the initial limit operator on X determined by the source

$$j : X \longrightarrow \mathsf{CAp}(\mathsf{CAp}(X, \mathbb{P}^{\#}), \mathbb{P}^{\#})$$

then $\mu \leq \lambda_X$. Now, since they both are pseudo-approach limits, in order to coincide we just have to verify the other inequality on ultrafilters. Suppose therefore on the contrary that there exists $a \in X$ and an ultrafilter $\mathscr{H} \in \mathsf{U}(X)$ such that $\mu\mathscr{H}(a) < \lambda_X\mathscr{H}(a)$. For all $W \in \mathscr{H}$ put

$$\widetilde{W} := \{k \in \mathsf{CAp}(X, \mathbb{P}^{\#}) \mid k(X \backslash W) = \infty_{\mathbb{P}}\}.$$

In particular then, for any $\gamma \in \mathbb{P}$, the two-valued function

$$k_\gamma^W : X \longrightarrow \mathbb{P}^\# : x \mapsto \begin{cases} \infty_\mathbb{P} & x \in X\backslash W, \\ \gamma & x \in W, \end{cases}$$

is a contraction which belongs to $\widetilde{W}$. It follows that $\{\widetilde{W} \mid W \in \mathscr{H}\}$ is a filterbasis on $\mathsf{CAp}(X, \mathbb{P}^\#)$. Let us denote by Ψ the filter generated by this base. Further, also consider the two-valued contraction

$$f : X \longrightarrow \mathbb{P}^\# : x \mapsto \begin{cases} \infty_\mathbb{P} & x \neq a, \\ \lambda_X \mathscr{H}(a) & x = a. \end{cases}$$

Now let $\mathscr{U} \in \mathsf{U}(X)$. If $\mathscr{U} \neq \mathscr{H}$ then $\operatorname{stack} \Psi(\mathscr{U}) = \operatorname{stack} \infty_\mathbb{P}$ and therefore

$$\lambda_\mathbb{P}^\# \operatorname{stack} \Psi(\mathscr{U})(f(a)) = 0.$$

If $\mathscr{U} = \mathscr{H}$ then $\operatorname{stack} \Psi(\mathscr{U}) \neq \operatorname{stack} \infty_\mathbb{P}$ and then

$$\lambda_\mathbb{P}^\# \operatorname{stack} \Psi(\mathscr{U})(f(a)) = \lambda_\mathbb{P} \operatorname{stack} \Psi(\mathscr{U})_{|\mathbb{P}}(f(a)) \leq f(a).$$

Thus it follows that for all $\mathscr{U} \in \mathsf{U}(X)$

$$\lambda_\mathbb{P}^\# \operatorname{stack} \Psi(\mathscr{U})(f(a)) \leq \lambda_X \mathscr{U}(a).$$

Moreover, since $f(x) = \infty_\mathbb{P}$ for $x \neq a$ we finally have

$$\lambda_\mathbb{P}^\# \operatorname{stack} \Psi(\mathscr{U}) \circ f \leq \lambda_X \mathscr{U}.$$

Now from the arbitrariness of $\mathscr{U}$ and upon applying 12.4.5 we can conclude that

$$\lambda_H^\mathbb{P} \Psi(f) = 0.$$

If we now put $\alpha := \mu \mathscr{H}(a)$ then it is clear that Ψ and f satisfy all the conditions in 12.4.2. Since $\alpha < f(a)$ we can thus choose $\beta \in [0, f(a) - \alpha[$ and then it follows from 12.4.2 that there exists $H \in \mathscr{H}$ such that $H^{<\beta>} \in \Psi$. Then let $W \in \mathscr{H}$ be such that $\widetilde{W} \subseteq H^{<\beta>}$. Since $k_\beta^W \in \widetilde{W}$ it follows that

$$W \cap H = \{x \mid k_\beta^W(x) \in [0, \beta]\} \cap H = \emptyset$$

which is a contraction, and hence we are finished. □

12.5 The Quasi-topos Hull of App

The quasi-topos hull of a category $\mathscr{C}$ (shortly denoted by QT($\mathscr{C}$)), if it exists, is the smallest quasi-topos $\mathscr{B}$ in which $\mathscr{C}$ is finally dense. Given a quasi-topos $\mathscr{A}$ in which $\mathscr{C}$ is finally dense, the quasi-topos hull of $\mathscr{C}$ is the full subcategory of $\mathscr{A}$ with those objects C for which there exists an initial source $(f_i : C \longrightarrow [A_i, B_i^\#])_{i \in I}$, such that $\forall i \in I$: A_i and B_i are in $\mathscr{C}$. In short, the quasi-topos hull of $\mathscr{C}$ is the initial (or bireflective) hull in $\mathscr{A}$ of the power-objects of type $[A, B^\#]$ for A and B in $\mathscr{C}$. See Herrlich (1987) and Schwarz (1989).

The quasi-topos hull of a category can be obtained by a two-step process. First one makes the extensional topological hull and then one makes the cartesian closed topological hull, precisely:

$$\mathrm{QT}(\mathscr{C}) = \mathrm{CCTH}(\mathrm{ETH}(\mathscr{C}))$$

It was observed by Schwarz in 1989 that the order of taking hulls on the right-hand side can not be interchanged (Schwarz 1989).

12.5.1 Proposition PsAp *is extensional.*

Proof This is analogous to the proof of the fact that CAp is extensional and we leave this to the reader. □

12.5.2 Theorem PsAp *is the quasi-topos hull of* PrAp.

Proof By 12.5.1 and 12.4.9 PsAp is a quasi-topos. Consequently it follows from 12.4.9 that it is the quasi-topos hull of PrAp. □

12.5.3 Theorem PsAp *is the quasi-topos hull of* App.

Proof By 12.2.2 App is finally dense in PsAp and by 12.3.6 PrAp is the extensional hull of App. Consequently it follows from 12.5.2 that PsAp is the quasi-topos hull of App. □

12.5.4 Corollary App *is finally dense in* PsAp *and hence also concretely reflective in* PsAp.

12.5.5 Theorem PsTop *is stable in* PsAp.

Proof Given a pseudo-approach space (X, λ) it is easily seen that λ_* and λ^* as defined in 12.2.4 are pseudo-approach limits and determine respectively the concrete reflection and coreflection. □

12.6 The Cartesian Closed Topological Hull of App

In this section we will construct the cartesian closed topological hull of App. We do this by identifying it with a subcategory of PsAp, the category of pseudo-approach spaces, which was shown to be the quasi-topos hull, QTH(App), of App.

Whereas the objects of PsAp can be described by axioms quite similar to those characterizing the objects of QTH(Top), the situation for CCTH(App) is somewhat different. In this case it are also the metric aspects of the theory which will play a prominent role.

12.6.1 Definition Given (X, λ) in PsAp, we define

$$\dot{F}^{\rho} := \{y \in X \mid \exists x \in F : \delta_{\bar{\lambda}}(x, \{y\}) \leq \rho\}.$$

The family $((\dot{-})^{\varepsilon})_{\varepsilon \in \mathbb{R}^+}$ is (clearly) not a tower in general. However, we could define

$$F^{\bullet\rho} := \{y \in X \mid \forall \rho' > \rho, \exists x \in F : \delta_{\bar{\lambda}}(x, \{y\}) \leq \rho'\}$$

such that $((-)^{\bullet\varepsilon})_{\varepsilon \in \mathbb{R}^+}$ does constitute a tower. Furthermore, one could observe in the sequel that consistently replacing $\dot{F}^{\rho}$ by $F^{\bullet\rho}$ would not make an essential change (other than the fact that the first form is nicer to work with, whereas the latter form has conceptual advantages). Further we define

$$\mathsf{d}_X : \mathsf{F}(X) \times \mathsf{F}(X) \longrightarrow \mathbb{P} : (\mathcal{F}, \mathcal{G}) \mapsto \mathsf{d}_X(\mathcal{F}, \mathcal{G}) := \inf\{\rho \geq 0 \mid \dot{\mathcal{G}}^{\rho} \subseteq \mathcal{F}\},$$

where $\dot{\mathcal{G}}^{\rho} := \text{stack}\{\dot{G}^{\rho} \mid G \in \mathcal{G}\}$.

Note that if $\mathsf{d}_X(\mathcal{F}, \mathcal{G}) < \alpha$ and $\mathsf{d}_X(\mathcal{G}, \mathcal{H}) < \beta$ then $\dot{\mathcal{G}}^{\alpha} \subseteq \mathcal{F}$ and $\mathcal{I} := \dot{\mathcal{H}}^{\beta} \subseteq \mathcal{G}$, and consequently, $\dot{\mathcal{H}}^{\alpha+\beta} \subseteq \dot{\mathcal{I}}^{\alpha} \subseteq \dot{\mathcal{G}}^{\alpha} \subseteq \mathcal{F}$. Hence d_X is a quasi-metric. Obviously d_X can attain the value ∞, by definition it is clearly not symmetric and for instance $\mathsf{d}_X(\dot{\mathcal{F}}^0, \mathcal{F}) = 0$.

We already know that in an approach space, for any $\mathcal{F} \in \mathsf{F}(X)$, the function $\lambda\mathcal{F}$ is a contraction. However we can also consider the function λ with two variables, filters on X and points of X. The foregoing definition of a quasi-metric on the set of all filters now makes it possible to consider contraction and continuity properties of this function of two variables.

From 12.2.3 we know that App is concretely reflective in PsAp, and given a pseudo-approach space $(\underline{X}, \lambda)$ we denote the limit operator associated with its concrete reflection in App by $\bar{\lambda}$. We also recall that we denote $\mathcal{T}_{\mathbb{P}} = \{]a, \infty] \mid a \in \mathbb{P}\} \cup \{\mathbb{P}\}$ (see 2.2.7).

12.6.2 Definition We define EpiAp to be the full subcategory of PsAp whose objects (X, λ) satisfy the following property.

(C) $\lambda : (\mathsf{U}(X), \mathcal{T}_{\mathsf{d}_X}) \times (X, \mathcal{T}_{\bar{\lambda}}) \longrightarrow (\mathbb{P}, \mathcal{T}_{\mathbb{P}})$ is a continuous map.

The following illustrates why, as usual, we could restrict ourselves to ultrafilters in the foregoing definition.

12.6.3 Proposition (C) *is equivalent to*

(C)′ $\lambda : (\mathsf{F}(X), \mathcal{T}_{\mathsf{d}_X}) \times (X, \mathcal{T}_{\bar{\lambda}}) \longrightarrow (\mathbb{P}, \mathcal{T}_{\mathbb{P}})$ is a continuous map.

Proof One implication is obvious.

Conversely, assume that $\lambda : (\mathsf{U}(X), \mathcal{T}_{\mathsf{d}_X}) \times (X, \mathcal{T}_{\bar{\lambda}}) \longrightarrow (\mathbb{P}, \mathcal{T}_{\mathbb{P}})$ is continuous. Now let $\mathcal{F} \in \mathsf{F}(X)$ and $x \in X$ be such that

$$K < \lambda(\mathcal{F})(x) = \sup_{\mathcal{U} \in \mathsf{U}(\mathcal{F})} \lambda(\mathcal{U})(x).$$

Then we can we find some ultrafilter $\mathcal{U} \supset \mathcal{F}$ such that $\lambda(\mathcal{U})(x) > K$, and hence also $V \in \mathcal{V}_{\bar{\lambda}}(x)$ and $\delta > 0$ such that $\mathsf{d}_X(\mathcal{U}, \mathcal{W}) \leq \delta$ (where $\mathcal{W} \in \mathsf{U}(X)$) and $y \in V$ implies that $\lambda(\mathcal{W})(y) > K$. We now have to consider some $\mathcal{G} \in \mathsf{F}(X)$ and $y \in V$ such that $\mathsf{d}_X(\mathcal{F}, \mathcal{G}) < \delta$ and $y \in V$. Since $\dot{\mathcal{G}}^{\delta} \subseteq \mathcal{F} \subseteq \mathcal{U}$, we find some $\mathcal{W} \in \mathsf{U}(\mathcal{G})$ such that $\dot{\mathcal{W}}^{\delta} \subseteq \mathcal{U}$. Indeed, assume otherwise that

$$\forall \mathcal{W} \in \mathsf{U}(\mathcal{G}), \exists W \in \mathcal{W} : \dot{W}^{\delta} \notin \mathcal{U}.$$

Then by 1.1.4 we can find $W_1, \ldots, W_n$ such that $\dot{W_i}^{\delta} \notin \mathcal{U}$ $(1 \leq i \leq n)$ and $W_1 \cup \ldots \cup W_n \in \mathcal{G}$. However, since $\mathcal{U} \supset \mathcal{F} \supset \dot{\mathcal{G}}^{\delta} \ni (W_1 \cup \ldots \cup W_n)^{\delta} = \dot{W_1}^{\delta} \cup \ldots \cup \dot{W_n}^{\delta}$, we find that $\dot{W_i}^{\delta} \in \mathcal{U}$ for some $1 \leq i \leq n$. Consequently, this is a contradiction and therefore there exists some $\mathcal{W} \in \mathsf{U}(\mathcal{G})$ such that $\dot{\mathcal{W}}^{\delta} \subseteq \mathcal{U}$, meaning that $\mathsf{d}_X(\mathcal{U}, \mathcal{W}) \leq \delta$. By previous choices, we then find that $\lambda(\mathcal{W})(x) > K$, hence also $\lambda(\mathcal{G})(x) \geq \lambda(\mathcal{W})(x) > K$. □

We are now in a position to state the main result of this section, which we will prove in several steps.

12.6.4 Theorem EpiAp *is the cartesian closed topological hull of* App.

STEP 1: We first show that App $\subseteq$ EpiAp.

12.6.5 Lemma *Let* (X, λ) *be an approach space and let* $\mathcal{F}, \mathcal{G} \in \mathsf{F}(X)$ *and* $\rho \geq 0$, *then the following properties hold.*

1. $\lambda \dot{\mathcal{F}}^{\rho} \leq \lambda \mathcal{F} + \rho$.
2. $\lambda \mathcal{F} \leq \lambda \mathcal{G} + \mathsf{d}_X(\mathcal{F}, \mathcal{G})$.

Proof 1. Let $U \in \sec \dot{\mathscr{F}}^{\rho}$, we then claim that $U^{(\rho)} \in \sec \mathscr{F}$. Indeed, let $F \in \mathscr{F}$, then we find $z \in U$ such that also $z \in \dot{F}^{\rho}$, meaning $\delta(y,\{z\}) \leq \rho$ for some $y \in F$. Hence, also $\delta(y,U) \leq \rho$, i.e. $y \in F \cap U^{(\rho)}$. If we now recall from 1.2.1 that

$$\lambda \dot{\mathscr{F}}^{\rho}(x) = \sup_{U \in \sec \dot{\mathscr{F}}^{\rho}} \delta(x,U) \text{ and } \lambda \mathscr{F}(x) = \sup_{U \in \sec \mathscr{F}} \delta(x,U),$$

then the foregoing clearly demonstrates what was required.

2. Let $\mathrm{d}_X(\mathscr{F},\mathscr{G}) < \alpha$, hence $\dot{\mathscr{G}}^{\alpha} \subseteq \mathscr{F}$. Then, by the first claim, it follows that

$$\lambda \mathscr{F} \leq \lambda \dot{\mathscr{G}}^{\alpha} \leq \lambda \mathscr{G} + \alpha.$$

By the arbitrariness of α, we conclude that $\lambda \mathscr{F} \leq \lambda \mathscr{G} + \mathrm{d}_X(\mathscr{F},\mathscr{G})$. □

The foregoing lemma shows that in an approach space, for any $x \in X$, also the function $\lambda(\cdot)(x)$ is a contraction. However we can show more.

12.6.6 Proposition *Let X be an approach space and let $(\mathscr{A}(x))_{x \in X}$ be the approach system. If we put*

$$\mathscr{B}_{\oplus}(\mathscr{F},x) := \{\mathrm{d}_X(\mathscr{F},\cdot) + \varphi \mid \varphi \in \mathscr{A}(x)\},$$

then $(\mathscr{B}_{\oplus}(\mathscr{F},x))_{(\mathscr{F},x) \in \mathsf{F}(X) \times X}$ is an approach basis on $\mathsf{F}(X) \times X$ and

$$\lambda : (\mathsf{F}(X) \times X, \widehat{\mathscr{B}_{\oplus}}) \longrightarrow \mathbb{P} \text{ is a contraction.}$$

Proof Let $\mathscr{B}_{\mathbb{P}}(x) := \{\varphi \in \mathbb{P}^{\mathbb{P}} \mid \varphi \leq d_{\mathbb{P}}(x,\cdot)\}$ if $x < \infty$ and let $\mathscr{B}_{\mathbb{P}}(\infty) := \{\theta_{]a,\infty]} \mid 0 \leq a < \infty\}$. We then know that this approach basis generates the approach system associated with $\delta_{\mathbb{P}}$ (see 1.2.62).

Now let $(\mathscr{F},x) \in \mathsf{F}(X) \times X$ and first assume that $\lambda \mathscr{F}(x) < \infty$. Also let $0 < \omega < \infty$ and $0 < \varepsilon$ be fixed. Since $\lambda \mathscr{F} : X \longrightarrow \mathbb{P}$ is a contraction we can find $\varphi \in \mathscr{A}(x)$ such that

$$(\lambda \mathscr{F}(x) - \lambda \mathscr{F}) \wedge \omega \leq \varphi + \varepsilon.$$

We then find that, for any $y \in X$:

$$\begin{aligned}(\lambda \mathscr{F}(x) - \lambda \mathscr{G}(y)) \wedge \omega &\leq (\lambda \mathscr{F}(x) - \lambda \mathscr{F}(y)) \wedge \omega + (\lambda \mathscr{F}(y) - \lambda \mathscr{G}(y)) \wedge \omega \\ &\leq \varphi(y) + \varepsilon + (\lambda \mathscr{G}(y) + \mathrm{d}_X(\mathscr{F},\mathscr{G}) - \lambda \mathscr{G}(y)) \wedge \omega \\ &\leq \varphi(y) + \varepsilon + \mathrm{d}_X(\mathscr{F},\mathscr{G}).\end{aligned}$$

Hence, by the arbitrariness of ω and ε we are finished for this case.

Now we assume that $\lambda \mathscr{F}(x) = \infty$. Then we then need to show that for arbitrary $\varepsilon > 0$ and $0 \leq K, \omega < \infty$, there exists $\varphi \in \mathscr{A}(x)$ such that

$$\forall \mathscr{G} \in \mathsf{F}(X) : \theta_{]K,\infty]}(\lambda \mathscr{G}) \wedge \omega \leq \mathrm{d}_X(\mathscr{F},\mathscr{G}) + \varphi + \varepsilon.$$

Since $\lambda\mathscr{F} : X \longrightarrow \mathbb{P}$ is a contraction, we can find $\varphi \in \mathscr{A}(x)$ such that

$$\theta_{]K+\omega,\infty]}(\lambda\mathscr{F}) \wedge \omega \leq \varphi + \varepsilon.$$

We now claim that for any $y \in X$:

$$\theta_{]K,\infty]}(\lambda\mathscr{G}(y)) \wedge \omega \leq \mathsf{d}_X(\mathscr{F}, \mathscr{G}) + \varphi(y) + \varepsilon.$$

If $\mathsf{d}_X(\mathscr{F}, \mathscr{G}) > \omega$ or $\lambda\mathscr{G}(y) > K$, this is clearly satisfied, so let us assume that $\mathsf{d}_X(\mathscr{F}, \mathscr{G}) \leq \omega$ and $\lambda\mathscr{G}(y) \leq K$. By the previous lemma, we then find that $\lambda\mathscr{F}(y) \leq \lambda\mathscr{G}(y) + \mathsf{d}_X(\mathscr{F}, \mathscr{G}) \leq K + \omega$, hence

$$\omega = \theta_{]K+\omega,\infty]}(\lambda\mathscr{F}(y)) \wedge \omega \leq \varphi(y) + \varepsilon \leq \mathsf{d}_X(\mathscr{F}, \mathscr{G}) + \varphi(y) + \varepsilon,$$

which proves our claim. □

This now allows us to draw the conclusion which we require.

12.6.7 Proposition App ⊆ EpiAp.

Proof Using notations as before, it is easily seen that the Top-coreflection of $(\mathsf{F}(X) \times X, \widehat{\mathscr{B}_{\oplus}})$ is $(\mathsf{F}(X) \times X, \mathscr{T}_{\mathsf{d}_X} \times \mathscr{T}_{\bar{\lambda}})$ from which the conclusion follows. □

STEP 2: Our next goal is to show that EpiAp is a cartesian closed topological category.

12.6.8 Proposition *Let $f : X \longrightarrow Y$ be a contraction between* PsAp*-objects, then $\bar{f} : (\mathsf{F}(X), \mathsf{d}_X) \longrightarrow (\mathsf{F}(Y), \mathsf{d}_Y) : \mathscr{F} \mapsto f(\mathscr{F})$ is also a contraction.*

Proof Since $f : X \longrightarrow Y$ is a contraction, we find that

$$\begin{aligned} f(\dot{F}^{\rho}) &= f(\{y \in X \mid \exists x \in F : \delta_{\bar{\lambda}_X}(x, \{y\}) \leq \rho\}) \\ &\subseteq \{y \in Y \mid \exists x \in f(F) : \delta_{\bar{\lambda}_Y}(x, \{y\}) \leq \rho\} = f(F)^{\cdot\rho} \end{aligned}$$

and, hence, $f(\mathscr{F})^{\cdot\rho} \subseteq f(\dot{\mathscr{F}}^{\rho})$ (for all $\mathscr{F} \in \mathsf{F}(X)$ and $\rho \geq 0$). Consequently, $\dot{\mathscr{G}}^{\rho} \subseteq \mathscr{F}$ implies that $f(\mathscr{G})^{\cdot\rho} \subseteq f(\dot{\mathscr{G}}^{\rho}) \subseteq f(\mathscr{F})$, which means that $\mathsf{d}_Y(f(\mathscr{F}), f(\mathscr{G})) \leq \mathsf{d}_X(\mathscr{F}, \mathscr{G})$. □

12.6.9 Theorem EpiAp *is concretely reflective in* PsAp*, in particular,* EpiAp *is a topological category.*

Proof Let $(f_i : (X, \lambda) \longrightarrow (X_i, \lambda_i))_{i \in I}$ be initial in PsAp, where all (X_i, λ_i) are in EpiAp. To show that (X, λ) satisfies (C), assume that $A \in \mathscr{T}_{\mathbb{P}}$, then it follows that

$$\lambda^{-1}(A) := \bigcup_{i \in I} \left(\lambda_i \circ (\bar{f}_i \times f_i)\right)^{-1}(A).$$

From the contractivity of all f_i, $i \in I$, the foregoing result and the fact that all λ_i, $i \in I$, satisfy (C) it follows that $\lambda^{-1}(A)$ is open. Hence λ is continuous and (X, λ) is in EpiAp. □

12.6.10 Proposition *Let X and Y be* PsAp*-objects, and let $\mathscr{G}$ be a filter on X, then the map $\bar{\mathscr{G}} : \mathsf{F}(\mathsf{PsAp}(X,Y)) \longrightarrow \mathsf{F}(Y) : \Psi \mapsto \Psi(\mathscr{G})$ is a contraction, i.e. for any pair Φ and Ψ of filters on* $\mathsf{PsAp}(X,Y)$: $\mathsf{d}_Y(\Phi(\mathscr{G}), \Psi(\mathscr{G})) \leq \mathsf{d}_{\mathsf{PsAp}(X,Y)}(\Phi, \Psi)$.

Proof Since $\mathrm{ev}_x : [X,Y] \longrightarrow Y$ is a contraction for any $x \in X$, it follows that

$$\begin{aligned}
\mathrm{ev}(G \times \dot{\varphi}^{\rho}) &= \bigcup_{x\in G} \mathrm{ev}_x(\dot{\varphi}^{\rho}) \\
&\subseteq \bigcup_{x\in G} (\mathrm{ev}_x(\varphi))^{\cdot\rho} \\
&\subseteq (\bigcup_{x\in G} \mathrm{ev}_x(\varphi))^{\cdot\rho} \\
&= (\mathrm{ev}(G \times \varphi))^{\cdot\rho},
\end{aligned}$$

and, hence, $\Phi(\mathscr{F})^{\cdot\rho} \subseteq \dot{\Phi}^{\rho}(\mathscr{F})$ (for all $\Phi \in \mathsf{F}(\mathsf{PsAp}(X,Y))$ and $\mathscr{F} \in \mathsf{F}(X)$).

Consequently, $\dot{\Psi}^{\rho} \subseteq \Phi$ implies that $\Psi(\mathscr{G})^{\cdot\rho} \subseteq \dot{\Psi}^{\rho}(\mathscr{G}) \subseteq \Phi(\mathscr{G})$, which means that $\mathsf{d}_Y(\Phi(\mathscr{G}), \Psi(\mathscr{G})) \leq \mathsf{d}_{\mathsf{PsAp}(X,Y)}(\Phi, \Psi)$. □

12.6.11 Theorem EpiAp *is closed under the formation of power-objects in* PsAp. *Moreover, if X is in* PsAp *and Y is in* EpiAp, *then $[X,Y]$ is in* EpiAp. *In particular,* EpiAp *is a cartesian closed category.*

Proof Let λ be the limit-operator of $[X,Y]$. To show that $[X,Y]$ satisfies (C), assume that $A \in \mathscr{T}_{\mathbb{P}}$, then it follows from the formula of λ (as a function of two variables) that

$$\lambda^{-1}(A) := \bigcup_{(\mathscr{F},x)\in \mathsf{F}(X)\times X} \Big(\big(\lambda_Y \circ (\bar{\mathscr{F}} \times \mathrm{ev}_x)\big)^{-1}\big(A \cap]\lambda_X\mathscr{F}(x), \infty]\big)\Big).$$

Hence it follows from the fact that all ev_x, $x \in X$, are contractions, the foregoing proposition and the fact that λ_Y satisfies (C) that $\lambda^{-1}(A)$ is open. Hence $[X,Y]$ is in EpiAp. □

STEP 3: We now turn to showing that proper "density" conditions are satisfied.

12.6.12 Theorem App *is finally dense in* EpiAp.

Proof This follows from 12.2.2. □

12.6.13 Corollary App *is a concretely reflective subcategory of* EpiAp.

To show the other required density, we first indicate the following lemma.

12.6.14 Lemma *Let* $\mathscr{F} \in \mathsf{F}(\mathbb{P})$ *and* $\varepsilon \geq 0$ *and* $0 \leq y < \infty$. *Then*

$$\lambda_{\mathbb{P}}(\mathscr{F})(y) \leq \varepsilon \Leftrightarrow \forall \beta > \varepsilon :]y - \beta, \infty] \in \mathscr{F}.$$

Proof This follows from the description of $\lambda_{\mathbb{P}}$ in 1.2.62. □

In the following we assume without restriction that $X \neq \emptyset$.

12.6.15 Proposition *Let* X *be in* EpiAp, *then the map*

$$j : X \longrightarrow [[X, \mathbb{P}], \mathbb{P}]$$

defined by $j(x)(f) = f(x)$ *is an initial contraction.*

Proof We first give the following diagram for clarity and mention that $j := \mathrm{ev}^*_{(X,\lambda),\mathbb{P}}$ is the map which makes the following diagram commute:

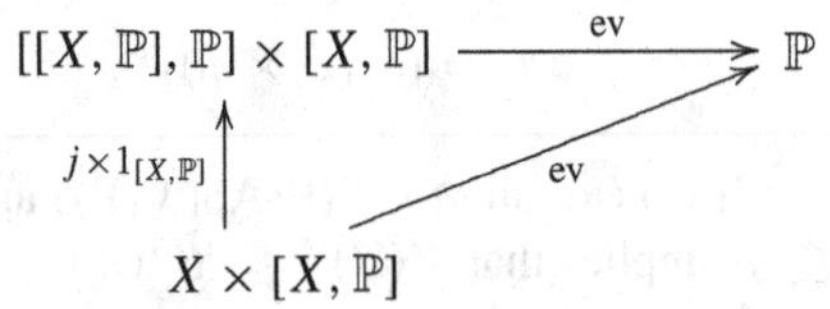

Hence, by properties of power-objects, j is a contraction.

In the following we also let

$$\lambda_H \text{ be the limit-operator on } [X, \mathbb{P}],$$
$$\lambda_{HH} \text{ be the limit-operator on } [[X, \mathbb{P}], \mathbb{P}],$$
$$\text{and } \hat{A}^{\delta} := \{f \in \mathsf{EpiAp}(X, \mathbb{P}) \mid f(A) \subseteq]\delta, \infty]\} \ (A \subseteq X, 0 \leq \delta < \infty).$$

To prove that j is initial, we will show for every ultrafilter $\mathscr{U}$ on X, $a \in X$ and $0 < K < \infty$ that $\lambda\mathscr{U}(a) > K$ implies that $\lambda_{HH}(j(\mathscr{U}))(j(a)) \geq K$. By 1.3.12, we then find that j is initial.

By 12.6.3 and the fact that X is in EpiAp, we find $V \in \mathscr{V}_{\bar{\lambda}}(a)$ and $\delta' > 0$ such that for all $\mathscr{F} \in \mathsf{F}(X)$ with $\mathsf{d}_X(\mathscr{U}, \mathscr{F}) < \delta'$ and $x \in V$, we have $\lambda\mathscr{F}(x) > K$.

Consider the map

$$g_0 : X \longrightarrow \mathbb{P} : \alpha \mapsto \delta_{\bar{\lambda}}(\alpha, X \setminus V),$$

which is a contraction by 1.3.4. As $V \in \mathscr{V}_{\bar{\lambda}}(a)$, we find that $\delta_1 := g_0(a) > 0$.

First assume that $g_0(a) < K$, then define $g_1 := g_0 + (K - \delta_1)$ and finally $g := g_1 \wedge K$ and $\delta'' := \delta_1$. We then find that $g \in \mathsf{EpiAp}(X, \mathbb{P})$, $g \leq K$, $g(a) = K$ and $\{g > K - \delta''\} \subseteq V$.

If however $g_0(a) \geq K$, then define $g := g_0 \wedge K$ and choose $0 < \delta'' < K$, then g and δ'' also fulfil the foregoing properties.

Now choose $0 < \delta < \delta' \wedge \delta''$ and define Ψ to be the filter on $\mathsf{EpiAp}(X, \mathbb{P})$ generated by the filterbasis

$$\{\hat{F}^{\delta} \mid F \neq \emptyset, \dot{F}^{\delta} \notin \mathscr{U}\}.$$

(It is clear that this is a filterbasis, as $\mathscr{U}$ is an ultrafilter and the constant ∞-function belongs to every set in this collection. Furthermore, if no such F were to exist, it would suffice to define $\Psi := \dot{0}$).

We now prove that $\lambda_H \Psi(g) \leq K - \delta$. Let $\mathscr{F}$ be a filter on X and $x \in X$ such that $\lambda_{\mathbb{P}}(\Psi(\mathscr{F}))(g(x)) > K - \delta$. Hence $K - \delta < \lambda_{\mathbb{P}}(\Psi(\mathscr{F}))(g(x)) \leq g(x)$, and consequently $g(x) > K - \delta$ and therefore $x \in V$. We also find that $\dot{\mathscr{F}}^{\delta} \subseteq \mathscr{U}$ (meaning that $\mathrm{d}_X(\mathscr{U}, \mathscr{F}) \leq \delta < \delta'$). If this were not the case, then we could find $F \in \mathscr{F}$ such that $\dot{F}^{\delta} \notin \mathscr{U}$, implying $\hat{F}^{\delta} \in \Psi$, hence $]\delta, \infty] \in \Psi(\mathscr{F})$. As $g(x) \leq K$, this implies in particular that

$$\forall \beta > K - \delta :]g(x) - \beta, \infty] \in \Psi(\mathscr{F}).$$

By the previous lemma, this means that $\lambda_{\mathbb{P}}(\Psi(\mathscr{F}))(g(x)) \leq K - \delta$, hence we have a contradiction.

By previous choices, we then find that $\lambda \mathscr{F}(x) > K$. Also

$$\lambda_{\mathbb{P}}(\Psi(\mathscr{F}))(g(x)) \leq g(x) \leq K,$$

and consequently $\lambda_{\mathbb{P}}(\Psi(\mathscr{F}))(g(x)) \leq \lambda \mathscr{F}(x)$. By definition of λ_H, we can thus conclude that $\lambda_H \Psi(g) \leq K - \delta$.

Now we show that $\lambda_{HH}(j(\mathscr{U}))(j(a)) \geq K$, by demonstrating that

$$\lambda_{\mathbb{P}}(j(\mathscr{U})(\Psi))(j(a)(g)) = \lambda_{\mathbb{P}}(\Psi(\mathscr{U}))(g(a)) = K > K - \delta = \lambda_H \Psi(g).$$

Let us assume the contrary, i.e. $\lambda_{\mathbb{P}}(\Psi(\mathscr{U}))(g(a)) \leq K - \varepsilon$, where $0 < \varepsilon < K$. This implies that $\forall \beta > K - \varepsilon : (g(a) - \beta, \infty] \in \Psi(\mathscr{U})$, hence

$$\forall \beta > K - \varepsilon : \exists U \in \mathscr{U} : \hat{U}^{K-\beta} \in \Psi.$$

In particular, for some $\gamma \geq 0$, $U \in \mathscr{U} : \hat{U}^{\gamma} \in \Psi$. Consequently, $\hat{F}^{\delta} \subseteq \hat{U}^{\gamma}$ for some $F \neq \emptyset$, $\dot{F}^{\delta} \notin \mathscr{U}$. However, this implies that if $z \notin \dot{F}^{\delta}$, then $\delta_{\bar{\lambda}}(-, \{z\}) \in \hat{F}^{\delta}$, thus $\delta_{\bar{\lambda}}(-, \{z\}) \in \hat{U}^{\gamma}$, and hence $z \notin U$ which shows that $U \subseteq \dot{F}^{\delta}$. This is a contradiction. □

STEP 4: Now we are in a position to combine all previous results and to prove the final step.

12.6.16 Theorem EpiAp *is the cartesian closed topological hull of* App.

Proof By the previous result, for any X in EpiAp we have an initial map $j : X \longrightarrow [[X, \mathbb{P}], \mathbb{P}]$ and since the functor $[-, \mathbb{P}]$:EpiAp $\longrightarrow$ EpiAp transforms final epi-sinks into initial sources (see Herrlich and Nel 1977) (and by 12.6.12, we can obtain $[X, \mathbb{P}]$ as a final lift of an epi-sink involving App-objects), we find that the class $\{\mathsf{App}(X,Y) \mid X, Y \in \mathsf{App}\}$ is initially dense in EpiAp and we are finished. □

We now show that EpiTop = CCTH(Top) has a nice relation to EpiAp. To this end, we first recall some facts regarding EpiTop introduced in Bourdaud (1975).

12.6.17 Definition Let (X,q) be a pseudotopological space. We denote its Top-reflection by $(X, \bar{q})$ and define the *point-operator* (with respect to (X,q)) as

$$\bullet : 2^X \longrightarrow 2^X : A \mapsto A^\bullet := \{x \in X \mid \mathrm{cl}_{\bar{q}}(\{x\}) \cap A \neq \emptyset\}.$$

Note that the point-operator determines a topological space, i.e. it is a topological closure operator.

12.6.18 Definition A pseudotopological space X is called an *Antoine space* or *epi-topological space* if and only if it satisfies the following properties (where $\mathcal{F}^\bullet$ is the filter generated by $\{F^\bullet \mid F \in \mathcal{F}\}$).

1. $\forall \mathcal{F} \in \mathsf{F}(X) : \lim \mathcal{F}$ is closed in $(X, \bar{q})$ (closed-domainedness).
2. $\forall \mathcal{F} \in \mathsf{F}(X) : \lim \mathcal{F} = \lim \mathcal{F}^\bullet$ (point-regularity).

The full subcategory of PsTop consisting of Antoine spaces is denoted by EpiTop and it was shown by work of Machado (1973) and Bourdaud (1975) that EpiTop = CCTH(Top).

12.6.19 Proposition *Let* $(X,q) \in \mathsf{PsTop}$, *then* $(X, \lambda_{\bar{q}}) = (X, \overline{\lambda_q})$.

Proof We will prove this by showing that $(X, \lambda_{\bar{q}})$ is also the App-reflection of (X, λ_q). To this end, let $f : (X, \lambda_q) \longrightarrow (X, \delta)$ be a contraction, where (X, δ) is in App. But then also $f : (X, \lambda_q) \longrightarrow (X,q_\delta)$ is a contraction, where we recall that the latter space is the PsTop-coreflection of (X, δ). Since we observed earlier that the Top-coreflection in App is just the restriction of the PsTop-coreflection, it follows that (X,q_δ) is a topological space, hence $f : (X, \lambda_{\bar{q}}) \longrightarrow (X,q_\delta)$ is a contraction. Consequently, $f : (X, \lambda_{\bar{q}}) \longrightarrow (X, \delta)$ is a contraction. □

12.6.20 Theorem EpiAp ∩ PsTop = EpiTop.

Proof If (X, λ) is in PsTop, one easily finds that condition (C) is equivalent to

(T) $\forall \mathcal{F} \not\longrightarrow a, \exists V \in \mathcal{V}_{(X,\mathcal{T}_{\bar{\lambda}})}(a) : \dot{\mathcal{G}}^0 \subseteq \mathcal{F}$ and $x \in V \Rightarrow \mathcal{G} \not\longrightarrow x$.

Also observe that in this case $\dot{\mathcal{G}}^0 = \mathcal{G}^\bullet$. Now assume that (T) holds (i.e. (X, λ) is at the same time in EpiAp and in PsTop).

Since $\dot{\mathcal{F}}^0 \subseteq \mathcal{F}$ (for all $\mathcal{F} \in \mathsf{F}(X)$), we find that for all $\mathcal{F} \not\longrightarrow a$ there exists $V \in \mathcal{V}_{(X,\mathcal{T}_{\bar{\lambda}})}(a)$ such that $V \subseteq (X \setminus \lim \mathcal{F})$, meaning $\lim \mathcal{F}$ is closed in $(X, \mathcal{T}_{\bar{\lambda}})$. Also, by letting $\mathcal{F} = \dot{\mathcal{H}}^0$, we find that $\dot{\mathcal{H}}^0 \not\longrightarrow a$ implies $\mathcal{H} \not\longrightarrow a$, hence $\lim \mathcal{H} = \lim \dot{\mathcal{H}}^0$ (for all $\mathcal{H} \in \mathsf{F}(X)$). Consequently, (X, λ) is a closed-domained, point-regular pseudotopological space (i.e. an Antoine space, see Antoine 1966a, b, c; Bourdaud 1974, 1975, 1976).

Conversely, assume $(X, \lambda) \in \mathsf{EpiTop}$ and $\mathcal{F} \not\longrightarrow a$. Let $V := X \setminus \lim \mathcal{F} \in \mathcal{V}_{(X,\mathcal{T}_{\bar{\lambda}})}(a)$ (as (X, λ) is closed-domained) and suppose $\dot{\mathcal{G}}^0 \subseteq \mathcal{F}$ and $x \in V$. We then find that $\mathcal{G} \not\longrightarrow x$, for if this were not the case, then also $\dot{\mathcal{G}}^0 \longrightarrow x$ (as (X, λ) is point-regular), implying $\mathcal{F} \longrightarrow x$, which is a contradiction. □

12.6.21 Theorem EpiTop *is stable in* EpiAp.

Proof Reflectivity is clear. As for coreflectivity, let (X, λ) be in EpiAp, then we show that (X, λ'), the PsTop-coreflection of (X, λ), is in EpiTop.

To this end, assume that $\mathcal{F} \overset{q_{\lambda'}}{\not\longrightarrow} x$, i.e. $\lambda\mathcal{F}(x) > 0$. Since $(X, \lambda) \in \mathsf{EpiAp}$, we find $V \in \mathcal{V}_{(X,\mathcal{T}_{\bar{\lambda}})}$ and $\delta > 0$ such that for $y \in V$ and $\mathsf{d}_{(X,\lambda)}(\mathcal{F}, \mathcal{G}) < \delta$, we have that $\lambda\mathcal{G}(y) > 0$.

As $1_X : (X, \lambda') \longrightarrow (X, \lambda)$ is a contraction, we find that $V \in \mathcal{V}_{(X,\mathcal{T}_{\bar{\lambda}'})}$ and that $\mathsf{d}_{(X,\lambda')}(\mathcal{F}, \mathcal{G}) < \delta$ implies that $\mathsf{d}_{(X,\lambda)}(\mathcal{F}, \mathcal{G}) < \delta$ (by 12.6.8).

Putting things together, we therefore obtain:

$$\forall \mathcal{F} \overset{q_{\lambda'}}{\not\longrightarrow} x, \exists V \in \mathcal{V}_{(X,\mathcal{T}_{\bar{\lambda}'})} : (y \in V \text{ and } \mathcal{G}^\bullet \subseteq \mathcal{F}) \Rightarrow \lambda(\mathcal{G})(y) > 0 \Rightarrow \mathcal{G} \overset{q_{\lambda'}}{\not\longrightarrow} y.$$

Consequently, (X, λ') satisfies (T) and is in PsTop, hence it is in EpiTop. □

The situation of the various hulls of the foregoing results are depicted in the following diagram.

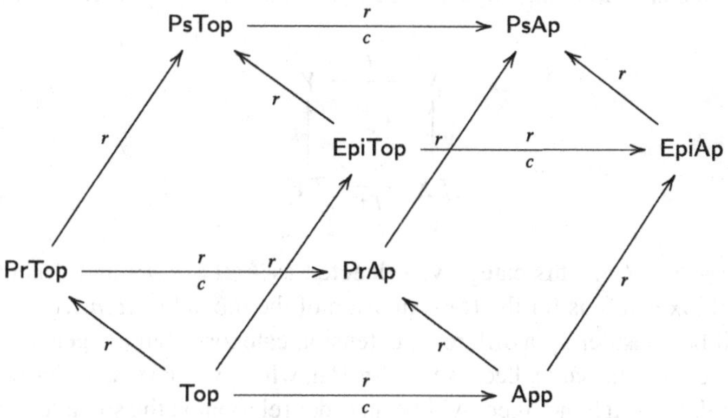

Fig. 12.1 The extensional, cartesian closed and quasi-topos hulls of App

12.7 A Lax-Algebraic Characterization of App

There is an interesting way to see that App can be viewed as the category of lax algebras associated with the ultrafilter monad (see Clementino and Hofmann 2003; Clementino et al. 2004).

In Clementino and Hofmann (2003) the proof that the category of lax algebras is isomorphic to App is somewhat circuitous, since a detour is made via distances. However, this is not necessary, and here we will present a new and more straightforward proof which makes direct links, between the lax algebraic structures for the ultrafilter monad and limit operators.

We recall the general principles for a lax setting with Rel as the "extension category". We suppose given a monad (T, e, m) (where T : Set $\longrightarrow$ Set and where e and m are respectively the unit and the multiplication) which can be extended to Rel, meaning that there is a lax-functor (denoted by the same symbol) T : Rel $\longrightarrow$ Rel which extends the original Set-Set functor such that the usual lax-diagrams hold (see e.g. Hofmann et al. 2014). A lax algebra for the monad is a pair (X, a) where $X \in$ Set and $a : TX \longrightarrow X$ is a relation such that

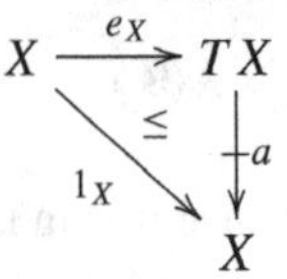

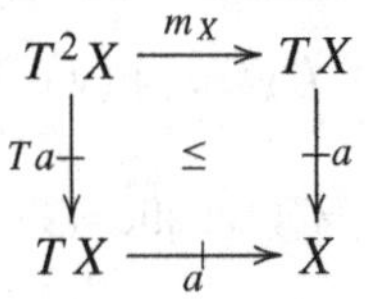

$$
\begin{array}{ccc}
X \xrightarrow{\;e_X\;} TX & \qquad & T^2X \xrightarrow{\;m_X\;} TX \\
\quad\searrow{\scriptstyle\le}\;\;\Big\downarrow{\scriptstyle a} & & Ta\Big\downarrow\quad{\scriptstyle\le}\quad\Big\downarrow{\scriptstyle a} \\
\quad\;{\scriptstyle 1_X}\quad X & & TX \xrightarrow[\;a\;]{} X
\end{array}
$$

i.e. $1_X \le a \circ e_X$ and $a \circ Ta \le a \circ m_X$.

These conditions are respectively called the *reflexivity* and the *transitivity* condition. Morphisms from (X, a) to (Y, b) are functions $f : X \longrightarrow Y$ satisfying:

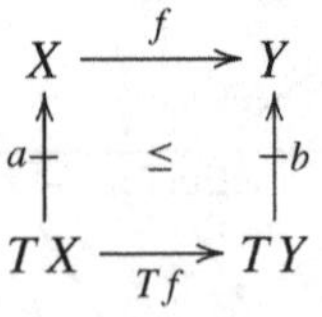

$$
\begin{array}{ccc}
X & \xrightarrow{\;f\;} & Y \\
a\Big\uparrow & {\scriptstyle \le} & \Big\uparrow b \\
TX & \xrightarrow[\;Tf\;]{} & TY
\end{array}
$$

i.e. $f \circ a \le b \circ Tf$. This category is denoted as $\mathrm{Alg}(T, e, m)$ and it is called the category of lax algebras for the Rel-extension of the monad (T, e, m).

We will be considering a different "extension category" but the general idea and principles remain the same, because, as for Rel, where we have an order relation on the set of relations at hand, there will be an order relation on the so-called numerical relations (which are $\mathbb{P}$-valued functions).

Precisely, we consider the *ultrafilter monad*, given by the following data. First we have the functor

$$\mathsf{U} : \mathsf{Set} \longrightarrow \mathsf{Set} : \begin{cases} X \mapsto \mathsf{U}(X) \\ f \mapsto \mathsf{U}(f) \end{cases}$$

where $\mathsf{U}(f)(\mathscr{U}) := \{A \mid f^{-1}(A) \in \mathscr{U}\}$ (also generated by $\{f(U) \mid U \in \mathscr{U}\}$). Since we have always denoted this extension of the map f simply by f we will continue to do so in the sequel.

Further we have the unit and multiplication

$$e_X : X \longrightarrow \mathsf{U}(X) : x \mapsto \dot{x} \quad \text{and} \quad m_X : \mathsf{U}(\mathsf{U}(X)) \longrightarrow \mathsf{U}(X) : \mathfrak{X} \mapsto \bigcup_{\mathscr{A} \in \mathfrak{X}} \bigcap_{\mathscr{U} \in \mathscr{A}} \mathscr{U}.$$

This monad has a lax extension $\mathsf{U} : \mathsf{Rel} \longrightarrow \mathsf{Rel}$ where for any sets X and Y, any relation $r : X \relbar\!\!\!\!+\!\!\!\rightarrow Y$ and any ultrafilters $\mathscr{U} \in \mathsf{U}(X)$ and $\mathscr{W} \in \mathsf{U}(Y)$ we have

$$\mathscr{U}\,\mathsf{U}(r)\mathscr{W} \Leftrightarrow \forall W \in \mathscr{W} : \{x \in X \mid \exists y \in W : xry\} \in \mathscr{U}.$$

(Note that a relation is a subset $r \subseteq X \times Y$ but that in the present context this is usually denoted $r : X \relbar\!\!\!\!+\!\!\!\rightarrow Y$, a practice to which we will adhere).

However, instead of considering Rel a new, numerical version is introduced, namely $\mathbb{P}\mathsf{Rel}$, the objects of which are sets and the morphisms of which are so-called *numerical relations*. These are functions $d : X \times Y \longrightarrow \mathbb{P}$ which, in a similar vain as in the Rel-case, we will denote as $d : X \relbar\!\!\!\!\twoheadrightarrow Y$. The identity of X is $\theta_{\Delta(X)}$ where $\Delta(X)$ stands for the diagonal of $X \times X$. For any set X, $\mathbb{P}\mathsf{Rel}(X)$ is equipped with the pointwise order and the composition of two numerical relations $d : X \relbar\!\!\!\!\twoheadrightarrow Y$ and $e : Y \relbar\!\!\!\!\twoheadrightarrow Z$ is defined as

$$e \circ d(x,z) := \inf_{y \in Y}(e(x,y) + d(y,z)).$$

The lax-extension of the ultrafilter monad to $\mathbb{P}\mathsf{Rel}$ is defined as follows. For each $d : X \relbar\!\!\!\!\twoheadrightarrow Y$ and for each $\alpha \in \mathbb{P}$, we have the relation $d_\alpha : X \relbar\!\!\!\!+\!\!\!\rightarrow Y$ given by

$$x d_\alpha y \Leftrightarrow d(x,y) \leq \alpha.$$

For any subset $A \subseteq X$ and any subset $\mathscr{A} \subseteq 2^X$ we set

$$d_\alpha(A) := \{y \in Y \mid \exists x \in A : x d_\alpha y\} \quad \text{and} \quad d_\alpha(\mathscr{A}) := \{d_\alpha(A) \mid A \in \mathscr{A}\}.$$

Then the assignment

$$\mathsf{U}(d) : \mathsf{U}(X) \times \mathsf{U}(Y) \longrightarrow \mathbb{P} : (\mathscr{U}, \mathscr{W}) \mapsto \inf\{\alpha \in \mathbb{P} \mid d_\alpha(\mathscr{U}) \subseteq \mathscr{W}\}$$

determines a lax extension of the ultrafilter monad to $\mathbb{P}\mathsf{Rel}$. For details we refer to Clementino and Hofmann (2003).

A lax algebra for this monad is a pair (X, a) where X is a set and $a : \mathsf{U}(X) \twoheadrightarrow X$. Reflexivity of a means that

$$\forall x \in X : a(\dot{x}, x) = 0,$$

and transitivity means that

$$\forall \mathfrak{X} \in \mathsf{U}^2(X), \forall \mathscr{U} \in \mathsf{U}(X), \forall x \in X : a(m_X(\mathfrak{X}), x) \leq \mathsf{U}(a)(\mathfrak{X}, \mathscr{U}) + a(\mathscr{U}, x).$$

That the category of lax algebras for the $\mathbb{P}$Rel-extension of the ultrafilter monad is isomorphic to App was proved in Clementino and Hofmann (2003) going via distances. We will give a completely new and straightforward proof identifying the lax-algebraic structures as limit operators via our simplified two-axiom characterization of limit operators.

We recall from our previous investigations, in particular 1.1.11, that we can indeed already characterize the structure of an approach space X by a limit operator λ defined for ultrafilters, satisfying two axioms, namely (L1) which says that

$$\forall x \in X : \lambda \dot{x}(x) = 0,$$

and (LU*) which says that for any set J

$$\forall \psi : J \longrightarrow X, \forall \sigma : J \longrightarrow \mathsf{U}(X), \forall \mathscr{F} \in \mathsf{U}(J) : \lambda \Sigma\sigma(\mathscr{F}) \leq \lambda\psi(\mathscr{F}) + \inf_{F \in \mathscr{F}} \sup_{j \in F} \lambda\sigma(j)\psi(j).$$

In order to be able to handle the transitivity formula we first prove the following lemma.

12.7.1 Lemma *For all $\mathfrak{X} \in \mathsf{U}^2(X)$ and $\mathscr{U} \in \mathsf{U}(X)$ we have*

$$\mathsf{U}(a)(\mathfrak{X}, \mathscr{U}) = \sup_{\mathscr{A} \in \mathfrak{X}} \sup_{U \in \mathscr{U}} \inf_{\mathscr{W} \in \mathscr{A}} \inf_{x \in U} a(\mathscr{W}, x).$$

Proof Let $\mathsf{U}(a)(\mathfrak{X}, \mathscr{U}) < \varepsilon$. Then there exists $\alpha < \varepsilon$ such that for all $\mathscr{A} \in \mathfrak{X}$, $a_\alpha(\mathscr{A}) \in \mathscr{U}$. Now take $\mathscr{A} \in \mathfrak{X}$ and $U \in \mathscr{U}$ then it follows that $U \cap a_\alpha(\mathscr{A}) \neq \emptyset$ and hence we can choose $y \in U \cap a_\alpha(\mathscr{A})$. Consequently there exists $\mathscr{W} \in \mathscr{A}$ such that $a(\mathscr{W}, y) \leq \alpha$. Hence

$$\inf_{\mathscr{Z} \in \mathscr{A}} \inf_{z \in U} a(\mathscr{Z}, z) \leq \alpha < \varepsilon.$$

Conversely, suppose that

$$\sup_{\mathscr{A} \in \mathfrak{X}} \sup_{U \in \mathscr{U}} \inf_{\mathscr{Z} \in \mathscr{A}} \inf_{z \in U} a(\mathscr{Z}, z) < \varepsilon.$$

Take $\mathscr{A} \in \mathfrak{X}$ and consider $a_\varepsilon(\mathscr{A}) = \{y \mid \exists \mathscr{W} \in \mathscr{A} : a(\mathscr{W}, y) \leq \varepsilon\}$. Suppose that $a_\varepsilon(\mathscr{A}) \notin \mathscr{U}$ then there exist $\mathscr{Z} \in \mathscr{A}$ and $z \notin a_\varepsilon(\mathscr{A})$ such that $a(\mathscr{Z}, z) < \varepsilon$. However if $z \notin a_\varepsilon(\mathscr{A})$ then $a(\mathscr{Z}, z) > \varepsilon$ which is a contradiction. Hence $a_\varepsilon(\mathscr{A}) \in \mathscr{U}$ and we are finished. □

12.7.2 Theorem *The category of lax algebras* $\mathsf{Alg}(\mathsf{U}, e, m)$ *for the* $\mathbb{P}\mathsf{Rel}$*-extension of the ultrafilter monad is isomorphic to* App.

Proof Reflexivity clearly is equivalent to (L1). So all that remains to be shown is that transitivity is equivalent to (LU*).

Let λ be a limit operator on X and let $\mathfrak{X} \in \mathsf{U}^2(X)$ and $\mathscr{U} \in \mathsf{U}(X)$. Put

$$\varepsilon := \mathsf{U}(a)(\mathfrak{X}, \mathscr{U}) = \sup_{\mathscr{A} \in \mathfrak{X}} \sup_{U \in \mathscr{U}} \inf_{\mathscr{W} \in \mathscr{A}} \inf_{x \in U} a(\mathscr{W}, x).$$

Let $\rho > 0$ and put

$$J := \{(\mathscr{G}, y) \in \mathsf{U}(X) \times X \mid \lambda \mathscr{G}(y) \leq \varepsilon + \rho\},$$

and consider the projections

$$\begin{array}{ccc} J & \xrightarrow{\psi := \mathrm{pr}_2} & X \\ {\scriptstyle \sigma := \mathrm{pr}_1}\downarrow & & \\ \mathsf{U}(X) & & \end{array}$$

Note that, by definition of ε and ρ, $\mathfrak{X} \times \mathscr{U}$ has a trace on J and consequently we can choose an ultrafilter $\mathscr{R} \in \mathsf{U}(J)$ finer than $\mathfrak{X} \times \mathscr{U}$. It then follows that

$$\mathfrak{X} = \mathrm{pr}_1(\mathscr{R}) = \sigma(\mathscr{R}) \text{ and } \mathscr{U} = \mathrm{pr}_2(\mathscr{R}) = \psi(\mathscr{R}),$$

and because of (LU*) we obtain, for any $x \in X$

$$\lambda \Sigma \sigma(\mathscr{R})(x) \leq \lambda \psi(\mathscr{R})(x) + \sup_{R \in \mathscr{R}} \inf_{z \in R} \lambda \sigma(z)(\psi(z))$$

and thus

$$\begin{aligned} \lambda m_X(\mathfrak{X})(x) &\leq \lambda \mathscr{U}(x) + \sup_{R \in \mathscr{R}, R \subseteq J} \inf_{(\mathscr{G}, y) \in R} \lambda \mathscr{G}(y) \\ &\leq \lambda \mathscr{U}(x) + \varepsilon + \rho. \end{aligned}$$

Consequently, by arbitrariness of ρ and the definition of ε it follows that λ satisfies the transitivity axiom.

Conversely let $a : \mathsf{U}(X) \rightarrow\!\!\!\!\!\rightarrow X$ satisfy the transitivity axiom and let J be a set, $\psi : J \longrightarrow X$, $\sigma : J \longrightarrow \mathsf{U}(X)$ and $\mathscr{F} \in \mathsf{U}(J)$. Put

$$\mathfrak{X} := \sigma(\mathscr{F}) \text{ and } \mathscr{U} := \psi(\mathscr{F}).$$

Then it follows that, for any $x \in X$

$$\begin{aligned}
a(m_X(\sigma(\mathscr{F})), x) &\leq a(\psi(\mathscr{F}), x) + \sup_{\mathscr{A} \in \sigma(\mathscr{F})} \sup_{U \in \psi(\mathscr{F})} \inf_{\mathscr{V} \in \mathscr{A}} \inf_{y \in U} a(\mathscr{V}, y) \\
&\leq a(\psi(\mathscr{F}), x) + \sup_{F \in \mathscr{F}} \inf_{\mathscr{V} \in \sigma(F)} \inf_{y \in \psi(F)} a(\mathscr{V}, y) \\
&\leq a(\psi(\mathscr{F}), x) + \sup_{F \in \mathscr{F}} \inf_{z \in F} a(\sigma(z), \psi(z)) \\
&= a(\psi(\mathscr{F}), x) + \inf_{F \in \mathscr{F}} \sup_{z \in F} a(\sigma(z), \psi(z))
\end{aligned}$$

which shows that a satisfies (LU*).

That via the identification of lax algebraic structures on the one hand with limit operators on the other hand, the morphisms in both categories coincide is an immediate consequence of the characterization of contractions via ultrafilters and the definition of morphisms in $\mathsf{Alg}(\mathsf{U}, e, m)$. □

12.8 Comments

1. **Stable subcategories of** App

The material in the first section of this chapter contains a correction to a result in Lowen (1997). In there it was namely stated that each stable subcategory equals App_Γ for some semigroup $\Gamma = \{0\} \cup [m, \infty]$, $m \in \mathbb{P}$. Although these semigroups, as seen from 12.1.7, do indeed generate stable subcategories, not all stable subcategories are generated by semigroups of this type, and the combined results of 12.1.6 and 12.1.7 contain the correct statement.

The interested reader will be able to verify that analogous results can be shown to characterize the stable subcategories of several of the other categories considered in this chapter.

2. **Stable subcategories of** PrAp

In a similar way as what we did for App it is possible to determine all stable subcategories of PrAp. With basically the same definitions, notations and concepts but now considering closed subsets of $\mathbb{P}$ which contain $\{0, \infty\}$ and without any semigroup requirements one can prove that the collection of stable subcategories of PrAp is given by the set $\{\mathsf{PrAp}_\Gamma \mid \{0, \infty\} \subseteq \Gamma \subseteq \mathbb{P}, \Gamma \text{ closed}\}$ where PrAp_Γ is the subcategory of PrAp with objects those (X, δ) for which $\mathrm{Im}\delta \subseteq \Gamma$.

Furthermore one can show that if Γ and Γ' are closed subsets of $\mathbb{P}$ containing $\{0, \infty\}$, then the subcategories PrAp_Γ and $\mathsf{PrAp}_{\Gamma'}$ are concretely isomorphic if and only if there exists an increasing bijection $\Gamma \to \Gamma'$. Finally one can also show that the set of stable subcategories of PrAp has cardinality $\mathfrak{c}$, and that there is a set of the same cardinality of non-isomorphic such subcategories.

3. **Further cartesian closed subcategories of** CAp

Bourdaud (1976) indicated the existence of a "family" of cartesian closed topological categories in Conv, the category of convergence spaces and continuous maps, where this "family" depended on certain choices of functors and of which the cartesian closed topological hull of Top is a particular instance. Also the CCT hull of Creg, is a specific instance of this family (see Bourdaud 1976). Such a family of CCT subcategories in CAp also exists with the CCT hull of App and the CCT hull of UAp as specific instances of this family. This, and more, can be found in the PhD thesis of Mark Nauwelaerts (2000).

4. **Premetric spaces in** PrAp

In Colebunders and Lowen (1988), it is shown that also the category of premetric spaces and non-expansive maps is embedded as a full and concretely coreflective category in PrAp. A premetric is a map measuring the distance between pairs of points with only condition that it has to be zero on the diagonal. The formula to embed such a space in PrAp is precisely the same as the one for metric spaces in App, i.e. given the premetric space (X, d) this is embedded in PrAp as the space (X, λ_d) where

$$\lambda_d(\mathscr{F})(x) := \inf_{F\in\mathscr{F}} \sup_{y\in F} d(x, y).$$

5. **Alternative lax-algebraic descriptions of** App

What we have denoted as $\mathsf{Alg}(\mathsf{U}, e, m)$ in the foregoing section, in Hofmann et al. (2014) is denoted as $(\beta, \mathbb{P}_+)$-Cat where β stands for the ultrafilter monad. However, since we were not dealing with other quantales than $\mathbb{P}$ and other extensions than $\mathbb{P}$Rel we preserved the original notation.

The first isomorphic description of approach spaces as lax algebras using a monad extension to Rel was given in Lowen and Vroegrijk (2008), based on the notion of functional ideals.

In Colebunders et al. (2011) an alternative way to obtain App as a category of lax algebras was described making use, for any set X, of certain functions from $\mathbb{P}_b^X$ to $\mathbb{P}$.

Meanwhile, several other isomorphic descriptions of App were obtained (Hofmann et al. 2014): first as $(\mathbb{F}, \mathbb{P}_+)$-Cat where $\mathbb{F}$ is the filter monad (this is based on Kleisli monoids and the fact that $\mathbb{F}$ is power-enriched (Hofmann et al. 2014)), second as $(\mathbb{J}, 2)$-Cat where $\mathbb{J}$ is a monad similar to what was used in Colebunders et al. (2011).

In a forthcoming paper by Colebunders, Lowen and Van Opdenbosch (2014) a power-enriched monad $\mathbb{I}$ is described and used to characterize App as $(\mathbb{I}, 2)$-Cat making extensive use of functional ideal convergence as given in the first chapter.

For a thorough study of lax algebraic theories and many more interesting results, also concerning the theory of approach spaces, we refer to Hofmann et al. (2014).

Appendix A
Formulas

For easy reference, we recall the transition formulas as well as the various formulas for initial and final structures which we proved throughout the text. Some formulas here are given in a concise form, a more general form can be found in the text (especially involving bases).

1. **Transition formulas from a distance δ**

   $\lambda\mathscr{F}(x) = \sup_{A\in\sec(\mathscr{F})} \delta(x, A).$
   $\alpha\mathscr{F}(x) = \sup_{F\in\mathscr{F}} \delta(x, F).$
   $\mathscr{A}(x) = \{\varphi \in \mathbb{P}^X \mid \forall A \subseteq X : \inf_{y\in A} \varphi(y) \leq \delta(x, A)\}.$
   $\mathscr{G} = \{d \in q\mathsf{Met}(X) \mid \forall A \subseteq X : \inf_{a\in A} d(\cdot, a) \leq \delta_A\}.$
   $\mathscr{G} = \{d \in q\mathsf{Met}(X) \mid \delta_d \leq \delta\}.$
   $\mathfrak{t}_\varepsilon(A) = A^{(\varepsilon)} = \{x \in X \mid \delta(x, A) \leq \varepsilon\}.$
   $\mathfrak{l}(\mu)(x) = \sup_{\omega<\infty} \sup_{\varepsilon>0} \inf_{i=1}^{n(\omega,\varepsilon)} (m_i^{\omega,\varepsilon} + \delta(x, M_i^{\omega,\varepsilon}))$ where, for each finite ω,

   $$(\inf_{i=1}^{n(\omega,\varepsilon)} (m_i^{\omega,\varepsilon} + \theta_{M_i^{\omega,\varepsilon}}))_{\varepsilon>0}$$

   is a development for $\mu \wedge \omega$.

2. **Transition formulas from a limit operator λ**

   $\delta(x, A) = \inf_{\mathscr{U}\in\mathsf{U}(A)} \lambda\mathscr{U}(x).$
   $\alpha\mathscr{F}(x) = \inf_{\mathscr{U}\in\mathsf{U}(\mathscr{F})} \lambda\mathscr{U}(x)\,.$
   $\mathscr{A}(x) = \{\varphi \in \mathbb{P}^X \mid \forall\mathscr{U} \in \mathsf{U}(X) : \sup_{U\in\mathscr{U}} \inf_{y\in U} \varphi(y) \leq \lambda\mathscr{U}(x)\}.$
   $\mathscr{G} = \{d \in q\mathsf{Met}(X) \mid \forall\mathscr{U} \in \mathsf{U}(X) : \sup_{U\in\mathscr{U}} \inf_{y\in U} d(\cdot, y) \leq \lambda\mathscr{U}\}.$

R. Lowen, *Index Analysis*, Springer Monographs in Mathematics,
DOI 10.1007/978-1-4471-6485-2

$\mathcal{G} = \{d \in q\mathsf{Met}(X) \mid \lambda_d \leq \lambda\}$.
$\mathfrak{t}_\varepsilon(A) = \{x \in X \mid \exists \mathcal{F} \in \mathsf{F}(A) : \lambda\mathcal{F}(x) \leq \varepsilon\}$.
$\mathfrak{J} \rightarrowtail x$ if and only if for all $\alpha \in [c(\mathfrak{J}), \infty[: \lambda \mathfrak{f}_\alpha \mathfrak{J}(x) \leq \alpha$.
$\mathfrak{U} \rightarrowtail x$ if and only if $\lambda \mathfrak{f}\mathfrak{U}(x) \leq c(\mathfrak{U})$ in case $\mathfrak{U}$ is prime.

3. **Transition formulas from an approach system** $\mathcal{A}$

$\delta(x, A) = \sup_{\varphi \in \mathcal{A}(x)} \inf_{y \in A} \varphi(y)$.
$\mathcal{G} = \{d \in pq\mathsf{M}^\infty(X) | \forall x \in X : d(x, \cdot) \in \mathcal{A}(x)\}$.
$\lambda\mathcal{F}(x) = \sup_{\varphi \in \mathcal{A}(x)} \inf_{F \in \mathcal{F}} \sup_{y \in F} \varphi(y)$.
$\alpha\mathcal{F}(x) = \sup_{\varphi \in \mathcal{A}(x)} \sup_{F \in \mathcal{F}} \inf_{y \in F} \varphi(y)$.
$\mathfrak{l}(\mu)(x) = \sup_{\varphi \in \mathcal{A}(x)} \inf_{y \in X} (\mu + \varphi)(y)$.
$\mathfrak{u}(\mu)(x) = \inf_{\varphi \in \mathcal{A}(x)} \sup_{y \in X} (\mu - \varphi)(y)$.
$\mathfrak{J} \rightarrowtail x$ if and only if $\mathcal{A}_b(x) \subseteq \mathfrak{J}$.

4. **Transition formulas from a gauge** $\mathcal{G}$

$\delta(x, A) = \sup_{d \in \mathcal{G}} \inf_{y \in A} d(x, y)$ or $\delta = \sup_{d \in \mathcal{G}} \delta_d$.
$\mathcal{A}(x) = \{\varphi \in \mathbb{P}^X \mid \{d(x, \cdot) \mid d \in \mathcal{G}\}$ dominates $\varphi\}$.
$\lambda\mathcal{F}(x) = \sup_{d \in \mathcal{G}} \inf_{F \in \mathcal{F}} \sup_{y \in F} d(x, y)$ or $\lambda = \sup_{d \in \mathcal{G}} \lambda_d$.
$\alpha\mathcal{F}(x) = \sup_{d \in \mathcal{G}} \sup_{F \in \mathcal{F}} \inf_{y \in F} d(x, y)$ or $\alpha = \sup_{d \in \mathcal{G}} \alpha_d$.
$\mathfrak{l}(\mu)(x) = \sup_{d \in \mathcal{G}} \inf_{y \in X} (\mu(y) + d(x, y))$ or $\mathfrak{l} = \sup_{d \in \mathcal{G}} \mathfrak{l}_d$.
$\mathfrak{u}(\mu)(x) = \inf_{d \in \mathcal{G}} \sup_{y \in X} (\mu(y) - d(x, y))$ or $\mathfrak{u} = \inf_{d \in \mathcal{G}} \mathfrak{u}_d$.

5. **Transition formulas from a lower regular function frame** $\mathfrak{L}$

$\delta(x, A) = \sup\{\rho(x) \mid \rho \in \mathfrak{L}, \rho_{|A} = 0\}$.
$\mathfrak{l}(\mu) = \sup\{\nu \in \mathfrak{L} \mid \nu \leq \mu\}$.
$\mathfrak{U} =< \{\alpha \ominus \mu \mid \mu \in \mathfrak{L}, \sup \mu < \alpha < \infty\} >$.

6. **Transition formulas from an upper regular function frame** $\mathfrak{U}$

$\mathcal{G} = \{d \mid \forall x \in X, \forall \omega < \infty : d(x, \cdot) \wedge \omega \in \mathfrak{U}\}$.
$\mathcal{A}(x) =< \{\mu \in \mathfrak{U} \mid \mu(x) = 0\} >$.
$\mathfrak{u}(\mu) = \inf\{\nu \in \mathfrak{U} | \mu \leq \nu\}$.
$\mathfrak{L} =< \{\alpha \ominus \mu \mid \mu \in \mathfrak{U}, \sup \mu < \alpha < \infty\} >$.

7. **Transition formulas from a tower** $\mathfrak{t}$

$\delta(x, A) = \inf\{\varepsilon \in \mathbb{R}^+ \mid x \in \mathfrak{t}_\varepsilon(A)\}$.
$\lambda\mathcal{F}(x) = \sup_{A \in \sec(\mathcal{F})} \inf\{\varepsilon \in \mathbb{R}^+ \mid x \in \mathfrak{t}_\varepsilon(A)\}$.

$\alpha\mathscr{F}(x) = \sup_{F\in\mathscr{F}} \inf\{\varepsilon \in \mathbb{R}^+ \mid x \in \mathfrak{t}_\varepsilon(F)\}$.

$\mathscr{A}(x) = \{\varphi \in \mathbb{P}^X \mid \forall A \subseteq X, \forall\varepsilon > 0 : x \in \mathfrak{t}_\varepsilon(A) \Rightarrow \inf_{y\in A} \varphi(y) \leq \varepsilon\}$ (closure-tower).

$\mathscr{A}(x) = \{\varphi \in \mathbb{P}^X \mid \forall\varepsilon \in \mathbb{R}^+, \forall\gamma > \varepsilon : \{\varphi < \gamma\} \in \mathscr{V}_\varepsilon(x)\}$ (neighbourhood-tower).

$\mathscr{G} = \{d \in q\mathsf{Met}(X) \mid \forall A \subseteq X : \mathfrak{t}_\varepsilon(A) \subseteq \{\inf_{y\in A} d(\cdot, y) \leq \varepsilon\}\}$.

$\mathscr{G} = \{d \in q\mathsf{Met}(X) \mid \mathfrak{t}_\varepsilon \leq \mathfrak{t}^d_\varepsilon\}$.

8. **Transition formulas from a lower hull operator $\mathfrak{l}$**

$\delta(x, A) = \mathfrak{l}(\theta_A)(x)$.

$\mathscr{A}(x) = \{\varphi \in \mathbb{P}^X \mid \forall\mu \in \mathbb{P}^X : \inf_{y\in X} (\mu + \varphi)(y) \leq \mathfrak{l}(\mu)(x)\}$.

$\mathscr{G} = \{d \in q\mathsf{Met}(X) \mid \forall\mu \in \mathbb{P}^X : \inf_{y\in X} (\mu(y) + d(\cdot, y)) \leq \mathfrak{l}(\mu)\}$.

$\mathscr{G} = \{d \in q\mathsf{Met}(X) \mid \mathfrak{l}_d \leq \mathfrak{l}\}$.

$\mathfrak{L} = \{\mu \in \mathbb{P}^X \mid \mathfrak{l}(\mu) = \mu\}$.

9. **Transition formulas from an upper hull operator $\mathfrak{u}$**

$\mathscr{G} := \{d \in q\mathsf{Met}(X) \mid \forall\omega < \infty\, \forall x \in X : \mathfrak{u}(d(x, \cdot) \wedge \omega)(x) = 0\}$.

$\mathfrak{U} = \{\mu \in \mathbb{P}^X_b \mid \mathfrak{u}(\mu) = \mu\}$.

$\mathscr{A}(x) = \{\varphi \in \mathbb{P}^X \mid \forall\omega < \infty : \mathfrak{u}(\varphi \wedge \omega)(x) = 0\}$.

10. **Transition formulas from a functional ideal convergence $\rightarrowtail$**

$\lambda\mathscr{F}(x) = \inf\{\alpha \mid \omega(\mathscr{F}) \oplus \alpha \rightarrowtail x\}$.

$\mathscr{A}_b(x) = \bigcap\{\mathfrak{J} \in \mathfrak{F}X \mid \mathfrak{J} \rightarrowtail x\}$.

11. **Gauge bases**

$\{d^\mu \mid \mu \in \mathfrak{U}\}$ where $d^\mu(x, y) := \mu(y) \ominus \mu(x)$.

$\{d_\mu \mid \mu \in \mathfrak{L}\}$ where $d_\mu(x, y) := \mu(x) \ominus \mu(y)$.

$\{d^\zeta_Z \mid Z \subseteq X, \zeta < \infty\}$ where

$$d^\zeta_Z(x, y) := (\delta(x, Z) \wedge \zeta) \ominus (\delta(y, Z) \wedge \zeta)$$

is not a gauge basis but does generate the associated distance.

12. **Initial structures** $(f_j : X \longrightarrow X_j)_{j\in J}$
In App.

$$\mathscr{G} = \widehat{\left\{ \sup_{j\in K} d_j \circ (f_j \times f_j) | K \in 2^{(J)}, \forall j \in K : d_j \in \mathscr{H}_j \right\}}.$$

$$\mathscr{A}(x) = \widehat{\left\{ \sup_{j\in K} \xi_j \circ f_j | K \in 2^{(J)}, \forall j \in K : \xi_j \in \mathscr{B}_j(f_j(x)) \right\}}.$$

$$\delta(x, A) = \sup_{\mathscr{P} \in \mathfrak{P}(A)} \min_{P \in \mathscr{P}} \sup_{j \in J} \delta_j(f_j(x), f_j(P))$$

($\mathfrak{P}(A)$ = the set of finite covers of A with subsets of A)

$\lambda\mathscr{F} = \sup_{j \in J} \lambda_j(f_j(\mathscr{F})) \circ f_j.$

$\mathfrak{L} := \{\mu \circ f_j \mid j \in J, \mu \in \mathfrak{L}_j\}^{\wedge\bigvee}.$

$\mathfrak{U} := \{\mu \circ f_j \mid j \in J, \mu \in \mathfrak{U}_j\}^{\vee\bigwedge}.$

$\mathfrak{J} \rightarrowtail x \Leftrightarrow \forall j \in J : f_j(\mathfrak{J}) \rightarrowtail f_j(x).$

In UG.

$$\mathscr{H} := \left\{ \sup_{j \in K} d_j \circ (f_j \times f_j) | K \in 2^{(J)}, \forall j \in K : d_j \in \mathscr{H}_j \right\}^{\sim}.$$

13. **Final structures** $(f_j : X_j \longrightarrow X)_{j \in J}$

$\mathfrak{L} := \{\mu \in \mathbb{P}^X \mid \forall j \in J : \mu \circ f_j \in \mathfrak{L}_j\}.$

$\mathfrak{U} := \{\mu \in \mathbb{P}_b^X \mid \forall j \in J : \mu \circ f_j \in \mathfrak{U}_j\}.$

Appendix B
Symbols

General

$\mathbb{P}$	$[0, \infty]$ either as set or as space
$\mathbb{P}_{\mathbb{E}}$	$\mathbb{P}$ equipped with the Euclidean structure
$\mathbb{P}_b^X$	Set of all bounded functions in $\mathbb{P}^X$
$\mathsf{F}(X)$	Set of filters on X
$\mathsf{U}(X)$	Set of ultrafilters on X
$\mathsf{F}(\mathscr{F})$	Set of filters finer that $\mathscr{F}$
$\mathsf{U}(\mathscr{F})$	Set of ultrafilters finer than $\mathscr{F}$
$\operatorname{stack}\mathscr{A}$	All supersets of sets in $\mathscr{A}$
$\sec\mathscr{F}$	Union of all ultrafilters finer than $\mathscr{F}$
$\Sigma\sigma(\mathscr{F})$	Diagonal filter of σ with respect to $\mathscr{F}$
$\widehat{\mathscr{A}}$	(local) saturation of $\mathscr{A}$
θ_A	Indicator of A
$\mathsf{Ind}(X)$	Set of all indicator functions on X
$\mathsf{Fin}(X)$	Set of functions in $\mathbb{P}^X$ taking a finite number of values
$a \ominus b$	$(a - b) \vee 0$
$c(\mathfrak{J})$	Characteristic value of the functional ideal $\mathfrak{J}$
$\mathfrak{f}_\alpha(\mathfrak{J})$	α-level filter associated with the functional ideal $\mathfrak{J}$
$\mathfrak{f}(\mathfrak{J})$	Filter associated with the functional ideal $\mathfrak{J}$
$\mathsf{i}(\mathscr{F})$	Functional ideal associated with the filter $\mathscr{F}$
$\mathfrak{J} \oplus \alpha$	α-translation of the functional ideal $\mathfrak{J}$
$\mathfrak{Z}$	Improper functional ideal
$\mathfrak{F}(X)$	Set of functional ideals on X
$\mathfrak{P}(X)$	Set of prime functional ideals on X
$\mathfrak{P}(\mathfrak{J})$	Set of prime functional ideals finer than $\mathfrak{J}$
$\mathfrak{P}_m(\mathfrak{J})$	Set of minimal prime functional ideals finer than $\mathfrak{J}$
$\Sigma\mathfrak{s}(\mathfrak{J})$	Diagonal functional ideal of σ with respect to $\mathfrak{J}$
d_Z^ζ	Quasi-metric $(x, y) \mapsto (\delta(x, Z) \wedge \zeta) \ominus (\delta(y, Z) \wedge \zeta)$
d_μ	Quasi-metric $(x, y) \mapsto \mu(x) \ominus \mu(y)$
d^μ	Quasi-metric $(x, y) \mapsto \mu(y) \ominus \mu(x)$

R. Lowen, *Index Analysis*, Springer Monographs in Mathematics,
DOI 10.1007/978-1-4471-6485-2

$d_{\mathbb{E}}$	Metric $(x, y) \mapsto \lvert x - y\rvert$
$d_{\mathbb{P}}$	Quasi-metric $(x, y) \mapsto x \ominus y$
cl	Usual notation for various closure operators
d^-	Quasi-metric $(x, y) \mapsto d(y, x)$
d^*	Quasi-metric $d \vee d^-$
$\mathscr{H}^s$	Symmetric saturation of $\mathscr{H}$
$\mathscr{K}(X)$	$\mathbb{R}$-valued contractions
$\mathscr{K}^*(X)$	Bounded $\mathbb{R}$-valued contractions
$\mathscr{U}(\mathscr{G})$	Uniform structure generated by the gauge $\mathscr{G}$
$\widetilde{\mathscr{D}}$	Uniform saturation of $\mathscr{D}$
$\omega_{\mathscr{H}}(\mathscr{F})$	$\mathscr{H}$-width of $\mathscr{F}$
$\mathrm{diam}_d(A)$	d-diameter of A
$\lim(\mathscr{F})$	Set of limit points of $\mathscr{F}$
$\mathrm{adh}(\mathscr{F})$	Set of adherence points of $\mathscr{F}$
$\mathscr{M}_{\mathscr{F}}$	Smallest Cauchy filter coarser than $\mathscr{F}$
$\mathscr{C}(X, Y)$	Set of all $\mathscr{C}$-morphisms between $\mathscr{C}$-objects X and Y
$[X, Y]$	$\mathscr{C}$-object on $\mathscr{C}(X, Y)$ in a cartesian closed topological category
$\beta^* X$	Approach Čech-Stone compactification of X
$\Delta_{\mathscr{D}}$	Proximity generated by $\mathscr{D}$
$\sigma(E, E')$	Weak topology on a normed space E
$\sigma(E', E)$	Weak* topology on dual space E'

Approach structures

δ	Distance
$A^{(\varepsilon)}$	The ε-enlargement of A
λ	Limit operator
$(\mathscr{A}(x))_{x\in X}$	Approach system
$(\mathscr{A}_b(x))_{x\in X}$	Bounded approach system
$\mathscr{G}$	Gauge or uniform gauge
$\mathscr{G}_b$	Bounded gauge or bounded uniform gauge
$(\mathrm{t}_\varepsilon)_\varepsilon$	Tower
$\mathfrak{L}$	Lower regular function frame
$\mathfrak{U}$	Upper regular function frame
$\mathfrak{l}$	Lower hull operator
$\mathfrak{u}$	Upper hull operator
$\rightarrowtail$	Functional ideal convergence
α	Adherence operator
$\delta_{\mathbb{E}}$	The "Euclidean" distance on $\mathbb{P}$
$\delta_{\mathbb{P}}, \lambda_{\mathbb{P}}, \mathscr{A}_{\mathbb{P}}$	The intrinsic approach structures on $\mathbb{P}$
$\lambda_{(\mathscr{F}, f)}$	Limit operator of filter approach space
$\delta_{(E,E')}$	Weak distance on a normed space E
$\delta_{(E',E)}$	Weak* distance on the dual space E'
$\lambda_{(E,E')}$	Weak limit operator on a normed space E
$\lambda_{(E',E)}$	Weak* limit operator on the dual space E'

$\mathscr{A}_{\mathscr{N}}$	Approach system derived from a local pre-norm system
δ_w	Weak distance on probability measures
λ_w	Weak limit operator on probability measures
δ_p	Distance of convergence in probability
λ_p	Limit operator of convergence in probability
δ_c	Distance of the continuity approach structure
λ_c	Limit operator of the continuity approach structure
δ_{W_d}	Wijsman distance on hyperspaces
$\delta_{prox(\Delta,d)}$	Proximity distance on hyperspaces
$\delta_{prox(d)}$	Proximal distance on hyperspaces
$\delta_{bprox(d)}$	Bounded proximal distance on hyperspaces
$\mathfrak{L}^{\wedge}$	Iinf-Vietoris lower regular function frame on hyperspaces
$\mathfrak{L}^{\vee}$	Sup-Vietoris lower regular function frame on hyperspaces
$\mathfrak{L}_v$	Vietoris lower regular function frame on hyperspaces

Indices

δ	Distance = index of closure
λ	Limit operator = index of convergence
α	Adherence operator = index of adherence
χ_c	Index of contractivity in case of functions
χ_c	Index of compactness in case of spaces
χ_{ce}	Index of closed expansiveness
χ_{oe}	Index of open expansiveness
χ_p	Index of properness
χ_{rc}	Index of relative compactness
χ_{sc}	Index of sequential compactness
χ_{rsc}	Index of relative sequential compactness
χ_{cc}	Index of countable compactness
χ_l	Lindelöf index
χ_{lc}	Index of local compactness
χ_{cn}	Index of connectedness
χ_{uc}	Index of uniform contractivity
χ_{pc}	Index of precompactness
χ_{cy}	Cauchy index
χ_{lcy}	Local Cauchy index
χ_{ec}	Index of equicontractivity
χ_{uec}	Index of uniform equicontractivity
χ_{wt}	Weak index of tightness
χ_{st}	Strong index of tightness
χ_{Lin}	Lindeberg index

Categories

$\mathsf{Alg}(\mathsf{U}, e, m)$	Lax algebras for the $\mathbb{P}$Rel-extension of ultrafilter monad
App	Approach spaces and contractions
App_0	T_0 approach spaces
App_1	T_1 approach spaces
App_2	T_2 approach spaces
App_{Rg}	Regular approach spaces
App_{Wa}	Weakly adjoint approach spaces
App_{Γ}	Γ-valued approach spaces
ApVec	Approach vector spaces and linear contractions
Bor	Bornological spaces and bounded maps
CAp	Convergence-approach spaces and contractions
Conv	Convergence spaces and continuous maps
CReg	Completely regular topological spaces
$c\mathsf{UAp}_2$	Complete T_2 uniform approach spaces
EpiAp	Epi-approach spaces and contractions
Fin	Finitely generated topological spaces
$k\mathsf{UAp}_2$	Compact T_2 uniform approach spaces
lcApVec	Locally convex approach spaces and linear contractions
lcTopVec	Locally convex topological spaces and linear continuous maps
Met	Metric spaces and non-expansive maps
MetVec	Metric vector spaces and linear non-expansive maps
$\mathbb{P}$Rel	Sets and numerical relations
PrAp	Pre-approach spaces and contractions
PrTop	Pretopological spaces and continuous maps
PsAp	Pseudo-approach spaces and contractions
PsTop	Pseudotopological spaces and continuous maps
$q\mathsf{Met}$	Quasi-metric spaces and non-expansive maps
$qm\mathsf{Top}$	Quasi-metrizable topological spaces
$qs\mathsf{Met}$	Quasi-semi-metric spaces and non-expansive maps
$q\mathsf{UG}$	Quasi-uniform gauge spaces and uniform contractions
$q\mathsf{Unif}$	Quasi-uniform spaces and uniformly continuous functions
Rel	Sets and relations
Set	Sets and functions
sNorm	Seminormed spaces and linear non-expansive maps
Top	Topological spaces and continuous maps
TopVec	Topological vector spaces and linear continuous maps
UAp	Uniform approach spaces
UG	Uniform gauge spaces and uniform contractions
Unif	Uniform spaces and uniformly continuous functions
Vec	Vector spaces and linear maps

References

Akhmerov, R.R., Kamenskii, M.I., Potapov, A.S., Rodkina, A.E., Sadovskii, B.N.: Measures of Noncompactness and Condensing Operators Operator Theory: Advances and Applications, vol. 55. Birkhäuser, Basel (1992)

Alestalo, P., Trotsenko, D.A., Väisälä, J.: Isometric approximation. Israel J. Math. **125**, 61–82 (2001)

Amir, D., Mach, J., Saatkamp, K.: Existence of Chebyshev centers, best n-nets and best compact approximants. Trans. Am. Math. Soc. **271**, 513–520 (1982)

Antoine, Ph: Étude élémentaire d'ensembles structurés. Bull. Soc. Math. Belg. **18**, 142–164 (1966a)

Antoine, Ph: Étude élémentaire des catégories d'ensembles structurés. II Bull. Soc. Math. Belg. **18**, 387–414 (1966b)

Antoine, Ph: Extension minimale de la catégorie des espaces topologiques. C. R. Acad. Sci. Paris Ser. A-B **262**, A1389–A1392 (1966c)

Arhangel'skii, A., Wiegandt, R.: Connectedness and disconnectedness in topology. Gen. Topol. Appl. **5**, 9–33 (1975)

Atsuji, M.: Uniform continuity of continuous functions of metric spaces. Pacific. J. Math. **8**, 11–16 (1958)

Aull, C.E., Lowen, R. (eds.) Handbook of the History of General Topology, vol. 3. Kluwer Academic Publishers, Dordrecht (2001)

Baboolal, D., Pillay, P.: On uniform Lipschitz-connectedness in metric spaces. Appl. Categor. Struct. **17**, 487–500 (2009)

Baekeland, R.: Measures of Compactness and their Application to Convergence of Probability Measures. Ph.D. Vrije Universiteit Brussel, Belgium (1992)

Baekeland, R., Lowen, R.: Measures of compactness in approach spaces. Comment. Math. Univ. Carol. **36**, 327–345 (1995)

Banaś, J.: Applications of measures of weak noncompactness and some classes of operators in the theory of functional equations in the Lebesgue space. Nonlinear Anal. **30**, 3283–3293 (1997)

Banaś, J., Goebel, K.: Measures of Noncompactness in Banach Spaces. Lecture Notes in Pure and Applied Mathematics, vol. 60. Marcel Dekker, New York (1980)

Banaschewski, B., Lowen, R., Van Olmen, C.: Sober approach spaces. Topol. Appl. **153**, 3059–3070 (2006)

Banaschewski, B., Lowen, R., Van Olmen, C.: Regularity in approach theory. Acta Math. Hung. **115**(3), 183–196 (2007)

Banaschewski, B., Lowen, R., Van Olmen, C.: Compactness in approach frames. Order **29**, 105–118 (2012)

Barbour, A.D., Chen, L.H.Y.: An Introduction to Stein's Method. Singapore University Press, Singapore; World Scientific Publishing Co, Singapore/Hackensack (2005)

Barbour, A.D., Hall, P.: Stein's method and the Berry-Esseen theorem. Aust. J. Stat. **26**, 8–15 (1984)

R. Lowen, *Index Analysis*, Springer Monographs in Mathematics,
DOI 10.1007/978-1-4471-6485-2

Beer, G.: Convergence of continuous linear functionals and their level sets. Arch. Math. **52**, 482–491 (1989)

Beer, G.: Mosco convergence and weak topologies for convex sets and functions. Mathematika **38**, 89–104 (1991)

Beer, G.: Topologies on Closed and Closed Convex Sets. Kluwer Academic Publishers, Dordrecht (1993)

Beer, G., Lechicki, A., Levi, S., Naimpally, S.: Distance functionals and suprema of hyperspace topologies. Annali di Matematica pura et applicata **CLXII**, 367–381 (1992)

Beer, G., Luchetti, R.: Weak topologies for the closed subsets of a metrizable space. Trans. Am. Math. Soc. **335**, 805–822 (1993)

Benavides, D.T.: Some properties of the set and ball measures of noncompactness and applications. J. Lond. Math. Soc. **34**(2), 120–128 (1986)

Bentley, H.L., Herrlich, H., Hušek M.: The Historical Development of Uniform, Proximal and Nearness Concepts in Topology in Handbook of the History of General Topology, pp. 577–630. Kluwer Academic Publishers, Dordrecht (1998)

Bentley, H.L., Herrlich, H., Lowen-Colebunders, E.: The category of Cauchy spaces is cartesian closed. Topol. Appl. **27**, 105–112 (1987)

Bentley, H.L., Herrlich, H., Lowen, R.: Improving Constructions in Topology in Category Theory at Work. Heldermann, Berlin (1991)

Benyamini, Y.: Asymptotic centers and best approximation of compact operators into $C(K)$. Constr. Approx. **1**, 217–229 (1985)

Berckmoes, B.: Approach structures in probability theory. Ph.D. Thesis. Universiteit, Antwerpen (2014)

Berckmoes, B., Lowen, R., Van Casteren, J.: Approach theory meets probability theory. Topol. Appl. **158**, 836–852 (2011a)

Berckmoes, B., Lowen, R., Van Casteren, J.: Distances on probability measures and random variables. J. Math. Anal. Appl. **374**, 412–428 (2011b)

Berckmoes, B., Lowen, R., Van Casteren, J.: An isometric study of the Lindeberg Feller central limit theorem via Steins method. J. Math. Anal. Appl. **405**(2), 484–498 (2013)

Berckmoes, B., Lowen, R., Van Casteren, J.: Stein's method and a quantitative Lindeberg CLT for the Fourier transforms of random vectors (submitted for publication)

Bergström, H.: On the central limit theorem in the case of not equally distributed random variables Skand. Aktuarietidskr **32**, 37–62 (1949)

Billingsley, P.: Convergence of Probability Measures. Wiley, New York (1968)

Birkhoff, G.: Lattice theory. Am. Math. Soc. **XXV** (Colloquium Publications, 1967)

Booth, P., Tillotson, J.: Monoidal closed, cartesian closed and convenient categories of topological spaces. Pacific J. Math. **88**, 35–53 (1980)

Börger, R.: Connectivity spaces and component categories. In: Proceedings of the International Conference on Categorical Topology. Toledo 1983, pp. 71–89. Heldermann, Berlin (1984)

Borges, C.R.: On stratifiable spaces. Pacific J. Math. **17**, 1–16 (1966)

Bourbaki, N.: Topologie Générale, chapitres 3 et 4: Groupes topologiques, nombres réels. Hermann, Paris (1960)

Bourbaki, N.: Topologie Générale, chapitre 10: Espaces fonctionnels. Hermann, Paris (1961)

Bourbaki, N.: Espaces Vectoriels Topologiques. Hermann, Paris (1964)

Bourdaud, G.: Structures d'Antoine associées aux semi-topologies et aux topologies. Compt. Rendus Acad. Sci. Paris **279**, 591–594 (1974)

Bourdaud, G.: Espaces d'Antoine et semi-espaces d'Antoine Cahiers. Topol. Géom. Diff. Cat. **16**, 107–133 (1975)

Bourdaud, G.: Some cartesian closed topological categories of convergence spaces in categorical topology. In: Proceedings of Mannheim 1975. Lecture Notes in Mathematics, vol. 540, pp. 93–108. Springer, Berlin (1976)

Bourgin, D.G.: Approximate isometries. Bull. Am. Math. Soc. **52**, 704–714 (1946)

Brandenburg, H., Hušek, M.: A remark on cartesian closedness. In: Proceeding of Gummersbach Conference on Category Theory, 1981, pp. 33–38. Springer, New York (1982)
Brezis, H.: Analyse fonctionelle, Théorie et applications. In: Collection Mathématiques appliqués pour la maitrise. Masson, Paris (1983)
Brezis, H.: Functional analysis, Sobolev spaces and partial differential equations. In: Universitext. Springer, New York (2011)
Brezis, H., Lieb, E.: A relation between pointwise convergence of functions and convergence of functionals. Proc. Am. Math. Soc. **88–3**, 486–490 (1983)
Brock, P., Kent, D.C.: Approach spaces, limit tower spaces and probabilistic convergence spaces. Appl. Categor. Struct. **5**, 99–110 (1997a)
Brock, P., Kent, D.C.: Regularity for approach spaces and probabilistic convergence spaces. Int. J. Math. Math. Sci. **20**, 637–646 (1997b)
Brock, P., Kent, D.C.: On convergence approach spaces. Appl. Categor. Struct. **6**, 117–125 (1998)
Brümmer, G.C.L., Giuli, E.: A categorical concept of completion of objects. Comment. Math. Univ. Carol. **33**, 131–147 (1992)
Brümmer, G.C.L., Giuli, E., Herrlich, H.: Epireflections which are completions. Cah. Top. Géom. Diff. Catégor. **33**, 71–73 (1992)
Brümmer, G.C.L., Sioen, M.: Asymmetry and bicompletion of approach spaces. Topol. Appl. **153**, 3101–3112 (2006)
Cantor, G.: Über unendliche, lineare Punktmannigfaltigkeiten. Math. Ann. **21**, 545–591 (1883)
Charalambos, A.D., Border, K.C.: Infinite Dimensional Analysis. Springer, Berlin (2007)
Chavez, M.A.: Spaces where all continuity is uniform. Am. Math. Mon. **92**, 487–489 (1985)
Choquet, G.: Convergences. Ann. Univ. Grenoble **23**, 57–112 (1947)
Claes, V.: Initially dense objects for metrically generated theories. Topol. Appl. **156**, 2082–2087 (2009)
Claes, V., Colebunders, E., Gerlo, A.: On the epimorphism problem and cowellpoweredness for metrically generated theories. Acta Math. Hung. **114**, 133–152 (2007)
Clementino, M.M., Hofmann, D.: Topological features of lax algebras. Appl. Categor. Struct. **11**, 267–286 (2003)
Clementino, M.M., Hofmann, D., Tholen, W.: One setting for all: metric, topology, uniformity and approach structures. Appl. Categor. Struct. **12**, 127–154 (2004)
Clementino, M.M., Giuli, E., Tholen, W.: A Functional Approach to Topology. In: Pedicchio, M.C., Tholen, W. (eds.) Categorical Foundations Special Topics in Order, Topology, Algebra, and Sheaf Theory. Cambridge University Press, Cambridge (2003)
Colebunders, E.: Function Classes of Cauchy Continuous Maps Monographs and Textbooks in Pure and Applied Mathematics, vol. 123. Marcel Dekker, New York (1989)
Colebunders, E., De Wachter, S., Lowen, R.: Intrinsic approach spaces on domains. Topol. Appl. **158**, 2343–2355 (2011)
Colebunders, E., De Wachter, S., Schellekens, M.: Approach spaces for complexity analysis. Appl. Categor. Struct. **22**(1), 119–136 (2014)
Colebunders, E., Gerlo, A.: Firm reflections generated by complete metric spaces. Cah. Topol. Geom. Diff. Categor. **48**, 243–260 (2007)
Colebunders, E., Gerlo, A., Sonck, G.: Function spaces and one point extensions for the construct of metered spaces. Topol. Appl. **153**, 3129–3139 (2006)
Colebunders, E., Lowen, R.: Metrically generated theories. Proc. Am. Math. Soc. **133**, 1547–1556 (2005)
Colebunders, E., Lowen, R.: Bornologies and metrically generated theories. Topol. Appl. **156**, 1224–1233 (2009)
Colebunders, E., Lowen, R.: Supercategories of Top and the inevitable emergence of categorical topology. In: Handbook of the History of General Topology, vol. 3, pp. 969–1026. Kluwer Academic Publishers, New York (2001)
Colebunders, E., Lowen, R., Rosiers, W.: Lax algebras via initial monad morphisms: app top. Met Ord. Topol. Appl. **158**, 882–903 (2011)

Colebunders, E., Lowen, R., Vandersmissen, E.: Uniqueness of completeness in metrically generated theories. Topol. Appl. **155**, 39–55 (2007)
Colebunders, E., Lowen, R., Van Geenhoven, A.: Local metrically generated theories. Topol. Appl. **159**, 2320–2330 (2012)
Colebunders, E., Lowen, R., Van Opdenbosch, K.: Approach spaces, functional ideal convergence and Kleisli monoids preprint (to appear)
Colebunders, E., Mynard, F., Trott, W.: Function spaces and contractive extensions in approach theory: the role of regularity. Appl. Categor. Struct. **22**(3), 551–564 (2014)
Colebunders, E., Vandersmissen, E.: Bicompletion of metrically generated constructs. Acta Math. Hung. **128**, 239–264 (2010)
Colebunders, E., Verbeeck, C.: Exponential objects in coreflective or quotient reflective subcategories: a comparison. Appl. Categor. Struct. **8**, 247–256 (2000)
Costantini, C., Levi, S., Zieminska, J.: Metrics that generate the same hyperspace convergence. Set-Valued Anal. **1**, 141–157 (1993)
Császár, A.: Foundations of General Topology. Pergamon Press, Oxford (1963)
Day, B.J., Kelley, G.M.: On topological quotient maps preserved by pullbacks or products. Proc. Camb. Phil. Soc. **67**, 553–558 (1970)
de la Harpe, P.: Topics in Geometric Group Theory. University of Chicago Press, Chicago (2000)
de Malafosse, B., Rakočević, V.: Applications of measure of noncompactness in operators on the spaces s_α, s_α^0, s_α^c, l_α^p. J. Math. Anal. Appl. **323**, 131–145 (2006)
de Pagter, B., Schep, A.R.: Measures of noncompactness of operators in Banach lattices. J. Funct. Anal. **78**(1), 3155 (1988)
Dieudonné, J.: Sur les espaces uniformes complets. Ann. Ecol. Norm. Supér. **56**, 276–291 (1939)
Dikranjan, D., Giuli, E.: Closure operators. Topol. Appl. **27**, 129–143 (1987)
Dikranjan, D., Tholen, W.: Categorical Structure of Closure Operators. Kluwer Academic Publishers, Dordrecht (1995)
Di Maio, G., Holà, L.: On hit-and-miss topologies. Rend. Acc. Sc. Fis. Mat. Napoli **LXII**, 103–124 (1997)
Dominguez Benavides T., Lorenzo P. Asymptotic centers and fixed points for multivalued nonexpansive mappings. Annales Universitatis MariaE Curie-Sklodowska. **LVIII**, 37–45 (2004)
Dugundji, J.: Topology. Allyn and Bacon, Boston (1967)
Edalat, A., Heckmann, R.: A computational model for metric spaces. Theor. Comput. Sci. **193**, 53–73 (1998)
Edelstein, M.: The construction of an asymptotic center with a fixed-point property. Bull. Am. Math. Soc. **78**, 206–208 (1972)
Edelstein, M.: Fixed point theorems in uniformly convex Banach spaces. Proc. Am. Math. Soc. **44**, 369–374 (1974)
Edwards, D.: The Structure of Superspace Studies in Topology. Academic Press, New York (1975)
Fletcher, P., Lindgren, W.F.: Quasi-uniform Spaces. Lecture Notes in Pure and Applied Mathematics, vol. 77. Marcel Dekker, New York (1982)
Gerlo, A.: Separation, completeness and compactness in metrically generated theories. Ph.D. (2007)
Gierz, G., Hofmann, K.H., Keimel, K., Lawson, J.D., Mislove, M., Scott, D.S.: Continuous Lattices and Domains Encyclopedia of Mathematics and its applications, vol. 93. Cambridge University Press, Cambridge (2003)
Gillman, L., Jerison, M.: Rings of Continuous Functions. Springer, Berlin (1976)
Giuli, E.: Two families of cartesian closed topological categories. Quaestiones Math. **6**, 353–362 (1983)
Grnicki, J.: Nonlinear ergodic theorems for asymptotically nonexpansive mappings in Banach spaces satisfying Opial's condition. J. Math. Anal. Appl. **161**, 440–446 (1991)
Herrlich, H.: Topologische Reflexionen und Coreflexionen. Lecture Notes in Mathematics, vol. 78. Springer, Berlin (1968)
Herrlich, H.: Categorical topology. Gen. Top. Appl. **1**, 1–15 (1971)
Herrlich, H.: On a concept of nearness. Gen. Topol. Appl. **4**, 191–212 (1974a)

Herrlich, H.: Topological structures. Math. Centre Tracts **52**, 59–122 (1974b)
Herrlich, H.: Categorical topology 1971–1981. In: Proceedings Fifth Prague Topology Symposium 1981, pp. 279–383. Heldermann, Berlin (1983)
Herrlich, H.: Einführung in die Topologie. Heldermann, Berlin (1986)
Herrlich, H.: Topological improvements of categories of structured sets. Topol. Appl. **27**, 145–155 (1987)
Herrlich, H.: Hereditary topological constructs. General topology and its relations to modern analysis and algebra VI. In: Proceedings of Sixth Prague Topological Symposium 1986, pp. 249–262. Heldermann (1988a)
Herrlich, H.: On the representability of partial morphisms in Top and in related constructs categorical algebra and its applications. Proceedings of Louvain-La-Neuve 1987. Lecture Notes in Mathematics, vol. 1348, pp. 143–153. Springer, Berlin (1988b)
Herrlich, H., Lowen-Colebunders, E., Schwarz, F.: Improving Top : PrTop and PsTop. In: Herrlich, H., Porst, H.E. (eds.) Category Theory at Work, pp. 21–34. Heldermann, Berlin (1991)
Herrlich, H., Nel, L.D.: Cartesian closed topological hulls. Proc. Am. Math. Soc. **62**, 215–222 (1977)
Herrlich, H., Strecker, G.E.: Category Theory. Allyn and Bacon, Boston, (1973) (reprint, 1979. Heldermann, Berlin)
Herrlich, H., Zhang, D.: Categorical properties of probabilistic convergence spaces. Appl. Categor. Struct. **6**, 495–513 (1998)
Hindman, N.: Ultrafilters and Combinatorial Number Theory. Lecture Notes in Mathematics, pp. 119–184. Springer, Berlin (1979)
Hofmann, D., Seal, G.J., Tholen, W. (eds.): Monoidal Topology: A Categorical Approach to Order, Metric and Topology. Cambridge University Press, Cambridge (2014)
Holgate, D., Sioen, M.: Approach structures and measures of connectedness. Quaestiones Math. **30**, 1–14 (2007)
Horvath, C.: Measure of noncompactness and multivalued mappings in complete topological vector spaces. J. Math. Anal. Appl. **108**, 403–408 (1985)
Hušek, M.: Reflective and coreflective subcategories of Unif and Top. Seminar Uniform Spaces, pp. 113–126 (1973)
Hušek, M.: Generalized proximity and uniform spaces I. Comment. Math. Univ. Carol. **5**, 247–266 (1964a)
Hušek, M.: Generalized proximity and uniform spaces II. Comment. Math. Univ. Carol. **6**, 119–139 (1964b)
Hyers, D.H., Ulam, S.M.: On approximate isometries. Bull. Am. Math. Soc. **51**, 288–292 (1945)
Hyers, D.H., Ulam, S.M.: On approximate isometries of the space of continuous functions. Ann. Math. **48**, 285–289 (1947)
Ioffe, A.: Nonsmooth analysis: differential calculus of nondifferentiable functions. Trans. Am. Math. Soc. **255**, 1–55 (1981)
Ioffe, A.: Proximal analysis and approximate subdifferentials. J. Lond. Math. Soc. **41**, 175–192 (1990)
Ioffe, A.: Metric regularity and subdifferential calculus. Russ. Math. Surv. **55**(3), 501–558 (2000)
Isbell J. R. Uniform spaces. Mathematical Surveys, vol. 12. American Mathematical Society, Providence (1964)
Isiwata, T.: Compact and realcompact κ-metrizable extensions. Topol. Proc. **10**, 95–102 (1985)
Isiwata, T.: Metrization of additive κ-metric spaces. Proc. Am. Math. Soc. **100**, 164–168 (1987)
Isiwata, T.: Continuity of additive κ-metric functions and metrization of κ-metric spaces. Proc. Am. Math. Soc. **104**, 988–992 (1988)
Jacka, S.D., Roberts, G.O.: On strong forms of weak convergence. Stoch. Process. Appl. **67**, 41–53 (1997)
Jaeger, G.: A note on neighbourhoods for approach spaces. Haceteppe J. Math. Stat. **41**, 283–290 (2012)

Jaeger, G.: Extensions of contractions and uniform contractions on dense subspaces. Quaestiones Mathematicae (2014) (to appear)
Jung, H.W.E.: Uber die kleinste Kugel die eine räumliche Figur einschliest. J. Reine Angew. Math. **123**, 241–257 (1901)
Kannan, V.: Reflexive cum coreflexive subcategories in topology. Math. Ann. **195**, 168–174 (1972)
Karlovitz, L.: On nonexpansive mappings. Proc. Am. Math. Soc. **55**, 321–325 (1976)
Kelley, G.J.: General Topology. Van Nostrand, New York (1955)
Kent, D.C.: Convergence functions and their related topologies. Fundam. Math. **54**, 125–133 (1964)
Kim, I.-S., Väth, M.: Some remarks on measures of noncompactness and retractions onto spheres. Topol. Appl. **154**, 3056–3069 (2007)
Kirk, W.A., Massa, S.: Remarks on asymptotic and Chebyshev centers. Houston J. Math. **16**(3), 357–364 (1990)
Kirk, A., Sims, B.: Handbook of Metric Fixed Point Theory. Kluwer Academic Publishers, Boston (2001)
Kowalsky, J.: Limesräume und Komplettierung. Math. Nachr. **12**, 301–340 (1954)
Kuczumow, T.: Weak convergence theorems for nonexpansive mappings and semi-groups in Banach spaces with Opial s property. Proc. Am. Math. Soc. **93**, 430–432 (1985)
Künzi, H.-P.: Nonsymmetric distances and their associated topologies: About the origins of basic ideas in the area of asymmetric topology. In: Aull, C.E., Lowen, R. (eds.) Handbook of the History of General Topology, vol. 3, pp. 853–968. Kluwer Acadamic Publisher, Dordrecht (2001)
Künzi, H.-P., Vajner, V.: Weighted quasi-metrics. Ann. N. Y. Acad. Sci. **728**, 64–77 (1994)
Kuratowski, C.: Sur les espaces complets. Fundam. Math. **15**, 301–309 (1930)
Kuratowski, C.: Topology. Academic Press, New York (1966)
Lami Dozo, E.: Multivalued nonexpansive mappings and opials condition. Proc. Am. Math. Soc. **38**, 286–292 (1973)
Lami Dozo, E.: Asymptotic Centers in Particular Spaces. Lecture Notes in Mathematics, vol. 886, pp. 199–207. Springer, Berlin (1981)
Lawson, J.D.: Spaces of maximal points. Math. Struct. Comput. Sci. **7**(5), 543–555 (1997)
Lechicki, A., Levi, S.: Wijsman convergence in the hyperspace of a metric space. Bull. Un. Mat. Ital. **5–B**, 435–452 (1987)
Lee, R.: The category of uniform convergence spaces is cartesian closed. Bull. Aust. Math. Soc. **15**, 461–465 (1976)
Lieb, E.H., Loss, M.: Analysis, 2nd edn. Graduate Studies in Mathematics, vol. 14. American Mathematical Society, Providence, RI (2001)
Lim, T.-C.: On asymptotic centers and fixed points of nonexpansive mappings. Can. J. Math. **32**, 421–430 (1980)
Liu, Q.H.: Iterative sequences for asymptotically quasi-nonexpansive mappings. J. Math. Anal. Appl. **259**, 17 (2001)
Lo, K.C.: Epistemic conditions for agreement and stochastic independence of ε-contaminated beliefs. Math. Soc. Sci. **39**, 207–234 (2000)
Lowen-Colebunders, E., Lowen, R.: A quasi-topos containing Conv and Met as full subcategories. Int. J. Math. Math. Sci. **11**, 417–438 (1988)
Lowen-Colebunders, E., Sonck, G.: Exponential objects and Cartesian closedness in the construct PrTop. Appl. Categor. Struct. **1**, 345–360 (1993)
Lowen-Colebunders, E., Sonck, G.: On the largest coreflective Cartesian closed subconstruct of PrTop. Appl. Categor. Struct. **4**, 69–79 (1996)
Lowen, R.: Approach Spaces: the Missing Link in the Topology-uniformity-metric Triad. Oxford Mathematical Monographs. Oxford University Press, Oxford (1997)
Lowen, R.: An Ascoli theorem in approach theory. Topol. Appl. **137**, 207–213 (2004)
Lowen, R., Sioen, M.: Proximal hypertopologies revisited. Set Valued Anal. **6**, 1–9 (1998)
Lowen, R., Sioen, M.: The Wijsman and Attouch-Wets topologies on hyperspaces revisited. Topol. Appl. **70**, 179–197 (1996)

Lowen, R., Sioen, M.: Weak representations of quantified hyperspace structures. Topol. Appl. **104**, 169–179 (2000a)
Lowen, R., Sioen, M.: A Wallman-Shanin-type compactification for approach spaces. Rocky Mt. J. Math. **30**, 1381–1419 (2000b)
Lowen, R., Sioen, M.: A note on separation in App. Appl. Gen. Topol. **4**, 475–486 (2003)
Lowen, R., Sioen, M., Vaughan, D.: Completing quasi-metric spaces—an alternative approach. Houston J. Math. **29**, 113–136 (2003)
Lowen, R., Vaughan, D.: A non-quasi-metric completion for quasi-metric spaces. Rend. Inst. Mat. Univ. Trieste **30**, 145–163 (1999)
Lowen, R., Verwulgen, S.: Approach vector spaces. Houston J. Math. **30**, 1127–1142 (2004)
Lowen, R., Vroegrijk, T.: A new lax algebraic characterisation of approach spaces. Quaderni di Mathematica **22**, 137–170 (2008)
Lowen, R., Windels, B.: AUnif : A common supercategory of Unif and Met. Int. J. Math. Math. Sci. **21**, 1–18 (1998)
Lowen, R., Windels, B.: Approach groups. Rocky Mt. J. Math. **30**, 1057–1073 (2000)
Lowen, R., Wuyts, P.: The Vietoris hyperspace structure for approach spaces. Acta Math. Hung. **139**(3), 286–302 (2013)
Machado, A.: Espaces d'Antoine et pseudo-topologies. Cah. Top. Géom. Diff. Categor. **14**, 309–327 (1973)
Malkowsky, E., Parashar, S.D.: Matrix transformations in spaces of bounded and convergent difference sequences of order m. Analysis **17**, 8797 (1997)
Malkowsky, E., Rakočecić, V.: An introduction into the theory of sequence spaces and measures of noncompactness (preprint)
Malkowsky, E., Rakočević, V.: The measure of noncompactness of linear operators between certain sequence spaces. Acta Sci. Math. (Szeged) **64**, 151–170 (1998)
Malkowsky, E., Rakočević, V.: Measure of noncompactness of linear operators between spaces of sequences that are $\overline{N}$, q summable or bounded. Czech. Math. J. **51**, 505–522 (2001)
Marny, T.: On epireflective subcategories of topological categories. Gen. Top. Appl. **10**, 175–181 (1979)
Martin K.: A foundation for computation. Ph.D. Thesis Tulane University (2000a)
Martin, K.: The measurement process in domain theory. Lect. Notes Comput. Sci. **1853**, 116–126 (2000b)
Matthews, S.G.: Partial metric topology in proceedings of the 8th summer conference on general topology and applications. Ann. N. Y. Acad. Sci. **728**, 183–197 (1994)
Mémoli, F.: Gromov-Hausdorff distances in Euclidean spaces. In: Computer Vision and Pattern Recognition Workshop CVPRW'08 IEEE Computer Society, pp. 1–8 (2008)
Michael, E.: Topologies on spaces of subsets. Trans. Am. Math. Soc. **71**, 152–182 (1951)
Morgenthaler, S.: A survey of robust statistics. Stat. Meth. Appl. **15**, 271–293 (2007)
Mrowka, S., Pervin, W.J.: On uniform connectedness. Proc. Am. Math. Soc. **15**, 446–449 (1964)
Nagata, J.: A survey of metrization theory II. Q & A Gen. Topol. **10**, 15–30 (1992)
Naimpally, S.A.: What is a hit-and-miss topology? Topol. Comment **8**, (2003)
Naimpally, S.A.: Proximity Approach to Problems in Topology and Analysis. Oldenbourg Wissenschaftsverlag GmbH, Munich (2009)
Naimpally, S.A., Pareek, C.M.: Generalized metric spaces via annihilators (Preprint)
Naimpally, S.A., Warrack, B.D.: Proximity Spaces. Cambridge University Press, Cambridge (1970)
Nauwelaerts, M.: Categorical hulls in approach theory. Ph.D. University of Antwerp (2000)
Nel, L.D.: Cartesian closed coreflective hulls. Quaestiones Math. **2**, 269–383 (1977)
Oltra, S., Valero, O.: Banach's fixed point theorem for partial metric spaces. Rend. Inst. Mat. Univ. Trieste **36**, 17–26 (2004)
Opial, Z.: Weak convergence of the sequence of successive approximations for nonexpansive mappings. Bull. Am. Math. Soc. **73**, 591–597 (1967)
Parthasarathy, K.R.: Probability Measures on Metric Spaces. Academic Press, New York (1967)

Parthasarathy, K.R., Steerneman, T.: A tool in establishing total variation convergence. Proc. Am. Math. Soc. **94**, 626–630 (1985)
Pettis, B.J.: Cluster sets of nets. Proc. Am. Math. Soc. **22**, 386–391 (1969)
Plotkin, G.: $\mathbb{T}^{\omega}$ as a universal domain. J. Comput. Syst. Sci. **17**, 209–230 (1978)
Preuss, G.: E-zusammenhängende Räume. Manuscr. Math. **3**, 331–342 (1970)
Preuss, G.: Theory of Topological Structures. Kluwer Academic Publishers, Dordrecht (1987)
Preuss, G.: Foundations of Topology. Kluwer Academic Publishers, Dordrecht (2002)
Prokhorov, A.V.: Borel-Cantelli lemma in Hazewinkel. Michiel. Encyclopaedia of Mathematics. Springer, Berlin (2001)
Pumplün, D., Röhrl, H.: Banach spaces and totally convex spaces I. Commun. Algebra **12**, 953–1019 (1984)
Pumplün, D., Röhrl, H.: Banach spaces and totally convex spaces II. Commun. Algebra **13**, 1047–1113 (1985)
Pumplün, D.: Absolutely convex modules and Saks spaces. J. Pure Appl. Algebra **155**, 257–270 (2001)
Rachev, S.T.: Probability Metrics and the Stability of Stochastic Models. Applied Probability and Statistics. Wiley Series in Probability and Mathematical Statistics. Wiley, Chichester (1991)
Robeys, K.: Extensions of products of metric spaces. Ph.D. Universiteit Antwerpen, RUCA (1992)
Schaefer, H.H.: Topological vector spaces. In: Macmillan Series in Advanced Mathematics and Theoretical Physics. Macmillan, New York (1966)
Schellekens, M.P.: The Smyth completion: a common foundation for denotational semantics and complexity analysis. Electr. Notes Theor. Comput. Sci. **1**, 535–556 (1995)
Schellekens, M.P.: A characterisation of partial metrizability: domains are quantifiable. Theoret. Comput. Sci. **305**, 409–432 (2003)
Schwarz, F.: Description of the topological universe hull. Lect. Notes Comp. Sci. **393**, 325–332 (1989)
Schwarz, F., Weck-Schwarz, S.: Internal description of hulls: a unifying approach. In: Herrlich, H., Porst, H.-E. (eds.) Category Theory at Work, pp. 35–45. Heldermann, Berlin (1991)
Scott, D.S.: Continuous lattices. Lect. Notes Math. **274**, 97–136 (1972)
Shchepin, E.V.: On topological products, groups, and a new class of spaces more general than metric spaces. Sov. Math. Dokl. **17**, 152–155 (1976a)
Shchepin, E.V.: Topology of limit spaces of uncountable inverse spectra. Russ. Math. Surv. **31**, 155–191 (1976b)
Shchepin, E.V.: On κ-metrizable spaces. Math. USSR Izv. **14**, 407–440 (1980)
Singer, I.: Some relations between dualities, polarities, coupling functionals and conjugations. J. Math. Anal. Appl. **115**, 1–22 (1986)
Sioen, M.: On the epireflectivity of the Wallman-Shanin-type compactification for approach spaces. Appl. Categor. Struct. **8**, 607–637 (2000)
Sioen, M., Verwulgen, S.: Quantified functional analysis and seminormed spaces: a dual adjunction. J. Pure Appl. Algebra **207**, 675–686 (2006)
Sioen, M., Verwulgen, S.: Quantified functional analysis: recapturing the dual unit ball. Result. Math. **45**, 359–369 (2004)
Sioen, M., Verwulgen, S.: Locally convex approach spaces. Appl. Gen. Topol. **4**, 263–279 (2003)
Stein, C.: A bound for the error in the normal approximation to the distribution of a sum of dependent random variables. In: Proceedings of the Sixth Berkeley Symposium on Mathematical Statistics and Probability (Univ. California, Berkeley, California, 1970/1971), vol. 2, pp. 583–602. Probability Theory. University of California Press, Berkeley (1972)
Stein, C.: Approximate Computation of Expectations. Institute of Mathematical Statistics Lecture Notes-Monograph Series, 7. Institute of Mathematical Statistics, Hayward (1986)
Suzuki, J., Tamano, K., Tanaka, Y.: κ -metrizable spaces, stratifiable spaces and metrization. Proc. Am. Math. Soc. **105**, 500–509 (1989)
Topsøe F.: Topology and Measure. Lecture Notes in Mathematics, vol. 133. Springer, Berlin (1970)
Väisälä, J.: A survey of nearisometries. arXiv: math/0201098v1 (2002)

Valentine, F.: Convex Sets. McGraw-Hill, New York (1965)
Vandersmissen, E.: Firm Completions in Metrically Generated Constructs. Ph.D., VUB (2008)
Vandersmissen, E., Van Geenhoven, A.: Function spaces in metrically generated theories. Topol. Appl. **156**(12), 2088–2100 (2009)
Van Douwen, E.: The Čech-Stone compactification of a discrete groupoid. Topol. Appl. **39**, 43–60 (1991)
Van Geenhoven, A.: Local metrically generated theories. Ph.D. University of Antwerp (2010)
Van Olmen, C.: A study of the interaction between frame theory and approach theory. Ph.D. University of Antwerp (2005)
Van Olmen, C., Verwulgen, S.: Every Banach space is reflexive. Appl. Categor. Struct. **14**, 123–134 (2006)
Vaughan, J.E.: Total Nets and Filters in Lecture Notes in Pure and Applied Mathematics, vol. 24, pp. 259–265. Marcel Dekker, New York (1976a)
Vaughan, J.E.: Products of topological spaces. Gen. Topol. Appl. **8**, 207–217 (1976b)
Verwulgen, S.: A categorical approach to quantified functional analysis. Ph.D. University of Antwerp (2003)
Verwulgen, S.: An isometric representation of the dual of $\mathscr{C}(X, \mathbb{R})$. Appl. Categor. Struct. **14**, 111–121 (2007a)
Verwulgen, S.: Duality, vector spaces and absolutely convex modules. Appl. Categor. Struct. **15**, 647–653 (2007b)
Vietoris, L.: Bereiche zweiter Ordnung. Monatsh f. Math. u. Phys. **32**, 258–280 (1922)
Vietoris, L.: Kontinua zweiter Ordnung. Monatsh. f. Math. u. Phys. **33**, 49–62 (1923)
Waszkiewicz, P.: Quantitative continuous domains. Appl. Categor. Struct. **11**, 41–67 (2003)
Weck-Schwarz, S.: Cartesian closed topological and monotopological hulls: A comparison. Topol. Appl. **38**, 263–274 (1991)
Weil A.: Sur les espaces à structure uniforme et sur la topologie générale. Actual. Sci. Ind. **551**, 3–40 (1937)
Windels, B.: Uniform Approach Theory. Ph.D. University of Antwerp (1997)
Wiśnicki, A., Wośko, J.: On relative Hausdorff measures of noncompactness and relative Chebyshev radii in Banach spaces. Proc. Am. Math. Soc. **124**, 2465–2474 (1996)
Wyler, O.: Are there topoi in topology? In: Binz, E., Herrlich, H. (eds.) Categorical Topology, pp. 699–719. Springer, Berlin (1976)
Wyler, O.: Lecture Notes on Topoi and Quasitopoi. World Scientific, Singapore (1991)
Zhang, D.: Tower extension of topological constructs. Comment. Math. Univ. Carol. **41**, 41–51 (2001)
Zsilinszky, L.: Topological games and hyperspace theories. Set Valued Anal. **6**, 187–207 (1998)
Zolotarev, V.M.: Probability metrics. Teor. Veroyatnost. i Primenen. **28**(2), 264–287 (1983). (Russian)

Further Reading

Alexandroff, P.: Über die Metrisation der im kleinen kompakten topologische Räume. Math. Ann. **92**, 294–301 (1924)
Amir, D.: On Jung's constant and related constants. In: Normed Linear Spaces in Texas Functional Analysis Seminar 1982–1983 (Austin, Tex.), Longhorn Notes, pp. 143–159. University of Texas Press, Austin (1983)
Anderson, C.L., Hyams, W.H., McKnight, C.K.: Center points of nets. Can. J. Math. **27**, 418–422 (1975)
Atsuji, M.: A space in which every continuous real function is uniformly continuous (general case). Solution of problem 6.2.16. Sûgaku 8, 211–213 (1956/1957)

Atsuji, M.: Uniform extension of uniformly continuous functions. Proc. Jpn Acad. **37**, 10–13 (1961)
Aull, C.E., Lowen, R. (eds.): Handbook of the History of General Topology, vol. 1. Kluwer Academic Publishers, Dordrecht (1997)
Aull, C.E., Lowen, R. (eds.): Handbook of the History of General Topology, vol. 2. Kluwer Academic Publishers, Dordrecht (1998)
Baekeland, R., Lowen, R.: Measures of Lindelöf and separability in approach spaces. Int. J. Math. Math. Sci. **17**, 597–606 (1994)
Beer, G.: A polish topology for the closed subsets of a polish space. Proc. Am. Math. Soc. **113**, 1123–1133 (1991)
Beer, G., DiConcilio, A.: Uniform convergence on bounded sets and the Attouch-Wets topology. Proc. Am. Math. Soc. **112**, 235–243 (1991)
Bentley, H.L., Herrlich, H.: Convergence Lowen-Colebunders. E. J. Pure Appl. Algebra **68**, 27–45 (1990)
Bergström, H.: Weak convergence of measures Probability and Mathematical Statistics. Academic Press, Inc. (Harcourt Brace Jovanovich, Publishers), New York-London (1982)
Bose, S.C.: Weak convergence to the fixed point of an asymptotically nonexpansive map. Proc. Am. Math. Soc. **68**(3), 305–308 (1978)
Bourbaki, N.: Topologie Générale, chapitre 9: Utilisation des nombres réels en topologie générale. Hermann, Paris (1958)
Bourbaki, N.: Topologie Générale, chapitre 1 et 2: Structures topologiques, structures uniformes. Hermann, Paris (1965)
Brodski, M.S., Mil'man, D.P.: On the center of a convex set. Dokl. Akad. Nauk SSSR (N.S.) 59, 837–840 (1948) (Russian)
Brümmer, G.C.L.: Topological categories. Topol. Appl. **18**, 27–41 (1984)
Burke, K.D., Lutzer, D.J.: Recent advances in the theory of generalized metric spaces. Topology. Lecture Notes in Pure and Applied Mathematics, vol. 24, pp. 1–70. Marcel Dekker, New York (1976)
Cagliari, F.: Disconnectednesses cogenerated by Hausdorff spaces. Cah. Top. Géom. Diff. Catégor. **29**, 3–8 (1985)
Cagliari, F., Cicchese, M.: Disconnectednesses and closure operators. Rend. Circ. Mat. Palermo **11**, 15–23 (1985)
Cartan, H.: Filtres et ultrafiltres. Compt. Rend. **205**, 777–779 (1937a)
Cartan, H.: Théorie des filtres. Compt. Rend. **205**, 595–598 (1937b)
Chen, L.H.Y., Shao, Q.-M.: A non-uniform Berry-Esseen bound via Stein's method. Probab. Theor. Relat. Fields **120**, 236–254 (2001)
Činčura, J.: Cartesian closed coreflective subcategories of the category of topological spaces. Topol. Appl. **41**, 205–212 (1991)
Činčura, J.: Products in Cartesian closed subcategories of the category of topological spaces. Topol. Appl. **59**, 195–200 (1994)
Clementino, M.M.: Hausdorff separation in categories. Appl. Categor. Struct. **1**, 285–296 (1993)
Clementino, M.M.: On categorical notions of compact objects. Appl. Categor. Struct. **4**, 15–29 (1996)
Clementino, M.M., Giuli, E., Tholen, W.: Topology in categories: compactness. Portugal. Math. **53**, 397–433 (1996)
Clementino, M.M., Tholen, W.: Separated and connected maps. Appl. Categor. Struct. **6**, 373–401 (1998)
Clementino, M.M., Hofmann, D.: Effective descent morphisms in categories of lax algebras. Appl. Categor. Struct. **12**, 413–425 (2004)
Cobos, F., Peetre, J.: Interpolation of compactness using Aronszajn-Gagliardo functors. Israel. J. Math. **68**, 220–240 (1989)
Cobos, F., Manzano, A., Martinez, A.: Interpolation theory and measures related to operator ideals. Quart. J. Math. Oxf. Ser. **50**(2), 401–416 (1999)

Colebunders, E., Lowen, R., Verbeeck, F.: Exponential objects in the construct PrAp. Cah. Top. Gom. Diff. Cat. **38**, 259–276 (1997)

Colebunders, E., Verbeeck, F.: A unifying theory on exponential objects in the setting of Čech closures and distances. In: Simon, P. (ed.) Proceedings of the Eighth Prague Topological Symposium, pp. 270–279, August (1996)

Colebunders, E., Lowen, R., Nauwelaerts, M.: The Cartesian closed hull of the category of approach spaces. Cah. Topol. Géom. Diff. Catég **XLII**, 242–260 (2001)

Colebunders, E., Lowen, R., Wuyts, P.: A Kuratowski-Mrówka theorem in approach theory. Topol. Appl. **153**, 756–766 (2005)

Di Maio, G., Lowen, R., Naimpally, S.A., Sioen, M.: Gap functionals, proximities and hyperspace compactification. Topol. Appl. **153**, 924–940 (2005)

Dimov, G.D.: On κ-metrizable Hausdorff compactifications of κ-metrizable spaces and a new class of spaces, including all separable metrizable spaces. C. R. Acad. Bulg. Sci. **36**, 1257–1260 (1983)

Dubuc, E.J., Porta, H.: Convenient categories of topological algebras and their duality theory. J. Pure Appl. Algebra **1**, 281–316 (1971)

Dubuc, E.J.: Concrete quasitopoi. Lect. Notes Math. **753**, 239–254 (1979)

Dudley, R.M.: Distances of probability measures and random variables. Ann. Math. Stat. **39**(5), 1563–1572 (1968)

Efremovič, V.A.: Infinitesimal spaces. Dokl. Akad. Nauk. SSSR **76**, 341–343 (1951)

Efremovič, V.A.: The geometry of proximity. Mat. Sb. **31**, 189–200 (1952)

Efremovič, V.A., Švarc, A.S.: A new definition of uniform spaces, metrization of proximity spaces. Dokl. Acad. Nauk. SSSR **89**, 393–396 (1953)

Egbert, R.J.: Products and quotients in probabilistic metric spaces. Pacific J. Math. **24**, 437–455 (1968)

Engelking, R.: Outline of General Topology. PWW—Polish Scientific Publishers, Warszawa (1977)

Feller, W.: An introduction to probability theory and its applications, vol. II, 2nd edn. Wiley, New York (1971)

Fisher, H.: Limesraüme. Math. Ann. **137**, 269–303 (1959)

Fréchet, M.: De l'écart numérique à l'écart abstrait. Port. Math. **5**, 121–131 (1946)

Frölicher, A.: Kompakt erzeugte Räume und Limesräume. Math. Z. **129**, 57–63 (1972)

Frölicher, A.: Smooth structures. Lect. Notes Math. **962**, 69–81 (1982)

Gaal, S.A.: Point Set Topology. Academic Press, New York (1964)

Garkavi, A.L.: The best possible net and the best possible cross-section of a set in a normed space. Am. Math. Soc. Transl. **39**, 111–132 (1964)

Gerlo, A., Vandersmissen, E., Van Olmen, C.: Sober approach spaces are firmly reflective for the class of epimorphic embeddings. Appl. Categor. Struct. **14**, 251–258 (2006)

Goebel, K., Kirk, W.A.: Topics in Metric Fixed Point Theory Cambridge Studies in Advanced Mathematics, vol. 28. Cambridge University Press, Cambridge (1990)

Grothendieck, A.: Produit tensoriels topologiques et espaces nucléaires. Mem. Am. Math. Soc. **16**, 140 (1955)

Gruenhage, G.: Generalized Metric Spaces. In: Handbook of Set-Theoretic Topology, pp. 423–501. North-Holland, Amsterdam (1984)

Gruenhage, G.: Generalized Metric Spaces and Metrization in Recent Progress in General Topology, pp. 239–274. North-Holland, Amsterdam (1992)

Gutev, V., Valov, V.: Sections, selections and Prohorov's theorem. J. Math. Anal. Appl. **360**, 377–379 (2009)

Hausdorff, F.: Grundzüge der Mengenlehre. Veit und Comp., Leipzig, (1914) (reprint, p. 1949. Chelsea, New York)

Harvey, J.M.: T_0-separation in topological categories. Quaestiones Math. **2**, 177–190 (1977)

Herrlich, H.: E-kompakte räume. Math. Zeitschr. **96**, 228–255 (1967)

Herrlich, H.: Limit-operators and topological coreflections. Trans. Am. Math. Soc. **46**, 203–210 (1969)

Herrlich, H.: Topological functors General. Topol. Appl. **4**, 125–142 (1974)

Herrlich, H.: Initial completions. Math. Zeitschrift **150**, 101–110 (1976)

Herrlich, H.: Compactness and Hausdorffness—old and new observations. Forum BMG **18**, 40–49 (2011)

Herrlich, H., Lowen, R.: On simultaneously reflective and coreflective subcategories. In G. C. L. Brümmer Festschrift, pp. 121–129 (1999)

Herrlich, H., Hušek, M.: Categorical topology. In: Recent Progress in General Topology, pp. 369–403. Elsevier Publishing Co., Amsterdam (1992)

Herrlich, H., Hušek, M.: Some open categorical problems in Top. Appl. Categor. Struct. **1**, 1–19 (1993)

Herrlich, H., Strecker, G.E.: Coreflective subcategories. Trans. Am. Math. Soc. **157**, 205–226 (1971)

Holà, L., Lucchetti, R.: Equivalence among hypertopologies (preprint)

Holmes, R.B.: A course in optimization and best approximation. Lecture Notes in Mathematics, vol. 257. Springer, Berlin (1972)

Huber, P.J., Ronchetti, E.M.: Robust statistics. Wiley Series in Probability and Statistics, 2nd edn. Wiley, Hoboken (2009)

Hušek, M.: Categorial connections between generalized proximity spaces and compactifications. In: Proceedings Symposium Berlin 1967 on Extension Theory of Topological Structures, pp. 127–132 (1967)

Hušek, M., Tozzi, A.: Generalized reflective cum coreflective classes in Top and Unif. Appl. Categor. Struct. **4**, 57–68 (1996)

Jaeger, G.: Gähler's neighbourhood condition for convergence approach spaces. Acta Math. Hungar. **139**, 19–31 (2013)

Johnstone, P.T.: Stone Spaces. In: Cambridge Studies in Advanced Mathematics, vol. 3. Cambridge University Press, Cambridge (1982)

Kallenberg, O.: Foundations of modern probability. In: Probability and Its Applications (New York), 2nd edn. Springer, New York (2002)

Kannan, V.: On a problem of Herrlich. J. Madurai Univ. **6**, 101–104 (1976)

Kawabe, J.: Convergence of compound probability measures on topological spaces. Colloquium Mathematicum **LXVII**, 161–176 (1994)

Kuczumow, T., Prus, S.: Asymptotic centers and fixed points of multivalued nonexpansive mappings. Houston J. Math. **16**, 465–468 (1990)

Künzi, H.-P.: Functorial admissible quasi-uniformities on topological spaces. Topol. Appl. **43**, 27–36 (1992)

Kushner, H.J.: Approximation and weak convergence methods for random processes, with applications to stochastic systems theory. In: MIT Press Series in Signal Processing, Optimization, and Control, vol. 6. MIT Press, Cambridge (1984)

Lee, Y.J., Lowen, R.: Approach theory in merotopic, Cauchy and convergence spaces I. Acta Math. Hung. **83**, 189–207 (1999a)

Lee, Y.J., Lowen, R.: Approach theory in merotopic, Cauchy and convergence spaces II. Acta Math. Hung. **83**, 209–229 (1999b)

Linz, P.: A critique of numerical analysis. Bull. Am. Math. Soc. **19**, 407–416 (1988)

Lowen-Colebunders, E., Lowen, R.: Topological quasi-topos hulls of categories containing topological and metric objects. Cah. Top. Géom. Diff. Catégor. **30**, 213–228 (1989)

Lowen, R.: Kuratowski's measure of noncompactness revisited. Q. J. Math. Oxf. **39**, 235–254 (1988)

Lowen, R.: Approach spaces: a common supercategory of Top and Met. Math. Nachr. **141**, 183–226 (1989a)

Lowen, R.: A topological category suited for approximation theory. J. Approx. Theor. **56**, 108–117 (1989b)

Lowen, R.: The shortest note on the smallest coreflective subcategories of Top. Quaestiones Math. **12**, 475–477 (1989c)

Lowen, R.: Cantor-connectedness revisited. Comment. Math. Univ. Carol. **33**, 525–532 (1992)

Lowen, R.: Approximation of weak convergence. Math. Nachr. **160**, 299–312 (1993)

Lowen, R.: Approach spaces. In: Encyclopedia of General Topology, pp. 293–298. North Holland (2003)

Lowen, R., Robeys, K.: Completions of products of metric spaces. Q. J. Math. Oxf. **43**, 319–338 (1991)

Lowen, R., Robeys, K.: Compactifications of products of metric spaces and their relation to Smirnov and Čech-Stone compactifications. Topol. Appl. **55**, 163–183 (1994)

Lowen, R., Sagiroglu, S.: Convex closures, weak topologies and feeble approach spaces. J. Convex Anal. **21**, 581–600 (2014)

Lowen, R., Sioen, M.: A unified look at completion in Met, Unif and App. Appl. Categor. Struct. **8**, 447–461 (2000a)

Lowen, R., Sioen, M.: Approximations in functional analysis. Result. Math. **37**, 345–372 (2000b)

Lowen, R., Sioen, M.: On the multitude of monoidal closed structures in UAp. Topol. Appl. **137**, 215–223 (2004)

Lowen, R., Verbeeck, C.: Local compactness in approach spaces: Part I. Int. J. Math. Math. Sci. **21**, 429–438 (1998)

Lowen, R., Verbeeck, C.: Local compactness in approach spaces: Part II. Int. J. Math. Math. Sci. **2**, 109–117 (2003)

Lowen, R., Sioen, M., Verwulgen, S.: Categorical topology in Beyond Topology contemporary mathematics. Am. Math. Soc. **486**, 1–36 (2009)

Lowen, R., Van Olmen, C.: Approach theory in Beyond Topology. Contemporary mathematics. Am. Math. Soc. **486**, 305–323 (2009)

Lowen, R., Windels, B.: On the quantification of uniform properties. Comment. Math. Univ. Carol. **28**, 749–759 (1997)

Lowen, R., Windels, B.: Quantifying completion. Int. J. Math. Math. Sci. **23**, 729–739 (2000)

Lowen, R., Wuyts, P.: A complete classification of the simultaneously reflective and coreflective subcategory of App. Appl. Categor. Struct. **8**, 235–245 (2000)

Lowen, R., Wuyts, P.: Stable subconstructs: a correction and new results. Appl. Categor. Struct. **19**, 539–555 (2011)

Malkowsky, E., Rakočević, V.: The measure of noncompactness of linear operators between spaces of *m*th-order difference sequences. Stud. Sci. Math. Hung. **35**, 381–395 (1999)

Menger, K.: Statistical metrics. Proc. Natl. Acad. Sci. USA **28**, 535–537 (1942)

Menger, K.: Probabilistic theories of relations. Proc. Natl. Acad. Sci. USA **37**, 178–180 (1951)

Meziani, L.: Tightness of probability measures on function spaces. J. Math. Anal. Appl. **354**, 202–206 (2009)

Michor, P.: Funktoren zwischen kategorien von Banach-und Waelbroeck-rŁumen Sitzungsberichte sterreichische Akademie Wiss. Abt II **182**, 43–65 (1974)

Minkler, J., Minkler, G., Richardson, G.: Subcategories of filter tower spaces. Appl. Categor. Struct. **9**, 369–379 (2001)

Nachbin, L.: Sur les espaces topologiques ordonnés. C. R. Acad. Sci. Paris **226**, 381–382 (1948)

Naimpally, S.A., Pareek, C.M.: Characterisations of metric and generalized metric spaces by real valued functions. Q A Gen. Topol. **8**, 425–439 (1990)

Nauwelaerts, M.: Cartesian closed hull for quasi-metric spaces. Comment. Math. Univ. Carol. **41**, 559–573 (2000)

Nel, L.D.: Initially structured categories and Cartesian closedness. Can. J. Math. **27**, 1361–1377 (1975)

Nel, L.D.: Cartesian closed topological categories. Springer Lect. Notes Math. **540**, 439–451 (1976)

Newey, W.K.: Uniform convergence in probability and stochastic equicontinuity. Econometrica **59**, 1161–1167 (1991)

Nishiura, E.: Constructive methods in probabilistic metric spaces. Fundam. Math. **67**, 124–155 (1970)

Nussbaum, R.D.: A generalization of the Ascoli theorem and an application to functional differential equations. J. Math. Anal. Appl. **35**, 600–610 (1971)

Penon, J.: Sur les quasi-topos. Cah. Top. Géom. Diff. Categor. **18**, 181–218 (1977)

Pollard, D.: Convergence of Stochastic Processes. Springer, Berlin (1984)
Preuss, G.: Allgemeine Topologie. Springer, Berlin (1972)
Preuss, G.: Trennung und Zusammenhang. Monatshefte für Mathematik **74**, 70–87 (1970)
Prokhorov, Y.V.: Convergence of random processes and limit theorems in probability theory. Theory Probab. Appl. **1**, 157–214 (1956)
Pumplün, D.: Eilenberg-moore algebras revisited. Seminarberichte **29**, 57–144 (1988)
Pumplün, D., Röhrl, H.: Separated totally convex spaces. Manusc. Math. **50**, 145–183 (1985)
Pumplün, D., Röhrl, H.: The coproduct of totally convex spaces. Beitr. Algebra Geom. **24**, 249–278 (1987)
Pumplün, D., Röhrl, H.: Convexity theories V: Extensions of absolutely convex modules. Appl. Categor. Struct. **8**, 527–543 (2000)
Råde, L.: A conversation with Harald Bergström. Stat. Sci. **12**, 53–60 (1997)
Reichel, H.C.: Some results on distance functions. In: Proceedings of the Fourth Prague Topological Symposium, pp. 371–380 (1976)
Reichel, H.C., Ruppert, W.: Über Distanzfunktionen mit Werten in angeordneten Halbgruppen. Monatsh. Math. **83**, 223–251 (1977)
Réveillac, A.: Convergence of finite-dimensional laws of the weighted quadratic variations process for some fractional brownian sheets. Stochast. Anal. Appl. **27**, 51–73 (2009)
Richter, G.: More on exponential objects in categories of pretopological spaces. Appl. Categor. Struct. **5**, 309–319 (1997)
Rudin, W.: Real and complex analysis. In: McGraw-Hill Series in Higher Mathematics. McGraw-Hill, New York (1966)
Rudin, W.: Functional Analysis International Series in Pure and Applied Mathematics. McGraw-Hill, New York (1991)
Ryan, R.A.: Introduction to tensor products of banach spaces. In: Springer Monographs in Mathematics. Springer, New York (2002)
Salbany, S.: A bitopological view of topology and order in categorical topology. In: Sigma Series in Pure Mathematics, vol. 5, pp. 481–504. Heldermann, Berlin (1984)
Sancetta, A.: Strong law of large numbers for pairwise positive quadrant dependent random variables. Stat. Infer. Stoch. Process. **12**, 55–64 (2008)
Semadeni, Z.: Banach spaces of continuous functions. Monografie Matematyczne, vol. 55. PWN, Warszawa (1971)
Schwarz, F.: Powers and exponential objects in initially structured categories and applications to categories of limit spaces. Quaestiones Math. **6**, 227–254 (1983)
Schweizer, B., Sklar, A.: Statistical metric spaces. Pacific J. Math. **10**, 313–334 (1960)
Schweizer, B., Sklar, A.: Probabilistic Metric Spaces. North-Holland, Amsterdam (1983)
Sine, R.: On the converse of the nonexpansive map fixed point theorem for Hilbert space. Proc. Am. Math. Soc. **100**, 489–490 (1987)
Sioen, M.: Extension Theory for Approach Spaces. Ph.D. University of Antwerp (1997)
Sioen, M.: The Čech-Stone compactification for Hausdorff uniform approach spaces is of Wallman-Shanin type. Quest. Ans. Topol. **16**, 189–211 (1998)
Smirnov, J.M.: On proximity spaces in the sense of V. A. Efremovič. Dokl. Akad. Nauk. SSSR 84, 895–898 (1952a) (Russian).
Smirnov, J.M.: On proximity spaces. Mat. Sb. **31**, 543–574 (1952b) (Russian)
Smirnov, J.M.: On completeness of proximity spaces. Dokl. Akad. Nauk. SSSR **8**, 761–764 (1953a) (Russian)
Smirnov, J.M.: On completeness of uniform and proximity spaces. Dokl. Akad. Nauk. SSSR **91**, 1281–1284 (1953b) (Russian)
Smyth, M.B.: Quasi-uniformities: reconciling domains with metric spaces. In: Lecture Notes in Computer Science, vol. 298. Springer, Berlin (1987)
Steen, L.A., Seebach Jr, J.A.: Counterexamples in Topology. Rinehart and Winston Inc., New york (1970)

Strecker, G.E.: On Cartesian closed topological hulls. In: Categorical Topology. Heldermann, Berlin (1984)
Sydow, W.: On hom-fuctors and tensor products of topological vector spaces. Lect. Notes Math. **962**, 292–301 (1982)
Tietze, H.: Beiträge zur allgemeinen Topologie I. Math. Ann. **88**, 290–312 (1923)
Van Casteren, J.: Markov Processes, Feller Semigroups and Evolution Equations. World Scientific Publishers, Singapore (2010)
Vandersmissen, E.: Completion via nearness for metrically generated constructs. Appl. Categor. Struct. **15**, 633–645 (2007)
Van Olmen, C., Verwulgen, S.: Lower semicontinuous function frames. Appl. Categor. Struct. **15**, 199–208 (2007)
Van Rooij, A.C.M.: Non-Archimedean Functional Analysis. Pure and Applied Mathematics, vol. 51. Marcel Dekker, New York (1978)
Varadarajan, V.S.: Convergence of stochastic processes. Bull. Am. Math. Soc. **67**, 276–280 (1961)
Verbeeck, F.: The category of pre-approach spaces. Ph.D. University of Antwerp (2000)
Waszkiewicz, P.: Distance and measurement in domain theory. Electr. Notes Theor. Comp. Sci. **45**, 1–15 (2001)
Weck-Schwarz, S.: T_0-objects and separated objects in topological categories. Quaestiones Math. **14**, 315–325 (1991)
Whitt, W.: Convergence of probability measures on the function space $\mathscr{C}[0, \infty)$. Ann. Math. Stat. **41**, 939–944 (1970)
Wilansky, A.: Topology for analysis. Ginn, A Xerox Company, Waltham, Massachusetts (1970)
Willard, S.: General Topology. Addison-Wesley, Reading (1970)

Index

R. Lowen, *Index Analysis*, Springer Monographs in Mathematics,
DOI 10.1007/978-1-4471-6485-2

K

L

Printed in the United States
By Bookmasters